Infrared Thermography as a Tool for Assessing Animal Welfare and Its Usefulness in Veterinary Research

Infrared Thermography as a Tool for Assessing Animal Welfare and Its Usefulness in Veterinary Research

Guest Editor

Daniel Mota-Rojas

MDPI

Basel • Beijing • Wuhan • Barcelona • Belgrade • Novi Sad • Cluj • Manchester

Guest Editor
Daniel Mota-Rojas
Neurophysiology, Behavior
and Animal Welfare
Assessment
Universidad Autónoma
Metropolitana
Mexico City
Mexico

Editorial Office
MDPI AG
Grosspeteranlage 5
4052 Basel, Switzerland

This is a reprint of the Special Issue, published open access by the journal *Animals* (ISSN 2076-2615), freely accessible at: https://www.mdpi.com/journal/animals/special_issues/53987Z8328.

For citation purposes, cite each article independently as indicated on the article page online and as indicated below:

Lastname, A.A.; Lastname, B.B. Article Title. *Journal Name* **Year**, *Volume Number*, Page Range.

ISBN 978-3-7258-3357-3 (Hbk)
ISBN 978-3-7258-3358-0 (PDF)
https://doi.org/10.3390/books978-3-7258-3358-0

Contents

About the Editor

Daniel Mota-Rojas

Daniel Mota-Rojas has significant research expertise in applied animal behavior and welfare. He obtained a Bachelor of Science degree in Veterinary Medicine and Animal Husbandry at the Metropolitan Autonomous University (UAM) in Mexico City. He earned a Master's in Production and Animal Health Sciences from the National Autonomous University of Mexico (UNAM) and a Ph.D. in Biological Sciences from the UAM in Mexico City. Currently, he serves as the Commissioner for the Ph.D. program in Biological and Health Sciences at the UAM. Dr. Mota-Rojas served as the Leader in Biomedical Science, focusing on the Pathophysiology of Stress, Behavior, and Welfare of Domestic Animals. His research centers on evaluating the role of cognitive bias in animal emotions, the neurological modulation of facial expressions, quality of life, quality of death, mother–young bonding, euthanasia, thermal biology, and the behavior and welfare of domestic and wild animals. He is a guest lecturer and international speaker on Animal Behavior and Welfare. Dr. Mota-Rojas is a member of the National System of Researchers of CONACYT in Mexico and has authored over 250 scientific articles indexed in SCOPUS with more than 5000 international citations. He is a member of the Mexican Network of Animal Welfare, the Mexican Academy of Sciences, and the Mexican Veterinary Academy. He is the editor of 13 books, encompassing topics related to animal behavior, cognition, emotional states, attachment and welfare, thermoregulation, global warming, pain assessment, facial expression, neonatology, and maternal behavior in domestic animals, including the water buffalo. He is currently co-editing the eBook "Animal Communication: Neurobiological, Chemical, and Physiological Aspects" alongside Professor Temple Grandin, published by Frontiers, Switzerland. Additionally, he serves on the Editorial Boards of several journals, including *Frontiers in Veterinary Science* (Switzerland), *Journal of Buffalo Science* (Lifescience Global, Canada), *Dog Behavior* (Italy), *CABI Reviews* (UK), and the journal *Animals* (MDPI, Basel, Switzerland).

Preface

It is both an honor and a pleasure to introduce this scientific collection, which brings together original research and review manuscripts on infrared thermography.

This Special Issue, entitled "Infrared Thermography as a Tool for Assessing Animal Welfare and Its Usefulness in Veterinary Research", has been meticulously curated to present the latest advancements in infrared thermography (IRT), encompassing innovations in thermal imaging acquisition and methodologies that enhance the application of this technology for evaluating temperature variations in both domestic and wild animals. These thermal fluctuations are closely associated with a range of environmental stressors, including exposure to extreme temperatures, pain, fear, hunger, anxiety, and boredom, as well as natural physiological processes such as disease progression, growth, aging, metabolic fluctuations, exercise, reproductive behaviors, maternal care, and various stages of physiological development. These encompass estrus, gestation, parturition, and lactation. Additionally, the impact of global warming and its potential to disrupt thermal homeostasis is a key consideration in this collection.

Within these pages, readers will find invaluable insights into the utility of IRT for assessing an animal's thermal state by measuring radiation emitted from the body due to superficial microcirculatory changes, which may either amplify or diminish thermal emissions. Furthermore, IRT emerges as a promising screening tool in veterinary clinical research, facilitating the early detection of local inflammatory processes, wounds, neoplasms, pain, neuropathies, and emotional distress.

This collection highlights the extensive range of benefits that IRT offers for diagnosing injuries, diseases, productivity, behavior, exercise response, autonomic activity, hypothalamic neuromodulation, neonatal thermoregulation at birth, inflammatory processes, and heat stress—key topics in veterinary medicine and animal sciences. Its applications extend to critical fields such as ethology, pathology, animal production, orthopedics, reconstructive surgery, and rehabilitation. Moreover, IRT serves as an effective means of evaluating the impact of climate change, heat stress, and the intricate interplay between physiological and behavioral responses across diverse species, including livestock, companion animals, laboratory models, and wildlife.

As societal interest in animal welfare continues to grow, IRT has gained prominence in assessing production animals, laboratory species, companion animals, and zoo-housed wildlife. This technology detects heat radiated from the body surface, translating it into measurable temperature variations that provide insight into an animal's comfort and welfare. From both physiological and practical perspectives, the regulation of blood flow near the body surface—through vasodilation and vasoconstriction—plays a crucial role in thermoregulation. Specific anatomical regions, referred to as "thermal windows," are particularly important for heat exchange due to their dense vascular networks, presence of specialized arteriovenous structures, and minimal fur coverage. These regions offer non-invasive and stress-free means of evaluating an animal's physiological status, making IRT an increasingly valuable tool in veterinary practice. Beyond its ease of application, IRT presents an efficient and, in some cases, cost-effective alternative to conventional physiological assessment methods, including residual feed intake prediction.

From a thermophysiological standpoint, alterations in blood flow are fundamental to heat exchange between the skin and the external environment. These processes are mediated by the sympathetic noradrenergic vasomotor control of smooth muscle, enabling adaptive thermoregulatory responses across species.

Readers will also gain a deeper understanding of how endothermic animals regulate their body temperature through dynamic heat exchange mechanisms, metabolically balancing generated heat with external thermal influences. This delicate equilibrium is essential for sustaining body temperature during various physiological states and activities throughout an animal's life. Thermal adjustments are not only driven by environmental fluctuations but also by factors such as age, nutritional status, stress, and inflammatory conditions—each of which induces internal temperature shifts that subsequently manifest as surface thermal variations.

Another crucial aspect explored in this collection is the assessment of wildlife welfare under human care, which necessitates reliable monitoring tools. As a non-invasive technique, IRT allows researchers to evaluate an animal's thermal condition by measuring radiation from different anatomical regions. However, due to species-specific variations in skin properties, fur density, body coverage, and pigmentation, thermal windows in wildlife remain underexplored. This volume presents key findings on the suitability of anatomical regions for thermal assessments, health monitoring, and habitat optimization in captive wildlife.

A particularly noteworthy topic featured in this Special Issue is the early detection of pregnancy in mammals, a critical factor for maximizing reproductive efficiency. Traditional pregnancy diagnosis methods are often technically demanding, labor-intensive, and costly. In contrast, IRT is emerging as an accessible, efficient, and relatively straightforward approach capable of identifying surface temperature fluctuations linked to physiological changes in gestating animals.

Additionally, this collection delves into the current methodologies used to assess acute pain in domestic animals. Given the multifaceted nature of pain, its evaluation requires an integrative approach that considers physiological, behavioral, and neurobiological indicators. Thermography offers a novel means of assessing an animal's physiological response to nociceptive stimuli through heat production, providing a potential correlation with the affective dimension of acute pain. Although IRT is still in the early stages of validation as a pain assessment tool, this volume compiles the latest research on its application for detecting acute pain responses in animals. One particularly compelling study included in this collection demonstrates that IRT can serve as a valuable non-invasive technique for assessing the physiological impact of emotionally charged experiences on sheep.

Various studies featured in this volume also highlight IRT's utility in veterinary medicine as a tool for evaluating peripheral vascular integrity. By detecting variations in heat emission, thermography can aid in diagnosing tissue viability, assessing the extent of injuries, and guiding wound management. While further research is required to refine its application in monitoring vascular health, preliminary findings from studies on thromboembolisms, neuropathic thermal alterations, and burn injuries suggest that IRT holds significant promise for assessing peripheral vascular disorders in veterinary medicine.

Another significant subject addressed in this collection is the characterization of thermographic profiles across diverse animal production systems. A particularly compelling case study explores the use of IRT in the Brazilian Amazon, where extreme environmental conditions—including high temperatures and humidity—pose challenges to animal welfare and productivity. By evaluating thermal variations within production environments, IRT facilitates a more comprehensive understanding of the climatic factors influencing livestock performance, enabling the implementation of management strategies to optimize health and productivity.

Closely linked to the broader implications of climate change, heat stress represents a serious concern for both domestic and wild animals. When animals experience heat stress, their autonomic nervous system activates a series of physiological and behavioral mechanisms to restore thermal balance. One such response is peripheral vasodilation, which promotes heat dissipation via the skin. As a result, IRT has been proposed as a viable method for assessing an animal's thermal status and predicting rectal temperature non-invasively. However, challenges remain in establishing precise predictive models, as the relationship between rectal temperature and IRT measurements does not always follow a direct linear correlation.

Several compelling topics have been highlighted in this Special Issue, but I would also like to mention additional valuable subjects such as associations between surface and rectal temperature in piglets; the importance of IRT for health assessment in killer whales; thermoregulation during field exercise in horses; the effect of density and lineage on the dorsal surface temperature of broilers; events with different emotional valence affect the eye's lacrimal caruncle temperature changes in sheep; and the relationship between the infrared eye temperature of beef cattle and associated biological responses at high environmental temperatures, among other topics. All these studies offer significant insights and have broad relevance.

Finally, it is important to emphasize that IRT is solidifying its role as a key tool for the study of animal physiology and veterinary research, complementing valuable fields such as pharmacology, ethology, algology, and pathology, among others—all of which are dedicated to promoting animal welfare. We hope that this collection not only expands knowledge on the application of IRT across various species and contexts but also inspires new research and strategies to enhance animal welfare on a global scale.

I deeply appreciate the trust and unwavering support of the authors in sharing their experiments and scientific findings in this Special Issue on *Infrared Thermography*.

Daniel Mota-Rojas
Guest Editor

animals

MDPI

Review

Thermal Imaging to Assess the Health Status in Wildlife Animals under Human Care: Limitations and Perspectives

Daniel Mota-Rojas [1,*], Alfredo M. F. Pereira [2], Julio Martínez-Burnes [3], Adriana Domínguez-Oliva [1], Patricia Mora-Medina [4], Alejandro Casas-Alvarado [1], Jennifer Rios-Sandoval [1], Ana de Mira Geraldo [2] and Dehua Wang [5]

1 Neurophysiology, Behavior and Animal Welfare Assessment, Department of Agricultural and Animal Production, Universidad Autónoma Metropolitana (UAM) Unidad Xochimilco, Mexico City 04960, Mexico
2 Mediterranean Institute for Agriculture, Environment and Development (MED), Institute for Advanced Studies and Research, Universidade de Évora, 7006-554 Évora, Portugal
3 Facultad de Medicina Veterinaria y Zootecnia, Universidad Autónoma de Tamaulipas, Victoria City 87000, Mexico
4 Facultad de Estudios Superiores Cuautitlán, Universidad Nacional Autónoma de México (UNAM), Cuautitlan Izcalli 54714, Mexico
5 School of Life Sciences, Shandong University, Qingdao 266237, China
* Correspondence: dmota100@yahoo.com.mx

Simple Summary: Ensuring the welfare of wildlife under human care requires tools to monitor their health and well-being. Infrared thermography is a non-invasive technique for assessing thermal states that measure the radiation emitted from the skin in distinct anatomical areas, known as thermal windows—anatomical regions with abundant capillaries and arteriovenous anastomosis that facilitate heat exchange with the environment. However, thermal windows for wildlife species have not yet been established due to the different characteristics of their skin, coats, fur, or coloring. This review discusses published information on the usefulness of the ocular, nasal, thoracic, abdominal, and podal anatomical regions as thermal windows for evaluating these animals' thermal responses and health status and monitoring habitat design. Another aspect that must be considered for wildlife under human care is the limitations of distinct species due to differences between animals and critical factors. Future studies should focus on establishing a precise application for each thermal window according to the specific characteristics of distinct animal species.

Abstract: Promoting animal welfare in wildlife species under human care requires the implementation of techniques for continuously monitoring their health. Infrared thermography is a non-invasive tool that uses the radiation emitted from the skin of animals to assess their thermal state. However, there are no established thermal windows in wildlife species because factors such as the thickness or color of the skin, type/length of coat, or presence of fur can influence the readings taken to obtain objective, sensitive values. Therefore, this review aims to discuss the usefulness and application of the ocular, nasal, thoracic, abdominal, and podal anatomical regions as thermal windows for evaluating zoo animals' thermal response and health status. A literature search of the Web of Science, Science Direct, and PubMed databases was performed to identify relevant studies that used IRT with wild species as a complementary diagnostic tool. Implementing IRT in zoos or conservation centers could also serve as a method for determining and monitoring optimal habitat designs to meet the needs of specific animals. In addition, we analyze the limitations of using IRT with various wildlife species under human care to understand better the differences among animals and the factors that must be considered when using infrared thermography.

Keywords: infrared thermography; thermal status; pain; thermal window

Citation: Mota-Rojas, D.; Pereira, A.M.F.; Martínez-Burnes, J.; Domínguez-Oliva, A.; Mora-Medina, P.; Casas-Alvarado, A.; Rios-Sandoval, J.; de Mira Geraldo, A.; Wang, D. Thermal Imaging to Assess the Health Status in Wildlife Animals under Human Care: Limitations and Perspectives. *Animals* **2022**, *12*, 3558. https://doi.org/10.3390/ani12243558

Academic Editor: Francesca Arfuso

Received: 17 November 2022
Accepted: 12 December 2022
Published: 15 December 2022

1. Introduction

Infrared thermography (IRT) is considered a non-invasive, real-time technique used as a complementary diagnostic tool for several physiological and pathological processes in domestic and wildlife species under human care [1–5]. The control of heat exchange between body surfaces and the external environment plays a crucial role in regulating body temperature during different physiological phases and/or activities throughout the lifetime of homeotherms. Thermoregulatory adjustments can be induced by changes in environmental temperature and numerous physiological factors including age, fasting and food intake, stressful circumstances, and inflammation status, which can cause changes in internal temperature that result in variations in body surface temperature. The evaluation of surface temperature using infrared thermography—a non-invasive technique that captures images of a specific body region at a distance—represents a valuable tool for monitoring the animals' physiological status, welfare, and stress responses [6,7]. The reactions of animals under stressful conditions primarily involve activating the sympathetic system and hypothalamic–hypophysis–adrenal axis (HPA) through the release of effector hormones, such as catecholamines, and glucocorticoid production, respectively. One result of this activation is stress-induced hyperthermia, which consists of an increased core body temperature with consequent temperature changes [8].

As a non-invasive tool, remote evaluation is considered advantageous because it minimizes handling or the use of chemical restraints [9] and prevents the development of processes like stress hyperthermia in small rodents and birds [10]. However, the wide diversity of species under human care has distinct anatomical, morphological, and physiological traits. For example, feathers are thermal insulators that maintain temperatures within normal ranges, but their presence limits the use of IRT, which requires zones with bare skin. Therefore, the eye region and legs are the preferred thermal windows in birds for evaluating superficial temperature changes [10]. In the case of animals with sparse hair, such as elephants, rhinoceroses, and hippopotami, or those that do not have long hair, like giraffes, or equids like zebras, IRT can be used in the abdomen, thorax, or ears. In contrast, for carnivores or species with winter coats, the recommendation is to evaluate windows located in the joints, feet (on the caudal side), or facial region (Figure 1) [1,11].

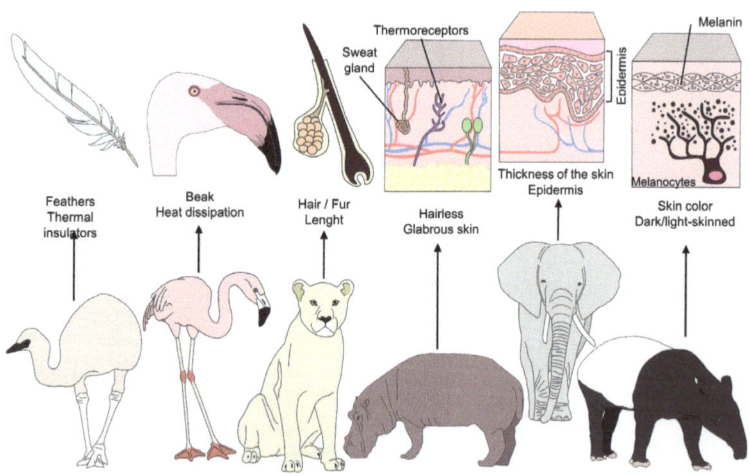

Figure 1. Morphological differences among furred, feathered, and bare-skinned animals. Thermoregulatory strategies in animals depend on various morphological elements. For example feathers serve as a thermal insulator in birds, and beaks facilitate heat dissipation. In contrast, the presence of hair and piloerection form a layer that conserves warm air to prevent any sudden drop in temperature. A thick layer of hair and dark hair color conserve heat and reduce radiation. Finally, skin color and the thickness of glabrous skin influence the amount of heat radiation in different species.

In addition to the features mentioned above, factors both external (environmental temperature, capture angle, distance to the target species, wind speed, and humidity) and internal (physiological state, feeding, metabolism, etc.) must also be considered to determine the validity and reliability of each anatomical region described as a potential thermal window. Therefore, the aim of this review, is to discuss the usefulness and application of the ocular, nasal, thoracic, abdominal, and podal regions as thermal windows to assess thermal responses and health status in animals. The limitations of IRT are also analyzed to better understand the species-specific characteristics that must be considered when using IRT with wild animals that are under human care. A literature search was performed to develop this review using the Web of Science, Science Direct, and PubMed databases to identify recent studies on the applicability of IRT with wild animals. The search keywords included "infrared thermography", "zoo animals", "wildlife under human care", "zoo animal disease", "zoo animal inflammation", and "non-invasive zoo animal health". The inclusion criteria were studies that used IRT as a complementary tool for diagnosing inflammatory, physiological, and/or pathological processes by evaluating thermal states in wildlife species. No year limit was placed on inclusion.

2. Advantages of IRT for Assessing the Thermal Status of Zoo Animals

IRT uses physical laws and surface properties to determine the superficial temperature of animals by measuring the amount of infrared radiation they emit (Figure 2) [12]. One of the main advantages that IRT offers is real-time, non-contact measuring of the surface temperatures of animals from distances of less than 1 m to over 1000 m, depending on species and the specific anatomical region selected [13]. This means that animals do not require extensive handling, physical restraint, or sedation, which is especially important when handling wild animals or those in zoos [1,14,15]. The IRT technique uses a camera with specialized thermal imaging lenses with diamond or germanium optical components and film to capture infrared radiation. A sensor processes this radiation to produce a digital radiometric image converted into a color pattern reflecting the temperatures detected [16].

The use of IRT with wild animals has focused primarily on detecting inflammatory processes, but other applications include aspects of thermoregulatory physiology in different taxa, population counts and comparisons, and evaluating the stages of pregnancy. IRT is helpful in the case of pregnancy because increased blood irrigation causes an increase in the surface temperature at the level of the flank in pregnant females. Additionally, it can help determine the association of certain emotional states with temperature changes and indirectly evaluate the habitat or housing of those animals [11,17].

For these reasons, IRT has been proposed as a method for monitoring enclosures and their characteristics to determine whether they comply with legal regulations and satisfy the basic biological needs of the species (e.g., shaded areas, bodies of water, adequate water supply) [18]. For example, alterations in the surface temperatures of elephants have been reported when the micro- or macroclimate of the installations are inappropriate for this species [19]. Similarly, in bears [20] and Bengal tigers (*Panthera tigris*) [21], IRT has recorded surface cutaneous temperature values and associated them with certain thermoregulatory behaviors, such as piloerection, seeking refuge in shaded areas, or rolling in the dirt to achieve thermoneutrality. This technique has also been used to assess the characteristics of the housing provided, which may favor or inhibit thermal comfort. Figure 3 shows the importance of providing sources of natural shade in enclosures for wildlife under human care since direct solar radiation without a proper resting area can cause heat stress in zoo animals.

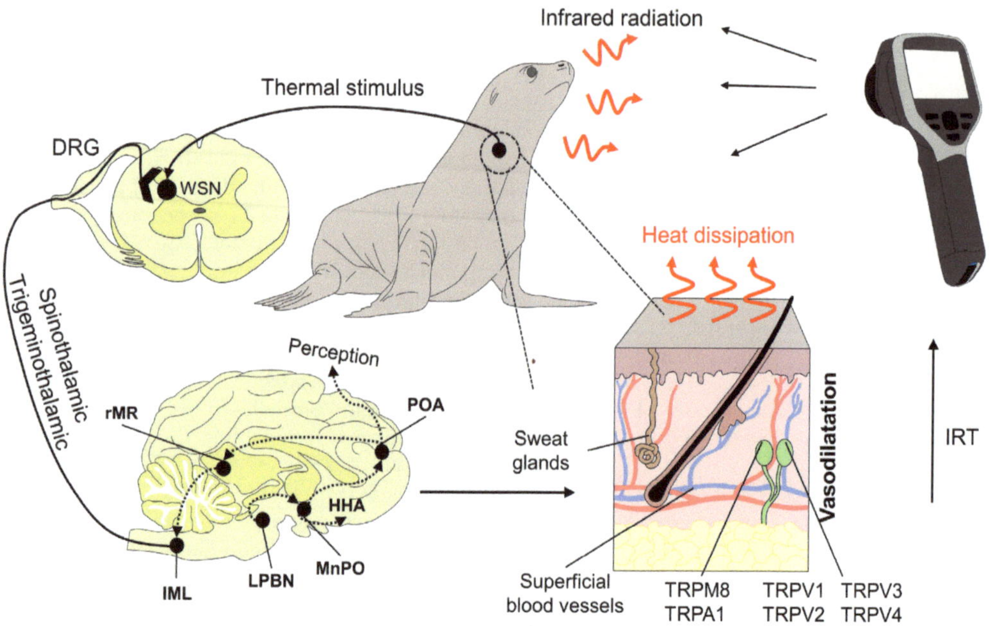

Figure 2. Physiological basis of infrared thermography and mechanisms of heat loss. Peripheral receptors in the dermal tissue, such as transient receptor potential vanilloid (TRPV, TRPV1, TRPV2, TRPV3, and TRPV4) or melastatin-related transient receptors (TRPM8), respond to heat and cold environmental stimuli, respectively. TRP ankyrin (A1) is postulated to have a function in noxious cold sensation and mechanosensation. These receptors initiate the transduction of the stimulus through primary peripheral nerve endings connected to secondary neurons in the spinal cord. Upon reaching this structure, warm-sensitive neurons (WSN) are activated and project the signal along the spinothalamic and trigeminothalamic pathways. Both of these tracts are connected to higher brain structures, such as the lateral parabrachial nucleus (LPBN) and median preoptic nucleus (MnPO) in the preoptic area of the hypothalamus (POA). DRG: dorsal root ganglion. HHA: hypothalamo-hypophyseal-adrenocortical. IRT: Infrared thermography. Their role is to generate behavioral, autonomic, and neuroendocrine responses. One of the autonomic responses to heat stimuli is peripheral vasodilatation by sympatho-adrenergic neurons in the rostral medullary raphe (rMR) and the intermediolateral nucleus of the spinal cord (IML), which increases heat dissipation. This response is captured by the lens of the infrared thermographic camera to generate a radiometric image.

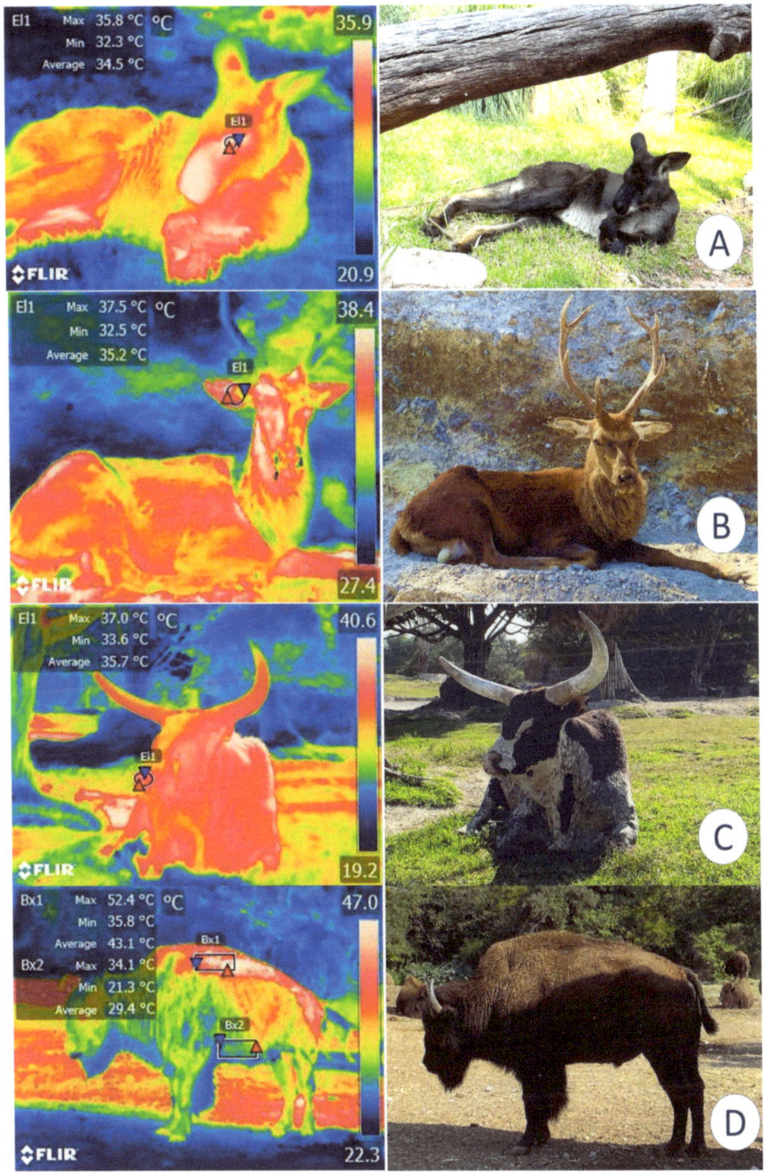

Figure 3. Importance of natural or artificial shade for the thermoregulatory behavior of various species. (**A**). Although this kangaroo (*Osphranter rufus*) is resting under natural shade, the average temperature of the lacrimal caruncle (El1) is 34.5 °C. (**B**). A sika deer (*Cervus nippon*) under shade shows an average temperature of 35.2 °C in the auricular region (El1). (**C**). The average nasal temperature of 35.7 °C of an ankole-watusi (*Bos taurus taurus ankole*) under the shade of a tree is depicted. (**D**). In an American bison (*Bison bison*), a comparison between the dorsal (Bx1) and ventral (Bx2) regions can be observed. An average temperature difference of 13.7 °C was registered between them. Therefore, shaded areas are a key element for ensuring the thermal comfort of wild animals under human care (Radiometric images and photographs by the authors). Maximum temperature (red triangle), and minimum (blue triangle). Thermal images obtained using a FLIR thermal camera.

Concerning thermal physiology, IRT recognizes several thermal windows that can be used for different species, including the African elephant (*Loxodonta africana*) [22], otters (*Lutra lutra* and *Pteronura brasiliensis*) [23], mole rats (*Fukomys mechowii* and *Heliophobius argenteocinereus*) [24], ground hornbills (*Bucorvidae*) [25], and koalas (*Phascolarctos cinereus*) [26]. Clinically, changes in thermal patterns, such as abnormal or asymmetrical thermal distribution, have been associated with pathologies, alterations in blood circulation, and inflammatory processes. In this area of study, Hilsberg-Merz [27] used IRT to evaluate problems affecting the joints and feet of elephants, rhinoceroses, giraffes, and hippopotami using IRT, while Church et al. [13] diagnosed a respiratory condition in an elephant related to a poorly positioned molar that was affecting the sinus area. In another study, Hurley-Sanders et al. [28] utilized IRT with fluorescein angiography to confirm a compromised vascular system and injury to the right wing in a Chilean flamingo (*Phoenicopterus chilensis*) that had a clinical history of acute weakness in that limb.

Thermography has also been used for the early diagnosis of diseases such as rabies in raccoons [29] and foot-and-mouth disease in mule deer (*Odocoileus hemionus*) [30]. Hurley-Sanders et al. [31] used it to monitor the healing process of an injury to the left forelimb of Gregory's wolf (*Canis rufus gregoryi*) after a skin graft. Results showed that IRT could aid in the early detection of a poor blood supply to the graft based on the stability of the vascularization of the transplanted skin. Recently, IRT has begun to be used in a field of research where emotional states derived from various stimuli have been found to correlate with thermal responses, especially in primates like rhesus monkeys (*Macaca mulatta*) [32], gorillas [33] and chimpanzees [34].

However, applying the IRT technique more widely in medical attention for zoo animals and wildlife species under human care requires not only a deep understanding of the anatomy of a wide range of species but also and evaluation of numerous other elements: the fact that humans do not have absolute control over the animal (movement, position relative to the sun, body zones, etc.), and environmental factors (e.g., direct radiation, humidity, rain, wind) that impact the accuracy and interpretation of the temperature readings obtained [1]. Specific body regions have been proposed for using IRT with individual species considering their characteristics and challenges and each one's morphological properties.

3. Facial Windows

3.1. Ocular Window (Regio Ocularis)

The ocular window assesses circulatory changes and thermal responses in the eye and the proximal area. Its usefulness has been reported in bovines [35], pigs [36], and dogs [37]. This window is surrounded by abundant periocular capillaries branching from the supra and infraorbital arteries. The latter irrigates the lacrimal canaliculi. Because its capillaries have autonomic innervation, they are sensitive to autonomic activation, especially during periods of acute stress. The area of the ocular window and its irrigation are represented in Figure 4.

The lacrimal carunclеis known to have sympathetic-innervated blood vessels that can respond to the activation of the HPA when exposed to a stressor. This is due to the secretion of catecholamines and the vasoconstriction response by its capillaries [38]. This effect has been reported in domesticated species such as sheep [2] and horses [39] during handling and transport, respectively. In those events, the temperature of the lacrimal caruncle has been seen to increase by up to 1 °C. In the case of wildlife species, Narayan et al. [26] evaluated the temperature of the lacrimal caruncle, abdomen, ears, and paws of three koalas born in captivity. They observed that the surface temperature of the ocular region had the greatest consistency and showed a significant increase compared to the ear region (31.6 °C vs. 23 °C, respectively), with a coefficient of variation of 4.83% in contrast to the values obtained from the ear (19.68%). This shows the window's usefulness for evaluating the body temperatures of animals. Their conclusions are similar to the findings of Melero et al. [40], who compared the thermal windows of five species of marine mammals: two cetaceans, a beluga whale (*Delphinapterus leucas*) and bottlenose dolphin (*Tursiops truncatus*), and three pinnipeds, a Patagonian sea lion (*Otaria flavescens*), a harbor seal (*Phoca vitulina*), and a Pacific walrus (*Odobenus rosmarus divergens*). The ocular region was considered appropriate for determining body temperature in the pinnipeds, the ocular region was considered appropriate for determining body temperature. In contrast, in the cetaceans, the internal mucosal region of the open blowhole was suggested as the most stable thermal window for monitoring body temperature non-invasively, as it showed differences of 1 °C compared to the normal rectal temperature of these species.

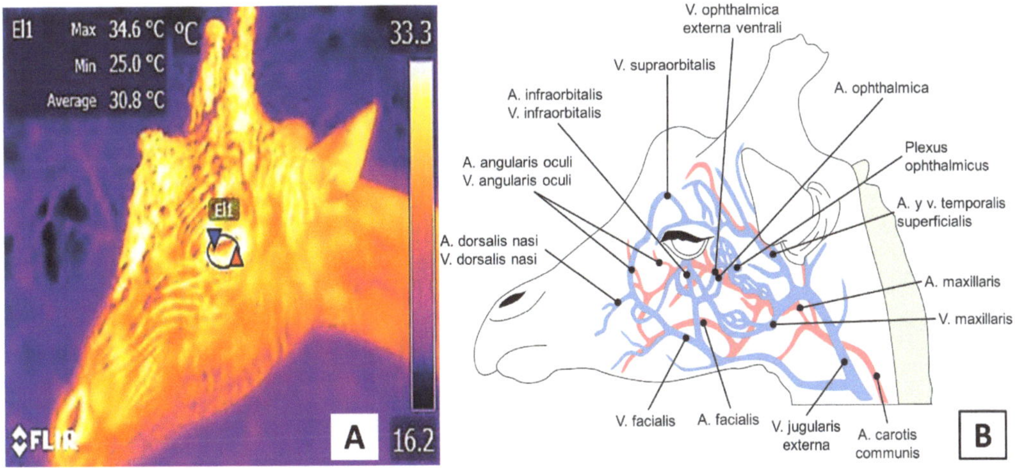

Figure 4. Depiction of the ocular window of a giraffe (*Giraffa camelopardis*). (**A**). A circle, approximately 3 cm in diameter encompasses the periocular region and a small portion of the eyelid. (**B**). Blood irrigation in this window comes from the supraorbital and infraorbital arteries, whose sympathetic innervation proceeds from branches of the facial nerve.

The studies described above are associated with the clinical usefulness of thermography focused on the ocular region for determining the thermal status of animals, which can be confirmed by similarity with the values obtained by conventional methods, such as rectal temperature. However, the effectiveness of IRT has also been compared to other approaches. Stryker's [21], for example, compared the thermal response of the eye to standard methods (rectal and axillary readings) and thermoregulatory behaviors in four species of large felines, namely, lions (*Panthera leo*), jaguars (*P. onca*), pumas (*Puma concolor*), and snow leopards (*P. uncia*). Using IRT, he identified that the surface temperature of the eye was similar to rectal values (39.9 °C and 39 °C, respectively). A second important observation was that the warm-climate felines (lions, jaguars) spent more time lying down (40–84%) than the cold-climate species (pumas, snow leopards) and had relatively less time performing active behaviors (5–10%). That study showed that temperature readings could provide information on the welfare of animals in enclosures when associated with different species' typical behaviors for adapting to their environment.

A study by South [41] reported a similar response in the European hedgehog (*Erinaceus europaeus*). There, IRT was applied to monitor hypothermia during hibernation using ocular temperature as a reference in 55 animals (23 males, 32 females). The temperature obtained with an axillary thermometer was significantly lower (1.3 ± 0.18 °C) than that of the periocular region (33.7 °C and 35.5 °C, respectively). Likewise, the ocular reading diagnosed hypothermia more precisely than relying on behavioral and clinical signs, suggesting that IRT is an accurate tool that can aid in evaluating the thermal states of animals and associate them with specific pathological and physiological conditions, thus contributing to the rescue of these species.

In contrast to the reliability and usefulness of IRT described above, a study of 39 Guianan squirrel monkeys (*Saimiri sciureus*) showed that facial temperature correlated poorly with rectal recordings ($r^2 = -0.10$, 95% IC= −0.27–0.07), as the average difference between the methods was 3.4 °C [42]. These results show that the precision of IRT depends greatly on choosing an adequate region for recording surface temperatures. Similar to descriptions of domesticated species, ocular IRT readings can be used to determine stress levels [2,43]. A study of giraffes (*Giraffa camelopardis*) evaluated the thermal response of the ocular and auricular regions under the assumption that, for these animals, feeding could be a stressful event. Results showed a correlation between blood cortisol levels ($p = 0.269$) and temperature that did not change before or after feeding. A positive relationship was also observed between the eye's and the ear's thermal response (Spearman rank $p = 0.759$, $p < 0.0001$). The authors attributed the absence of any ocular thermal response to a potentially enjoyable emotional state, suggesting that IRT could be useful for identifying emotional states in animals [44]. Vaz et al. [45] found a similar response in African lions (*P. leo*), where the IRT response at the ocular level was used to understand large felines' personality and stress physiology, when coordinated with ethological studies and fecal glucocorticoid levels. The authors identified three personality types by studying 22 African lions in two locations, (dominant, gentle, and neurotic). The animals had differences in their fecal cortisol metabolite levels, but these did not influence their thermal responses, so it was impossible to correlate body surface temperature with the personality type or age of the animals. One important element is the standardized distance of 1 m between the thermal camera and the ocular region. Another is lens angle, which must be perpendicular (90°) to the sagittal plane to ensure valid readings [46,47].

As observed in zoo species, unlike in domestic animals, measuring the surface temperature at the ocular level is a remote method to assess body temperature without handling the animals. However, its application to assess the stress degree or ANS activation has yet to be well established. Therefore, as reported in large ruminants, future studies must consider each species' emotional factor and anatomical differences, as reported in large ruminants [38].

Finally, it is important to consider that the ocular window can be evaluated by other techniques, such as optical coherence tomography (OCT). This method could help evaluate in vivo vascular responses by identifying the reactions of surface ocular blood vessels. Meleppat et al. [48] used this approach with aerial, aquatic, and terrestrial species like the Long Evans rats, gray short-tailed opossums (*Monodelphis domestica*), white sturgeons (*Acipenser transmontanus*), and great horned owls (*Bubo virginianus*). OCT has proven effective in identifying retinal and choroidal vascular structures and retinal layers [49]. Therefore, it can be suggested as a novel tool for diagnosing ophthalmologic diseases since the responses recorded by IRT may differ when animals are exposed to stressors.

3.2. Auricular Window (Regio Auricularis)

The auricular window is the most important region for thermographic evaluation in elephants. The ears of elephants represent approximately 20% of the total body surface area [50] and have a dense subcutaneous vasculature provided by the auricular arteries and veins (*arteria auricularis caudalis*) on the caudal surface [51] (Figure 5). The elephants' ears dissipate as much as 8% of body heat [52], though Hilsberg-Merz [1] mentions that African elephants—the species with the largest ears—can lose up to 30% of excess heat by radiation and convection through the ears. In elephants, the extension of the auricular pavilion compensates for the lack of sweat glands [22].

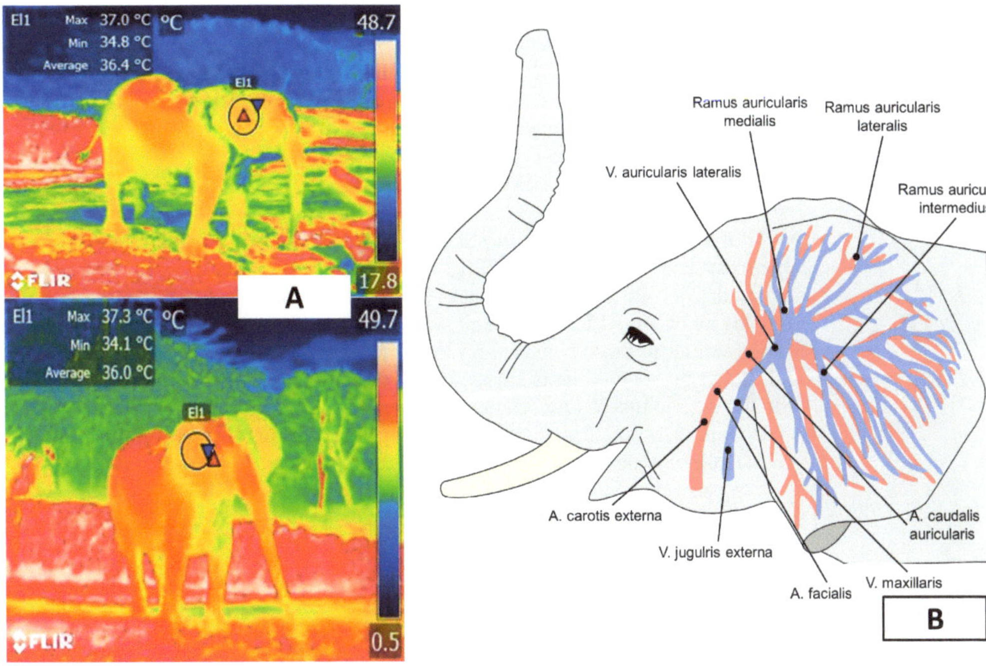

Figure 5. Auricular window of an African elephant (*Loxodonta africana*). (**A**). A circle around the external auditory canal represents the window. (**B**). The blood supply to the ear comes from the branches of the lateral auricular artery and the medial and marginal areas of the ear.

Because of these characteristics, IRT can assess the auricular region to determine the thermostability of the species and the influence that environmental temperature and/or enclosure design exert on it. Phillips and Heath [50] studied the heat exchange mechanism in four female African elephants (*L. africana*), finding that surface temperatures of the pinna change in response to environmental temperature through vasomotor action. They further stated that readings proximal to the external auditory meatus were higher than those registered at the extremities of the pinna (39 °C and 18 °C, respectively, at an ambient temperature of 18 °C). Parallel to changes in auricular microcirculation, ear-flapping is another method of thermoregulation that is believed to transfer heat by rotational oscillations [53]. This process involves the rotator, *auriculo occipitalis*, *postauriculatis*, *platysma myoides*, and *sphinctor profundus* muscles in the Loxodonta genus, and the rotator, *auriculo occipitalis*, *postauricularis*, and *zygomatico auricularis* muscles in the Elephas genus [54]. Therefore, the vasculature and motor control of the ears allow heat dissipation through radiative heat and convection, facilitate heat exchange in that region, and are considered a useful window for evaluating the thermal comfort zone of these species. Ear-flapping dissipates heat by forced convection, thus increasing the rate of heat loss.

A study of six African elephants (*L. africana*) by Weissenbock et al. [22] reported similar results. They took 325 thermograms of six body regions at times of 5, 10, and 30 m: head, proboscis, torso, forelimb, hindlimb, and ear. The ear lobe and distal areas of the pinna delimited the latter. IRT in this region increased in a range of 33.8–35.3 °C in warm weather after the animals returned to their indoor enclosure. Seventeen minutes after returning, the average ear temperature was 17.1 °C, while after 68 min it increased, revealing areas of maximum and minimum temperatures at 33 °C and 22.2 °C [22]. This area remained the body region with the lowest surface temperature. Intriguingly, the authors recorded different thermal patterns in the two ears of the same animal. This difference has also been reported in leporids, where sympathetic innervation can control the temperature of each pinna independently. In contrast to the variability and absence of any established pattern in other thermal windows, auricular IRT associates the thermoregulatory response of the animals with their housing conditions. Therefore, it is essential to (i) provide water for frequent bathing and (ii) consider other morphological characteristics, such as the elephant's wrinkled skin, which increases the surface area and ensures more efficient heat loss due to water evaporation [50].

The design of the enclosures also influences thermostability. Here, IRT can help evaluate the thermal responses of animals. Figure 6 illustrates how this thermal window was used to assess the thermal states of African elephants exposed to heat stress as they performed natural thermoregulating behaviors, such as ear-flapping, bathing and splashing in mud. The authors of a study of Asian elephants that had recently been moved to a new enclosure observed lethargy in all subjects. Their IRT temperature readings of the ears were similar to those recorded for the body. They considered this an abnormal pattern since the ears of elephants are structures designed for heat dissipation. However, upon analyzing the macro environment of the pachyderms, the authors found high levels of humidity (95%) and insufficient access to sources of water, two conditions that resulted in heat stress and higher ear temperatures as compensatory mechanisms [19]. In that case, the vapor partial pressure was close to the saturation pressure, thus reducing the evaporation rate and decreasing evaporative heat loss. The only possible response by those animals was to increase the peripheral temperature (increased auricular IRT) to maximize heat loss through radiation and convection. An important point to emphasize in that study is that the entire group showed the same pattern of increased auricular IRT. This made it possible to non-invasively discern between a pathological process in a single individual and a general, simultaneous response by the herd [1].

IRT has also been utilized to diagnose inflammatory pathologies in Asian elephants (*E. maximus*). Avni-Magen et al. [55] found that assessing regions like the head, ears, body, and limbs at a distance of 2–3 m identified inflammatory processes with a sensitivity and specificity of 89.2 and 83.4%, respectively, and a negative predictive value (NPPV) of 99.4%. In that study, increases of up to 2° in the delta temperature of the ear surface made it possible to detect inflammatory processes before clinical diagnoses were made due to visible signs. However, due to the low positive predictive value (PPV) of 19.3%, the authors concluded that IRT could be recommended as a complementary method that requires other confirmatory diagnostic techniques.

The effects of pathologies and thermostability have been studied in other zoo species, such as equids, large felines, large ruminants, and wild birds, among others. Da Costa et al. [56] used IRT with 18 captive canines, including the crab-eating fox (*Cerdocyon thous*), maned wolf (*Chrysocyon brachyurus*), and hoary fox (*Pseudalopex vetulus*) in a study that lasted almost a year. Readings showed that the temperature of the left ear increased, compared to the right ear (24.6 and 21.3 °C, respectively). This effect was attributed to an abscess that was not perceivable in a remote inspection. In addition, it was also possible to verify post-treatment efficacy since the thermographic values were equal in both ears after 9 days of treatment. A study of giant sable antelopes (*Hippotragus niger*) and mishmi takins (*Budorcas taxicolor taxicolor*) used thermography to evaluate the effects of heat stress and how it influenced the animals' behavior and thermoregulation mechanisms [1]. In contrast, in a study of a feline patient (*Leopardus tigrinus*) with renal failure, IRT did not show diagnostic validity or reveal any abnormal thermographic patterns in the animal's pads due to the thickness of the dermis, which impeded taking accurate infrared reading [56]. Similarly, the ear region of Koalas, unlike the studies of elephants mentioned above, proved to be the least consistent window (compared to the ocular, abdominal, dorsal, and paws), as it showed a coefficient variation of only 19.68% and an average temperature of 22.11 °C, as opposed to the ocular IRT readings, which had a coefficient variation of 4.83% [26].

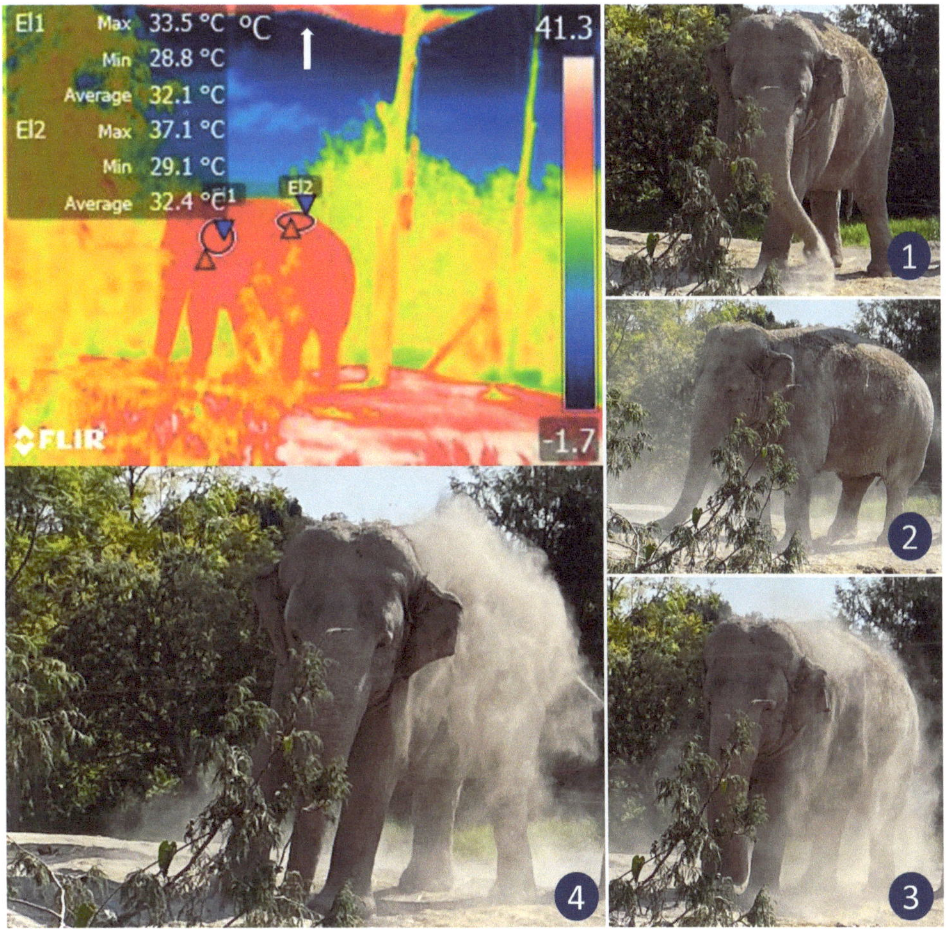

Figure 6. Thermoregulatory behavior in an African elephant (*Loxodonta africana*) under heat stress. The sequence of digital images shows dustbathing in an elephant exposed to high temperatures that triggered thermal imbalance (see images 1, 2, 3, and 4). The radiometric image shows the average temperature of the auricular (El1) and dorsal (El2) thermal windows. Under normal conditions, the ear temperature in elephants oscillates between 25 and 31 °C. However, the thermal image shows a temperature in the auricular region as high as 33.5 °C. IRT imaging reveals that although the animal covers its back with dust to thermoregulate, this behavior might be insufficient to achieve thermostability in this environment because the maximum temperature of the dorsum was 37.1 °C. Although the enclosure has a shade mesh as a roof to reduce direct sunlight (white arrow), it is not enough to reduce thermal stress in this species. (Radiometric images and photographs by the authors).

Finally, the application of IRT with captive wildlife species could be adapted for use with animals living in the wild. For example, auricular IRT has been evaluated in chimpanzees concerning behavioral changes, such as whimpers and aggressive barks, to assess the effect of emotions in response to conspecific vocalizations. In those test animals, after listening to aggressive bark vocalizations, the auricular surface temperature increased ($p < 0.05$) by an average of 0.7 ± 0.25 °C, but the temperature increase was greater during neutral calls due to an increase in blood flow that improved the performance of the animals' auditory sensitivity to the vocalizations of their conspecifics [57]. These data demonstrate the usefulness of the auricular/pinna region for assessing various animal processes, especially the limits of thermoneutrality within which species held in captivity must be maintained to avoid behavioral and physio-metabolic complications triggered by heat stress.

3.3. Nasal Window (Regio Nasalis)

The nasal window is an area of particular interest due to the sympathetic branch's ability to reflect ANS activity during events involving stress, excitement, physical activity, and disease processes, all of which cause vasoconstriction of peripheral blood vessels [58,59]. Figure 7 shows the vasomotor action derived from branches of the maxillary and sphenopalatine arteries, which are innervated by the facial nerve [60].

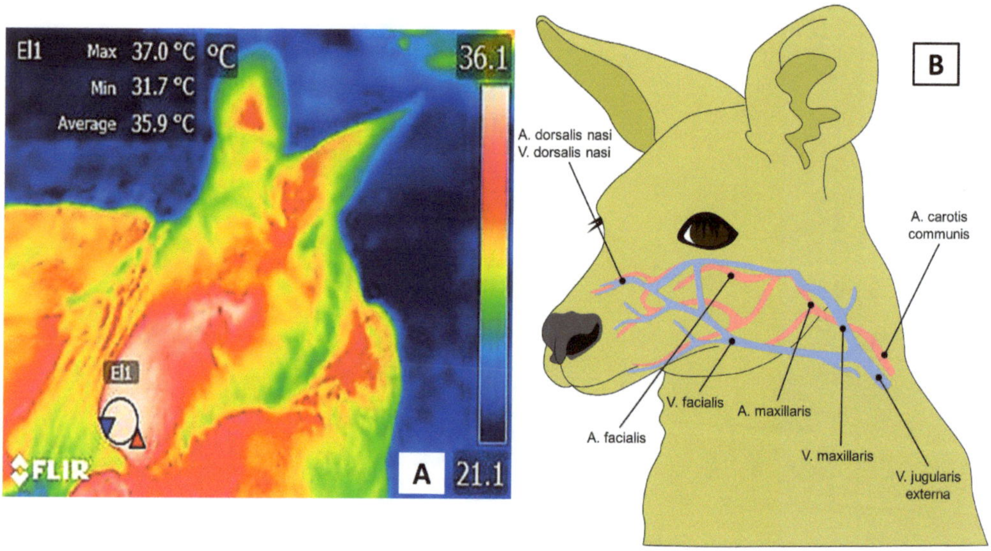

Figure 7. Nasal window in a kangaroo (*Osphranter rufus*). (**A**). A circle around the nasal region, including both nostrils, delimits the nasal window. (**B**). Circulation in the nose depends on branches of the maxillary artery and the *a. facialis* and *a.* and *v. dorsalis nasi*, which are innervated by the facial nerve.

Ross et al. [34] recorded nasal thermograms from 29 chimpanzees (*Pan troglodytes*) to identify the emotional arousal of those animals according to the species' behavioral repertoire. In the 538 thermographic images taken in that study, the temperature of the nasal region decreased significantly in the individuals considered active compared to those classified as showing inactive behaviors. The authors concluded that the nasal region is a thermal window of great importance in this species but that needs to be evaluated or compared with other anatomical areas, such as the forehead, to achieve validation. In another study, the emotional responses and changes in cutaneous peripheral perfusion in the nasal area were assessed in 17 common marmoset monkeys (*Callithrix jacchus*) exposed to conditions with distinct emotional valence. The presence of food (considered an exciting event) increased nasal temperature by 0.8 ± 0.7 °C. In contrast, the presentation of negative stimuli led to a 0.5 ± 0.8 °C decrease in the temperature of this area. Decreased nasal IRT readings were also associated with aggressive vocalizations and piloerection during aversive events [61]. The physiological mechanism that explains this thermal response under exposure to a potentially aversive stimulus is sympathetic activation and the consequent secretion of catecholamines, which are responsible for triggering peripheral capillary vasoconstriction and, therefore, inducing a decrease in the temperature radiated at the surface level [16].

In other non-human primates, IRT has been applied to evaluate the thermal response of chimpanzees (*P. troglodytes*) to adverse auditory stimuli (vocalizations). According to the results, nasal temperatures decreased significantly (0.16 $\pm$ 0.18 °C, $p < 0.001$) compared to thermal responses in the ear canal—where the opposite effect was observed—as temperatures increased (0.06 $\pm$ 0.09 °C, $p < 0.001$) [57]. This same response was reported in a study of rhesus monkeys (Macaca mulatta), where a decrease in temperature at the nasal level (0.2 $\pm$ 0.1 °C) was observed when the animals were exposed to a threatening physical agent. In addition, and proportionally to the decrease in temperature, facial movements such as bared teeth, mouth opening, and constant lip licking were observed [32], all of which form part of the facial language this species employs to communicate mental states [32,62].

A study of gorillas (*Gorilla gorilla*) by Heintz et al. [33] assessed the temperature of the nasal region and its correlation with cortisol and oxytocin levels during training with positive reinforcement and cognitive tasks. Observations showed that the nasal temperature decreased significantly during both moments ($p < 0.014$), while salivary cortisol and oxytocin also decreased after the sessions. In contrast, a higher frequency of self-directed behaviors was observed during the controlled conditions. Those results show the inconsistency and difficulty of validating the nasal window as a method for evaluating emotions, though other authors consider it a reliable area for identifying emotional states in non-human primates under human care [63].

In contrast, McFarland et al. [64] evaluated thermal responses in different body regions of vervet monkeys (*Chlorocebus pygerythrus*), including the dorsum and face, especially the nasal region. Results revealed that although the temperature of the nasal region presented significant changes, they were not related to body temperature and thus could not be considered an effective approach to measure vasoconstriction accurately. The authors attributed this effect to the influence of environmental factors that can alter thermographic readings. Their observations were similar to those reported by Thompson et al. [65], who evaluated the nasal area of howler monkeys (*Alouatta palliata*). In that study, the nasal region was not found to reflect internal processes because autonomous processes regulate body temperature for the production and loss of heat. This observation allows us to understand why this window does not represent a relevant region for heat transfer in red-rumped agouti (*Dasyprocta leporina*), a species in which cooling due to water evaporation in the nasal mucosa can affect the temperature of this region [66].

Given these findings, it is evident that this thermal window can help identify and understand emotional states, especially in primates. Nevertheless, the contradictions and difficulties in associating microvascular vasomotor responses to a specific emotional event have made it necessary to explore its validity more widely and whether the response in primates could be found in other species. Without a doubt, the use of these thermal windows to evaluate the health status of zoo animals is a field that requires further studies to determine their usefulness. Preliminary results of the authors' work with gazelles and deer, for example, have found that animals with clinical signs of respiratory disease have alterations in the surface temperature of the nose, as shown in Figure 8.

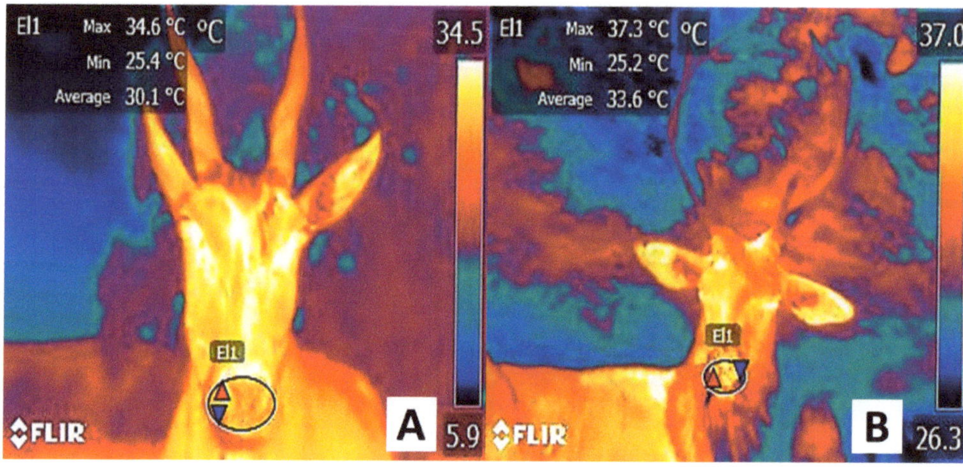

Figure 8. Comparison of the thermal state of the nasal region in healthy and sick animals. (**A**). Temperature of the nasal region in a gazelle (*Gazelle gazella*). This animal presents no health alterations or clinical signs, and its overall health condition is good, as shown by the maximum temperature (red triangle) of 34.6 °C in the nasal region (El1) and an average value of 30.1 °C. (**B**). Temperature of the nasal region in a deer (*Odocoileus virginianus*). In this case, an adynamic animal with upper respiratory disease and nasal discharge presents a maximum nasal temperature (El1) (red triangle) of 37.3 °C and an average temperature of 33.6 °C. Compared to the healthy animal, a difference between 2.7° and 3.5 °C can be seen. This may be due to a pathological process that includes the release of cytokines and prostaglandin E_2, which can cause a feverish state and promote the activation of thermoregulation mechanisms such as vasodilation to effectuate heat loss.

4. Body Windows

4.1. Thoracic Region (Regio Costalis) (Regio Pectoris)

Evidence from facial regions indicates that they can be used as thermal windows to determine physiological parameters such as body temperature and respiratory rate, which can help verify individuals' health status [67–69]. One example is the case of water buffaloes, a species for which the thoracic window has been proposed to evaluate thermal comfort [70,71]. Body windows, like the thoracic region, have been mainly used to identify and follow up on inflammatory processes and infectious diseases, where IRT can assist by monitoring lesions and achieving timely diagnoses of hyperemia and hypothermia disorders [70,71].

The anatomical characteristics of this region and the network of blood vessels that supply muscles like the pectoral fascia (*fascia pectoralis*), ventral thoracic serratus (*serratus ventralis thoracis*), transversus thoracis (*transversus thoracis*), and thoracic rectum (*rectus thoracis*) during inflammatory or pathological processes generate the secretion of proinflammatory mediators, such as prostaglandins, serotonin, histamine, and interleukins. These substances cause local vasodilation of the capillaries that irrigate the muscles and, consequently, irradiate greater heat, which is detectable using IRT [72].

Menzel et al. [58] observed that the temperature in the thoracic region was 2.4 °C higher than in the abdominal region when determining the health status of the lungs of domesticated species such as pigs. This increase correlated positively with the thickness of the lung tissue layer and local temperature increases ($r_S = 0.50$, $p = 0.03$). In this context, a later study carried out with the same species related to febrile responses detected by IRT in animals inoculated with *Actinobacillus pleuropneumoniae* found a positive correlation between rectal and surface temperatures ($r^2 = 0.52$, $p < 0.001$), as the former was significantly higher than the latter (0.5 ± 0.3 °C vs. 0.3 ± 0.3 °C) ($p < 0.0001$), though this did not represent a meaningful difference ($p > 0.05$) [73]. Both of those studies showed the clinical capacity of

IRT to detect local pulmonary inflammatory processes [74,75], though it performed poorly in identifying febrile states, as results were similar to those found in studies of the ocular thermal window.

What remains to be determined is the specificity of the thoracic region for detecting increases in surface temperature as a result of local pathological processes involving prostaglandins F2α (PGF2α) or interleukin-1 (IL-1) that cause inflammation of the surrounding lung tissue [76]. As da Costa et al. reported [56], IRT has been utilized in this area to evaluate injuries generated by trauma. Those authors analyzed the IRT of 42 felines and canines, including the maned wolf (*Chrysocyon brachyurus*), crab-eating fox (*Cerdocyon thous*), hoary fox (*Pseudalopex vetulus*), jaguar (*Panthera onca*), hoar puma (*Puma concolor*), and ocelot (*Leopardus pardalis*). In the dorsal region of the thorax and limbs, temperatures increased significantly by 3.1–3.3 °C, indicating an inflammatory process accompanied by clinical signs and confirmed by radiographic diagnoses. Although these findings helped detect conditions in the corresponding area, the authors reported that the density and thickness of the animals' coats could affect thermal readings. The authors determined that because thicker coats trap more stable air, it is impossible to measure the radiation emitted by the epidermis accurately.

As reported by Hilsberg [1] in a review of IRT findings in different species of wild animals under human care, the thoracic window has been suggested for evaluating local flank lesions in Asian elephants (*E. maximus*) and white (*Ceratotherium simum*) and black rhinoceroses (*Rhinoceros unicornis*). In these animals, modifications of their daily activities were associated with lesions in the thoracic area of the right flank, making it possible to achieve early diagnoses with no need to immobilize the subjects. The author further stated that, unlike domestic species (such as the water buffalo and domestic cattle), the thorax region is not considered an effective window for determining thermal comfort, thus, this is a limitation that arises when validating this region [38].

Subsequently, the thoracic thermal window was proposed as a potential region that could assist in identifying pathological conditions, such as infectious diseases, and recognizing hyperthermia derived from the secretion of proinflammatory substances. Moreover, during the development of lesions in this area, vasodilatation and increases in local temperature allow the non-invasive detection of possible affected areas and, based on the information obtained by IRT, implement medical imaging or laboratory techniques to confirm diagnoses. However, species-specific factors, such as hair density and color, must be considered when using or validating this window because they can affect readings.

4.2. Abdominal (Regio Abdominis)

The abdominal region of zoo animals is used less frequently as a window than previously mentioned regions. However, the irrigation that flows from the caudal aorta artery (*arteria aorta abdominalis*) and its branches to the abdominal organs or the dermal areas of the *regio abdominis cranialis*, *medius*, and *caudalis* quadrants provide the anatomical basis for evaluating physiological states such as gestation in some species [60] (Figure 9).

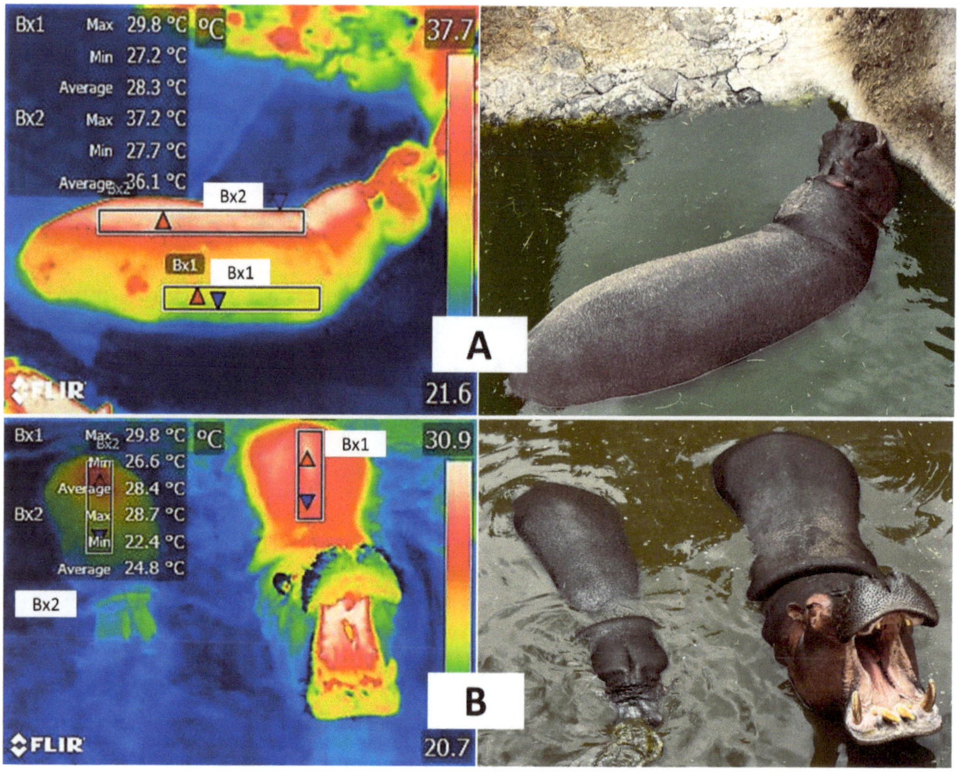

Figure 9. Importance of ponds or pools to improve the comfort level of hippopotami housed in zoos. (**A**). Dorsal and lateral body window with total immersion in a pond. A difference of 7.8 °C can be seen between the mean temperatures of the lateral (Bx1) and dorsal regions (Bx2). This is due to the effect of solar radiation that falls directly on the dorsal region, which, in contact with the lateral region with the water, can dissipate heat more efficiently and maintain lower temperatures. (**B**). Effect of immersion in water on the thermoregulation of a mother and her calf. A decrease of 3.6° °C is observed in the mean temperature in the offspring's dorsal region (Bx2) in contrast to that of the mother (Bx1). This temperature drop is associated with the fact that the calf's skin was moist when the image was taken. Both cases reflect the importance of ponds or pools to help maintain thermoneutrality in animals with poor thermoregulation mechanisms, like the hippopotamus (*Hippopotamus amphibius*). (Radiometric images and photographs by the authors).

In rhinoceroses, IRT has detected increases in *regio abdominis medius* and *regio abdominis lateralis* temperatures from the end of the third trimester of gestation, with increases of up to 6 °C in the uterine zone compared to maternal body temperatures and decreases in the temperature of this zone (of approximately 2.6 °C) after giving birth [9]. Similarly, in a babirusa (*Babyrousa babyrussa*) housed at the Wilhelma Botanical and Zoological Garden, IRT was utilized as a complementary technique during gestation from August 2016 to January 2017. Krueger et al. [77] delimited a thermal window as a rectangular area enclosing the hindleg, and foreleg, spine, abdomen, and teats. The maximum temperatures of the right and left side of that thermal window correlated with the weeks of gestation (R = 0.66, R = 0.61, $p < 0.05$, respectively), with changes detected from the third trimester, but no differences between the two sides ($p > 0.05$). More specifically, the flank temperature reached 29 °C five days before parturition, and teat temperature increased from 27 °C in the fourth week of gestation to 32 °C five days before calving. This increase in the irradiated temperature of the teats was more evident than in other areas of the thermal window because the teats have glabrous skin due to an increase in energy demand for lactation. The authors also mentioned that one advantage of this window is that prolonged exercise or activity does not affect thermographic readings and, therefore, does not interfere with interpretations [9].

As has occurred in domesticated animals such as horses, in which pregnant mares show increases of 2 °C in the abdominal region compared to non-pregnant females [78], in larger zoo species, such as the white (*Ceratotherium simum*) and black rhinoceroses (*Diceros bicornis*), giraffe, and elephant, gestation can be diagnosed by superficial dermal changes during the first trimester [1]. Furthermore, IRT is suggested for use with rhinoceroses as a tool for detecting inflammatory processes and temperature alterations in animals with dermal pathologies [79].

Another application of IRT in the abdominal region is in evaluating the thermal needs of species such as big captive cats, including lions (*P. leo*), tigers (*P. tigris*), jaguars (*P. onca*), amur tigers (*P. tigris altaica*), pumas (*P. concolor*), and snow leopards (*P. uncia*). In these cases, IRT has been used to determine the influence of the conditions of captivity and the microenvironment on the animals' thermoregulatory behavior [21]. Thermography was also used to observe the functionality and effectiveness of heat dissipation behaviors in nine Bengal tigers (*Panthera tigris tigris*), especially lying in a dorsal decubitus posture with exposure to the loins. Although this behavior was not prevalent in felines (between <4 and 13.1%), the surface temperature of the area increased as a vasomotor mechanism to dissipate heat during hot weather. This finding led the author to suggest the *regio pubica* as a promising field of study and recommended determining its validity as a thermal window and thermal threshold in this species. In addition, that study mentions the importance of considering key design factors of the installations, such as the use of glass in exhibits, which reflects solar radiation and can heat the enclosure due to short- and long-wavelength radiation, thus impacting the thermal comfort of captive animals [80]. In other studies, conducted with wild felines such as the jaguar (*Panthera onca*), jaguarundi (*Herpailurus yagouaroundi*), and oncilla (*Leopardus tigrinus*), IRT has been applied to detect conditions such as dermatitis, pododermatitis, and periodontitis. The authors of those reports affirm that IRT has a high sensitivity for evaluating facial (*faciei regions*) and appendicular regions (*membri thoracici and membri pelvini regions*), but low diagnostic sensitivity in regions such as the trunk (*regio pectoris*) and abdomen (*regio abdominis*), due to the presence of thick fur that interferes with thermographic readings [56]. Other windows suggested for these species are at the facial level (eye, mouth, tongue) [11]. They could help assess thermoregulation in wild felines [13].

In contrast to previous reports on koalas (*Phascolarctos cinereus*), Narayan et al. [26] studied the body regions of three captive individuals to determine the most suitable thermal window. In addition to the eyes, one window they could validate, the abdominal area had the highest consistency for measuring surface temperatures, with a coefficient variation of 11.01%, an average temperature of 22.13 °C, and visibility of 48.54% in radiometric images. This can be suggested as an option for this marsupial and other zoo animals when the aim is to evaluate their heat exchange efficiency, as well as for physiological states such as gestation, during which higher estrogen levels promote increases in irradiated temperatures, thus facilitating identification by IRT.

5. Podal Windows (Regio Carpi, Regio Tarsi)

The podal or phalangeal thermal window has been utilized mainly for joint pathologies or inflammatory injuries. Figure 10 shows the irrigation of this region by arteries such as the dorsal metatarsal (*metatarsea dorsalis*), medial tarsal (*tarsea medialis*), and *tarsea perforans* in the case of the pelvic limb, or the dorsal carpal (*carpeus palmaris*) and palmar (*palmaris*) arterial branches in the case of the thoracic limb [60]. When an inflammatory process develops due, for example, to the presence of bacteria, immune cells (neutrophils or macrophages) trigger the release of proinflammatory substances (e.g., serotonin, histamine, prostaglandin, interleukins) that cause local hyperemia [81], an effect that can be detected by IRT [1,27].

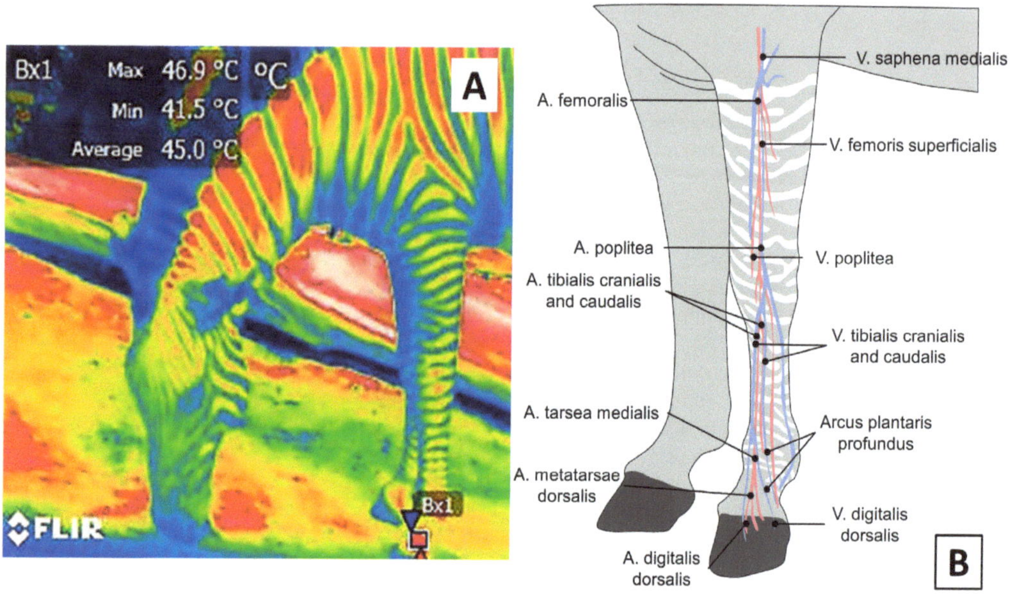

Figure 10. Podal window. (**A**). The podal region assessed using a rectangle that proximally encompasses the metacarpophalangeal joint down to the distal phalanx. (**B**). This window has been used to detect inflammatory processes in paws that generate the secretion of pro-inflammatory substances and, consequently, greater heat radiation.

In large herbivores such as giraffes, rhinoceroses, elephants, and hippopotami, it is often difficult to observe the exact location of injured areas on limbs, especially when no clinical signs are visible [27]. In these cases, IRT can aid in identifying injured areas [19]. Dunbar et al. [29] evaluated changes in the surface temperature of infected areas in mule deer (*Odocoileus hemionus*) infected experimentally with the foot-and-mouth disease virus. In their subjects, it was possible to identify an increase in the maximum temperature in the podal region of 2 °C two days post-infection ($p = 0.017$). Given this initial observation, the authors proposed that detecting thermal body changes may make it possible to identify inflammatory responses that would help diagnose podal and joint pathologies and monitor claudication and pododermatitis in the species mentioned above [13,27,55,82].

Working with Asian elephants (*E. maximus*), Avni-Magen et al. [55] evaluated the sensitivity and specificity of IRT in the foot region to identify inflammatory conditions. The regions evaluated were divided into three groups (one negative, one positive, and one pre-inflammatory). In the group of pre-inflammatory areas, temperatures were 2 °C higher than in the healthy areas. The authors also reported that IRT showed a sensitivity of 89.2% and a specificity of 83.4%, with a predictive value of 99.4%. These results demonstrate the usefulness of this window for this type of inflammatory disorder in large mammals. Similarly, in bird species, Duncan et al. [83] evaluated the foot region of 67 penguins for 15 months, including specimens of king (*Aptenodytes patagonicus*), macaroni (*Eudyptes chrysolophus*), and southern rockhopper penguins (*Eudyptes chrysocome*). They found that when an active lesion was present, the affected area showed a temperature 1.5 °C higher than in inactive or uninjured areas, with a positive correlation between foot and lesion site temperature (rho= 0.94, $p = 0.0001$). In addition, Porter-Blackwell et al. [84] evaluated 10 Falconiformes of the red-tailed (*Buteo jamaicensis*), Harris' (*Parabuteo unicinctus*), and Swainson's hawks (*Buteo swainsoni*) species, to study tarsal and plantar thermal responses in limbs with injuries in these areas. They found a significant correlation between the metatarsal and plantar temperatures when abrasive lesions were present (r = 0.81).

At the foot level, IRT has been suggested as an effective method for evaluating thermoregulation processes in other species. Phillips and Sanborn [85], for example, evaluated the thermoregulatory capacity of birds such as the ostrich (*Struthio camelus*), emu (*Dromaius novaehollandiae*), and double-wattled cassowary (*Casuarius casuarius*), species in which the legs constitute a high percentage of the total anatomy and are involved in heat regulation and exchange. It has been reported that the feet and toes of those birds make up 12–17.5% of the total body surface area and that 40% of the metabolic heat they produce is eliminated through these surfaces; thus, they should be considered important sources of heat dissipation. In a study of otters, Kuhn and Meyer [23] found that vasomotor responses in the feet and tail are the main mechanisms for regulating heat loss. The sparsely-furred paws of the Eurasian otter (*L. lutra*) are the only body parts where significant heat loss may occur. When otters are in the water or at low temperatures, vasoconstriction in their skin prevents excessive heat loss, while vasodilation occurs to dissipate excess heat favored, as well, by the interdigital membranes that increase the surface area available for heat exchange.

As a result, in animals where diagnoses of foot or footpad lesions are common, IRT can identify potential lesions that could affect locomotor activity and the quality of life of individuals, as well as the thermoregulatory role of the legs in some species. To date, however, its sensitivity and specificity have only been determined in elephants. Given this scarce evidence, the possibility of validating this window is limited, complicated by the reality that factors such as a generalized inflammatory response may impede precisely identifying the affected areas.

6. Limitations on the Use of IRT in Wildlife Species under Human Care

Scientific evidence has shown that IRT has clinical and diagnostic applications with zoo animals [64]. However, effectively implementing this tool requires considering factors that can influence recordings and their interpretation. Hilsberg-Merz [1] mentions that the quality and length of an animal's coat can prevent elements such as mud or dust from interfering with thermographic readings. Other elements to be considered are species-specific thermoregulation mechanisms, morpho-anatomical characteristics such as coat color patterns, hair length, dermal thickness, location of glands, size of ears or appendages such as antlers, and the presence or absence of hair, fur, or feathers [55]. Coat colors and patterns can influence readings and heat loss during short-wave solar radiation interactions. In this case, darker coats have higher absorptivity and generate higher IRT readings, though this only corresponds to higher transmissivity in certain cases. The primary effect is that, as the coat assumes a higher temperature, there is a less favorable thermal gradient for heat exchange through radiation between the epidermis and the coat. Likewise, as Mota Rojas et al. [86] mentioned, hair characteristics such as length and color variation also influence heat loss or gain. Figure 11 shows how the color of the hair in bicolor animals like zebras generates a distinct heat radiation pattern that helps maintain thermostability in these species.

Figure 11. Effect of coat color on the surface temperature of a zebra (*Equus quagga*). It has been suggested that the zebra's stripes play a role in thermoregulation. IRT images show that the average temperature of the white stripes in the neck is 35.9 °C (Bx1), while the average temperature of the zone with black hair is 47.9 °C (Bx2), a difference of 11 °C. This evidence reaffirms the importance of considering hair color in some species when using IRT since it can influence the amount of solar radiation that may be absorbed. (Radiometric images and photographs by the authors).

Similarly, the presence of feathers in birds and their color and length can alter the body's ability to dissipate heat [87]. Feathers are known as a thermal insulating mechanism that maintains the temperature of birds within an average value of 40 °C. Therefore, when applying IRT techniques to these species, the role of feathers must be considered to interpret their thermoregulatory capacity accurately [88]. Figure 12 presents thermal images of different bird species to exemplify the importance of this factor.

Figure 12. Differences in the thermoregulatory performance of birds with different types of feathers. (**A**). Radiometric image of a military macaw (*Ara militaris*) at rest on a perch. In this species, the ocular window can serve to evaluate thermal responses by detecting average ocular temperatures (El1) (35.2 °C). (**B**). The dorsal region (EL1) of flamingos (*Phoenicopterus ruber*) covered by feathers registered an average value of 21.5 °C despite direct exposition to solar radiation. (**C**). In contrast to the light-colored feathers of the flamingo, the dorsal area (El1) of a black vulture (*Coragyps atratus*) shows an average temperature of 62.8 °C with a maximum value of 68.2 °C. In the digital image of the vulture, a spread-wing posture is observed. This is a thermoregulatory behavior that helps dissipate heat when birds experience temperatures above their thermoneutral range. (**D**). Infrared reading of an emu (*Dromaius novaehollandiae*) where the temperature in the dorsal (Bx1) and ventral regions (Bx2) can be compared, observing an average difference of 6.2 °C. This difference may be attributable to the presence of long feathers that aid in dissipating heat. (Radiometric images and photographs by the authors). Maximum temperature (red triangle), and minimum (blue triangle). Thermal images obtained using a FLIR thermal camera.

Despite thermography's clinical and diagnostic applications, climatic factors that may influence readings, such as exposure to direct solar radiation, precipitation, wind, and humidity, must all be factored in [89,90]. One example of this is a study of large felines in distinct habitats, where the presence of water influenced readings by as much as 1 °C. Findings like those confirm the need to acclimatize the animals for up to two hours before taking radiometric images [21].

The corpus of evidence reviewed suggests that implementing IRT in zoos and conservation centers could serve as a method for selecting and monitoring ideal habitat design, which must meet the needs of each particular species [91]. In practical terms, using IRT with these species could generate information on thermal comfort levels with respect to the characteristics of their microenvironments and help determine the correct features for promoting their welfare and appropriate behaviors.

7. Perspectives for Thermography Applied to Zoo Animals, and Other Windows Suggested

Although various windows have been identified for most species of animals in captivity, some studies propose specific anatomical regions for each animal. In the case of capybaras (*Hydrochoerus hydrochaeris*), IRT was used to monitor the health of adult and young animals by taking thermograms of the facial, joints, and head regions. In these animals, the temperature increases registered in the supranasal gland of resting adults made it possible to improve preventive medicine measures with no need for handling or containment [92].

IRT has been used with the genital region of chimpanzees, as well as with the animals' facial, auricular, and nasal areas, and their hands, neck, and feet, to identify the adaptive behaviors that females have developed to conceal gestation and avoid infanticidal behavior, a common practice of the males in the group. Observations of these females showed that, in contrast to the monitoring of gestation in rhinoceroses, pregnant chimpanzees had similar temperatures to non-pregnant females in the genital tumescence period during the swelling stage, with a maximum difference of approximately 1 °C during the other stages. Their findings led the authors to conclude that in this species, IRT is not a reliable tool for detecting gestation due to the adaptive behaviors of females [57]. In contrast, the vaginal sheath of elephants and rhinoceroses is suggested as an adequate region for detecting stages of the reproductive cycle, such as estrus. Other authors have recorded temperature increases in relation to reproductive behaviors such as mounts by males [1].

In Malayan sun bears (*Helarctos malayanus*) and polar bears (*Ursus maritimus*) at temperatures of 5–30 °C and 0–20 °C, respectively, the surface temperature of their eyes, snout, face, trunk, shoulder region, inner leg, and outer leg were compared to determine the range of the thermoneutral zone by identifying an absence of thermal exchange on the body surface, a phenomenon that is necessary for both species. It has been suggested that IRT could help assess the thermoneutrality zone in Malayan bears by relating the thermal aspect to autonomic and behavioral responses. The authors found a proportional increase in IRT when ambient temperatures increased in all windows tested except the nasal window. A range of 24–28 °C was registered as ideal for those bears. In other results, when environmental temperatures reached 15 °C (a value below the minimum range), numerous behaviors were altered, including resting, foraging, nesting, social play, physical activity, and standing [93]. The study of polar bearsfound that the test animals showed signs of heat stress at temperatures of 12–19 °C, with temperature increases of 18–24.5 °C in the shoulders and back. Their findings allowed the authors to establish a range in which body and environmental temperatures were close (24–28 °C) and to suggest that the zone of thermoneutrality for this species is limited. Finally, they ascertained that, in polar bears, the extremities and shoulder region are the main areas of heat loss [20]. Thermal images of black bears (*Ursus americanus*) taken by the authors (Figure 13) have proven useful in identifying differences in heat radiation associated with the characteristics of the enclosure.

In that case, providing natural shade prevents direct solar radiation from reaching the animals, thus reducing the presentation of heat stress.

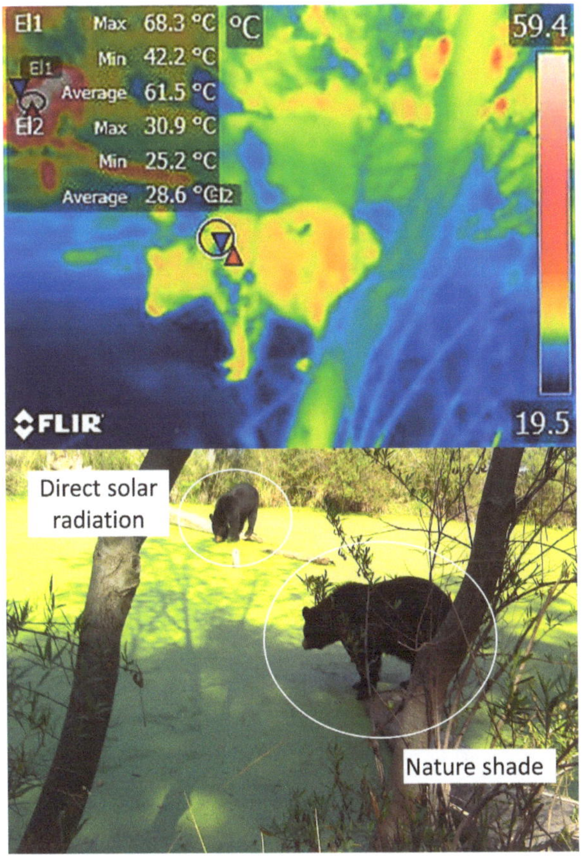

Figure 13. Comparison of the surface temperature in the dorsal area of two American black bears (*Ursus americanus*). Upon assessing the influence of shaded areas on the thermal state of these animals, it is clear that the bear standing under the shade (right side of the thermogram) has an average dorsal superficial temperature of 28.6 °C. In comparison, the individual exposed to direct solar radiation recorded an average value of 61.5 °C. The difference of 32.9 °C between these two animals is an important trait that needs to be considered when designing enclosures for wildlife under human care (Radiometric images and photographs by the authors).

In addition, somewhat unconventional applications have been explored in other windows, such as the nasal one. For example, Dunbar et al. [29], evaluated temperature variations in regions such as the dorsal, ocular, and nasal windows in raccoons (*Procyon lotor*) inoculated with the rabies virus. The authors found an increase of 30.4 ± 3.5 °C in the nasal window of the infected animals before the onset of the prodromic phase of the disease ($p < 0.0001$). For this reason, thermographic studies of this kind have been proposed as indicators contributing to both animal and public health.

In light of these findings, future study trends should focus on establishing precise applications for each thermal window according to the characteristics of individual animal species. The objective is to implement non-invasive methods to monitor and optimize thermal responses, health, and environmental conditions of wild animals under human care [9].

8. Conclusions

IRT is a tool with potential applications in wild animals under human care. Its non-invasiveness and capacity for remote evaluation are attractive qualities for assessing the health status of wild species with no need for means of chemical or physical restraint, mainly when used in conjunction with other techniques.

In several species of felines, herbivores, birds, canines, and non-human primates, IRT has identified infectious pathologies, inflammatory processes derived from injuries, and physiological states such as gestation. Efforts have also been undertaken to use IRT to understand the emotional states of animals better. However, as we know from experience with humans and domesticated animals, IRT has individual, environmental, and technical limitations that must be evaluated when interpreting the thermal responses of animals.

The importance of implementing these techniques in zoo animals is not limited to animal welfare, though this is a priority for wildlife species under human care. For veterinarians and zookeepers, IRT can also provide information about the conditions of enclosures and how their design can alter the thermoregulatory mechanisms of diverse species by providing environments that better satisfy their biological needs, prevent heat stress, and facilitate preventive health monitoring. Another benefit of applying IRT in zoos is the potential contributions that findings could make to our scientific knowledge of wild animals in captivity that would be impossible to obtain in wild conditions. Finally, the information obtained by utilizing IRT in zoos can be applied to wild populations. Approaches of this kind would support the role of zoos as wildlife research and conservation centers, where techniques such as IRT are available to evaluate, objectively and non-invasively, the daily activities of animals, aid in preserving positive mental states, and, in that way, ensure proper clinical, behavioral, physiological, habitat, and emotional welfare.

Author Contributions: All authors have contributed to the conceptualization, writing, reading, and approval of the final manuscript. All authors have read and agreed to the published version of the manuscript.

Funding: This research received no external funding.

Institutional Review Board Statement: Not applicable.

Informed Consent Statement: Not applicable.

Data Availability Statement: Not applicable.

Conflicts of Interest: The authors declare that they have no conflict of interest.

References

1. Hilsberg-Merz, S. Infrared Thermography in Zoo and Wild Animals. In *Zoo and Wild Animal Medicine*; Elsevier: Amsterdam, The Netherlands, 2008.
2. Cannas, S.; Palestrini, C.; Canali, E.; Cozzi, B.; Ferri, N.; Heinzl, E.; Minero, M.; Chincarini, M.; Vignola, G.; Dalla Costa, E. Thermography as a Non-Invasive Measure of Stress and Fear of Humans in Sheep. *Animals* **2018**, *8*, 146. [CrossRef] [PubMed]
3. Leite, R.F.; Rodrigues, L.A.; Freitas, N.M.A.; Nogueira, F.R.B.; da Silva, L.J.; De Souza, B.B.; de Miranda Neto, E.G. Principles and Cautions in the Infrared Thermography Application in Sheep. *Multidiscip. Rev.* **2022**, *5*, e2022001. [CrossRef]
4. Elias, B.; Starling, M.; Wilson, B.; McGreevy, P. Influences on Infrared Thermography of the Canine Eye in Relation to the Stress and Arousal of Racing Greyhounds. *Animals* **2021**, *11*, 103. [CrossRef] [PubMed]
5. Dela Ricci, G.; Silva-Miranda, K.O.; Titto, C.G. Infrared Thermography as a Non-Invasive Method for the Evaluation of Heat Stress in Pigs Kept in Pens Free of Cages in the Maternity. *Comput. Electron. Agric.* **2019**, *157*, 403–409. [CrossRef]
6. Giannetto, C.; Di Pietro, S.; Falcone, A.; Pennisi, M.; Giudice, E.; Piccione, G.; Acri, G. Thermographic Ocular Temperature Correlated with Rectal Temperature in Cats. *J. Therm. Biol.* **2021**, *102*, 103104. [CrossRef] [PubMed]
7. Arfuso, F.; Rizzo, M.; Giannetto, C.; Giudice, E.; Fazio, F.; Piccione, G. Age-Related Changes of Serum Mitochondrial Uncoupling 1, Rumen and Rectal Temperature in Goats. *J. Therm. Biol.* **2016**, *59*, 47–51. [CrossRef]
8. Arfuso, F.; Acri, G.; Piccione, G.; Sansotta, C.; Fazio, F.; Giudice, E.; Giannetto, C. Eye Surface Infrared Thermography Usefulness as a Noninvasive Method of Measuring Stress Response in Sheep during Shearing: Correlations with Serum Cortisol and Rectal Temperature Values. *Physiol. Behav.* **2022**, *250*, 113781. [CrossRef]

9. Hilsberg-Merz, S. Infrared Thermography in Zoo Animals: New Experiences with This Method, Its Use in Pregnancy and Inflammation Diagnosis and Survey of Environmental Influences and Thermoregulation in Zoo Animals. *Eur. Assoc. Zoo Wildl. Vet.* **1998**, *3*, 397–409.
10. Jerem, P.; Herborn, K.; McCafferty, D.; McKeegan, D.; Nager, R. Thermal Imaging to Study Stress Non-Invasively in Unrestrained Birds. *J. Vis. Exp.* **2015**, *2015*, 53184. [CrossRef]
11. Mota-Rojas, D.; Martínez-Burnes, J.; Casas-Alvarado, A.; Gómez-Prado, J.; Hernández-Ávalos, I.; Domínguez-Oliva, A.; Lezama-García, K.; Jacome-Romero, J.; Rodríguez-González, D.; Pereira, A.M.F. Clinical Usefulness of Infrared Thermography to Detect Sick Animals: Frequent and Current Cases. *CABI Rev.* **2022**, *22*, 1–27. [CrossRef]
12. Speakman, J.; Ward, S. Infrared Thermography: Principles and Applications. *Zool. complex Syst.* **1998**, *101*, 224–232.
13. Church, J.; Cook, N.; Schaefer, A. *Recent Applications of Infrared Thermography for Animal Welfare and Veterinary Research: Everything from Chicks to Elephants*; InfraMation: Las Vegas, NV, USA, 2009.
14. Whitham, J.; Miller, L. Using Technology to Monitor and Improve Zoo Animal Welfare. *Anim. Welf.* **2016**, *25*, 395–409. [CrossRef]
15. McCafferty, D.J. The Value of Infrared Thermography for Research on Mammals: Previous Applications and Future Directions. *Mamm. Rev.* **2007**, *37*, 207–223. [CrossRef]
16. Gómez-Prado, J.; Pereira, A.M.F.; Villanueva-García, D.; Wang, D.; Domínguez-Oliva, A.; Mora-Medina, P.; Hernández-Avalos, I.; Martínez-Burnes, J.; Casas-Alvarado, A.; Olmos-Hernández, A.; et al. Thermoregulation mechanisms and perspectives for validating thermal windows in pigs with hypothermia and hyperthermia: An overview. *Front. Vet. Sci.* **2022**, *9*, 1023294. [CrossRef]
17. Mota-Rojas, D.; Wang, D.; Titto, C.G.; Gómez-Prado, J.; Carvajal-de la Fuente, V.; Ghezzi, M.; Boscato-Funes, L.; Barrios-García, H.; Torres-Bernal, F.; Casas-Alvarado, A.; et al. Pathophysiology of Fever and Application of Infrared Thermography (IRT) in the Detection of Sick Domestic Animals: Recent Advances. *Animals* **2021**, *11*, 2316. [CrossRef]
18. Rekant, S.I.; Lyons, M.A.; Pacheco, J.M.; Arzt, J.; Rodriguez, L.L. Veterinary Applications of Infrared Thermography. *Am. J. Vet. Res.* **2016**, *77*, 98–107. [CrossRef]
19. Hilsberg-Merz, S. Aspekte Zur Klinischen Anwendung Der Infrarot-Thermographie in Der Zoo- Und Wildtiermedizin. Ph.D. Thesis, Universität Leipzig, Leipzig, Germany, 2000.
20. Schneider, M. Behavioural and Autonomic Thermoregulation in Malayan Sun Bears (*Helarctos Malayanus*) and Polar Bears (*Ursus Maritimus*). Ph.D. Thesis, Universität zu Köln, Köln, Germany, 2015.
21. Stryker, J.A. Thermoregulatory Behaviour Assessment and Thermal Imaging of Large Felids. Ph.D. Thesis, University of Guelph, Guelph, ON, Canada, 2016.
22. Weissenböck, N.M.; Weiss, C.M.; Schwammer, H.M.; Kratochvil, H. Thermal Windows on the Body Surface of African Elephants (*Loxodonta Africana*) Studied by Infrared Thermography. *J. Therm. Biol.* **2010**, *35*, 182–188. [CrossRef]
23. Kuhn, R.; Meyer, W. Infrared Thermography of the Body Surface in the Eurasian Otter Lutra Lutra and the Giant Otter Pteronura Brasiliensis. *Aquat. Biol.* **2009**, *6*, 143–152. [CrossRef]
24. Okrouhlík, J.; Burda, H.; Kunc, P.; Knížková, I.; Šumbera, R. Surprisingly Low Risk of Overheating during Digging in Two Subterranean Rodents. *Physiol. Behav.* **2015**, *138*, 236–241. [CrossRef]
25. van Vuuren, A.K.; Kemp, L.V.; McKechnie, A.E. The Beak and Unfeathered Skin as Heat Radiators in the Southern Ground-hornbill. *J. Avian Biol.* **2020**, *51*, jav.02457. [CrossRef]
26. Narayan, E.; Perakis, A.; Meikle, W. Using Thermal Imaging to Monitor Body Temperature of Koalas (*Phascolarctos Cinereus*) in A Zoo Setting. *Animals* **2019**, *9*, 1094. [CrossRef] [PubMed]
27. Hilsberg-Merz, S. Clinical Application of Infrared-Thermography in Inflammation Diagnosis in Mega-Herbivores. In Proceedings of the European Association of Zoo- and Wildlife Veterinarians (EAZWV) 4th Scientific Meeting, Heidelberg, Germany, 8–12 May 2002; EAZWV: Heidelberg, Germany, 2002; pp. 315–320.
28. Hurley-Sanders, J.L.; Bowman, K.F.; Wolfe, B.A.; Nutter, F.B.; Sladky, K.K.; Stoskopf, M.K. Use of Thermography and Fluorescein Angiography in the Management of a Chilean Flamingo with Avascular Necrosis of the Wing. *J. Avian Med. Surg.* **2012**, *26*, 255–257. [CrossRef] [PubMed]
29. Dunbar, M.R.; MacCarthy, K.A.; Dunbar, M.R.; MacCarthy, K.A. Use of Infrared Thermography to Detect Signs of Rabies Infection in Raccoons (*Procyon Lotor*). *J. Zoo Wildl. Med.* **2006**, *37*, 518–523. [CrossRef] [PubMed]
30. Dunbar, M.R.; Johnson, S.R.; Rhyan, J.C.; McCollum, M. Use of Infrared Thermography to Detect Thermographic Changes in Mule Deer (*Odocoileus Hemionus*) Experimentally Infected with Foot-and-Mouth Disease. *J. Zoo Wildl. Med.* **2009**, *40*, 296–301. [CrossRef] [PubMed]
31. Hurley-Sanders, J.L.; Sladky, K.K.; Nolan, E.C.; Loomis, M.R. Use of Cortical Bone Fenestration, Autogenous Free Skin Graft, and Thermography for Wound Treatment and Monitoring in a Red Wolf (*Canis Rufus Gregoryi*). *J. Zoo Wildl. Med.* **2015**, *46*, 617–620. [CrossRef] [PubMed]
32. Nakayama, K.; Goto, S.; Kuraoka, K.; Nakamura, K. Decrease in Nasal Temperature of Rhesus Monkeys (*Macaca Mulatta*) in Negative Emotional State. *Physiol. Behav.* **2005**, *84*, 783–790. [CrossRef]
33. Heintz, M.R.; Fuller, G.; Allard, S. Exploratory Investigation of Infrared Thermography for Measuring Gorilla Emotional Responses to Interactions with Familiar Humans. *Animals* **2019**, *9*, 604. [CrossRef]
34. Ross, S.R.; Lake, B.R.; Fultz, A.; Hopper, L.M. An Evaluation of Thermal Imaging as a Welfare Monitoring Tool for Captive Chimpanzees. *Primates* **2021**, *62*, 919–927. [CrossRef]

35. Stewart, M.; Webster, J.; Schaeder, A. Infrared Thermography and Heart Rate Variability for Non-Invasive Assessment of Animal Welfare. *ANZCCART Hum. Sci. News* **2008**, *21*, 1–6.
36. Schaefer, A.L.; Cook, N.; Tessaro, S.V.; Deregt, D.; Desroches, G.; Dubeski, P.L.; Tong, A.K.W.; Godson, D.L. Early Detection and Prediction of Infection Using Infrared Thermography. *Can. J. Anim. Sci.* **2004**, *84*, 73–80. [CrossRef]
37. Lush, J.; Ijichi, C. A Preliminary Investigation into Personality and Pain in Dogs. *J. Vet. Behav.* **2018**, *24*, 62–68. [CrossRef]
38. Mota-Rojas, D.; Pereira, M.F.A.; Wang, D.; Martínez-Burnes, J.; Ghezzi, M.; Hernández-Ávalos, I.; Lendez, P.; Mora-Medina, P.; Casas, A.; Olmos-Hernández, A.; et al. Clinical Applications and Factors Involved in Validating Thermal Windows in Large Rumiants to Assess Health and Productivity. *Animals* **2021**, *11*, 2247. [CrossRef]
39. Butterfield, C.; Grumpelt, B.; Kimmel, D.; Patterson, R.; Jones, K.; Scott, S.L.; Schaefer, A. The Pretransport Management of Stress in Performance Horses. *J. Equine Vet. Sci.* **2018**, *69*, 145–148. [CrossRef]
40. Melero, M.; Rodríguez-Prieto, V.; Rubio-García, A.; García-Párraga, D.; Sánchez-Vizcaíno, J.M. Thermal Reference Points as an Index for Monitoring Body Temperature in Marine Mammals. *BMC Res. Notes* **2015**, *8*, 411. [CrossRef]
41. South, K.E. Infrared Thermography of *Erinaceus Europaeus*: Applications with Hypothermia and Hibernation. Ph.D. Thesis, University of Plymouth, Plymouth, UK, 2018.
42. Pizzi, R.; Dowling, A.; Brown, D.; Girling, S.; Pearson, S.; Bacon, H.; Pereira, Y.M. Facial Thermography Is Not Useful in Assessing Body Temperature in Common Squirrel Monkeys (*Saimiri Sciureus*) in Comparison to Rectal Temperatures. *J. Zoo Aquarium Res.* **2015**, *3*, 94–98.
43. Jansson, A.; Lindgren, G.; Velie, B.D.; Solé, M. An Investigation into Factors Influencing Basal Eye Temperature in the Domestic Horse (*Equus Caballus*) When Measured Using Infrared Thermography in Field Conditions. *Physiol. Behav.* **2021**, *228*, 113218. [CrossRef]
44. Heintz, M.R.; Fuller, G.A.; Allard, S.M. Welfare Assessment Using Infrared Thermography and Hormonal Measures During a Giraffe Feeding Program with Visitors. Available online: https://assets.speakcdn.com/assets/2332/heintz_giraffeirtaza2017poster.pdf (accessed on 12 September 2022).
45. Vaz, J.; Bartley, A.; Hunt, J. Personality Matters: Exploring the Relationship between Personality and Stress Physiology in Captive African Lions. *BMC Zoo.* **2021**, *7*, 30. [CrossRef]
46. Ijichi, C.; Evans, L.; Woods, H.; Yarnell, K. The Right Angle: Validating a Standardised Protocol for the Use of Infra-Red Thermography of Eye Temperature as a Welfare Indicator. *Anim. Welf.* **2020**, *29*, 123–131. [CrossRef]
47. Bukhari, S.S.U.H.; McElligott, A.G.; Parkes, R.S.V. Quantifying the Impact of Mounted Load Carrying on Equids: A Review. *Animals* **2021**, *11*, 1333. [CrossRef]
48. Meleppat, R.K.; Fortenbach, C.R.; Jian, Y.; Martinez, E.S.; Wagner, K.; Modjtahedi, B.S.; Motta, M.J.; Ramamurthy, D.L.; Schwab, I.R.; Zawadzki, R.J. In Vivo Imaging of Retinal and Choroidal Morphology and Vascular Plexuses of Vertebrates Using Swept-Source Optical Coherence Tomography. *Transl. Vis. Sci. Technol.* **2022**, *11*, 11. [CrossRef]
49. Meleppat, R.K.; Zhang, P.; Ju, M.J.; Manna, S.K.; Jian, Y.; Pugh, E.N.; Zawadzki, R.J. Directional Optical Coherence Tomography Reveals Melanin Concentration-Dependent Scattering Properties of Retinal Pigment Epithelium. *J. Biomed. Opt.* **2019**, *24*, 1. [CrossRef] [PubMed]
50. Phillips, P.K.; Heath, J.E. Heat Exchange by the Pinna of the African Elephant (*Loxodonta Africana*). *Comp. Biochem. Physiol. Part A Physiol.* **1992**, *101*, 693–699. [CrossRef] [PubMed]
51. Isaza, R.; Hunter, R. Drug Delivery to Captive Asian Elephants—Treating Goliath. *Curr. Drug Deliv.* **2004**, *1*, 291–298. [CrossRef] [PubMed]
52. Rees, P.A. Asian Elephants (*Elephas Maximus*) Dust Bathe in Response to an Increase in Environmental Temperature. *J. Therm. Biol.* **2002**, *27*, 353–358. [CrossRef]
53. Koffi, M.; Andreopoulos, Y.; Jiji, L.M. The Role of Pinnae Flapping Motion on Elephant Metabolic Heat Dissipation. *J. Heat Transfer* **2014**, *136*, 101101. [CrossRef]
54. Marchant, G.H.; Shoshani, J. Head Muscles of *Loxodonta Africana* and *Elephas Maximus* with Comments on *Mammuthus Primigenius* Muscles. *Quat. Int.* **2007**, *169–170*, 186–191. [CrossRef]
55. Avni-Magen, N.; Zaken, S.; Kaufman, E.; Kelmer, G. Use of Infrared Thermography in Early Diagnosis of Pathologies in Asian Elephants (*Elephas Maximus*). *Isr. J. Vet. Med.* **2017**, *72*, 22–27.
56. da Costa, A.L.M.; Rassy, F.B.; Cruz, J.B. da Diagnostic applications of infrared thermography in captive brazilian canids and felids. *Arch. Vet. Sci.* **2020**, *25*, 2. [CrossRef]
57. Dezecache, G.; Zuberbühler, K.; Davila-Ross, M.; Dahl, C.D. Skin Temperature Changes in Wild Chimpanzees upon Hearing Vocalizations of Conspecifics. *R. Soc. Open Sci.* **2016**, *4*, 160816. [CrossRef]
58. Diaz-Piedra, C.; Gomez-Milan, E.; Di Stasi, L.L. Nasal Skin Temperature Reveals Changes in Arousal Levels Due to Time on Task: An Experimental Thermal Infrared Imaging Study. *Appl. Ergon.* **2019**, *81*, 102870. [CrossRef]
59. Matsuno, S.; Mizuno, T.; Asano, H.; Mito, K.; Itakura, N. Estimating Autonomic Nerve Activity Using Variance of Thermal Face Images. *Artif. Life Robot.* **2018**, *23*, 367–372. [CrossRef]
60. International Committee on Veterinary Gross Anatomical Nomenclature. *Nomina Anatómica Veterinaria*; World Association of Veterinary Anatomists: Hanover, Germany, 2017; pp. 1–178.
61. Ermatinger, F.A.; Brügger, R.K.; Burkart, J.M. The Use of Infrared Thermography to Investigate Emotions in Common Marmosets. *Physiol. Behav.* **2019**, *211*, 112672. [CrossRef]

62. Parr, L.; Heintz, M. Facial Expression Recognition in Rhesus Monkeys, *Macaca Mulatta*. *Anim. Behav.* **2009**, *77*, 1507–1513. [CrossRef]
63. Chotard, H.; Ioannou, S.; Davila-Ross, M. Infrared Thermal Imaging: Positive and Negative Emotions Modify the Skin Temperatures of Monkey and Ape Faces. *Am. J. Primatol.* **2018**, *80*, 1–35. [CrossRef]
64. McFarland, R.; Barrett, L.; Fuller, A.; Hetem, R.S.; Porter, W.P.; Young, C.; Henzi, S.P. Infrared Thermography Cannot Be Used to Approximate Core Body Temperature in Wild Primates. *Am. J. Primatol.* **2020**, *82*, e23204. [CrossRef]
65. Thompson, C.L.; Williams, S.H.; Glander, K.E.; Teaford, M.F.; Vinyard, C.J. Body Temperature and Thermal Environment in a Generalized Arboreal Anthropoid, Wild Mantled Howling Monkeys (*Alouatta Palliata*). *Am. J. Phys. Anthropol.* **2014**, *154*, 1–10. [CrossRef]
66. De Queiroz, J.P.A.F.; de Souza, J.B.F.; de Morais Oliveira, V.R.; de Souza Castelo, T.; Tavares Dantas, M.R.; de Macedo Costa, L.L.; de Oliveira, M.F. Sensible Heat Transfer and Thermal Windows in *Dasyprocta Leporina* (Mammalia, Rodentia). *Biol. Rhythm Res.* **2019**, *50*, 232–244. [CrossRef]
67. Farrar, K.L.; Field, A.E.; Norris, S.L.; Jacobsen, K.O. Comparison of Rectal and Infrared Thermometry Temperatures in Anesthetized Swine (*Sus Scrofa*). *J. Am. Assoc. Lab. Anim. Sci.* **2020**, *59*, 221–225. [CrossRef]
68. Wang, F.-K.; Shih, J.-Y.; Juan, P.-H.; Su, Y.-C.; Wang, Y.-C. Non-Invasive Cattle Body Temperature Measurement Using Infrared Thermography and Auxiliary Sensors. *Sensors* **2021**, *21*, 2425. [CrossRef]
69. Lowe, G.; Sutherland, M.; Waas, J.; Schaefer, A.; Cox, N.; Stewart, M. Infrared Thermography—A Non-Invasive Method of Measuring Respiration Rate in Calves. *Animals* **2019**, *9*, 535. [CrossRef]
70. Mota-Rojas, D.; Napolitano, F.; Braghieri, A.; Guerrero-Legarreta, I.; Bertoni, A.; Martínez-Burnes, J.; Cruz-Monterrosa, R.; Gómez, J.; Ramírez-Bribiesca, E.; Barrios-García, H.; et al. Thermal Biology in River Buffalo in the Humid Tropics: Neurophysiological and Behavioral Responses Assessed by Infrared Thermography. *J. Anim. Behav. Biometeorol.* **2021**, *9*, 1–12. [CrossRef]
71. Mota-Rojas, D.; Wang, D.; Titto, C.G.; Martínez-Burnes, J.; Villanueva-García, D.; Lezama-García, K.; Domínguez, A.; Hernández-Avalos, I.; Mora-Medina, P.; Verduzco, A.; et al. Neonatal infrared thermography images in the hypothermic ruminant model: Anatomical-morphological physiological aspects and mechanisms for thermoregulation. *Front. Vet. Sci.* **2022**, *9*, 963205. [CrossRef] [PubMed]
72. Vainionpää, M. *Thermographic Imaging in Cats and Dogs Usability as a Clinical Method*; University of Helsinki: Helsinki, Finland, 2014.
73. Loughmiller, J.A.; Spire, M.F.; Dritz, S.S.; Fenwick, B.W.; Hosni, M.H.; Hogge, S.B. Relationship between Mean Body Surface Temperature Measured by Use of Infrared Thermography and Ambient Temperature in Clinically Normal Pigs and Pigs Inoculated with *Actinobacillus Pleuropneumoniae*. *Am. J. Vet. Res.* **2001**, *62*, 676–681. [CrossRef] [PubMed]
74. Menzel, A.; Beyerbach, M.; Siewert, C.; Gundlach, M.; Hoeltig, D.; Graage, R.; Seifert, H.; Waldmann, K.-H.; Verspohl, J.; Hennig-Pauka, I. Actinobacillus Pleuropneumoniae Challenge in Swine: Diagnostic of Lung Alterations by Infrared Thermography. *BMC Vet. Res.* **2014**, *10*, 199. [CrossRef] [PubMed]
75. Menzel, A.; Siewert, C.; Gasse, H.; Seifert, H.; Hoeltig, D.; Hennig-Pauka, I. Infrared Thermography of the Pig Thorax: An Assessment of Selected Regions of Interest by Computed Tomographical and Anatomical Parameters. *Anat. Histol. Embryol.* **2015**, *44*, 107–117. [CrossRef]
76. Petersson, M.; Uvnäs-Moberg, K.; Nilsson, A.; Gustafson, L.-L.; Hydbring-Sandberg, E.; Handlin, L. Oxytocin and Cortisol Levels in Dog Owners and Their Dogs Are Associated with Behavioral Patterns: An Exploratory Study. *Front. Psychol.* **2017**, *8*, 1796. [CrossRef]
77. Krueger, F.; Knauf-Witzens, T.; Getto, S. New Approach in Thermal Pregnancy Diagnosis: Teat's Heating in Babirusa (*Babyrousa Babyrussa*). *Theriogenology* **2019**, *133*, 144–148. [CrossRef]
78. Bowers, S.; Gandy, S.; Anderson, B.; Ryan, P.; Willard, S. Assessment of Pregnancy in the Late-Gestation Mare Using Digital Infrared Thermography. *Theriogenology* **2009**, *72*, 372–377. [CrossRef]
79. Miller, M.A.; Buss, P.E. Chapter 55—Rhinoceridae (Rhinoceroses). *Fowler's Zoo Wild Anim. Med.* **2015**, *8*, 538–547. [CrossRef]
80. Stryker, J.A.; Atkinson, J.L.; Brown, R.D.; Barney, D.; Robinson, J.A.B.; Duncan, J.; Finegan, E.J. Behavioral Repertoire Assessment of Bengal Tigers (*Panthera Tigris*) with Focus on Thermoregulatory Behavior. *Int. J. Biometeorol.* **2019**, *63*, 1369–1379. [CrossRef]
81. Galindo, F.; Broom, D.M. The Effects of Lameness on Social and Individual Behavior of Dairy Cows. *J. Appl. Anim. Welf. Sci.* **2002**, *5*, 193–201. [CrossRef]
82. Giggin, T.; Narayanan, M.K. Advancements in Diagnostic Imaging of Elephant Feet. In *Reviews of Veterinary Research-What Next?* Lulu Publication: Morrisville, NC, USA, 2021; pp. 152–176. ISBN 978-1-300-21732-9.
83. Duncan, A.E.; Torgerson-White, L.L.; Allard, S.M.; Schneider, T. An Evaluation of Infrared Thermography for Detection of Bumblefoot (Pododermatitis) in Penguins. *J. Zoo Wildl. Med.* **2016**, *47*, 474–485. [CrossRef]
84. Porter-Blackwell, R.; Paul-Murphy, J.R.; le Jeune, S.S.; Thermography, C.V.; Martínez-López, B.; Seibert, B.A. Pilot Study: Correlation of the Surface Skin Temperature Between the Leg and Foot Using Thermographic Imaging in Captive Hawks. *J. Avian Med. Surg.* **2020**, *34*, 164. [CrossRef]
85. Phillips, P.K.; Sanborn, A.F. An Infrared, Thermographic Study of Surface Temperature in Three Ratites: Ostrich, Emu and Double-Wattled Cassowary. *J. Therm. Biol.* **1994**, *19*, 423–430. [CrossRef]
86. Mota-Rojas, D.; Titto, C.G.; de Mira Geraldo, A.; Martínez-Burnes, J.; Gómez, J.; Hernández-Ávalos, I.; Casas, A.; Domínguez, A.; José, N.; Bertoni, A.; et al. Efficacy and Function of Feathers, Hair, and Glabrous Skin in the Thermoregulation Strategies of Domestic Animals. *Animals* **2021**, *11*, 3472. [CrossRef]

87. Graveley, J.M.F.; Burgio, K.R.; Rubega, M. Using a Thermal Camera to Measure Heat Loss Through Bird Feather Coats. *J. Vis. Exp.* **2020**, *1*, 160. [CrossRef]
88. Pap, P.L.; Osváth, G.; Daubner, T.; Nord, A.; Vincze, O. Down Feather Morphology Reflects Adaptation to Habitat and Thermal Conditions across the Avian Phylogeny. *Evolution* **2020**, *74*, 2365–2376. [CrossRef]
89. Vainionpää, M.; Tienhaara, E.-P.; Raekallio, M.; Junnila, J.; Snellman, M.; Vainio, O. Thermographic Imaging of the Superficial Temperature in Racing Greyhounds before and after the Race. *Sci. World J.* **2012**, *2012*, 182749. [CrossRef]
90. Fortin, D.; Larochelle, J.; Gauthier, G. The Effect of Wind, Radiation and Body Orientation on the Thermal Environment of Greater Snow Goose Goslings. *J. Therm. Biol.* **2000**, *25*, 227–238. [CrossRef]
91. Cilulko, J.; Janiszewski, P.; Bogdaszewski, M.; Szczygielska, E. Infrared Thermal Imaging in Studies of Wild Animals. *Eur. J. Wildl. Res.* **2013**, *59*, 17–23. [CrossRef]
92. da Silva, L.G.; Siqueira, N.M.C.; Gama, D.B.F.; de Souza Martins, J.V.; de Oliveira Lima, R.; de Oliveira, L.V.; Mateus, R.G.; Rita, P.H.S. Termografia de Capivaras (*Hydrochoerus Hydrochaeris*) Do Perímetro Urbano de Campo Grande-MS. *Pubvet* **2019**, *13*, 148. [CrossRef]
93. Schneider, M.; Ziegler, T.; Kolter, L. Thermoregulation in Malayan Sun Bears (*Helarctos Malayanus*) and Its Consequences for in Situ Conservation. *J. Therm. Biol.* **2020**, *91*, 102646. [CrossRef] [PubMed]

animals

Review

Assessment of Pain and Inflammation in Domestic Animals Using Infrared Thermography: A Narrative Review

Alexandra L. Whittaker [1], Ramon Muns [2], Dehua Wang [3], Julio Martínez-Burnes [4], Ismael Hernández-Ávalos [5], Alejandro Casas-Alvarado [6], Adriana Domínguez-Oliva [2,6] and Daniel Mota-Rojas [6,*]

1 School of Animal and Veterinary Sciences, Roseworthy Campus, University of Adelaide, Roseworthy, SA 5116, Australia
2 Agri-Food and Biosciences Institute, Hillsborough, Co Down BT 26 6DR, Northern Ireland, UK
3 School of Life Sciences, Shandong University, Qingdao 266237, China
4 Facultad de Medicina Veterinaria y Zootecnia, Universidad Autónoma de Tamaulipas, Victoria City 87000, Mexico
5 Clinical Pharmacology and Veterinary Anesthesia, Facultad de Estudios Superiores Cuautitlán, Universidad Nacional Autónoma de México (UNAM), Cuautitlán 54714, Mexico
6 Neurophysiology, Behaviour and Animal Welfare Assessment, DPAA, Xochimilco Campus, Universidad Autónoma Metropolitana, Mexico City 04960, Mexico
* Correspondence: dmota100@yahoo.com.mx

Simple Summary: Current acute pain assessment protocols in domestic animals require the use of a suite of methods to evaluate the multidimensional components of pain. Thermal imaging is a method that assesses the physiological response, via heat production, to a noxious stimulus. This may correlate with the affective experience of acute pain. Since the use of the method as a pain assessment tool is still in its relative infancy and has yet to be fully evaluated empirically, the present review brings together current knowledge on the use of infrared thermography to evaluate acute pain in animals. We also propose future directions in this area of research.

Abstract: Pain assessment in domestic animals has gained importance in recent years due to the recognition of the physiological, behavioral, and endocrine consequences of acute pain on animal production, welfare, and animal model validity. Current approaches to identifying acute pain mainly rely on behavioral-based scales, quantifying pain-related biomarkers, and the use of devices monitoring sympathetic activity. Infrared thermography is an alternative that could be used to correlate the changes in the superficial temperature with other tools and thus be an additional or alternate acute pain assessment marker. Moreover, its non-invasiveness and the objective nature of its readout make it potentially very valuable. However, at the current time, it is not in widespread use as an assessment strategy. The present review discusses scientific evidence for infrared thermography as a tool to evaluate pain, limiting its use to monitor acute pain in pathological processes and invasive procedures, as well as its use for perioperative monitoring in domestic animals.

Keywords: nociception; inflammatory response; invasive procedures; surgery; analgesia; castration; teeth clipping; infrared thermography (IRT)

Citation: Whittaker, A.L.; Muns, R.; Wang, D.; Martínez-Burnes, J.; Hernández-Ávalos, I.; Casas-Alvarado, A.; Domínguez-Oliva, A.; Mota-Rojas, D. Assessment of Pain and Inflammation in Domestic Animals Using Infrared Thermography: A Narrative Review. *Animals* **2023**, *13*, 2065. https://doi.org/10.3390/ani13132065

Academic Editor: Éric Troncy

Received: 21 May 2023
Revised: 17 June 2023
Accepted: 20 June 2023
Published: 22 June 2023

1. Introduction

Pain is a multidimensional sensory and emotional experience that it is now accepted that animals can perceive, regardless of their inability to verbalize the experience, as discussed in the latest definition proposed by the International Association for the Study of Pain [1]. The current approach to pain management in animals is through categorizing physiological and behavioral changes in an attempt to provide timely treatment and reduce welfare impact [2]. However, behavioral-based pain scales, one of the most common strategies to assess this negative state, are often subjective and dependent on the observer. This impacts the reliability and confidence we can have in the findings [3,4].

Infrared thermography (IRT) has been proposed as an objective method to evaluate acute pain, with a basis in the physiological response during pain perception [5,6]. Temperature variations can be measured using IRT after a nociceptive stimulus due to the activation of the nervous system and its vasomotor consequences, inducing an inflammatory process and an increase or decrease in heat radiation [7,8]. Therefore, theoretically, IRT can assist in monitoring the physiological function and pain status of animal patients. Moreover, the non-invasiveness of IRT makes it a tool that can be used in animals with complex management [9].

However, there is still controversy as to whether this temperature variation response is associated with pain level and if it is reliable as a guide to adjusting analgesic treatment during invasive procedures in animals. Likewise, selecting the anatomical region or thermal window is essential since the nociceptive process induces peripheral vasoconstriction or vasodilation depending on the evaluated region, influencing the amount of heat radiation [10,11]. Research on IRT is still limited; thus, this review will bring together current research on IRT applied to evaluate acute pain in domestic animals, discuss gaps in knowledge, and propose avenues for future investigation restricted to acute pain monitoring.

2. Search Methodology

For the present review, a literature search was performed in databases such as Web of Science, Scopus, and PubMed. The following keywords were used: "domestic animal infrared thermography", "animals inflammatory process", "animal surgery", "acute painful surgical procedures in animals" and "neurobiology of acute pain in animals". Inclusion criteria were that there was full-text access, they were written in English or Spanish, and there was a discussion of IRT in the context of evaluating pain during inflammatory diseases or surgical procedures. Likewise, those articles mentioning the neurobiology of pain and its physiological changes were used to describe the physiological basis of IRT. Articles were excluded if they used IRT to assess body temperature but did not establish any association with acute pain. In total, 147 references were cited in the present review.

3. Neurobiology of Pain and Its Association with the Superficial Thermal Response

Pain presence is highly relevant to the health and well-being status of animals. The pain experience leads to behavioral and physiological consequences which differ in magnitude based on intensity [3,12,13]. An understanding of the neurobiological process underlying pain perception is necessary to understand the relationship between pain and thermal response.

The neurobiological process of pain consists of five phases, which, according to authors such as Bell [14] and Lamont et al. [15], are (1) transduction, (2) transmission, (3) modulation, (4) projection, and (5) perception [16]. Transduction is the transformation of mechanical, thermal, and chemical stimuli to an electrical impulse due to nociceptors present in peripheral tissues [17,18]. Transient receptor potential (TRP), particularly TRPV1, TRPV2, TRPV3, TRPV4, TRPM8, and TRPA1, participates in this phase by transducing noxious stimuli [17,19,20]. These receptors alter the action potential when they detect temperature disturbances [21]. For example, Kozyreva et al. [22] evaluated anesthetized rats receiving two cold thermal stimuli types. They found that TRPM8 activation by cold, with the use of menthol as an agonist of this receptor, increased oxygen consumption and induced peripheral vasoconstriction to shift the heat and maintain core temperature. The use of in vitro models has shown that activating TRPV1 channels in response to hot thermal stimuli increase the entrance of K^+ or Na^+ ions to promote energy expenditure and thermogenesis by decreasing visceral fat accumulation [23,24]. These examples show that TRPs initiate the response to thermal disturbances and painful sensations that are subsequently transmitted by type Aδ and C nociceptive fibers [25].

The activation of nociceptors during the transduction and transmission of a painful stimulus triggers different thermal responses that can be local (inflammation) or systemic (autonomic response). To achieve this, the modulation of the signal in the dorsal horn of the spinal cord suppresses or enhances the projection and perception of pain in cerebral structures [14,26,27]. Different brain regions activate due to pain, including the primary and secondary somatosensory cortices, cingulate cortex, insular cortex, and amygdala [28,29].

The central nervous system (CNS) processing contributes to the thermal response since the nerve signal is transmitted through the spinothalamic tract, resulting in the neurosecretion of catecholamines (epinephrine and norepinephrine). This also occurs in the locus coeruleus and the thalamus, transmitting nociceptive information to the somatosensory cortex, hypothalamus, and hippocampus [30,31]. Through this mechanism, pain induces the activation of the hypothalamic-pituitary-adrenal axis, the sympathetic nervous system, and, consequently, the release of glucocorticoids [32,33]. On the other hand, activation of the sympathetic-adrenomedullary system alters cardiovascular function due to catecholamine secretion by activating adrenoceptors in the vascular endothelium [34,35]. This event causes hypertension, tachycardia, tachypnea, and changes in peripheral and central temperature [35]. Redaelli et al. [36] used IRT to monitor rodents with induced spinal cord injury. In these animals, the temperature of the interscapular region, where brown adipose tissue is located, increased by 5 °C. This effect could be explained by thermogenesis, and the increase in the availability of energy resources is due to gluconeogenesis, lipolysis, and proteolysis [37]. Therefore, these mechanisms may clarify the value of IRT as an assistance tool for pain recognition [11,38].

Finally, when tissue injury occurs, an inflammatory process arises due to the release of pro-inflammatory substances such as cytokines, interleukins, prostaglandin, and ions, which can alter the action of the nociceptors, thus facilitating the activation of TRP channels, a phenomenon known as peripheral sensitization [39,40]. Local heat is a consequence of the vasodilation caused by pro-inflammatory mediators, altering the amount of dissipated heat [41]. Through these mechanisms, it has been suggested that IRT can be used to recognize processes such as laminitis, mastitis, or tendon injuries [5,42]. This evidence strongly suggests that the neurobiological process of pain is related to the thermal response, as shown in Figure 1, and this reaction can be identified using IRT.

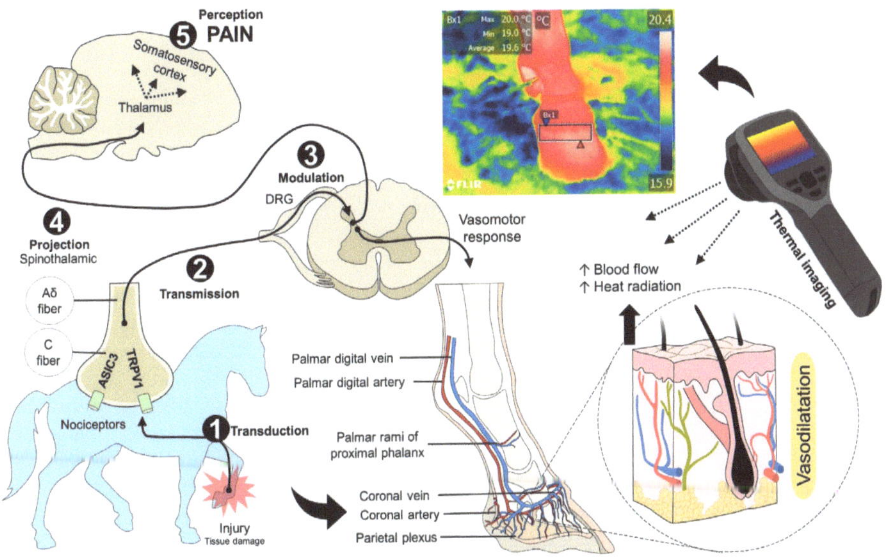

Figure 1. Nociceptive pathway during laminitis. The inflammatory process and tissue injury caused by laminitis trigger the first phase of nociception. (1) Transduction. The noxious stimulus is recognized

and transformed into an electrical signal by peripheral nerves (Aδ and C fibers), known as nociceptors. In these free nerve endings, receptors such as ASIC3 or TRPV1, among others, are activated to create action potentials that will be transmitted to the DRG of the spinal cord. (2) Transmission. Through Aδ and C fibers, the noxious input is transmitted to the spinal cord, to synapse with second-order neurons in the gray matter of the structure. (3) Modulation. Once the signal reaches the spinal cord, spinal interneurons are responsible for projecting or modulating the signal by releasing inhibitory or excitatory neurotransmitters. (4) Projection. Through the spinothalamic tract, the electrical signal reaches superior centers in the brain, mainly the thalamus. (5) Perception. From the thalamus, third-order neurons project to the somatosensory cortex, where the conscious recognition of pain is developed. Due to the interaction of the thalamus with other regions such as the hypothalamus, pain activates sympathetic centers, which causes physiological, endocrine, and behavioral responses. The vasomotor response, occurring as a result of pain and inflammation, causes vasodilation in the injured site to promote immune cells invasion to the injury site, with consequent healing. The increase in blood flow in the region also increases the amount of radiated heat from the skin. This element is captured by thermal cameras, helping to identify an inflammatory process in animals. ASIC: acid-sensing ion channel; DRG: dorsal root ganglion; TRPV1: transient potential receptor vanilloid 1.

4. Inflammatory Responses and Pain during the Pathological Process and Its Relation to Thermal Changes

Inflammatory responses associated with acute pain result from sudden or constant exposure to noxious stimuli and the interaction between pro- and anti-inflammatory mediators [43]. They serve as a mechanism for recovery from injury, trauma, sepsis, or infections. Nonetheless, mismanagement of acute inflammation can gradually lead to chronic disorders and persistent pain [44,45]. The labels of acute or chronic are generally based on a time construct, which typically ranges from a few days (acute) to weeks (chronic). There may also be differences in pain intensity associated with these different timelines; for example, chronic pain may be of a lesser intensity but prolonged.

Pain, as one of the cardinal signs of inflammation, is accompanied by an increase in core and surface temperature at the injury site—an element that can be monitored with IRT—due to the interaction between pro-inflammatory mediators and nociceptors. At the inflammatory site, prostaglandins, prostacyclin, interleukins (IL), leukotrienes, cytokines, histamine, serotonin, bradykinin, and alpha tumor necrosis factor (TNF-α), among others, induce vasodilatation and increase blow flow and permeability of blood vessels, emigration of leukocytes, migration by chemotaxis to the injury site, and the release of chemical mediators, causing signs such as redness, swelling, pain, loss of function, and local heat (Figure 2) [46–50].

Pain during an acute inflammatory process can be maintained by mast cells releasing histamine [51], bradykinin, and prostaglandins [52]. Pro-inflammatory substances such as Cytokines as IL-1β, IL-6, and TNF-α act on Toll-like receptors and TNF receptors [53], triggering various systemic changes, ranging from blood parameter alteration, organ failure, and even death, such as in the case of pneumonia in dogs (Figure 3) [54–57]. In the respiratory system, aspirated pathogens are removed by the mucociliary system and eliminated by alveolar macrophages and neutrophils. Alternatively, those reaching the lung by blood circulation are attacked by intravascular macrophages. In addition, phagocytosed pathogens by macrophages and dendritic cells are conducted to lymphoid tissues, allowing B and T Lymphocytes to contact pathogens and contributing to the humoral and cellular immune response. Antigen recognition in the lymphoid tissue leads to T cell proliferation and differentiation; hence, differentiated effector and memory T cells enter the circulation, migrate to the antigen site, activate macrophages, and kill ingested pathogens. However, activated cells in the bloodstream release cytokines producing cytokine storm and systemic inflammation [58].

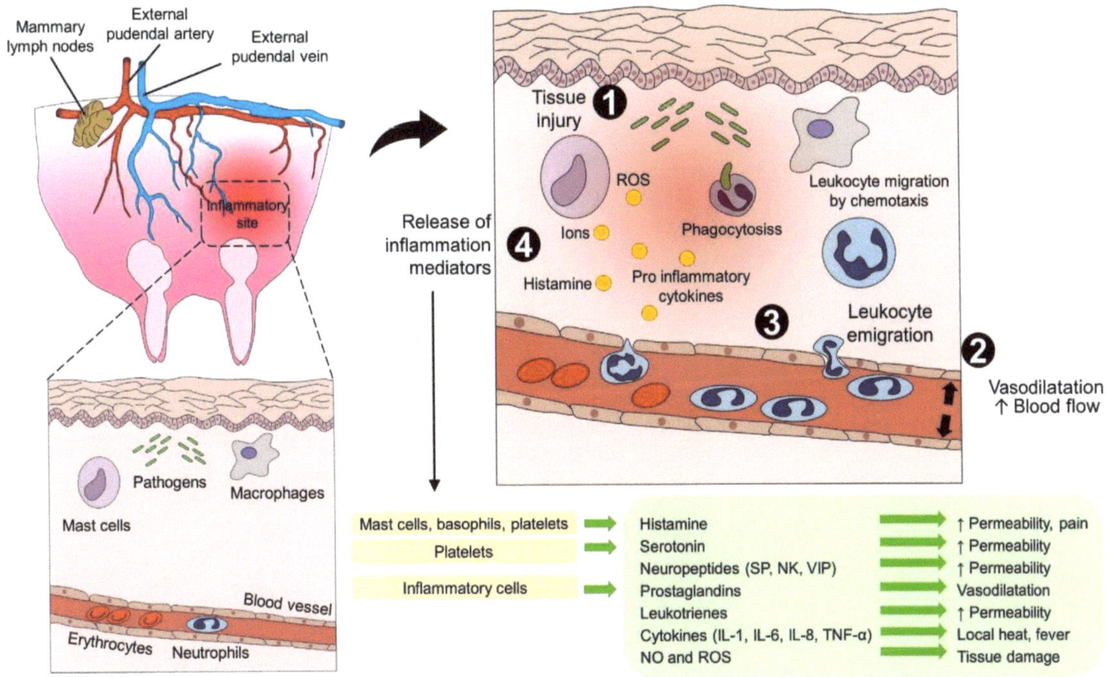

Figure 2. The local inflammatory response in the udder of a patient with mastitis. (1) The presence of a tissue injury or pathogens initiates inflammation by releasing local chemical compounds. (2) First-line mediators such as histamine, NO, ROS, and ions induce vasodilation to increase the permeability of the blood vessels and blood flow. (3) Leukocyte activation, emigration, and migration cause the presence and degranulation of neutrophils and other inflammatory cells into the injury site. (4) Release of pro-inflammatory substances such as prostaglandins, leukotrienes, cytokines, and neuropeptides, among others, exacerbates the inflammatory response to promote pathogens' phagocytosis, and healing IL: interleukin; NK: neurokinin; NO: nitric oxide; SP: substance P; VIP: vasoactive intestinal polypeptide.

Recently, IRT has been implemented in veterinary medicine. Infrared cameras can be divided into two types: cooled and uncooled array detectors. Cooled cameras have limited application for clinical use because they need to cool down before taking the image. Therefore, the process must be organized in advance [59]. On the other hand, uncooled thermal cameras use pyroelectric or ferroelectric materials for infrared radiation detection and pixel formation. Uncooled detectors constantly calibrate according to ambient temperature, being suitable for clinical application [60–63]. However, uncooled chamber detectors cannot provide a fixed absolute temperature due to constant calibration. Therefore, measuring the absolute temperature of the subject is not possible without compensatory programs for the internal temperature of the detector [64]. Moreover, IRT cameras also differ according to the number of lenses. Research-grade thermal cameras are characterized by three lenses that determine the focal length [65]. This is relevant because the capacity of the camera to detect thermal variability is higher in smaller spot-size lenses [65].

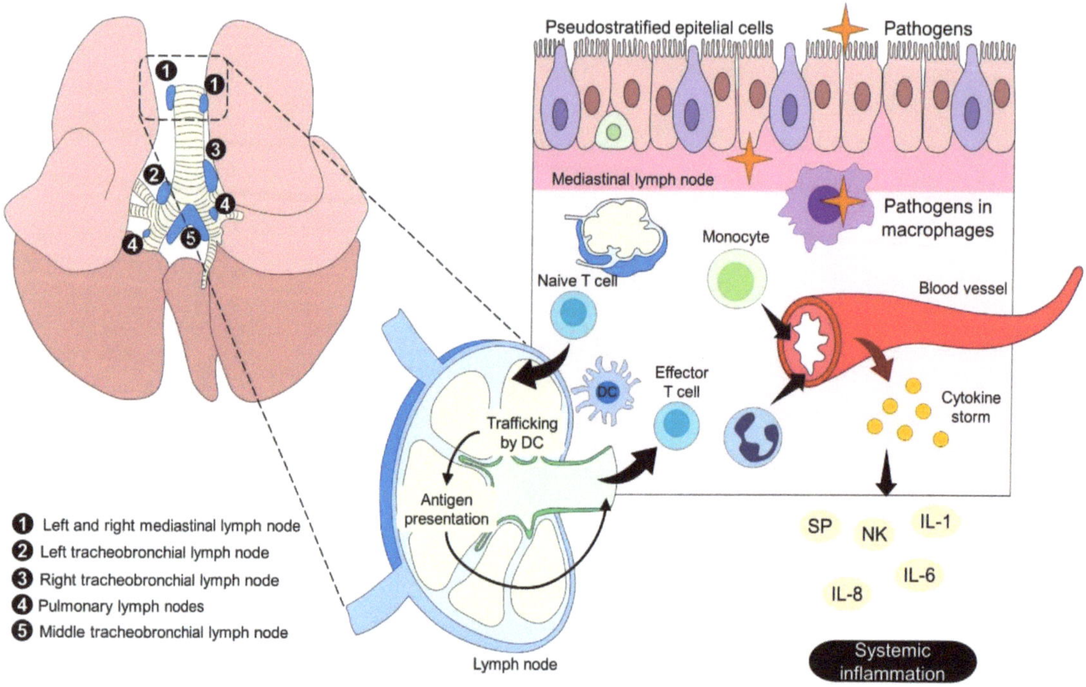

Figure 3. The systemic inflammatory response in pneumonia. Infection of the airway epithelium by pathogens leads to an inflammatory response where the mediastinal lymph node has a crucial role by participating in trafficking by DC, known as the take up of pathogens to the lymph node. Within the lymph node, antigen presentation takes place on naïve T cells to release effector T cells and induce the adaptative response of the organism. Degranulation of monocyte and polymorphic cells into the bloodstream causes a cytokine storm (SP, NK, IL), a reaction that promotes systemic inflammation and clinical signs. DC: dendritic cells; IL: interleukin; NK: neurokinin; SP: substance P.

These cameras have been used in animals to study diseases such as neoplasia [66], musculoskeletal disorders [42], vascular, metabolic disorders, and infectious conditions associated with fever [34,67]. For example, in equines, infectious colic accompanied by fever and pain has been assessed in preliminary studies. Figure 4 shows the thermal differences between a healthy mare and a mare with acute colic. In the sick horse, the inflammatory response arises from the presence of cytokines and vasoactive amines, causing vasomotor changes and altering the radiated heat detected by IRT. Contrary reactions were also found in these patients, with a decrease in ocular temperature, due to vasoconstriction, and an increase in caudal abdominal temperature due to vasodilation and cytokine release. In equines, common disorders such as colic are characterized by acute pain and local inflammation of the intestinal mucosa. Due to this event, as shown in Figure 3, IRT could help to identify these vascular changes. However, limited case reports have been published about this pathology and its association with IRT, limiting its use [68]. Subsequent studies by Schaefer et al. [69,70] have suggested that implementing IRT can help to diagnose febrile states in domestic animals due to infectious conditions. According to these same authors, in domestic cattle with Bovine Respiratory Disease Complex, they show an increase in maximum surface temperature due to an increase in neutrophils/lymphocytes [71]. Other studies reported an IRT efficiency in recognizing this disease between 71–80% when compared to standard diagnostic methods [72]. This would reinforce the recommendation of using IRT as an assistance tool for recognizing infectious states in farm animals [73–76], in addition to its application in humans to identify febrile patients in airports during the COVID-19 pandemic [77,78].

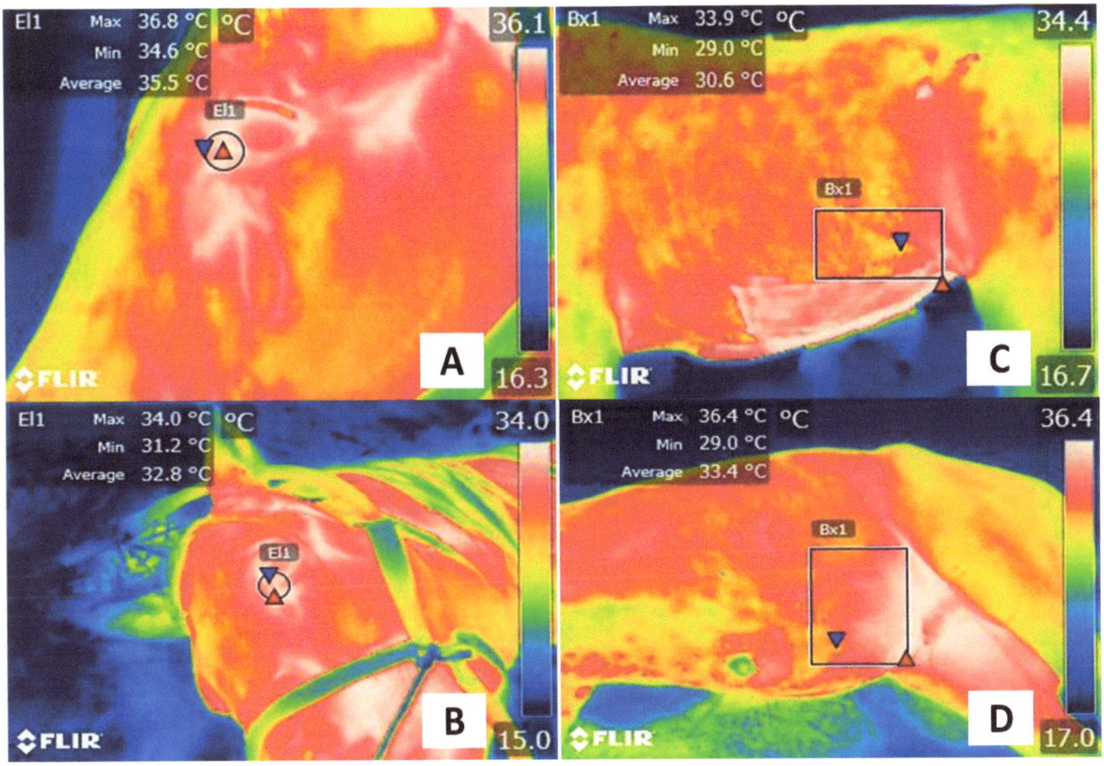

Figure 4. Superficial thermal response associated with acute pain in a horse with colic. (**A**,**C**). Healthy female quarter-horse patient. The maximum temperature in the lacrimal caruncle (El1) was 36.8 °C (red triangle), and the minimum was 34.6 °C (blue triangle). In the caudal abdominal region (Bx1), the surface temperature presented a maximum temperature of 33.9 °C (red triangle) and a minimum of 29 °C (blue triangle). (**B**) Thoroughbred female equine with acute pain associated with colic. It is observed that the maximum (red triangle) and minimum (blue triangle) temperatures of the lacrimal caruncle (El1) decreased by 2.8 °C and 3.4 °C, respectively, compared to the healthy horse. (**D**) Contrary to the thermal response in the lacrimal caruncle, in the caudal abdominal region (Bx1) of the sick animal, the maximum surface temperature increased by 2.5 °C. The thermal images were taken by the authors of the present review. Radiometric images were obtained using a T1020 FLIR thermal camera. Image resolution 1024 × 768; up to 3.1 MP with UltraMax. FLIR Systems, Inc. Wilsonville, OR, USA.

Mastitis is a prevalent disease in dairy cattle [79–81] that affects animals' health and causes economic loss due to a reduced milk yield [82,83]. It is a disease not always accompanied by clinical signs, although local inflammation and pain are commonly present in the udder. Therefore, the ability to diagnose this condition by IRT based on udder surface temperature would be valuable (Figure 5) [71]. Machado et al. [84] evaluated subclinical mastitis in 38 cows using thermal imaging of the udder. The authors observed a high correlation between the udder temperature and the somatic cell count. This relation was predominant in the posterior quarter of the udder, where the recorded temperature in the right quarter was 40.6 °C, while the left side recorded 39.8 °C, with correlation values of R2 = 0.97 and R2 = 0.88, respectively. This illustrates that the technique may not only be useful for binary evaluations based on the presence or absence of disease but may be able to detect severity.

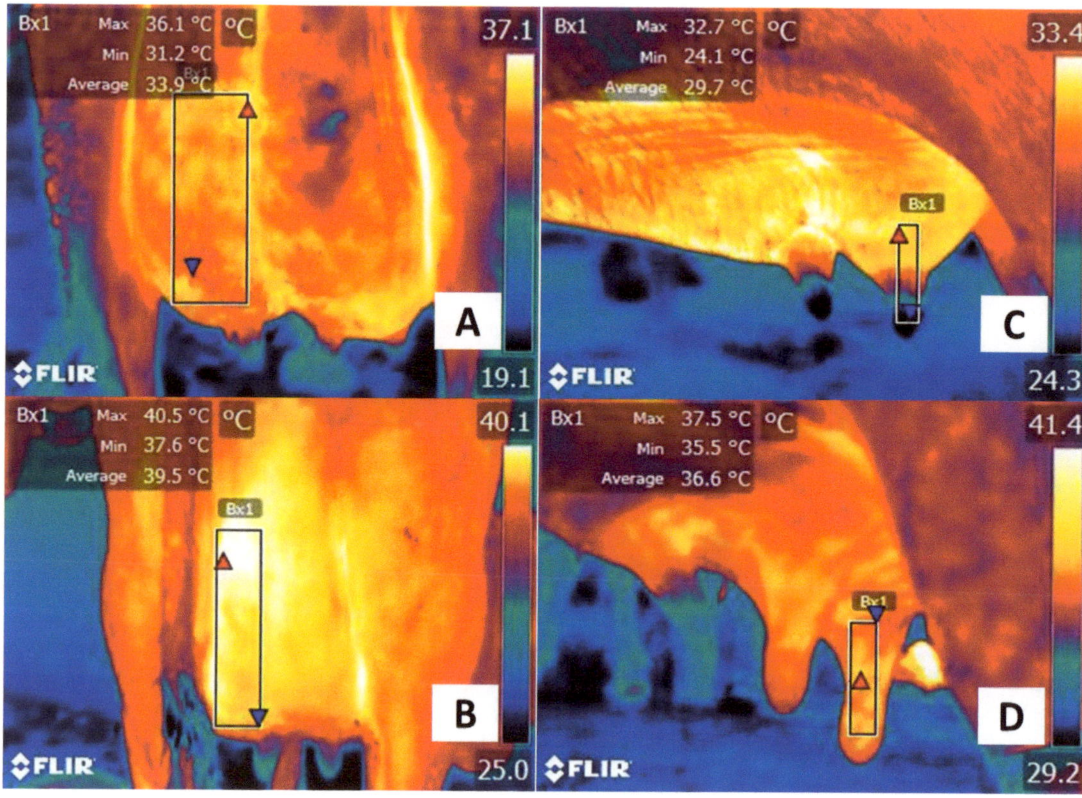

Figure 5. Thermographic evaluation of the thermal response of the mammary gland in dairy cows and buffaloes. (**A**) Healthy udder of a Holstein bovine. The surface temperature of the caudal quarter (Bx1) of the udder shows a maximum temperature of 36.1 °C (red triangle) and a minimum value of 31.2 °C (blue triangle). (**B**) Udder of a Holstein cow with mastitis. The maximum and minimum temperature of the caudal quarter (Bx1) is 4.4 °C and 6.4 °C higher, respectively. (**C**) Healthy udder in a Murrah water buffalo. The surface temperature of the healthy lateral caudal quarter (Bx1) recorded a maximum temperature of 32.7 °C (red triangle) and a minimum of 24.1 °C (blue triangle). (**D**) Udder with mastitis in dairy buffalo. Compared with the thermal response of the healthy udder, the temperature of the caudal quarter (Bx1) of the udder is 4.8 °C higher at the maximum temperature (red triangle) and 11.4 °C of the minimum temperature (blue triangle) compared to the temperature of the healthy udder. Infectious agents such as *Escherichia coli* or *Staphylococcus* sp. can trigger the release of several pro-inflammatory cytokines that cause vasodilation and increase heat radiation associated with acute pain. The thermal images were taken by the authors of the present review. Radiometric images were obtained using a T1020 FLIR thermal camera. Image resolution 1024 × 768; up to 3.1 MP with UltraMax. FLIR Systems, Inc. Wilsonville, OR, USA.

In another study, Zaninelli et al. [85] evaluated the udder health status of 310 dairy cows through somatic cell counts and IRT in 3 farms under field conditions. The authors observed that sick animals presented a maximum temperature of 36.08 ± 0.22 °C and a somatic cell count of $930.81 \pm 96.58 \times 10^3$. In contrast, healthy cows had a maximum temperature of 35.79 ± 0.15 °C and a somatic cell count of $592.38 \pm 71.40 \times 10^3$, finding a correlation between both variables. Thermographic research in infectious processes has highlighted that activation of the immune system can cause substances such as IL-1, IL6,

TNF-α, and PGE2 to cross the blood-brain barrier and, therefore, act on the anterior preoptic area of the hypothalamus, resulting in heat production and increasing the radiated energy, being able to differentiate with IRT the sick animal from the healthy ones [86,87].

Hoof and locomotion disorders, characterized by lameness (Figure 6), are other examples of chronic inflammatory conditions that have been evaluated with IRT. Weimer et al. [88] applied IRT to study bacterial chondronecrosis with osteomyelitis (BCO) secondary to laminitis in chickens. In the study, chickens with lameness had severe lesions caused by BCO diagnosed macroscopically. Lower temperatures were recorded in the hock joint, shank, tarsometatarsus, foot leg, and phalangeal regions. However, the investigators concluded that IRT was more accurate in identifying an increase in leg temperature in subclinical chondronecrosis lesions rather than a temperature drop in clinical lesions.

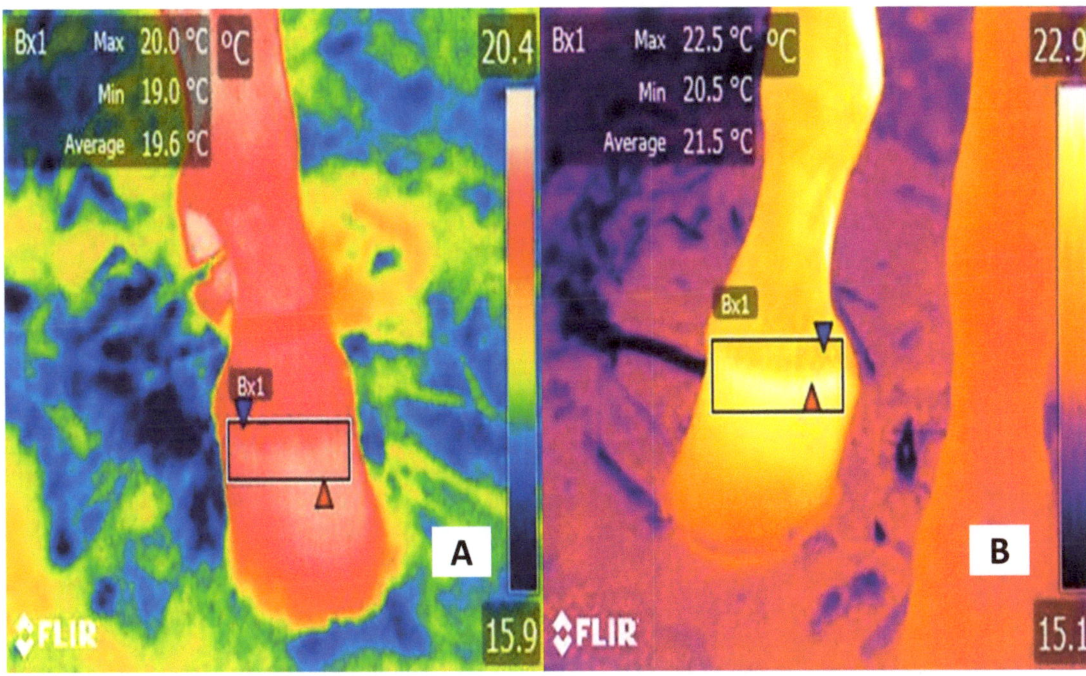

Figure 6. Thermographic evaluation of hoof inflammation due to laminitis. (**A**) Healthy hoof. The hoof of a quarter-horse male with no reported pathologies. A maximum temperature of 20 °C (red triangle) and a minimum of 19 °C (blue triangle) can be observed on the surface of the coronet band (Bx1). (**B**) Hoof with laminitis. The thermal response of the coronet band (Bx1) of a female horse of the Azteca breed recorded increases by 2.5 °C and 1.5 °C at the maximum (red triangle) and minimum temperature (blue triangle). This may be due to the invasion of infectious agents during laminitis, which can trigger the release of interleukin-1, interleukin-10, histamine, and prostaglandin F2 alpha, causing an increase of heat in the inflamed area. The thermal images were taken by the authors of the present review.

Fever is another sign of infection that IRT may be able to identify whilst minimizing animal handling [69,89]. IRT could help in the early diagnosis of diseases with generalized subcutaneous edema, nasal discharge, oral lesions, cyanotic tongue, dyspnea, and lameness [90,91]. This could be extremely valuable in the case of notifiable diseases such as bluetongue allowing early detection of fever and swift management and eradication. Pérez de Diego et al. [92] implemented IRT in 9-month-old male Merino sheep experimentally infected with bluetongue virus [93,94]. The surface temperature of the eye and the surrounding skin was taken, with blood samples being collected on the second-day post-inoculation. A moderate correlation between rectal temperature and IRT (r = 0.504) was found, although IRT registered lower values than rectal temperature. However, the difference was only 1.46 ± 0.05 °C, which is unlikely to be of clinical significance, i.e., IRT is a good proxy for rectal temperature. Another condition that can result in systemic inflammation is autoimmune encephalomyelitis. This disease is characterized by excessive cytokine production associated with fever, fatigue, sleep disturbances, myalgia, and cognitive disorders [95]. Faraji et al. [96] reported a high susceptibility to autoimmune encephalomyelitis among female mice. Females not only presented with tail weakness, paralysis, and increased levels of pro and anti-inflammatory cytokines but also had higher dorsal temperatures than males (females vs. males; 37.72 ± 1.63 °C vs. 30.61 ± 1.59 °C; t18 = 3.06, $p < 0.01$).

These studies demonstrate that IRT can detect relevant differences between healthy and sick animals and that it can serve as an additional tool to detect inflammation, infectious diseases, and pathological processes or to determine the severity of the disease when a greater thermal variation is observed.

5. IRT as an Assistance Tool in Invasive and Surgical Procedures

5.1. Surgical Procedures

Several husbandry interventions (e.g., disbudding, castration, teeth clipping, tail docking, among others) are known to cause pain in animals. This pain is exacerbated by inappropriate analgesic or anesthetic protocols [97]. A survey by Fajt et al. [98] reported that only 30.5–33.9% of calves under six months old receive analgesics during castration and 62.5–68.1% during dehorning. In piglets, a survey in 24 European countries found that 54% of animals are castrated without anesthesia and analgesia [99]. Similarly, only 42% of sheep producers in Australia use pain relief for castration and tail docking [100].

The lack of analgesia causes physiological and behavioral responses associated with acute pain perception. These responses have been monitored through behavior-based scales, pain biomarkers (e.g., cortisol, substance P), cardiorespiratory variables (heart rate and respiratory rate), and productive performance (body weight, daily food intake) [101]. Some studies have implemented IRT to assess the presence/absence of pain and the pharmacological efficacy of compounds proposed within analgesic protocols to promote welfare during these invasive procedures [102]. The following discusses these investigations.

Castration is a common management technique used across species to control breeding. The procedure consists of either the physical removal or chemical inactivation of the testicles. These events cause inflammation and pain due to the manipulation of soft tissues [103]. The inflammatory response generated has been studied in tomcats using superficial testicular IRT to determine the clinical viability of chemical castration—a combination of 20% $CaCl_2$ with 0.5% dimethyl sulfoxide—at 18-time points. Paranzini et al. [104] showed that this treatment caused a minimal local inflammatory reaction with only a decrease in IRT values 10 min after the injection due to temporary vasoconstriction. The unusual results with the opposite of the expected response may be reconciled since the cats demonstrated a normal behavioral repertoire, suggesting a lack of pain.

Unlike dogs and cats, farm animals rarely receive analgesia or anesthesia during castration. Reasons for this issue include that, in many countries, there are no approved products for pain management during surgical castration in species that will enter the food chain. If a drug can be used, it must be administered by aveterinarian [105]. This has led to a large body of research studying animal pain during or after on-farm procedures and the testing of alternative procedures or alternative agents to mitigate pain. Some of these have employed IRT. For example, CO_2 surgical lasers for piglet castration were tested, and IRT of the surgical wound was used to show that, in comparison to scalpel-castrated piglets, temperatures were significantly lower [106]. However, despite this finding, behavioral evaluations showed that laser-castrated animals displayed more pain-related behaviors (e.g., scratching, stiffness, trembling, and tail wagging), and greater tissue damage around the incision site was observed. The latter may explain the unexpected finding since damage can alter vascularization in the zone, influencing IRT reading. This finding highlights the need for further research on IRT and the factors influencing its response prior to widespread use.

Aside from castration, other invasive procedures, such as otectomy and gastrotomy, have been evaluated using the superficial temperature of the surgical wound. Saidu et al. [107] evaluated the temperature of surgical wounds in indigenous dogs subjected to castration, otectomy, and gastrotomy, registering increases within the first 12 h post-surgery (+2.34 °C, +0.58 °C, and +0.56 °C, respectively), and, at 48 h, the temperature in the first two groups was higher than those undergoing gastrotomy. Accordingly, the authors concluded that increased values could be associated with pain and inflammatory responses, an effect that was not observed in gastrotomy. However, the lack of significant changes in gastrotomy does not necessarily imply that this procedure is non-painful since IRT reading depends on blood supply from the skin. It is plausible that the lack of blood supply to the linea alba, where the incision would have been made, may have altered the results in this scenario.

5.2. Efficacy of Analgesics

IRT has also been used to evaluate the administration of drugs for pain management. In calves, Stewart et al. [108] evaluated the effect of surgical castration using maximum eye temperature and physiological variables such as heart rate variability, heart rate, and concentrations of cortisol, epinephrine, and norepinephrine. When comparing the effect of the surgical castration alone, a surgical procedure with local analgesia, and sham handling as the control group, eye temperature increased significantly in the patient's undergoing surgery, regardless of the analgesia. This effect also corresponded with cortisol increases. The authors concluded that IRT could assess the activation of the autonomous nervous system (ANS) after the perception of acute pain since both the endocrine and physiological responses were associated with increased superficial temperature. The efficacy of topical analgesics was assessed in Holstein calves receiving 3.33 mg/kg of transdermal flunixin meglumine applied to the dorsal midline during castration. Treated animals registered the highest ocular temperatures compared to placebo and previously castrated animals (35.4 °C, 34.5 °C, and 34.3 °C, respectively). Although cortisol concentrations were significantly lower in the treated animals, substance P concentrations were similar between all experimental groups. The authors concluded based on the results that transdermal flunixin had little analgesic effect [109]. A similar result was observed by Graves et al. [110] in a study on castrated mix breed goats receiving the same analgesic protocol. Scrotal IRT, heart rate, respiratory rate, and rectal temperature were measured with the finding that there was no difference between flunixin and placebo-treated groups for the physiological parameters measured However, there was some evidence of a flunixin effect as food intake was improved in the analgesic group. The authors reconciled that the analgesic may have had some effect but possibly needed dose refinement to show the change in physiological parameters expected.

Electroencephalography and ocular IRT were used to assess pain in another castration study on merino ram lambs [111]. Animals were allocated to [111] groups of castration with meloxicam and lignocaine, castration untreated and sham [111]. At 1 h post-surgery, untreated animals and those that received the drug recorded increases in ocular temperature of approximately 0.3 °C (basal = 38.6 °C vs. 1 h post-surgery = 38.8 °C). In the groups receiving meloxicam and lignocaine, the temperature increased 0.6 °C at 4 h post-surgery (basal = 38.6 °C vs. 4 h post-surgery = 39.2 °C), which was likely a consequence of waning analgesia. Given that there were abnormal lying, standing, and walking behavioral responses, it was suggested that the analgesic protocol was not efficacious [111]. In contrast, successful analgesic treatment in piglets via milk was reported by Bates et al. [112], who administered meloxicam orally to Yorkshire X Landrace sows—after farrowing—to test the transfer of the drug through milk. On post-farrowing day five, gilts and piglets were subjected to castration, tail docking, and iron injection. Treated animals had lower cortisol concentrations with no differences in mean plasma substance P levels. IRT at the lower cranial region (including cranial, left, and right ear, and snout temperature) was lower in control piglets (sham handling) than in treated animals, concluding that animals receiving meloxicam maintained a consistent temperature before and after the procedures, and the temperature values were similar to those registered in the control group.

An often-cited justification for the lack of analgesia provision to newborn or very young farm animals is the belief that their nociceptive pathways are not fully developed and are incapable of perceiving pain in the same way as mature animals [101,113]. Nonetheless, studies with animals of different ages dispute this assertion. Bergamasco et al. [97] evaluated unmitigated surgical castration in Holstein calves at different ages (six weeks, three, and six months) and after exposure to a simulated castration. In general, average surface eye temperature was higher in the castration groups but in both groups, a significant decrease was recorded. This change can be attributed to sympathetically mediated vasoconstriction and an adaptative response of the parasympathetic system to painful stimuli. However, there was an effect of age, where calves of six weeks had lower temperatures than older calves. Additionally, younger animals had lower cortisol concentrations, with peak values returning faster to baseline (within 50 min post-surgery) than in older animals (120 min). This information is critical since it does not imply that younger animals are not in pain and should be withheld analgesia but that analgesic protocols must consider age to guide appropriate drug choice and dose selection.

The effect of different castration techniques on pain generation has also been studied with IRT. In Holstein claves, castration by banding, cut-and-clamp, and cut-and-pull, plus the influence of bull calves' age was evaluated using ocular IRT and electroencephalography, electrodermal activity, and pain biomarkers (cortisol and substance P). The electroencephalogram recorded desynchronization in all treatments, representing general arousal but greater desynchronization in older calves, with higher cortisol concentrations and increased electrodermal activity. Maximum eye temperature had an apparent increase after treatment in all groups and ages, but average IRT temperature was affected by age, where bulls of 6 months recorded the highest temperatures [114]. Similarly, in beef cattle, castration by band and knife in animals of 3 different ages (1 week, 2 months, and 4 months old) was assessed by IRT in the scrotal area, together with salivary cortisol, substance P, haptoglobin concentrations, and activity levels [115]. This study concluded that pain treatment should be provided, particularly to calves older than 2 months old, where band castration registered the highest temperatures (37.2 °C), and inflammation was maintained for up to 14 days with swelling for up to 35 days. These indicators could suggest a chronic pain response and point to the potential utility of IRT to assess this more long-lived response [115]. Both examples show a response associated with the activation of the ANS and biomarker release, such as nitric oxide. Taken together, these findings suggest that castration can be performed at an early age, but with consideration for hyperalgesia effects and the importance of analgesic protocols in all groups of animals [114].

IRT has also been used to monitor routine procedures such as tail docking, teeth clipping, and castration in piglets to assess the physiological changes and the analgesic efficacy of different compounds. Tail docking is used to prevent tail-biting behavior in piglets [116]. Figure 7 shows the sequence of thermal images taken in 1-day-old piglets (White Large x Landrace) before, during, and after tail docking with side-cutter pliers. In general, an increase in temperature on the ocular surface and a decrease in the nasal region are observed, a reaction that could be attributed to acute pain perception. In the case of dairy cattle, the presence of chronic pain due to tail docking and increased sensitivity to heat and cold was shown by Eicher et al. [117] through thermal imaging of the tail. When comparing docked and non-docked animals, after the sensitivity test, tails from docked animals remained 1.43 °C warmer, showing the increasing sensitivity as a result of the procedure. This is similar to the phantom limb effect in humans. These studies show that IRT could also be applied to monitor acute and chronic pain.

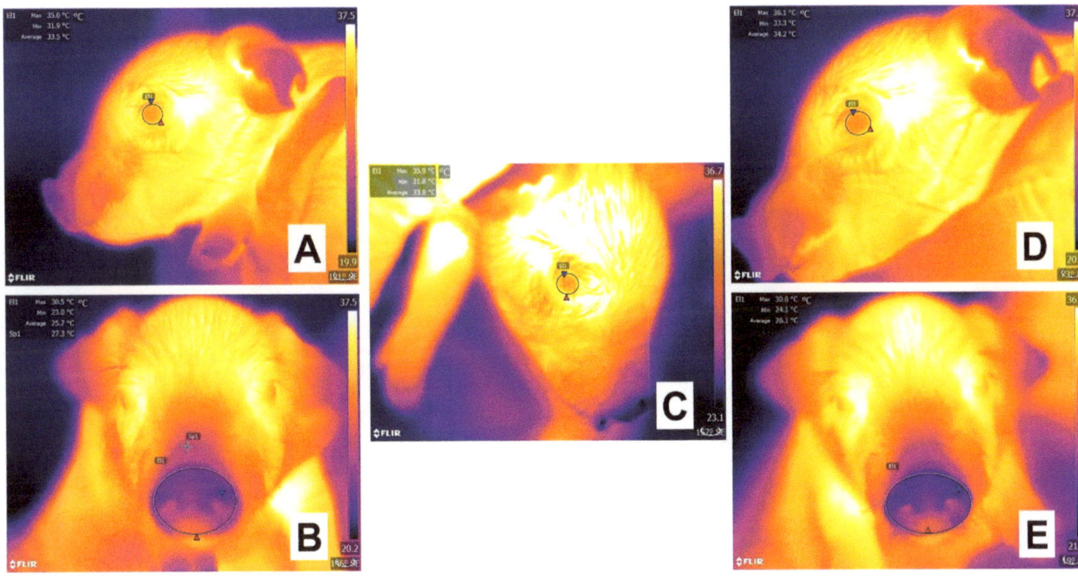

Figure 7. Evaluation of the thermal response of 1-day-old piglets during tail docking with side-cutter pliers. (**A**,**B**) Before tail docking, the ocular surface (**A**, El1) of the piglet registered a maximum temperature (red triangle) of 35.0 °C, while the nasal window (**B**, El1) had a maximum temperature (red triangle) of 30.5 °C. (**C**). During the procedure, the ocular surface window increased its maximum temperature by 0.9 °C. (**C**,**D**) When comparing basal values with the surface temperatures taken immediately after tail docking, the maximum temperature of the piglet at the ocular window (**D**, El1) increased by 1.1 °C. At the same time, the nasal region (**E**, El1) decreased by 0.5 °C. This ambivalent reaction could be a result of the HPA axis increasing blood flow to important organs such as the eye and limiting it to a peripheral region such as the piglet's nose. Sp1: default focal point of the software. The thermal images were taken by the authors of the present review. Radiometric images were obtained using a T1020 FLIR thermal camera. Image resolution 1024 × 768; up to 3.1 MP with UltraMax. FLIR Systems, Inc. Wilsonville, OR, USA.

Dehorning, another practice commonly performed on farm animals, was studied by Van der Saag et al. [118] in Hereford calves. The analgesic effect of buccal meloxicam and local anesthetics was studied using IRT on dehorning wounds, in the first seven days after dehorning. In general, there was no effect of treatment on behavioral and thermographic parameters. However, temperature increases were time-dependent with increases from day 1 (38.83 ± 0.42 °C) to days 3 and 7 (40.43 ± 0.35 °C, 40.30 ± 0.40 °C, respectively). In another study in dairy calves, cautery disbudding was associated with increased nociception as shown by ocular IRT, plasma cortisol, substance P concentrations, and pressure algometry. When comparing animals receiving a single oral dose of firocoxib (0.5 mg/kg) and placebo after a cornual block, the animals receiving placebo had higher cortisol concentrations and heart rates. However, no significant differences were found in ocular IRT between groups (both groups recorded 37.7 ± 0.1 °C), although a decrease by 0.12 °C and 0.15 °C was reported between 2 and 4 h after disbudding, respectively [119]. These results suggest that firocoxib dose might reduce nociception, particularly during the first 24 h post procedure, but IRT alone might not be suitable to determine its analgesic efficacy.

Teeth clipping is performed mainly in piglets to prevent damage to the sow's udder or injuries to littermates, such as tail biting. Studies aimed at implementing IRT to evaluate this practice are limited. Radaelli et al. [120] used tooth temperature during teeth grinding in piglets 16 h after birth. At peak moments, the average tooth temperature ranged from 50 °C to 90 °C. These increases lasted 2 s. In spite of this rise, IRT showed that soft tissue remained unaffected when the procedure was performed correctly. Preliminary results of a study made by the authors during teeth clipping with clippers in piglets are shown in Figure 8. In this figure, the increase in the ocular surface temperature and the upper lip was recorded, while values in the nasal thermal window dropped by 1.3 °C due to the potential perception of pain and the peripheral vasomotor changes.

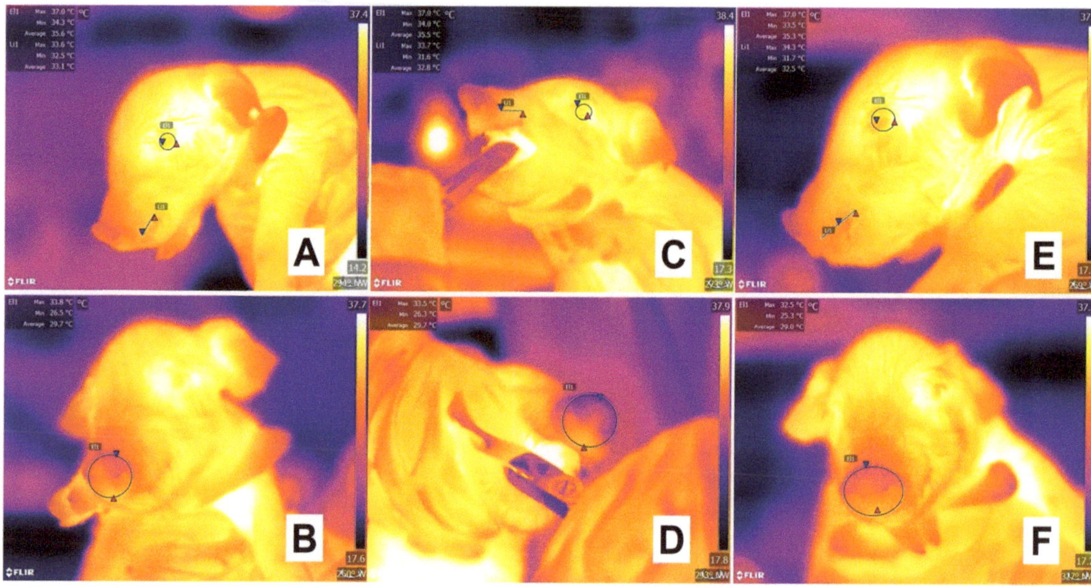

Figure 8. Evaluation of the thermal response in White large x Landrace piglets during teeth clipping. (**A**,**B**) Before teeth clipping in 1-day-old piglets, the thermal windows of the ocular surface (**A**, El1), upper lip (Li1), and nose (**B**, El1) show a maximum temperature (red triangle) of 37.0 °C, 33.6 °C, and 33.8 °C, respectively. (**C**,**D**) During the procedure performed with clippers, the ocular surface's maximum temperature (red triangle) (**C**, El1) maintained the same value as basal recordings. However, the maximum temperature of the upper lip (Li1) increased by 0.1 °C and in the nasal window (**D**, El1) decreased by 0.3 °C. (**E**,**F**) After the routine procedure, although the maximum temperature of the ocular surface (**E**, El1) did not change compared to basal values, a progressive increase was observed in the upper lip (Li1) by 0.7 °C. The drop in the maximum temperature (red triangle) of the nasal region (**F**, El1) by 1.3 °C shows the physiological response of animals when perceiving potential painful stimuli. The thermal images were taken by the authors of the present review. Radiometric images were obtained using a T1020 FLIR thermal camera. Image resolution 1024 × 768; up to 3.1 MP with UltraMax. FLIR Systems, Inc. Wilsonville, OR, USA.

Since teeth clipping is usually performed to avoid lesions in littermates, some studies adopting IRT as a method to record the association between lesion score due to tail biting by conspecifics and changes in temperature due to inflammation and infection were studied by Teixeira et al. [121] in pigs. IRT at the tail base and ear base recorded lower temperatures in animals with lesion scores of 0–1 (between approximately 32 and 33 °C). However, considering some of the conflicting findings with respect to IRT presented, herein, it would be wise to corroborate this finding with other behavioral and physiological variables to reliably establish the piglets' pain state welfare level before making any conclusions on the utility of IRT for monitoring in this procedure.

Other procedures that require analgesic protocols, such as laparotomy, have been studied by Viscardi et al. [122], who compared the analgesic effect of meloxicam and flunixin meglumine in sheep using pain-related facial expressions and IRT. The authors found no differences in facial expression scores or pain-associated behaviors assessed in animals treated with meloxicam and flunixin. Although no differences were observed between the temperature of the surgical site, both treatments reduced the temperature by 3.4 °C. The same has been reported in piglets, where the surface temperature of the cranial region and cortisol concentration decreased with the administration of meloxicam [112].

6. IRT as a Method to Monitor the Autonomic Response and Pain Intensity

A comprehensive evaluation of pain must consider the type of pain (acute and chronic) and its intensity (mild, moderate, and severe) [3,123]. Although current behavioral-based scales can be used to assess this, due to the multifactorial nature of pain and other factors such as masking of pain response in some species, the sensitivity of these tools is limited and often subjective [4,124]. For this reason, IRT has been suggested as a tool to aid in pain recognition [42]. Casas-Alvarado et al. [125] evaluated 21 bitches of different breeds undergoing ovariohysterectomy. After following an analgesic protocol of lidocaine alone or in combination with pure opioids, the thermal response of the lacrimal gland and eyelids did not show differences between treatments. Whilst this may suggest a lack of sensitivity of the IRT in determining pain intensity, given that the scores obtained on a validated pain assessment scale were also the same, a preferred interpretation may be that the analgesic effect of both drugs was similar.

During surgery or any procedure that causes acute pain, there is a need to evaluate analgesic treatments to ensure that the dose, selected drug, and protocol are adequate for the animals. IRT has the potential for use as part of this assessment. For example, pain management monitoring in a canine patient undergoing ovariohysterectomy is shown in Figure 9, where the intraoperative administration of intravenous meloxicam with rescue tramadol [126] restored the basal temperature of the patient in the lacrimal caruncle.

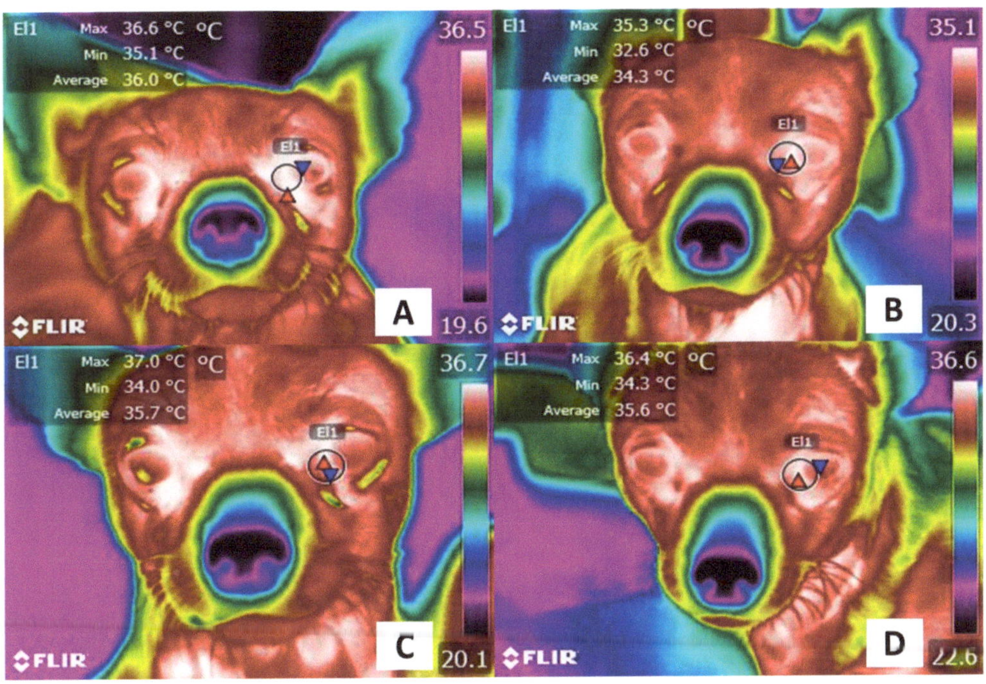

Figure 9. Thermal response associated with acute pain in dogs undergoing surgery. (**A**) Before surgery. A 1-year-old bitch of the Chihuahua breed was subjected to ovariohysterectomy under analgesic management with meloxicam (0.1 mg/kg IV). The maximum temperature of the lacrimal caruncle (El1) had values of 36.6 °C (red triangle), while the minimum was 35.1 °C (blue triangle). (**B**) 1 h after surgery. A decrease in the temperature of the lacrimal caruncle is observed, with maximum (red triangle) and

minimum (blue triangle) values of 35.3 °C and 32.6 °C, respectively. This represents a decrease of 1.3 °C and 2.5 °C, respectively. This could represent a sympathetic response to the effect of pain despite the use of analgesics. **(C)** 2 h after surgery. After administering tramadol as rescue analgesia, the maximum and minimum temperatures of the lacrimal caruncle (El1) (37.0 °C and 34.0 °C) were recorded above the basal values, due to the effect of the drug that reduces sympathetic tone. **(D)** 3 h post-surgery. Although it is observed that the maximum temperature of the lacrimal caruncle (El1) decreased by 0.6 °C (red triangle), the minimum increased by 0.3 °C (blue triangle), which could be associated with post-surgical stability. The thermal images were taken by the authors of the present review. Radiometric images were obtained using a T1020 FLIR thermal camera. Image resolution 1024 × 768; up to 3.1 MP with UltraMax. FLIR Systems, Inc. Wilsonville, OR, USA.

Tapper et al. [127] assessed pain intensity using pressure algometry and thermal sensitivity in 12 mixed-breed sows with lameness due to laminitis. They observed that the affected limbs tolerated less pressure and had a greater thermal response than healthy animals. Sodium salicylate and flunixin meglumine administration increased pressure tolerance and decreased the difference in thermal response between affected and healthy limbs. Thus, it was observed that analgesic treatment could reduce the radiation emitted due to an inflammatory event, as schematized in Figure 10 with the use of non-steroidal analgesics (NSAIDs). Likewise, Stubsjøen et al. [128] evaluated the infrared thermal response of the eye together with heart rate variability (HRV) to estimate the degree of pain in six sheep exposed to a forelimb tourniquet. The results showed that the stimulus decreased eye temperature by 0.47 °C, while heart rate and HRV increased. In another case report, an athlete-horse of the Amazon was diagnosed with septic arthritis, cursing with tachycardia (48 beats per minute) and tachypnea (60 breaths per minute) due to the perception of pain, by using IRT, ozone therapy applied to the injury site lowered by 1.9 °C the initial local temperature of 39.1 °C [129]. Accordingly, the discussed studies show that IRT can be considered a complementary tool to recognize acute pain and confirm the effectiveness of treatment.

Although there is a growing body of evidence suggesting that IRT may be useful as a pain assessment tool, it is clear from the above that there are conflicting findings. These conflicts may arise due to the measurements performed during research studies. The exclusive measurement of one parameter, such as maximum temperature, only provides limited data and cannot give an overall evaluation of the thermal state. Other elements such as thermal symmetry, chronology, and delta T, among others, are equally important for evaluation and contrast between studies. There is also controversy around the impact of various pharmacological agents on local temperatures and a lack of understanding as to the level of confounding this might create on results when trying to interpret in the context of pain. As an example, Holmes et al. [130] observed that local analgesics such as mepivacaine hydrochloride in distal joints in equine limbs did not change local temperature after administration. In contrast, reviews regarding the use of IRT argue that local analgesics such as lidocaine can cause hyperthermia due to the blockage of postganglionic fibers, which causes vasodilation and increased heat radiation [11,66]. As a result, further research is needed.

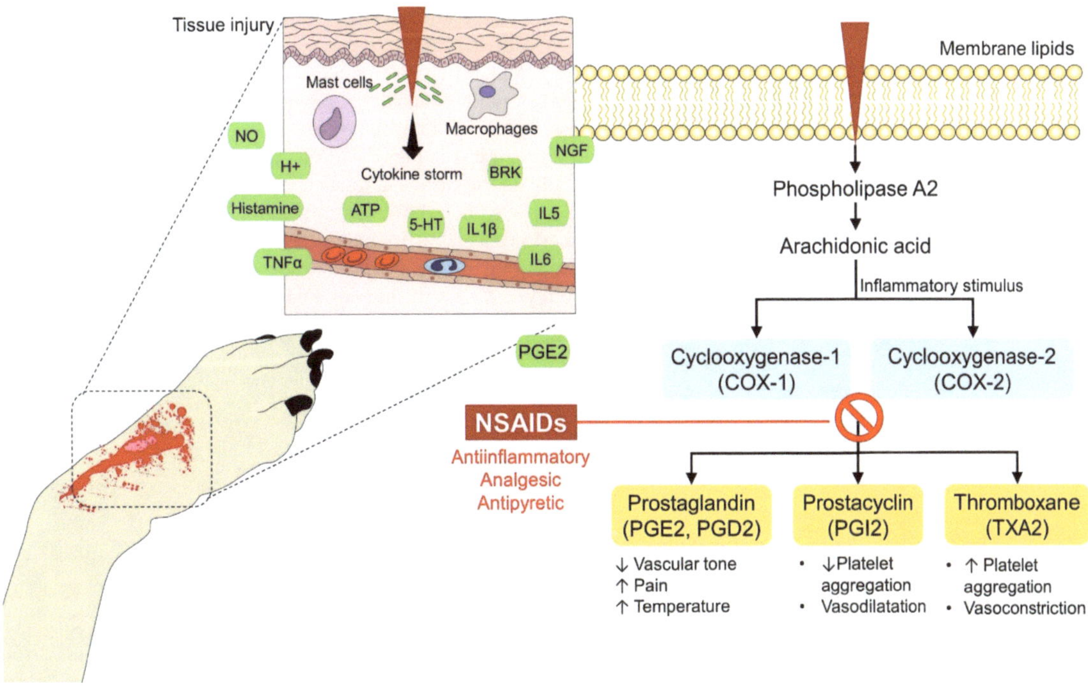

Figure 10. Mechanism of action and effect of analgesics in the nociceptive response and local inflammation. After tissue injury, the released pro-inflammatory mediators (e.g., NO, H+, histamine, 5-HT, among others) extend the local reaction to produce the five cardinal signs of inflammation. The administration of analgesic drugs, for example, NSAIDs, lessens this reaction by inhibiting the cyclooxygenase enzyme. In this way, reactions promoted by prostaglandin and prostacyclin, such as vasodilatation, pain, and local and systemic hyperthermia can be prevented. 5-HT: serotonin; ATP: adenosine triphosphate; BRK: bradykinin; H+: hydrogen; IL: interleukin; NGF: nerve growth factor; NO: nitric oxide; NSAIDs: non-steroidal anti-inflammatory drugs; PGE2: prostaglandin E2; TNFα: tumor necrosis factor alpha.

7. Future Research on IRT Application

Based on the research to date, it seems that IRT is beneficial as an additional method to recognize acute pain through the identification of superficial temperature changes under specific pathological conditions [7,131] such as tendonitis, laminitis, or mastitis [132,133]. However, it is necessary to establish the correlation between the thermal response and currently used tools to evaluate pain, such as pain scales [10] or behavioural-based measures of affective state [134]. Further, it is necessary to consider multiple measurements or features of the infra-red image to improve diagnostic accuracy, for example, symmetry and chronology. The utility of IRT is enhanced by increasing availability of complementary technologies, such as integration with parametric and non-parametric platforms, and the ability to use machine learning and artificial intelligence to automate data extraction and analysis.

Another aspect to be considered in future research is the correlation of autonomous activity through various tools such as IRT, infrared pupillometry, parasympathetic tone activity, heart rate variability, CARDEAn (cardiovascular depth analgesia), surgical plethysmography index (SPI), skin conductance, and photoplethysmography waveform amplitude (PPGA) [59]. For example, pupillometry has been applied to study the activation of the sympathetic nervous system after painful stimuli, modifying pupil diameter [135,136]. Regarding the parasympathetic tone activity index, its use with IRT could allow nociceptive monitoring of pain processes and their treatments by identifying hemodynamic instability, especially in patients under anesthesia [137,138].

An important field of research where IRT could help to refine and promote animal welfare is during the slaughter or euthanasia of farm, companion, and laboratory animals in order to prevent pain and negative emotional states [74,139,140]. Weschenfelder et al. [74] have used infrared ocular thermography in pigs during slaughter to associate the physiological response with meat quality. In the case of animals used in science, preliminary results of the authors of the present article have shown that IRT can be applied during exposure to different types of euthanasia (injectable, inhalational, and physical) to determine the physiological reaction of the organism and potentially identify painful states. Figure 11 contains thermal images from Wistar rats during euthanasia with CO_2.

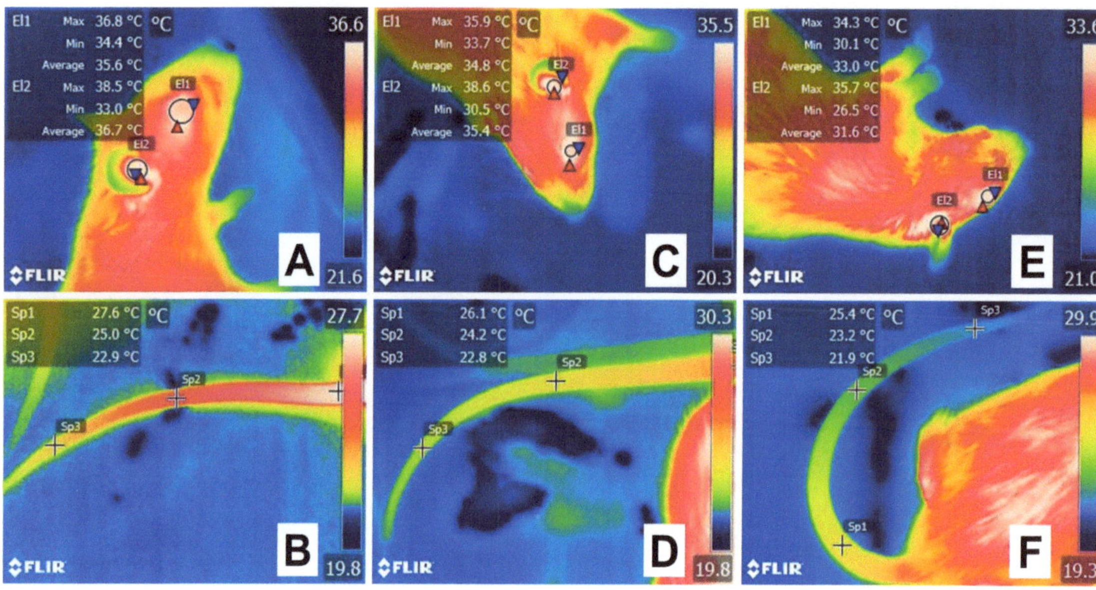

Figure 11. Thermal images in Wistar rats exposed to euthanasia with CO_2. (**A**,**B**) Before the application of the euthanasia method, the ocular surface (El1), the auricular window (El2), and the base of the tail (Sp1) had a maximum temperature of 36.8 °C, 38.5 °C, and 27.6 °C, respectively. (**C**,**D**) During the euthanasia of the rodent with CO_2 exposition inside a chamber, a general drop in the maximum temperature of the ocular surface (El1) and tail base (Sp1) by 0.9 °C and 1.5 °C was recorded. In contrast, an increase in the maximum temperature of the auricular region was reported (38.6 °C). (**E**,**F**) Within the first two minutes after exposition to CO_2 euthanasia, all thermal windows showed a significant drop in temperature. When compared to basal values, the maximum temperature of the ocular region (El1) decreased by 2.5 °C, the auricular region (El2) by 2.8 °C, and the tail base (Sp1) by 2.2 °C. The progressive temperature drops in the rats, assessed by IRT, could help to determine the vasodilation effect of the euthanasia drug and could even be associated with pain when correlated with other evaluation tools. Maximal temperature is indicated with a red triangle and the minimal with a blue triangle. The thermal images were taken by the authors of the present review. Radiometric images were obtained using a T1020 FLIR thermal camera. Image resolution 1024 × 768; up to 3.1 MP with UltraMax. FLIR Systems, Inc. Wilsonville, OR, USA.

The selection of thermal windows according to the species (e.g., farm [141], companion [66], laboratory [37], or wildlife under human care [7] is required to objectively categorize the changes in autonomic activity and the degree of pain. This could be applied during the transport of livestock, including water buffaloes, in whom IRT can detect changes in the surface temperature. This trait might have a repercussion on subproduct quality (Figure 12).

Similarly, the application of IRT to detect subtle temperature changes in neonates [142–144], stress [141,145,146], or to assess the health status of the animals [147,148] must consider its use together with other techniques to improve the sensitivity and specificity of IRT as a complementary diagnostic method. Finally, it is important to consider that the current focus of IRT is to evaluate and monitor acute pain. There is controversy about the use of IRT to assess chronic pain or processes where inflammation is not present. Finally, IRT is increasingly being considered as a technology amenable to be used in precision farming as part of automated, computerized systems that help increase livestock production and allow early intervention to safeguard the welfare state [149,150].

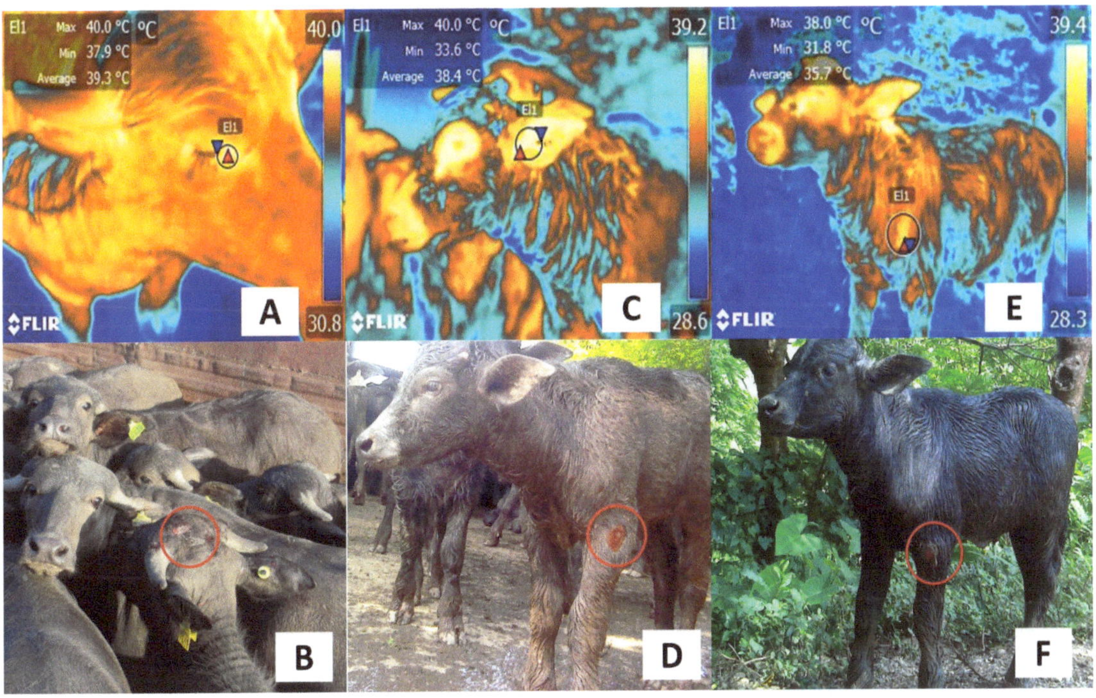

Figure 12. Recognition of inflammatory lesions in water buffaloes. (**A**,**B**) Thermal response due to lesion in the parietal region (red circle). An increase in the superficial temperature of the lacrimal caruncle (El1) can be associated with the lesion, with a maximum temperature of 40 °C (red triangle) and a minimum of 37.9 °C (blue triangle). (**C**,**D**) An ulcerative lesion on the elbow can be seen in the digital image. Through thermal imaging, the maximum temperature of the auricular region is 40 °C with a minimum of 33.6 °C. (**E**,**F**) The digital image shows a lesion in the shoulder region (red circle), whereas the radiometric image shows that the surface temperature of this region (El1) presented a maximum temperature of 38 °C (red triangle) and a minimum temperature of 31.8 °C (blue triangle), which would help to corroborate the presence of a lesion in the shoulder region as a result of the inflammatory process. The thermal images were taken by the authors of the present review.

8. Conclusions

Pain management in domestic animals has gained attention due to the increased recognition of its impact on welfare, production, and scientific model validity. Implementing tools in veterinary medicine to recognize pain, prevent said consequences, and establish proper anti-nociceptive and analgesic protocols is essential. As a non-invasive technique, IRT is an alternative for pain evaluation that has been used to detect acute processes during inflammatory states such as osteoarthritis or during routine surgical procedures such as castration, tail docking, and disbudding, where animals do not always receive analgesic drugs. According to the evidence, IRT applied to the ocular surface, the surgical wound, or the anatomical region where the inflammation occurs can identify temperature changes triggered by vasomotor changes in the dermis. IRT has helped to identify acute and chronic pain in companion and farm animals and has served as a complementary tool to evaluate the analgesic efficacy of different drugs.

At the current time it is sensible to use IRT with physiological parameters (e.g., heart rate, rectal temperature, heart rate variability), biomarkers (cortisol, substance P, and norepinephrine), activity levels (e.g., accelerometers), top-notch technologies (electroencephalography and electromyography), and behavioral-based scales to provide a comprehensive evaluation into the multidimensional components of pain.

Author Contributions: All authors contributed to the conceptualization, writing, and reading. All authors have read and agreed to the published version of the manuscript.

Funding: This research received no external funding.

Institutional Review Board Statement: Not applicable.

Informed Consent Statement: Not applicable.

Data Availability Statement: Not applicable.

Conflicts of Interest: The authors declare no conflict of interest.

References

1. Raja, S.N.; Carr, D.B.; Cohen, M.; Finnerup, N.B.; Flor, H.; Gibson, S.; Keefe, F.J.; Mogil, J.S.; Ringkamp, M.; Sluka, K.A.; et al. The revised International Association for the Study of Pain definition of pain: Concepts, challenges, and compromises. *Pain* **2020**, *161*, 1976–1982. [CrossRef] [PubMed]
2. Mainau, E.; Manteca, X. Pain and discomfort caused by parturition in cows and sows. *Appl. Anim. Behav. Sci.* **2011**, *135*, 241–251. [CrossRef]
3. Hernández-Avalos, I.; Mota-Rojas, D.; Mora-Medina, P.; Martínez-Burnes, J.; Casas Alvarado, A.; Verduzco-Mendoza, A.; Lezama-García, K.; Olmos-Hernandez, A. Review of different methods used for clinical recognition and assessment of pain in dogs and cats. *Int. J. Vet. Sci. Med.* **2019**, *7*, 43–54. [CrossRef] [PubMed]
4. Hugonnard, M.; Leblond, A.; Keroack, S.; Cadoré, J.; Troncy, E. Attitudes and concerns of French veterinarians towards pain and analgesia in dogs and cats. *Vet. Anaesth. Anal.* **2004**, *31*, 154–163. [CrossRef]
5. Rekant, S.I.; Lyons, M.A.; Pacheco, J.M.; Arzt, J.; Rodriguez, L.L. Veterinary applications of infrared thermography. *Am. J. Vet. Res.* **2016**, *77*, 98–107. [CrossRef]
6. Čebulj-Kadunc, N.; Frangež, R.; Kruljc, P. Infrared thermography in equine practice. *Vet. Stanica* **2020**, *51*, 425–432. [CrossRef]
7. Mota-Rojas, D.; Pereira, A.M.F.; Martínez-Burnes, J.; Domínguez-Oliva, A.; Mora-Medina, P.; Casas-Alvarado, A.; Rios-Sandoval, J.; de Mira Geraldo, A.; Wang, D. Thermal imaging to assess the health status in wildlife animals under human care: Limitations and perspectives. *Animals* **2022**, *12*, 3558. [CrossRef]
8. Bertoni, A.; Mota-Rojas, D.; Álvarez-Macias, A.; Mora-Medina, P.; Guerrero-Legarreta, I.; Morales-Canela, A.; Gómez-Prado, J.; José-Pérez, N.; Martínez-Burnes, J. Scientific findings related to changes in vascular microcirculation using infrared thermography in the river buffalo. *J. Anim. Behav. Biometeorol.* **2020**, *8*, 288–297. [CrossRef]
9. Zhang, Z.; Zhang, H.; Liu, T. Study on body temperature detection of pig based on infrared technology: A review. *Artif. Intell. Agric.* **2019**, *1*, 14–26. [CrossRef]
10. Lush, J.; Ijichi, C. A preliminary investigation into personality and pain in dogs. *J. Vet. Behav.* **2018**, *24*, 62–68. [CrossRef]
11. Casas-Alvarado, A.; Mota-Rojas, D.; Hernández-Ávalos, I.; Mora-Medina, P.; Olmos-Hernández, A.; Verduzco-Mendoza, A.; Reyes-Sotelo, B.; Martínez-Burnes, J. Advances in infrared thermography: Surgical aspects, vascular changes, and pain monitoring in veterinary medicine. *J. Therm. Biol.* **2020**, *92*, 102664. [CrossRef] [PubMed]
12. Matičić, D.; Stejskal, M.; Pećin, M.; Kreszinger, M.; Pirkić, B.; Vnuk, D.; Smolec, O.; Rumenjak, V. Correlation of pain assessment parameters in dogs with cranial cruciate surgery. *Vet. Arh.* **2010**, *80*, 597–609.
13. Sneddon, L. The impact of nociception and pain: Implications for animal welfare legislation. *Physiol. News Mag.* **2022**, 12–15. [CrossRef]
14. Bell, A. The neurobiology of acute pain. *Vet. J.* **2018**, *237*, 55–62. [CrossRef]
15. Lamont, L.A.; Tranquilli, W.J.; Grimm, K.A. Physiology of pain. *Vet. Clin. N. Am. Small Anim. Pract.* **2000**, *30*, 703–728. [CrossRef]
16. Youn, D.H.; Kim, T.W.; Cho, H.J. Pain in animals: Anatomy, physiology, and behaviors. *J. Vet. Clin.* **2017**, *34*, 347–352. [CrossRef]
17. Sneddon, L.U. Comparative physiology of nociception and pain. *Physiology* **2018**, *33*, 63–73. [CrossRef]
18. Rosenbaum, T.; Simon, S.A. TRPV1 Receptors and Signal Transduction. In *TRP Ion Channel Function in Sensory Transduction and Cellular Signaling Cascades*; Liedtke, W.B., Heller, S., Eds.; CRC Press/Taylor & Francis: Boca Ratón, MA, USA, 2007; pp. 69–84.

19. Marwaha, L.; Bansal, Y.; Singh, R.; Saroj, P.; Bhandari, R.; Kuhad, A. TRP channels: Potential drug target for neuropathic pain. *Inflammopharmacology* **2016**, *24*, 305–317. [CrossRef]
20. Sneddon, L.U.; Elwood, R.W.; Adamo, S.A.; Leach, M.C. Defining and assessing animal pain. *Anim. Behav.* **2014**, *97*, 201–212. [CrossRef]
21. Lezama-García, K.; Mota-Rojas, D.; Pereira, A.M.F.; Martínez-Burnes, J.; Ghezzi, M.; Domínguez, A.; Gómez, J.; de Mira Geraldo, A.; Lendez, P.; Hernández-Ávalos, I.; et al. Transient Receptor Potential (TRP) and thermoregulation in animals: Structural biology and neurophysiological aspects. *Animals* **2022**, *12*, 106. [CrossRef]
22. Kozyreva, T.V.; Kozaruk, V.P.; Tkachenko, E.Y.; Khramova, G.M. Agonist of TRPM8 channel, menthol, facilitates the initiation of thermoregulatory responses to external cooling. *J. Therm. Biol.* **2010**, *35*, 428–434. [CrossRef]
23. Zhang, F.; Jara-Oseguera, A.; Chang, T.-H.; Bae, C.; Hanson, S.M.; Swartz, K.J. Heat activation is intrinsic to the pore domain of TRPV1. *Proc. Natl. Acad. Sci. USA* **2018**, *115*, E317–E324. [CrossRef] [PubMed]
24. Tobita, N.; Makino, M.; Fujita, R.; Jyotaki, M.; Shinohara, Y.; Yamamoto, T. Sweet scent lactones activate hot capsaicin receptor, TRPV1. *Biochem. Biophys. Res. Commun.* **2021**, *534*, 547–552. [CrossRef] [PubMed]
25. Bourne, S.; Machado, A.G.; Nagel, S.J. Basic anatomy and physiology of pain pathways. *Neurosurg. Clin. N. Am.* **2014**, *25*, 629–638. [CrossRef] [PubMed]
26. Wiese, A.J.; Yaksh, T.L. Nociception and pain mechanism. In *Handbook of Veterinary Pain Management*; Gaynor, J.S., Muir, W.W., Eds.; Elsevier: St. Louis, MO, USA, 2009.
27. Muir, W.W. Physiology and pathopjysiology of pain. In *Handbook of Veterinary Pain Management*; Gaynor, J.S., Muir, W.W., Eds.; Mosby: St. Louis, MO, USA, 2009; pp. 13–49.
28. Nakata, H.; Sakamoto, K.; Kakigi, R. Meditation reduces pain-related neural activity in the anterior cingulate cortex, insula, secondary somatosensory cortex, and thalamus. *Front. Psychol.* **2014**, *5*, 1489. [CrossRef]
29. Bushnell, M.C.; Čeko, M.; Low, L.A. Cognitive and emotional control of pain and its disruption in chronic pain. *Nat. Rev. Neurosci.* **2013**, *14*, 502–511. [CrossRef]
30. Garland, E.L. Pain Processing in the Human Nervous System. *Prim. Care Clin. Off. Pract.* **2012**, *39*, 561–571. [CrossRef]
31. Voisin, D.L.; Guy, N.; Chalus, M.; Dallel, R. Nociceptive stimulation activates locus coeruleus neurones projecting to the somatosensory thalamus in the rat. *J. Physiol.* **2005**, *566*, 929–937. [CrossRef]
32. Mota-Rojas, D.; Miranda-Cortés, A.; Casas-Alvarado, A.; Mora-Medina, P.; Boscato-Funes, L.; Hernández-Ávalos, I. Neurobiología y modulación de la hipertermia inducida por estrés agudo y fiebre en los animales. *Abanico Vet.* **2021**, *11*, 1–17. [CrossRef]
33. Hernández-Avalos, I.; Mota-Rojas, D.; Mendoza- Flores, J.E.; Casas-Alvarado, A.; Flores-Padilla, K.; Miranda-Cortes, A.E.; Torres-Bernal, F.; Gómez-Prado, J.; Mora-Medina, P. Nociceptive pain and axiety in equines: Physiological and behavioral alterations. *Vet. World* **2021**, *14*, 2984–2995. [CrossRef]
34. Mota-Rojas, D.; Wang, D.; Titto, C.G.; Gómez-Prado, J.; Carvajal-de la Fuente, V.; Ghezzi, M.; Boscato-Funes, L.; Barrios-García, H.; Torres-Bernal, F.; Casas-Alvarado, A.; et al. Pathophysiology of fever and application of infrared thermography (IRT) in the detection of sick domestic animals: Recent advances. *Animals* **2021**, *11*, 2316. [CrossRef]
35. Muir, W.W. Pain: Mechanisms and management in horses. *Vet. Clin. N. Am. Equine Pract.* **2010**, *26*, 467–480. [CrossRef]
36. Redaelli, V.; Papa, S.; Marsella, G.; Grignaschi, G.; Bosi, A.; Ludwig, N.; Luzi, F.; Vismara, I.; Rimondo, S.; Veglianese, P.; et al. A refinement approach in a mouse model of rehabilitation research. Analgesia strategy, reduction approach and infrared thermography in spinal cord injury. *PLoS ONE* **2019**, *14*, e0224337. [CrossRef]
37. Verduzco-Mendoza, A.; Bueno-Nava, A.; Wang, D.; Martínez-Burnes, J.; Olmos-Hernández, A.; Casas, A.; Domínguez, A.; Mota-Rojas, D. Experimental applications and factors involved in validating thermal windows using infrared thermography to assess the health and thermostability of laboratory animals. *Animals* **2021**, *11*, 3448. [CrossRef]
38. Mota-Rojas, D.; Olmos-Hernández, A.; Verduzco-Mendoza, A.; Lecona-Butrón, H.; Martínez-Burnes, J.; Mora-Medina, P.; Gómez-Prado, J.; Orihuela, A. Infrared thermal imaging associated with pain in laboratory animals. *Exp. Anim.* **2021**, *70*, 1. [CrossRef] [PubMed]
39. Yam, M.; Loh, Y.; Tan, C.; Khadijah Adam, S.; Abdul Manan, N.; Basir, R. General pathways of pain sensation and the major neurotransmitters involved in pain regulation. *Int. J. Mol. Sci.* **2018**, *19*, 2164. [CrossRef] [PubMed]
40. Latremoliere, A.; Woolf, C.J. Central sensitization: A generator of pain hypersensitivity by central neural plasticity. *J. Pain* **2009**, *10*, 895–926. [CrossRef] [PubMed]
41. Närhi, M.; Kontturi-Närhi, V.; Hirvonen, T.; Ngassapa, D. Neurophysiological mechanisms of dentin hypersensitivity. *Proc. Finn. Dent. Soc.* **1992**, *88*, 15–22. [PubMed]
42. Mota-Rojas, D.; Martínez-Burnes, J.; Casas-Alvarado, A.; Gómez-Prado, J.; Hernández-Ávalos, I.; Domínguez-Oliva, A.; Lezama-García, K.; Jacome-Romero, J.; Rodríguez-González, D.; Pereira, A.M.F. Clinical usefulness of infrared thermography to detect sick animals: Frequent and current cases. *CABI Rev.* **2022**, *17*, 1–17. [CrossRef]
43. Medzhitov, R. Inflammation 2010: New Adventures of an Old Flame. *Cell* **2010**, *140*, 771–776. [CrossRef]
44. Isailovic, N.; Daigo, K.; Mantovani, A.; Selmi, C. Interleukin-17 and innate immunity in infections and chronic inflammation. *J. Autoimmun.* **2015**, *60*, 1–11. [CrossRef] [PubMed]
45. Zhou, Y.; Hong, Y.; Huang, H. Triptolide attenuates inflammatory response in membranous glomerulo-nephritis rat via downregulation of NF. *Kidney Blood Press. Res.* **2016**, *41*, 901–910. [CrossRef] [PubMed]

46. Nichols, S. Chronic Udder Abscess. In *Comparative Veterinary Anatomy*; Orsini, J.A., Grenager, N.S., de Lahunta, A., Eds.; Academic Press: London, UK, 2022; pp. 1236–1241.
47. Sugimoto, M.A.; Vago, J.P.; Perretti, M.; Teixeira, M.M. Mediators of the resolution of the inflammatory response. *Trends Immunol.* **2019**, *40*, 212–227. [CrossRef] [PubMed]
48. Goulart, D.B.; Mellata, M. *Escherichia coli* mastitis in dairy cattle: Etiology, diagnosis, and treatment challenges. *Front. Microbiol.* **2022**, *13*, 928346. [CrossRef]
49. Mohan, H. Inflammation and healing. In *Textobook of Pahotology*; Mohan, H., Ed.; Jaypee Brothers Medical Publishers: New Delhi, India, 2015; pp. 116–164.
50. Cai, Z.; Cui, J.; Yuan, H.; Cheng, M. Application and research progress of infrared thermography in temperature measurement of livestock and poultry animals: A review. *Comput. Electron. Agric.* **2023**, *205*, 107586. [CrossRef]
51. Gilfillan, A.M.; Austin, S.J.; Metcalfe, D.D. Mast cell biology: Introduction and overview. In *Mast Cell Biology*; Gilfillan, A.M., Metcalfe, D., Eds.; Advances in Experimental Medicine and Biology Book Series; Springer: Boston, MA, USA, 2011; pp. 2–12.
52. Hsieh, F.H. Primer to the immune response. *Ann. Allergy Asthma Immunol.* **2014**, *113*, 333. [CrossRef]
53. Kaminska, B. MAPK signalling pathways as molecular targets for anti-inflammatory therapy—From molecular mechanisms to therapeutic benefits. *Biochim. Biophys. Acta—Proteins Proteom.* **2005**, *1754*, 253–262. [CrossRef]
54. Czaja, A.J. Hepatic inflammation and progressive liver fibrosis in chronic liver disease. *World J. Gastroenterol.* **2014**, *20*, 2515. [CrossRef]
55. Liu, Z.; Wang, Y.; Wang, Y.; Ning, Q.; Zhang, Y.; Gong, C.; Zhao, W.; Jing, G.; Wang, Q. Dexmedetomidine attenuates inflammatory reaction in the lung tissues of septic mice by activating cholinergic anti-inflammatory pathway. *Int. Immunopharmacol.* **2016**, *35*, 210–216. [CrossRef]
56. Reed, S.G.; Ager, A. Immune responses to IAV infection and the roles of L-Selectin and ADAM17 in lymphocyte homing. *Pathogens* **2022**, *11*, 150. [CrossRef]
57. Forbester, J.L.; Humphreys, I.R. Genetic influences on viral-induced cytokine responses in the lung. *Mucosal Immunol.* **2021**, *14*, 14–25. [CrossRef] [PubMed]
58. Lopez, A.; Martinson, S.A. Respiratory System, Thoracic Cavities, Mediastinum, and Pleurae. In *Pathologic Basis of Veterinary Disease*; Zachary, J.F., Ed.; Elsevier: Amsterdam, The Netherlands, 2022; Chapter 9; pp. 547–642.
59. Tattersall, G.J. Infrared thermography: A non-invasive window into thermal physiology. *Comp. Biochem. Physiol. Part A Mol. Integr. Physiol.* **2016**, *202*, 78–98. [CrossRef] [PubMed]
60. Denoble, A.E.; Hall, N.; Pieper, C.F.; Kraus, V.B. Patellar Skin Surface Temperature by Thermography Reflects Knee Osteoarthritis Severity. *Clin. Med. Insights Arthritis Musculoskelet. Disord.* **2010**, *3*, CMAMD.S5916. [CrossRef] [PubMed]
61. Infernuso, T.; Loughin, C.A.; Marino, D.J.; Umbaugh, S.E.; Solt, P.S. Thermal Imaging of Normal and Cranial Cruciate Ligament-Deficient Stifles in Dogs. *Vet. Surg.* **2010**, *39*, 410–417. [CrossRef]
62. Douthit, T.L.; Bormann, J.M.; Bello, N.M. Assessing the Association Between Hoof Thermography and Hoof Doppler Ultrasonography for the Diagnosis of Lameness in Horses. *J. Equine Vet. Sci.* **2014**, *34*, 275–280. [CrossRef]
63. Figuereido, T.; Dzyekanski, B.; Pimpao, C.T.; Silveira, A.B.; Gapriglione, L.G.; Michelotto, P. V Use of Infrared Thermography to Detect Intrasynovial Injections in Horses. *J. Equine Vet. Sci.* **2013**, *33*, 257–260. [CrossRef]
64. Pedreros, F.; Pezoa, J.E.; Torres, S.N. Compensating internal temperature effects in uncooled microbolometer-based infrared cameras. In *Infrared Imaging Systems: Design, Analysis, Modeling, and Testing XXIII*; Holst, G.C., Krapels, K.A., Eds.; SPIE: London, UK, 2012; pp. 472–481.
65. Playà-Montmany, N.; Tattersall, G.J. Spot size, distance and emissivity errors in field applications of infrared thermography. *Methods Ecol. Evol.* **2021**, *12*, 828–840. [CrossRef]
66. Casas-Alvarado, A.; Martínez-Burnes, J.; Mora-Medina, P.; Hernández-Avalos, I.; Domínguez-Oliva, A.; Lezama-García, K.; Gómez-Prado, J.; Mota-Rojas, D. Thermal and circulatory changes in diverse body regions in dogs and cats evaluated by infrared thermography. *Animals* **2022**, *12*, 789. [CrossRef]
67. Mota-Rojas, D.; Titto, C.G.; Orihuela, A.; Martínez-Burnes, J.; Gómez-Prado, J.; Torres-Bernal, F.; Flores-Padilla, K.; Carvajal-de la Fuente, V.; Wang, D. Physiological and behavioral mechanisms of thermoregulation in mammals. *Animals* **2021**, *11*, 1733. [CrossRef]
68. Fleischmann, T.; Siewert, C.; Staszyk, C.; Schulze, M.; Stadler, P.; Seifert, H. Thermal Imaging as an Aid to the Diagnosis of Pain in Horses—First Results. In *IFMBE Proceedings, Proceedings of the World Congress on Medical Physics and Biomedical Engineering, Munich, Germany, 7–12 September 2009*; Dössel, O., Schlegel, W.C., Eds.; Springer: Berlin/Heidelberg, Germany, 2009; Volume 25/2.
69. Schaefer, A.L.; Cook, N.; Tessaro, S.V.; Deregt, D.; Desroches, G.; Dubeski, P.L.; Tong, A.K.W.; Godson, D.L. Early detection and prediction of infection using infrared thermography. *Can. J. Anim. Sci.* **2004**, *84*, 73–80. [CrossRef]
70. Schaefer, A.; Genho, D.; Clisdell, R.; von Gaza, H.; DesRoches, G.; Hiemer, L.; Pelech, G.; Grumpelt, B.; Patterson, R. 497 The automated and real time use of infrared thermography in the detection and correction of DFD and fevers in cattle. *J. Anim. Sci.* **2018**, *96*, 275. [CrossRef]
71. Schaefer, A.L.; Cook, N.J.; Bench, C.; Chabot, J.B.; Colyn, J.; Liu, T.; Okine, E.K.; Stewart, M.; Webster, J.R. The non-invasive and automated detection of bovine respiratory disease onset in receiver calves using infrared thermography. *Res. Vet. Sci.* **2012**, *93*, 928–935. [CrossRef] [PubMed]

72. Schaefer, A.L.; Cook, N.J.; Church, J.S.; Basarab, J.; Perry, B.; Miller, C.; Tong, A.K.W. The use of infrared thermography as an early indicator of bovine respiratory disease complex in calves. *Res. Vet. Sci.* **2007**, *83*, 376–384. [CrossRef] [PubMed]
73. McManus, R.; Boden, L.A.; Weir, W.; Viora, L.; Barker, R.; Kim, Y.; McBride, P.; Yang, S. Thermography for disease detection in livestock: A scoping review. *Front. Vet. Sci.* **2022**, *9*, 965622. [CrossRef] [PubMed]
74. Weschenfelder, A.V.; Saucier, L.; Maldague, X.; Rocha, L.M.; Schaefer, A.L.; Faucitano, L. Use of infrared ocular thermography to assess physiological conditions of pigs prior to slaughter and predict pork quality variation. *Meat Sci.* **2013**, *95*, 616–620. [CrossRef] [PubMed]
75. Church, J.S.; Hegadoren, P.R.; Paetkau, M.J.; Miller, C.C.; Regev-Shoshani, G.; Schaefer, A.L.; Schwartzkopf-Genswein, K.S. Influence of environmental factors on infrared eye temperature measurements in cattle. *Res. Vet. Sci.* **2014**, *96*, 220–226. [CrossRef]
76. Stewart, M.; Wilson, M.T.; Schaefer, A.L.; Huddart, F.; Sutherland, M.A. The use of infrared thermography and accelerometers for remote monitoring of dairy cow health and welfare. *J. Dairy Sci.* **2017**, *100*, 3893–3901. [CrossRef]
77. Khaksari, K.; Nguyen, T.; Hill, B.Y.; Perrault, J.; Gorti, V.; Blick, E.; Gonzalez Cano, T.; Shadgan, B.; Quang, T.; Malpani, R.; et al. Review of the efficacy of infrared thermography for screening infectious diseases with applications to COVID-19. *J. Med. Imaging* **2021**, *8*, 010901. [CrossRef]
78. Martinez-Jimenez, M.A.; Loza-Gonzalez, V.M.; Kolosovas-Machuca, E.S.; Yanes-Lane, M.E.; Ramirez-GarciaLuna, A.S.; Ramirez-GarciaLuna, J.L. Diagnostic accuracy of infrared thermal imaging for detecting COVID-19 infection in minimally symptomatic patients. *Eur. J. Clin. Investig.* **2021**, *51*, e13474. [CrossRef]
79. Bertoni, A.; Napolitano, F.; Mota-Rojas, D.; Sabia, E.; Álvarez-Macías, A.; Mora-Medina, P.; Morales-Canela, A.; Berdugo-Gutiérrez, J.; Guerrero-Legarreta, I.; Mendoza, A.B.; et al. Similarities and differences between river buffaloes and cattle: Health, physiological, behavioral and productivity aspects. *J. Buffalo Sci.* **2020**, *9*, 92–109. [CrossRef]
80. Mota-Rojas, D.; De Rosa, G.; Mora-Medina, P.; Braghieri, A.; Guerrero-Legarreta, I.; Napolitano, F.; Mota Rojas, D. Dairy buffalo behaviour and welfare from calving to milking. *CAB Rev. Perspect. Agric. Vet. Sci. Nutr. Nat. Resour.* **2019**, *14*, 1–9. [CrossRef]
81. Berry, R.J.; Kennedy, A.D.; Scott, S.L.; Kyle, B.L.; Schaefer, A.L. Daily variation in the udder surface temperature of dairy cows measured by infrared thermography: Potential for mastitis detection. *Can. J. Anim. Sci.* **2003**, *83*, 687–693. [CrossRef]
82. Pyorala, S.; Taponen, S. Coagulase-negative Staphylococci—Emerging mastitis pathogens. *Vet. Microbiol.* **2009**, *134*, 3–8. [CrossRef] [PubMed]
83. Macmillan, K.; Colazo, M.G.; Cook, N.J. Evaluation of infrared thermography compared to rectal temperature to identify illness in early postpartum dairy cows. *Res. Vet. Sci.* **2019**, *125*, 315–322. [CrossRef]
84. Machado, N.A.F.; Da Costa, L.B.S.; Barbosa-Filho, J.A.D.; De Oliveira, K.P.L.; De Sampaio, L.C.; Peixoto, M.S.M.; Damasceno, F.A. Using infrared thermography to detect subclinical mastitis in dairy cows in compost barn systems. *J. Therm. Biol.* **2021**, *97*, 102881. [CrossRef]
85. Zaninelli, M.; Redaelli, V.; Luzi, F.; Bronzo, V.; Mitchell, M.; Dell'Orto, V.; Bontempo, V.; Cattaneo, D.; Savoini, G. First evaluation of infrared thermography as a tool for the monitoring of udder health status in farms of dairy cows. *Sensors* **2018**, *18*, 862. [CrossRef]
86. Soszyński, D. The pathogenesis and the adaptive value of fever. *Postepy Hig. Med. Dosw.* **2003**, *57*, 531–554.
87. Walter, E.J.; Hanna-Jumma, S.; Carraretto, M.; Forni, L. The pathophysiological basis and consequences of fever. *Crit. Care* **2016**, *20*, 200. [CrossRef]
88. Weimer, S.L.; Wideman, R.F.; Scanes, C.G.; Mauromoustakos, A.; Christensen, K.D.; Vizzier-Thaxton, Y. The utility of infrared thermography for evaluating lameness attributable to bacterial chondronecrosis with osteomyelitis. *Poult. Sci.* **2019**, *98*, 1575–1588. [CrossRef]
89. Pérez de Diego, A.C.; Athmaram, T.N.; Stewart, M.; Rodríguez-Sánchez, B.; Sánchez-Vizcaíno, J.M.; Noad, R.; Roy, P. Characterization of protection afforded by a bivalent virus-like particle vaccine against bluetongue virus serotypes 1 and 4 in sheep. *PLoS ONE* **2011**, *6*, e26666. [CrossRef]
90. Schwartz-Cornil, I.; Mertens, P.P.C.; Contreras, V.; Hemati, B.; Pascale, F.; Bréard, E.; Mellor, P.S.; MacLachlan, N.J.; Zientara, S. Bluetongue virus: Virology, pathogenesis and immunity. *Vet. Res.* **2008**, *39*, 46. [CrossRef] [PubMed]
91. Maclachlan, N.J.; Drew, C.P.; Darpel, K.E.; Worwa, G. The pathology and pathogenesis of bluetongue. *J. Comp. Pathol.* **2009**, *141*, 1–16. [CrossRef] [PubMed]
92. Pérez de Diego, A.C.; Sánchez-Cordón, P.J.; Pedrera, M.; Martínez-López, B.; Gómez-Villamandos, J.C.; Sánchez-Vizcaíno, J.M. The use of infrared thermography as a non-invasive method for fever detection in sheep infected with bluetongue virus. *Vet. J.* **2013**, *198*, 182–186. [CrossRef] [PubMed]
93. Hofmann, M.A.; Renzullo, S.; Mader, M.; Chaignat, V.; Worwa, G.; Thuer, B. Genetic characterization of toggenburg orbivirus, a new bluetongue virus, from goats, Switzerland. *Emerg. Infect. Dis.* **2008**, *14*, 1855–1861. [CrossRef] [PubMed]
94. Maan, S.; Maan, N.S.; Nomikou, K.; Veronesi, E.; Bachanek-Bankowska, K.; Belaganahalli, M.N.; Attoui, H.; Mertens, P.P.C. Complete genome characterisation of a novel 26th bluetongue virus serotype from Kuwait. *PLoS ONE* **2011**, *6*, e26147. [CrossRef] [PubMed]
95. Heesen, C. Fatigue in multiple sclerosis: An example of cytokine mediated sickness behaviour? *J. Neurol. Neurosurg. Psychiatry* **2006**, *77*, 34–39. [CrossRef]

96. Faraji, J.; Bettenson, D.; Babatunde, S.; Gangur-Powell, T.; Yong, V.W.; Metz, G.A.S. Thermoregulatory dynamics reveal sex-specific inflammatory responses to experimental autoimmune encephalomyelitis in mice: Implications for multiple sclerosis-induced fatigue in females. *Brain Behav. Immun.—Health* **2022**, *23*, 100477. [CrossRef]
97. Bergamasco, L.; Edwards-Callaway, L.N.; Bello, N.M.; Mijares, S.H.; Cull, C.A.; Rugan, S.; Mosher, R.A.; Gehring, R.; Coetzee, J.F. Unmitigated surgical castration in calves of different ages: Cortisol concentrations, heart rate variability, and infrared thermography findings. *Animals* **2021**, *11*, 2719. [CrossRef]
98. Fajt, V.R.; Wagner, S.A.; Norby, B. Analgesic drug administration and attitudes about analgesia in cattle among bovine practitioners in the United States. *J. Am. Vet. Med. Assoc.* **2011**, *238*, 755–767. [CrossRef]
99. De Briyne, N.; Berg, C.; Blaha, T.; Temple, D. Pig castration: Will the EU manage to ban pig castration by 2018? *Porc. Health Manag.* **2016**, *2*, 29. [CrossRef]
100. Kynetec. *2017 Merino Husbandry Practices Survey*; Australian Wool Innovation Limited: Sydney, Australia, 2018; pp. 1–152.
101. Steagall, P.V.; Bustamante, H.; Johnson, C.B.; Turner, P.V. Pain management in farm animals: Focus on cattle, sheep and pigs. *Animals* **2021**, *11*, 1483. [CrossRef] [PubMed]
102. Godyń, D.; Herbut, E.; Walczak, J. Infrared thermography as a method for evaluating the welfare of animals subjected to invasive procedures—A review. *Ann. Anim. Sci.* **2013**, *13*, 423–434. [CrossRef]
103. Hoa, V.-B.; Song, D.-H.; Seol, K.-H.; Kang, S.-M.; Kim, H.-W.; Jang, S.-S.; Cho, S.-H. Half-castration is a newly effective method for increasing yield and tenderness of male cattle meat. *Anim. Biosci.* **2022**, *35*, 1258–1269. [CrossRef]
104. Paranzini, C.S.; Sousa, A.K.; Cardoso, G.S.; Perencin, F.M.; Trautwein, L.G.C.; Bracarense, A.P.F.R.L.; Martins, M.I.M. Effects of chemical castration using 20% CaCl 2 with 0.5% DMSO in tomcats: Evaluation of inflammatory reaction by infrared thermography and effectiveness of treatment. *Theriogenology* **2018**, *106*, 253–258. [CrossRef] [PubMed]
105. Baysinger, A.; Webb, S.R.; Brown, J.; Coetzee, J.F.; Crawford, S.; DeDecker, A.; Karriker, L.A.; Pairis-Garcia, M.; Sutherland, M.A.; Viscardi, A.V. Proposed multidimensional pain outcome methodology to demonstrate analgesic drug efficacy and facilitate future drug approval for piglet castration. *Anim. Health Res. Rev.* **2021**, *22*, 163–176. [CrossRef]
106. Viscardi, A.V.; Cull, C.A.; Kleinhenz, M.D.; Montgomery, S.; Curtis, A.; Lechtenberg, K.; Coetzee, J.F. Evaluating the utility of a CO2 surgical laser for piglet castration to reduce pain and improve wound healing: A pilot study. *J. Anim. Sci.* **2020**, *98*, skaa320. [CrossRef] [PubMed]
107. Saidu, A.M.; Olorunfemi, O.J.; Laku, D. Thermography following castration, otectomy, and gastrotomy in Nigerian indigenous dogs. *Sahel J. Vet. Sci.* **2023**, *20*, 50–56. [CrossRef]
108. Stewart, M.; Verkerk, G.A.A.; Stafford, K.J.J.; Schaefer, A.L.L.; Webster, J.R.R. Noninvasive assessment of autonomic activity for evaluation of pain in calves, using surgical castration as a model. *J. Dairy Sci.* **2010**, *93*, 3602–3609. [CrossRef]
109. Kleinhenz, M.D.; Van Engen, N.K.; Smith, J.S.; Gorden, P.J.; Ji, J.; Wang, C.; Perkins, S.C.B.; Coetzee, J.F. The impact of transdermal flunixin meglumine on biomarkers of pain in calves when administered at the time of surgical castration without local anesthesia. *Livest. Sci.* **2018**, *212*, 1–6. [CrossRef]
110. Graves, M.T.; Schneider, L.; Cox, S.; Caldwell, M.; Krawczel, P.; Lee, A.; Lear, A. Evaluation of the pharmacokinetics and efficacy of transdermal flunixin for pain mitigation following castration in goats. *Transl. Anim. Sci.* **2020**, *4*, txaa198. [CrossRef]
111. Harris, C.; White, P.J.; Hall, E.; Van der Saag, D.; Lomax, S. Evaluation of electroencephalography, behaviour and eye temperature in response to surgical castration in sheep. *Animals* **2021**, *11*, 637. [CrossRef]
112. Bates, J.L.; Karriker, L.A.; Stock, M.L.; Pertzborn, K.M.; Baldwin, L.G.; Coetzee, J.F. The impact of translactational delivered meloxicam analgesia on biomarkers of pain and distress after piglet processing. *Iowa State Univ. Anim. Indust. Rep.* **2014**, *11*, 1–13.
113. Guatteo, R.; Levionnois, O.; Fournier, D.; Guémené, D.; Latouche, K.; Leterrier, C.; Mormède, P.; Prunier, A.; Servière, J.; Terlouw, C.; et al. Minimising pain in farm animals: The 3S approach—'Suppress, Substitute, Soothe'. *Animal* **2012**, *6*, 1261–1274. [CrossRef] [PubMed]
114. Dockweiler, J.C.; Coetzee, J.F.; Edwards-Callaway, L.N.; Bello, N.M.; Glynn, H.D.; Allen, K.A.; Theurer, M.E.; Jones, M.L.; Miller, K.A.; Bergamasco, L. Effect of castration method on neurohormonal and electroencephalographic stress indicators in Holstein calves of different ages. *J. Dairy Sci.* **2013**, *96*, 4340–4354. [CrossRef]
115. Marti, S.; Meléndez, D.M.; Pajor, E.A.; Moya, D.; Heuston, C.E.M.; Gellatly, D.; Janzen, E.D.; Schwartzkopf-Genswein, K.S. Effect of band and knife castration of beef calves on welfare indicators of pain at three relevant industry ages: II. Chronic pain1. *J. Anim. Sci.* **2017**, *95*, 4367–4380. [CrossRef] [PubMed]
116. Sutherland, M.A.; Davis, B.L.; McGlone, J.J. The effect of local or general anesthesia on the physiology and behavior of tail docked pigs. *Animal* **2011**, *5*, 1237–1246. [CrossRef]
117. Eicher, S.D.; Cheng, H.W.; Sorrells, A.D.; Schutz, M.M. Short communication: Behavioral and physiological indicators of sensitivity or chronic pain following tail docking. *J. Dairy Sci.* **2006**, *89*, 3047–3051. [CrossRef] [PubMed]
118. Van der Saag, D.; Lomax, S.; Windsor, P.A.; Taylor, C.; White, P.J. Evaluating treatments with topical anaesthetic and buccal meloxicam for pain and inflammation caused by amputation dehorning of calves. *PLoS ONE* **2018**, *13*, e0198808. [CrossRef]
119. Stock, M.L.; Millman, S.T.; Barth, L.A.; Van Engen, N.K.; Hsu, W.H.; Wang, C.; Gehring, R.; Parsons, R.L.; Coetzee, J.F. The effects of firocoxib on cautery disbudding pain and stress responses in preweaned dairy calves. *J. Dairy Sci.* **2015**, *98*, 6058–6069. [CrossRef]

120. Redaelli, V.; Luzi, F.; Farish, M.; Nanni Costa, L. The use of thermography to assess the teeth temperature during resection by grinding in piglets. In Proceedings of the 11th International Conference on Quantitative InfraRed Thermography, Naples, Italy, 11–14 June 2012; pp. 1–2.
121. Teixeira, D.L.; Boyle, L.A.; Enríquez-Hidalgo, D. Skin temperature of slaughter pigs with tail lesions. *Front. Vet. Sci.* **2020**, *7*, 198. [CrossRef]
122. Viscardi, A.V.; Reppert, E.J.; Kleinhenz, M.D.; Wise, P.; Lin, Z.; Montgomery, S.; Daniell, H.; Curtis, A.; Martin, M.; Coetzee, J.F. Analgesic comparison of flunixin meglumine or meloxicam for soft-tissue surgery in sheep: A pilot study. *Animals* **2021**, *11*, 423. [CrossRef] [PubMed]
123. Reid, J.; Nolan, A.M.; Scott, E.M. Measuring pain in dogs and cats using structured behavioural observation. *Vet. J.* **2018**, *236*, 72–79. [CrossRef] [PubMed]
124. Ison, S.; Rutherford, K. Attitudes of farmers and veterinarians towards pain and the use of pain relief in pigs. *Vet. J.* **2014**, *202*, 622–627. [CrossRef] [PubMed]
125. Casas-Alvarado, A.; Mota- Rojas, D.; Hernández-Avalos, I.; Martínez-Burnes, J.; Rosas, M.; Miranda-Cortés, A.E.; Domínguez-Oliva, A.; Mora-Medina, P. Assessment of thermal response, cardiorespiratory parameters and postoperative analgesia in dogs undergoing ovariohysterectomy with different combinations of epidural anesthesia and isoflurane. *J. Anim. Behav. Biometeorol.* **2023**, *11*, e2023009. [CrossRef]
126. Domínguez-Oliva, A.; Casas-Alvarado, A.; Miranda-Cortés, A.E.; Hernández-Avalos, I. Clinical pharmacology of tramadol and tapentadol, and their therapeutic efficacy in different models of acute and chronic pain in dogs and cats. *J. Adv. Vet. Anim. Res.* **2021**, *8*, 404. [CrossRef]
127. Tapper, K.R.; Johnson, A.K.; Karriker, L.A.; Stalder, K.J.; Parsons, R.L.; Wang, C.; Millman, S.T. Pressure algometry and thermal sensitivity for assessing pain sensitivity and effects of flunixin meglumine and sodium salicylate in a transient lameness model in sows. *Livest. Sci.* **2013**, *157*, 245–253. [CrossRef]
128. Stubsjøen, S.M.; Flø, A.S.; Moe, R.O.; Janczak, A.M.; Skjerve, E.; Valle, P.S.; Zanella, A.J. Exploring non-invasive methods to assess pain in sheep. *Physiol. Behav.* **2009**, *98*, 640–648. [CrossRef]
129. Silva, L.K.; Martorano, L.G.; Da Silva, W.; Garcia, A.R.; Fernandes, G.B.; Reis, A.D.S.B.; Gomes, W.N.; Correa, F.R.A.; da Silva, F.P.; Beldini, T.P.; et al. Uso associado da termografia por infravermelho e ozonioterapia para diagnóstico e tratamento de um processo inflamatório em equino: Relato de caso. *Pubvet* **2020**, *14*, 1–9. [CrossRef]
130. Holmes, L.C.; Gaughan, E.M.; Gorondy, D.A.; Hogge, S.; Spire, M.F. The effect of perineural anesthesia on infrared thermographic images of the forelimb digits of normal horses. *Can. Vet. J. La Rev. Vet. Can.* **2003**, *44*, 392–396.
131. Gómez-Prado, J.; Pereira, A.M.F.; Wang, D.; Villanueva-García, D.; Domínguez-Oliva, A.; Mora-Medina, P.; Hernández-Avalos, I.; Martínez-Burnes, J.; Casas-Alvarado, A.; Olmos-Hernández, A.; et al. Thermoregulation mechanisms and perspectives for validating thermal windows in pigs with hypothermia and hyperthermia: An overview. *Front. Vet. Sci.* **2022**, *9*, 1023294. [CrossRef]
132. Grégoire, J.; Bergeron, R.; D'Allaire, S.; Meunier-Salaün, M.-C.; Devillers, N. Assessment of lameness in sows using gait, footprints, postural behaviour and foot lesion analysis. *Animal* **2013**, *7*, 1163–1173. [CrossRef] [PubMed]
133. Colak, A.; Polat, B.; Okumus, Z.; Kaya, M.; Yanmaz, L.E.; Hayirli, A. Short communication: Early detection of mastitis using infrared thermography in dairy cows. *J. Dairy Sci.* **2008**, *91*, 4244–4248. [CrossRef] [PubMed]
134. Whittaker, A.L.; Marsh, L.E. The role of behavioural assessment in determining 'positive' affective states in animals. *CABI Rev.* **2019**, 1–13. [CrossRef]
135. Larson, M.D.; Behrends, M. Portable infrared pupillometry. *Anesth. Anal.* **2015**, *120*, 1242–1253. [CrossRef]
136. Pouzot-Nevoret, C.; Junot, S.; Goffette, L.; Bonnet-Garin, J.-M.; Allaouchiche, B.; Magnin, M. Use of pupillometry for the evaluation of analgesia in dogs hospitalized in intensive care: A prospective study. *Res. Vet. Sci.* **2023**, *158*, 96–105. [CrossRef]
137. Gruenewald, M.; Ilies, C. Monitoring the nociception–anti-nociception balance. *Best Pract. Res. Clin. Anaesthesiol.* **2013**, *27*, 235–247. [CrossRef]
138. Mansour, C.; Mocci, R.; Santangelo, B.; Sredensek, J.; Chaaya, R.; Allaouchiche, B.; Bonnet-Garin, J.-M.; Boselli, E.; Junot, S. Performance of the Parasympathetic Tone Activity (PTA) index to predict changes in mean arterial pressure in anaesthetized horses with different health conditions. *Res. Vet. Sci.* **2021**, *139*, 43–50. [CrossRef]
139. Domínguez-Oliva, A.; Mota-Rojas, D.; Hernández-Avalos, I.; Mora-Medina, P.; Olmos-Hernández, A.; Verduzco-Mendoza, A.; Casas-Alvarado, A.; Whittaker, A.L. The neurobiology of pain and facial movements in rodents: Clinical applications and current research. *Front. Vet. Sci.* **2022**, *9*, 1016720. [CrossRef]
140. Mota-Rojas, D.; Domínguez-Oliva, A.; Martínez-Burnes, J.; Casas-Alvarado, A.; Hernández-Ávalos, I. Euthanasia and Pain in Canine Patients with terminal and chronic-degenerative diseases: Ethical and legal aspects. *Animals* **2023**, *13*, 1265. [CrossRef]
141. Mota-Rojas, D.; Pereira, M.F.A.; Wang, D.; Martínez-Burnes, J.; Ghezzi, M.; Hernández-Ávalos, I.; Lendez, P.; Mora-Medina, P.; Casas, A.; Olmos-Hernández, A.; et al. Clinical applications and factors involved in validating thermal windows in large ruminants to assess health and productivity. *Animals* **2021**, *11*, 2247. [CrossRef]
142. Villanueva-García, D.; Mota-Rojas, D.; Martínez-Burnes, J.; Olmos-Hernández, A.; Mora-Medina, P.; Salmerón, C.; Gómez, J.; Boscato, L.; Gutiérrez-Pérez, O.; Cruz, V.; et al. Hypothermia in newly born piglets: Mechanisms of thermoregulation and pathophysiology of death. *J. Anim. Behav. Biometeorol.* **2021**, *9*, 1–10. [CrossRef]

143. Lezama-García, K.; Martínez-Burnes, J.; Marcet-Rius, M.; Gazzano, A.; Olmos-Hernández, A.; Mora-Medina, P.; Domínguez-Oliva, A.; Pereira, A.M.F.; Hernández-Ávalos, I.; Baqueiro-Espinosa, U.; et al. Is the weight of the newborn puppy related to its thermal balance? *Animals* **2022**, *12*, 3536. [CrossRef] [PubMed]
144. Napolitano, F.; Bragaglio, A.; Braghieri, A.; El-Aziz, A.H.A.; Titto, C.G.; Villanueva-García, D.; Mora-Medina, P.; Pereira, A.M.F.; Hernández-Avalos, I.; José-Pérez, N.; et al. The effect of birth weight and time of day on the thermal response of newborn water buffalo calves. *Front. Vet. Sci.* **2023**, *10*, 1084092. [CrossRef] [PubMed]
145. Flores-Peinado, S.; Mota-Rojas, D.; Guerrero-Legarreta, I.; Mora-Medina, P.; Cruz-Monterrosa, R.; Gómez-Prado, J.; Guadalupe Hernández, M.; Cruz-Playas, J.; Martínez-Burnes, J. Physiological responses of pigs to preslaughter handling: Infrared and thermal imaging applications. *Int. J. Vet. Sci. Med.* **2020**, *8*, 71–84. [CrossRef] [PubMed]
146. Stewart, M.; Webster, J.R.; Verkerk, G.A.; Schaefer, A.L.; Colyn, J.J.; Stafford, K.J. Non-invasive measurement of stress in dairy cows using infrared thermography. *Physiol. Behav.* **2007**, *92*, 520–525. [CrossRef] [PubMed]
147. Sutherland, M.A.; Worth, G.M.; Dowling, S.K.; Lowe, G.L.; Cave, V.M.; Stewart, M. Evaluation of infrared thermography as a non-invasive method of measuring the autonomic nervous response in sheep. *PLoS ONE* **2020**, *15*, e0233558. [CrossRef]
148. Lowe, G.; Sutherland, M.; Waas, J.; Schaefer, A.; Cox, N.; Stewart, M. Infrared Thermography—A Non-invasive method of measuring respiration rate in calves. *Animals* **2019**, *9*, 535. [CrossRef] [PubMed]
149. Berkmans, D. *Advances in Precision Livestock Farming*; Burleing Dodds Science Publishing: Cambridge, UK, 2022.
150. Luzi, F.; Mitchell, M.; Nanni, C.L.; Redaelli, V. *Thermography: Current Status and Advances in Livestock Animals and in Veterinary Medicine*; Fondazione Iniziative Zooprofilattiche e Zootecniche: Brescia, Italy, 2013; 216p.

animals

MDPI

Case Report

Infrared Thermography of the Blowhole as a Potential Diagnostic Tool for Health Assessment in Killer Whales (*Orcinus orca*)

Jennifer P. Russell [1,2,*], Steve D. Osborn [2], Kelsey E. S. Herrick [1], Todd L. Schmitt [1] and Todd Robeck [3,4]

1 SeaWorld San Diego Zoological Department, 500 SeaWorld Drive, San Diego, CA 92109, USA
2 SeaWorld San Antonio Zoological Department, 10500 SeaWorld Drive, San Antonio, TX 78251, USA
3 SeaWorld Orlando Zoological Department, 7007 SeaWorld Drive, Orlando, FL 32821, USA
4 SeaWorld and Busch Gardens Species Preservation Laboratory, 2595 Ingraham St., San Diego, CA 92109, USA
* Correspondence: jennifer.russell@seaworld.com

Simple Summary: Non-invasive techniques, such as infrared thermography, are ideal diagnostic tools for evaluation of the health status of marine megafauna such as killer whales (*Orcinus orca*). Infrared thermography of the blowhole was implemented in the diagnostic work-up of two male killer whales (also known as 'orcas') under managed care, and accurately detected changes in blowhole mucosa emissivity consistent with pyrexia in both cases. Infrared thermography of the blowhole is a potentially useful addition for the health assessment of killer whales as a screening tool to identify pyrexic animals within a group. The aim of this case report is to describe the clinical presentation and diagnostic findings in two killer whales with signs consistent with inflammation and pyrexia, most notably that an increase in blowhole temperature obtained by infrared thermography aligned with increased rectal temperature.

Abstract: Killer whales (*Orcinus orca*) are experiencing increasing environmental pressures, with some ecotypes being identified as endangered, and the development and validation of non-invasive health assessment tools is critical for assessing the well-being of individuals within these endangered populations. Infrared thermography of the blowhole is a non-contact method of temperature measurement that was recently investigated in killer whales in managed care. Two male killer whales presenting with clinical signs at separate institutions had veterinary clinical health assessments performed, which included infrared thermography of the blowhole as well as concurrent rectal temperature measurement. The current case report is aimed at describing the clinical use of infrared thermography of the blowhole as a method to detect elevated body temperature in two killer whales. Both animals exhibited blowhole temperatures above the previously reported values (36.4 °C and 37.6 °C; the mean in healthy whales is reported to be 34.21 ± 1.47 °C) with concurrently elevated rectal temperatures, as well as clinicopathologic findings consistent with a systemic inflammatory response (e.g., neutrophilia, increased fibrinogen and erythrocyte sedimentation rate, hypoferritinemia). Following veterinary intervention, both animals' blowhole and rectal temperatures returned to baseline. Infrared thermography of the blowhole represents a promising tool for the identification of pyrexic animals and with further investigation may be considered as part of conservation health assessments for threatened free-ranging populations.

Keywords: infrared thermography; blowhole; killer whale; *Orcinus orca*

Citation: Russell, J.P.; Osborn, S.D.; Herrick, K.E.S.; Schmitt, T.L.; Robeck, T. Infrared Thermography of the Blowhole as a Potential Diagnostic Tool for Health Assessment in Killer Whales (*Orcinus orca*). *Animals* **2024**, *14*, 1867. https://doi.org/10.3390/ani14131867

Academic Editor: Daniel Mota-Rojas

Received: 16 May 2024
Revised: 19 June 2024
Accepted: 19 June 2024
Published: 25 June 2024

1. Introduction

As apex predators, killer whales (*Orcinus orca*) accumulate legacy chemical pollutants, such as polychlorinated biphenyl (PCBs) and mercury, and are one of the most contaminated marine mammals in the world [1–3]. While prey availability and anthropogenic factors are among the largest threats to wild populations, infectious disease is also a commonly reported cause of death, possibly due to the immunosuppressive effects of contaminants, malnutrition, or physiologic stress imposed by anthropogenic factors [1,4,5]. The validation of non-invasive methods for diagnostics and data collection is important due to the logistic difficulty of collecting direct physiologic samples or measurements from aquatic megafauna [6,7]. For animals that may be experiencing disease processes, non-invasive techniques are ideal. Stress associated with handling may compromise the individual further by disrupting social interactions, appetite, or willingness to participate in wellness exams [8,9]. Stress associated with handling naive or wild cetaceans can potentially result in stress-induced hyperthermia, electrolyte derangements, changes in circulating thyroid and adrenal hormones, or exertional rhabdomyolysis [9–13]. Non-invasive methods also circumvent any complications associated with the use of chemical restraint [14].

Health assessment of marine megafauna, such as killer whales, presents unique challenges to the veterinarian. These animals are often too large for traditional diagnostic imaging modalities (e.g., ultrasound tissue penetration is non-diagnostic at depths greater than 30 cm with commercially available equipment, and, similarly, radiographic imaging is limited by animal size and beam penetration [14–16]). Moreover, they live exclusively in an aquatic environment, where simple physical examination can be logistically difficult, and even something as simple as thoracic auscultation is inhibited by a thick blubber layer, rapid breath cycle, and water interfering with equipment [15]. This is in addition to the major constraint posed by unwell animals who may become non-participatory due to feelings of malaise, or free-ranging animals who are not directly accessible. Rectal temperature is used as part of the routine veterinary exam; however, this data can only be regularly collected if the animal is conditioned to accept a rectal probe and willing to participate. Until recently, no reference ranges for rectal temperature in killer whales were available [17].

Core body temperature is a physiological parameter that provides insight into clinical health status, and estimation of core body temperature (typically via rectal temperature measurement) is part of the minimum diagnostic database in veterinary medicine [18,19]. Infrared thermography can detect anatomical regions in many species that may serve as non-invasive proxies to core body temperature, known as thermal windows [20,21]. These are body surfaces that exhibit elevated heat exchange from the animal to the environment, and they typically have quantitative and qualitative differences in vascularization as compared to well-insulated areas of the body [21]. By identifying and using appropriate thermal windows, infrared imaging has the potential to recognize vasomotor changes consistent with disease processes that induce inflammation or pyrexia [22,23].

The blowhole of free-ranging and managed care cetaceans has been shown to provide a consistently detectable heat anomaly, given the lack of insulating blubber and large venous plexuses associated with the accessory sinus system [17,24–27]. This creates a potentially useful thermal window for infrared imaging. Infrared thermography of the blowhole was found to be nearly identical to rectal temperatures collected from bottlenose dolphins (*Tursiops truncatus*) and approximately 1 °C lower than rectal temperatures in beluga whales (*Delphinapterus leucas*) and 1.28 °C lower than rectal temperatures in killer whales [17,26]. Recent data comparing infrared thermography of the blowhole to rectal temperature in killer whales showed that the accuracy of the infrared blowhole reading improved as the mean temperature of the animal increased (i.e., as the mean rectal and blowhole temperatures increased, so did the accuracy of the infrared camera) [17]. Thus, it is possible that infrared imaging of the blowhole may have utility as a non-invasive diagnostic screening tool to identify pyrexic individuals in a population of whales.

A reliable non-contact method of evaluating body temperature, such as infrared thermography, may not only have implications for individual animals receiving veterinary care, but may also serve as a useful data collection tool when performing health assessments of wild killer whale populations [7,18,28,29]. Studies using infrared thermography in marine mammals are limited, and, to the authors' knowledge, there are no case reports describing the use of infrared thermography of the blowhole in killer whales as part of a clinical diagnostic work-up. It is the authors' objective to retrospectively report the clinical use of infrared thermography of the blowhole in two killer whale cases and discuss how detectable changes in blowhole temperature complement the assessment of health. The normality of clinicopathological results (i.e., hematology, biochemistry, and rectal and blowhole temperatures) was determined by comparisons with our previously reported killer whale age- and sex-specific reference intervals (RIs) [17,30,31], as well as comparisons with historical individual baseline data. Any value outside of these ranges was considered abnormal.

2. Case 1

A 10-year-old male killer whale living at Facility A, weighing 2561 kg, presented with slow movement through the habitat, halitosis, and regurgitation of food items. Appetite and general demeanor toward animal care staff were unchanged and husbandry behaviors were offered by the animal voluntarily. Veterinary physical examination and infrared thermography were performed on the same day as clinical signs were first reported. Infrared thermography of the oral mucosa was unremarkable. The whale was asked by husbandry staff to rest in ventral recumbency at the surface of the water and breathe normally while a handheld infrared camera (FLIR E40 Thermal Imaging Camera, Teledyne FLIR, 2700 SW Parkway Ave., Wilsonville, OR 97070, USA) was held directly above the blowhole at a distance of approximately 2 ft (0.61 m), as previously described (Figure 1a) [17]. The camera was calibrated prior to the exam using melting ice at 0 °C. An infrared video was taken of the blowhole opening and closing during a natural breath cycle, while a dynamic crosshair tracked the hottest area visible within the image frame. Blowhole temperature was 37.6 °C (Figure 2a). The reported mean observed blowhole range for healthy killer whales is 34.21 ± 1.47 °C [17]. The animal was then asked by trainers to rest in dorsal recumbency for insertion of a rectal temperature probe to 40 cm depth (Digi-Sense Temp Series Thermocouple thermometer, Cole-Palmer, 625 East Bunker Court, Vernon Hills, LI 60061, USA). Rectal temperature at presentation was 36.8 °C (the expected temperature range for male killer whales <4000 kg [ETR] is 35.10 to 35.50 °C [17]). A blood sample from the ventral fluke periarterial vascular rete (PAVR) revealed a decreased alkaline phosphatase (ALP) (497 IU/L, down from 531 IU/L), elevated erythrocyte sedimentation rate (ESR) (5 mm/h; 0 mm/h is typical for the species), increased fibrinogen (441 mg/dL, up from 227 mg/dL; RI 250.6 to 281.8 mg/dL), increased globulin (3.27 g/dL, up from 2.88 g/dL; RI 2.54 to 2.92 g/dL), decreased serum iron (26 μg/dL, down from 86 μg/dL; RI 69.3 to 86.6 μg/dL), and increased white blood cell count (WBC) ($6.60 \times 10^3/\mu L$ with 82% relative neutrophilia, up from $5.13 \times 10^3/\mu L$; RI 5.32–$6.07 \times 10^3/\mu L$). These changes are consistent with a systemic inflammatory response and were notably changed from a routine sample collected two weeks prior to presentation. The animal was started on empiric treatment with levofloxacin (5 mg/kg orally [PO] q 24 h), amoxicillin clavulanate (7 mg/kg PO q 12 h), voriconazole (0.27 mg/kg PO q 24 h), and famotidine (0.4 mg/kg PO q 24 h) [32]. Diagnostic imaging was performed, including voluntary dental radiographs (unremarkable) and abdominal and thoracic ultrasound (unremarkable except for a subjective decrease in intestinal motility from baseline). Ancillary diagnostic investigation included blowhole sputum sampling (for anaerobic, aerobic, fungal, and acid-fast culture and sensitivity and respiratory microbiome characterization [via next generation DNA sequencing] and avian influenza polymerase chain reaction [PCR] [negative]) as well as whole blood pan-bacterial and pan-fungal PCR (negative), fecal cytology and gram stain (unremarkable), enteric viral PCR (negative), and fecal microbiome characterization. Findings were supportive of a presumptive diagnosis of uncomplicated gastroenteritis, which responded well to continued supportive treatment (famotidine was discontinued, and omeprazole [360 mg total dose PO q 24 h] and sucralfate [20 g total dose PO q 8 h] were initiated [32]). Rectal temperature decreased back to baseline to 35.5 °C, and blowhole temperature was 33.7 °C three days after initiating treatment (Figure 2b). Full clinical resolution was achieved within 15 days of initial presentation with normalization of complete blood count (CBC), serum biochemistry, ESR, and fibrinogen, and medications were discontinued on day 18.

(a) (b)

Figure 1. (**a**) A veterinarian performing an infrared blowhole temperature reading in a killer whale resting at the surface of the water. (**b**) A digital photograph of the open blowhole of a killer whale, for reference.

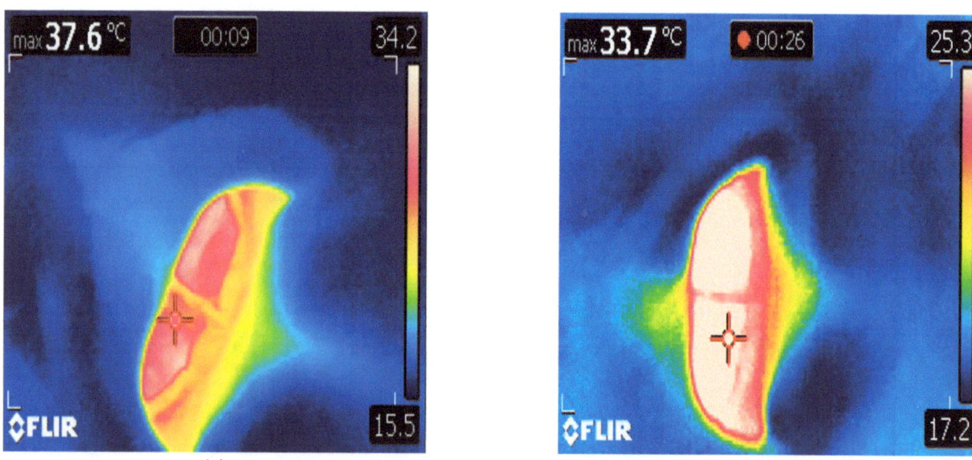

(a) (b)

Figure 2. (**a**) A still image from an infrared video clip of the open blowhole of a male killer whale (*Orcinus orca*) described in Case 1 on the day of presentation. The red crosshair, or 'hotspot', tracks the maximum temperature within the image frame and reports it in the top left corner (37.6 °C). The bar on the right depicts the range of pixel colors assigned to temperature values. (**b**) A still image from an infrared video clip of the blowhole of the same killer whale following veterinary intervention.

3. Case 2

A 21-year-old male killer whale living at Facility B, weighing 3672 kg, presented with decreased appetite and resting at the bottom of the pool. Veterinary physical exam was performed on the same day as clinical signs were reported. Infrared thermography of the oral mucosa revealed an increase in surface temperature at the lower right eighth tooth, which had associated periapical lucency on voluntary dental radiographs. The animal was asked by husbandry staff to rest in ventral recumbency at the surface of the water and breathe normally while a handheld infrared camera (the same model as in Facility A) was held directly above the blowhole at a distance of approximately 2 ft (0.61 m), as previously described [17]. The camera was calibrated using melting ice at 0 °C prior to examination. Blowhole temperature was 36.4 °C on the day of presentation (Figure 3a). The animal was then asked by trainers to rest by the side of the pool in dorsal recumbency for insertion of a rectal temperature probe to 40 cm depth. Rectal temperature was 37.4 °C. A blood sample collected the same day from the ventral fluke PAVR revealed decreased ALP (69 IU/L, down from 100 IU/L), markedly elevated ESR (23 mm/h; 0 mm/h is typical for the species) and fibrinogen (608 mg/dL, up from 265 mg/dL), globulin (4.4 g/dL, up from 3.7 g/dL), decreased iron (23 µg/dL, down from 75 µg/dL), and leukocytosis ($9.46 \times 10^3/\mu L$ characterized by an 86% neutrophilia, up from $6.08 \times 10^3/\mu L$). These changes are consistent with a systemic inflammatory response, are outside of the available reference intervals for this species [17,30,31], and were notably changed from a routine sample collected one week prior to presentation. This animal was diagnosed with a periapical tooth abscess and a dental extraction was subsequently performed. The animal responded well to antimicrobial therapy and analgesia (clindamycin 4.5 mg/kg PO q 12 h, voriconazole 0.27 mg/kg PO q 24 h, and aspirin 240 grain equine bolus total dose PO once). Rectal and blowhole temperatures returned to 35.5 °C and 33.6 °C, respectively, within one week of initiating treatment (Figure 3b). Normalization of CBC, serum biochemistry, ESR, and fibrinogen followed dental extraction performed 14 days later.

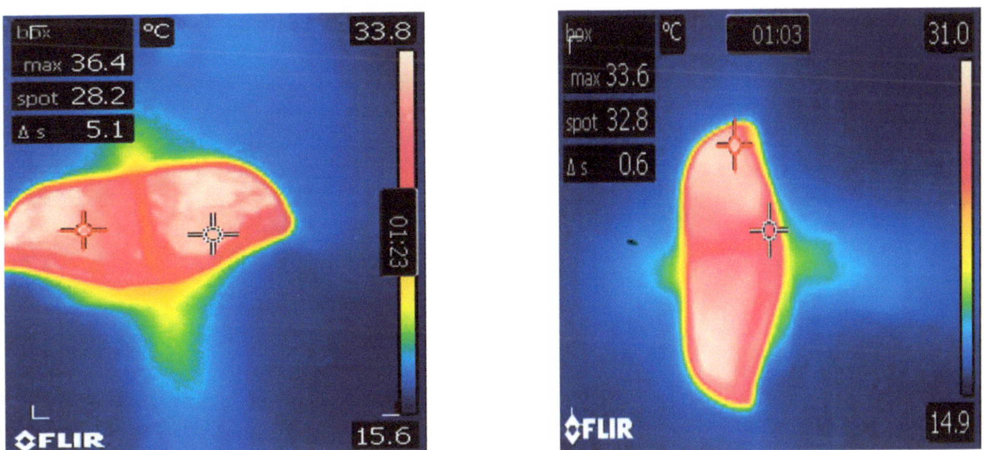

(a) (b)

Figure 3. (**a**) A still image from an infrared video of the open blowhole of a killer whale (*Orcinus orca*) described in Case 2. The camera in this case was in the 'Hotspot-Spot' setting, whereby a red crosshair dynamically tracks the hottest detectable pixel within the frame ('max') and a white crosshair remains static in the center of the screen ('spot'). The blowhole temperature in this case was recorded as 36.4 °C. (**b**) A still image from an infrared video clip of the blowhole of the same killer whale following veterinary intervention.

4. Discussion

In both cases, blowhole and rectal temperatures simultaneously increased outside of the reported observed healthy range for the species [17] and, in the context of concurrent diagnostic results, were interpreted as pyrexia for these individuals. The additional information provided by the blood analyses and diagnostic imaging supported working diagnoses of inflammatory processes resulting in peripheral vasodilation and increased thermal radiation consistent with a pyrexic state. Rigorous monitoring of the blood parameters and temperatures of these individuals was important to document both a positive response to treatment and clinical resolution.

Infrared thermography is not a direct measurement of temperature but rather the detection of infrared radiation emitted by a subject. This is translated into an image, which depicts each infrared energy level using arbitrarily assigned colors [21,33]. In these cases, infrared blowhole temperature measurement and rectal temperature via rectal probe were used as surrogates of core body temperature, as invasive absolute core body temperature measurement was not performed. In each case, different disease processes resulted in a similar systemic inflammatory response. In mammals, the release of pyrogens and proinflammatory cytokines alter the thermal 'set point' in the anterior hypothalamus and cause an increase in core body temperature. This, in turn, results in changes in peripheral vasodilation and heat exchange with the environment [21,22,34–36]. Pyrexia is distinct from nonfebrile hyperthermia, where the hypothalamus is not influenced, but is rather a function of heat gain vs. loss (e.g., during exercise, inadequate heat dissipation, etc.) [34,35,37]. This is a key concept to note if performing infrared thermography on animals where physical or chemical restraint is employed, as stress-induced hyperthermia may occur [8,9].

While case number is a limitation in the current report, infrared thermography of the blowhole as a complimentary diagnostic tool proved to be reliable at detecting systemic inflammation and pyrexia. It was well-tolerated and comfortable for the animals (who readily participated), was safe to perform for both animal and clinician, and provided informative insight into the thermoregulatory changes associated with the clinical state of the animals, allowing for timely and successful treatment to be initiated. In Case 2, the use of aspirin as an analgesic may have contributed to the reduction in rectal and blowhole temperature as it is also an antipyretic agent. Nevertheless, infrared thermography was able to detect this reduction in temperature.

Infrared temperature measurement of the blowhole was performed in these cases in a clinical setting and not under controlled experimental conditions. While these animals were participatory and conditioned to rest at the surface for blowhole and rectal temperature measurement, infrared thermography is a non-contact method that may theoretically be used in individuals that are not conditioned, non-participatory due to malaise or other factors, or not amenable to handling, such as in the case of free-ranging animals. Such assessments may potentially be incorporated into data collection performed via unmanned aerial systems (drones) as previously described [7,24,38–41]; however, further investigation is required, given that there are multiple factors that may influence the infrared temperature measurement of the blowhole of a free-ranging animal, including altitude or distance from the subject, air temperature, water temperature, wind speed, cloud cover, emittance of surrounding sea water, the animals' wet skin surface, or blowhole vapor interference [17,21,22,38,42]. It is also unknown whether blowhole temperatures will vary beyond what is observed here during swimming or exercise. Further investigation is required to determine if killer whales become hyperthermic with exercise and whether this is detectable via infrared imaging. In studies examining the rectal temperature of bottlenose dolphins, it was found that rectal temperatures obtained during exercise were 0.5 °C cooler than temperatures measured at rest [43].

Additional case report data are required to investigate whether infrared thermography of the blowhole can accurately detect clinically significant temperature elevation in various disease states beyond those described here, or whether it might be a useful measurement tool to detect decreases in core body temperature associated with parturition in females [44,45]. The two individual case examples reported herein may provide evidence to support the use of blowhole thermography for identifying pyrexia in animals resting at the surface of the water. However, it cannot be assumed that a systemically unwell animal will always have a blowhole temperature above the normal range. Given that the tendency for blowhole temperature readings is to be below those of rectal temperature in the healthy animal, and that accuracy improves with increasing mean temperature, it is unlikely that this temperature measurement will yield false positives; however, more clinical case data are needed to investigate this [17]. In Case 1, the blowhole temperature recorded was higher than the rectal temperature, which supports this concept. While rectal temperature is now performed daily as part of the routine clinical health assessment for killer whales in the authors' facilities, the utility of blowhole temperature has only recently been investigated in this species [17]. Only two clinical cases incorporating infrared temperature measurement of the blowhole have occurred within the last three years, and these are the only two animals to date that have had measured blowhole temperatures above the observed 'normal' range [17].

5. Conclusions

Infrared thermography of the blowhole was a reliable sensor for inflammation and pyrexia in two killer whales presenting with clinical signs. This non-invasive method was well-tolerated by the animals with no adverse effects. Infrared thermography of the blowhole has clinical application for managed-care individual animals receiving veterinary care (particularly animals that are not amenable to rectal temperature measurement) and may possibly also serve as a useful screening tool when performing health assessments of free-ranging killer whale populations via unmanned aerial systems, allowing the detection of pyrexic animals within a group. Additional case report data incorporating blowhole infrared thermography as part of the diagnostic toolkit will help to corroborate these findings.

The detection of pyrexic animals might be beneficial for epidemiologists investigating infectious or inflammatory disease prevalence within a population (e.g., animals identified as pyrexic may provide insight for serological surveys or stranding data as to whether animals sampled are demonstrating seroconversion due to previous disease exposure vs. active infection). It may benefit the clinical evaluation of live stranded animals to assist rehabilitative efforts or the identification of individuals in need of veterinary or other intervention (e.g., via direct or remote delivery of therapeutics [46]). It might assist in the identification of immunocompromised individuals in a critically endangered population or in the investigation of the physiologic effects of various environmental or anthropogenic influences, which may provide valuable data in support of species protective legislation or other influential conservation measures. Blowhole temperature data from managed-care and free-ranging killer whales and other cetacean species may therefore provide valuable contextual data that will compliment assessments of health status for wild populations.

Author Contributions: Investigation, J.P.R., S.D.O., K.E.S.H. and T.L.S.; writing—original draft preparation, J.P.R.; writing—review and editing, J.P.R., S.D.O., K.E.S.H., T.L.S. and T.R. All authors have read and agreed to the published version of the manuscript.

Funding: This research received no external funding.

Institutional Review Board Statement: The information provided in this report was part of a clinical diagnostic investigation and did not require ethical approval. All activities were conducted with direct supervision from the veterinary medical care teams at SeaWorld San Diego, SeaWorld San Antonio, and United Parks and Entertainment.

Informed Consent Statement: Not applicable.

Data Availability Statement: No new data were created for this report.

Acknowledgments: The authors thank Fran Rowley, Rebecca Rivera, Tara Klimek, Julie Sigman, Maria Love, Mary Nations, Amy Reese, Pete Peters, Jen Davenport, Amber Johnson-Reed, Caitlin Brush, Mike Felice, and the hospital, laboratory, and animal care staff at SeaWorld California and SeaWorld Texas. This is SeaWorld Parks & Entertainment Technical contribution 2024-6.

Conflicts of Interest: The authors declare no conflicts of interest.

References

1. Williams, R.; Lacy, R.C.; Ashe, E.; Barrett-Lennard, L.; Brown, T.M.; Gaydos, J.K.; Gulland, F.; MacDuffee, M.; Nelson, B.W.; Nielsen, K.A.; et al. Warning sign of an accelerating decline in critically endangered killer whales (*Orcinus orca*). *Commun. Earth Environ.* **2024**, *5*, 173. [CrossRef]
2. Andvik, C.; Jourdain, E.; Lyche, J.L.; Karoliussen, R.; Borgå, K. High Levels of Legacy and Emerging Contaminants in Killer Whales (*Orcinus orca*) from Norway, 2015 to 2017. *Environ. Toxicol. Chem.* **2021**, *40*, 1848–1858. [CrossRef] [PubMed]

3. Desforges, J.-P.; Hall, A.; McConnell, B.; Rosing-Asvid, A.; Barber, J.L.; Brownlow, A.; De Guise, S.; Eulaers, I.; Jepson, P.D.; Letcher, R.J.; et al. Predicting global killer whale population collapse from PCB pollution. *Science* **2018**, *361*, 1373–1376. [CrossRef] [PubMed]
4. Gaydos, J.K.; St. Leger, J.; Raverty, S.; Nollens, H.H.; Haulena, M.; Ward, E.J.; Emmons, C.K.; Hanson, M.B.; Balcomb, K.C.; Ellifrit, D.; et al. Epidemiology of skin changes in endangered Southern Resident killer whales (*Orcinus orca*). *PLoS ONE* **2023**, *18*, e0286551. [CrossRef] [PubMed]
5. Raverty, S.; St. Leger, J.; Noren, D.P.; Burek Huntington, K.; Rotstein, D.S.; Gulland, F.M.D.; Ford, J.K.B.; Hanson, M.B.; Lambourn, D.M.; Huggins, J.; et al. Pathology findings and correlation with body condition index in stranded killer whales (*Orcinus orca*) in the northeastern Pacific and Hawaii from 2004 to 2013. *PLoS ONE* **2020**, *15*, e0242505. [CrossRef]
6. Hunt, K.E.; Moore, M.J.; Rolland, R.M.; Kellar, N.M.; Hall, A.J.; Kershaw, J.; Raverty, S.A.; Davis, C.E.; Yeates, L.C.; Fauquier, D.A.; et al. Overcoming the challenges of studying conservation physiology in large whales: A review of available methods. *Conserv. Physiol.* **2013**, *1*, cot006. [CrossRef] [PubMed]
7. Rhodes, L.D.; Emmons, C.K.; Wisswaesser, G.S.; Wells, A.H.; Hanson, M.B. Bacterial microbiomes from mucus and breath of southern resident killer whales (*Orcinus orca*). *Conserv. Physiol.* **2022**, *10*, coac014. [CrossRef] [PubMed]
8. Atkinson, S.; Dierauf, L.A. Stress and Marine Mammals. In *CRC Handbook of Marine Mammal Medicine*, 3rd ed.; Gulland, F.M.D., Dierauf, L.A., Whitman, K.L., Eds.; CRC Press: Boca Raton, FL, USA, 2018; pp. 153–168.
9. Karaer, M.C.; Čebulj-Kadunc, N.; Snoj, T. Stress in wildlife: Comparison of the stress response among domestic, captive, and free-ranging animals. *Front. Vet. Sci.* **2023**, *10*, 1167016. [CrossRef] [PubMed]
10. Hao, Y.-J.; Zhao, Q.-Z.; Wu, H.-P.; Chen, D.-Q.; Gong, C.; Li, L.; Wang, D. Physiological responses to capture and handling of free-ranging male Yangtze finless porpoises (Neophocaena phocaenoides asiaeorientalis): Marine & Freshwater Behaviour & Physiology. *Mar. Freshw. Behav. Physiol.* **2009**, *42*, 315–327. [CrossRef]
11. Herráez, P.; Sierra, E.; Arbelo, M.; Jaber, J.R.; de Los Monteros, A.E.; Fernández, A. Rhabdomyolysis and myoglobinuric nephrosis (capture myopathy) in a striped dolphin. *J. Wildl. Dis.* **2007**, *43*, 770–774. [CrossRef]
12. Sierra, E.; Fernández, A.; Espinosa de Los Monteros, A.; Jaber, J.R.; Andrada, M.; Herráez, P. Complex polysaccharide inclusions in the skeletal muscle of stranded cetaceans. *Vet. J.* **2012**, *193*, 152–156. [CrossRef] [PubMed]
13. Roe, W.; Spraker, T.R. Capture-related myopathy in marine mammals and exertional rhabdomyolysis in horses: A possible link? *Vet. J.* **2012**, *193*, 10–11. [CrossRef] [PubMed]
14. Dold, C.; Ridgway, S. Cetaceans. In *Zoo Animal and Wildlife Immobilization and Anesthesia*; John Wiley & Sons, Ltd.: Oxford, UK, 2014; pp. 679–691. [CrossRef]
15. McBain, J.F. Cetacean Medicine. In *CRC Handbook of Marine Mammal Medicine*, 2nd ed.; CRC Press: Boca Raton, FL, USA, 2001; pp. 895–906.
16. Sweeney, J.C.; Ridgway, S.H. Procedures for the clinical management of small cetaceans. *J. Am. Vet. Med. Assoc.* **1975**, *167*, 540–545. [PubMed]
17. Russell, J.P.; St. Germain, M.; Osborn, S.D.; Schmitt, T.L.; Herrick, K.E.S.; Robeck, T. Comparison of infrared thermography of blowhole mucosa with rectal temperatures in killer whales (*Orcinus orca*). *Front. Mar. Sci.* **2024**, *11*, 1369287. [CrossRef]
18. McCafferty, D.J. The value of infrared thermography for research on mammals: Previous applications and future directions. *Mammal. Rev.* **2007**, *37*, 207–223. [CrossRef]
19. McCafferty, D.J.; Gallon, S.; Nord, A. Challenges of measuring body temperatures of free-ranging birds and mammals. *Anim. Biotelemetry* **2015**, *3*, 33. [CrossRef]
20. Andrade, D.V. Thermal windows and heat exchange. *Temperature* **2015**, *2*, 451. [CrossRef]
21. Tattersall, G.J. Infrared thermography: A non-invasive window into thermal physiology. *Comp. Biochem. Physiol. Part A Mol. Integr. Physiol.* **2016**, *202*, 78–98. [CrossRef]
22. Rekant, S.I.; Lyons, M.A.; Pacheco, J.M.; Arzt, J.; Rodriguez, L.L. Veterinary applications of infrared thermography. *Am. J. Vet. Res.* **2016**, *77*, 98–107. [CrossRef]
23. Mota-Rojas, D.; Pereira, A.M.F.; Martínez-Burnes, J.; Domínguez-Oliva, A.; Mora-Medina, P.; Casas-Alvarado, A.; Rios-Sandoval, J.; de Mira Geraldo, A.; Wang, D. Thermal Imaging to Assess the Health Status in Wildlife Animals under Human Care: Limitations and Perspectives. *Animals* **2022**, *12*, 3558. [CrossRef]
24. Cuyler, L.C.; Wiulsrød, R.; ØRitsland, N.A. Thermal Infrared Radiation from Free Living Whales. *Mar. Mammal. Sci.* **1992**, *8*, 120–134. [CrossRef]
25. Costidis, A.; Rommel, S.A. Vascularization of Air Sinuses and Fat Bodies in the Head of the Bottlenose Dolphin (*Tursiops truncatus*): Morphological Implications on Physiology. *Front. Physiol.* **2012**, *3*, 243. [CrossRef] [PubMed]
26. Melero, M.; Rodríguez-Prieto, V.; Rubio-García, A.; García-Párraga, D.; Sánchez-Vizcaíno, J.M. Thermal reference points as an index for monitoring body temperature in marine mammals. *BMC Res. Notes* **2015**, *8*, 411. [CrossRef] [PubMed]
27. Lonati, G.; Zitterbart, D.; Miller, C.; Corkeron, P.; Murphy, C.; Moore, M. Investigating the thermal physiology of Critically Endangered North Atlantic right whales Eubalaena glacialis via aerial infrared thermography. *Endanger. Species Res.* **2022**, *48*, 139–154. [CrossRef]
28. Bossart, G. Marine Mammals as Sentinel Species for Oceans and Human Health. *Vet. Pathol.* **2011**, *48*, 676–690. [CrossRef] [PubMed]

29. Barratclough, A.; Wells, R.S.; Schwacke, L.H.; Rowles, T.K.; Gomez, F.M.; Fauquier, D.A.; Sweeney, J.C.; Townsend, F.I.; Hansen, L.J.; Zolman, E.S.; et al. Health Assessments of Common Bottlenose Dolphins (*Tursiops truncatus*): Past, Present, and Potential Conservation Applications. *Front. Vet. Sci.* **2019**, *6*, 444. [CrossRef] [PubMed]
30. Nollens, H.H.; Robeck, T.R.; Schmitt, T.L.; Croft, L.L.; Osborn, S.; McBain, J.F. Effect of age, sex, and season on the variation in blood analytes of a clinically normal ex situ population of killer whales (*Orcinus orca*). *Vet. Clin. Pathol.* **2019**, *48*, 100–113. [CrossRef] [PubMed]
31. Gulland, F.M.D.; Dierauf, L.; Whitman, K.L. Appendix 1: Normal Hematology and Serum Chemistry Ranges. In *CRC Handbook of Marine Mammal Medicine*, 3rd ed.; CRC Press: Boca Raton, FL, USA, 2018; p. 1008.
32. Simeone, C.A.; Stoskopf, M.K. Pharmaceuticals and Formularies. In *CRC Handbook of Marine Mammal Medicine*, 3rd ed.; Gulland, F., Dierauf, L.A., Whitman, K.L., Eds.; CRC Press: Boca Raton, FL, USA, 2018; pp. 607–673.
33. Usamentiaga, R.; Venegas, P.; Guerediaga, J.; Vega, L.; Molleda, J.; Bulnes, F. Infrared Thermography for Temperature Measurement and Non-Destructive Testing. *Sensors* **2014**, *14*, 12305–12348. [CrossRef] [PubMed]
34. Black, V.L.; Whitworth, F.J.S.; Adamantos, S. Pyrexia in juvenile dogs: A review of 140 referred cases. *J. Small Anim. Pract.* **2019**, *60*, 116–120. [CrossRef]
35. Balli, S.; Shumway, K.R.; Sharan, S. Physiology, Fever. In *StatPearls*; StatPearls Publishing: Treasure Island, FL, USA, 2024.
36. Chow, A.W. Complications, Diagnosis, and Treatment of Odontogenic Infections. UpToDate 2024. Available online: https://www.uptodate.com/contents/complications-diagnosis-and-treatment-of-odontogenic-infections (accessed on 6 June 2024).
37. Flood, J. The Diagnostic Approach to Fever of Unknown Origin in Dogs. In *Compendium: Continuing Education for Veterinarians*; CompendiumVet.com: Irvine, CA, USA, 2009; pp. 14–21.
38. Burke, C.; Rashman, M.; Wich, S.; Symons, A.; Theron, C.; Longmore, S. Optimizing observing strategies for monitoring animals using drone-mounted thermal infrared cameras. *Int. J. Remote Sens.* **2019**, *40*, 439–467. [CrossRef]
39. Horton, T.W.; Hauser, N.; Cassel, S.; Klaus, K.F.; Fettermann, T.; Key, N. Doctor Drone: Non-invasive Measurement of Humpback Whale Vital Signs Using Unoccupied Aerial System Infrared Thermography. *Front. Mar. Sci.* **2019**, *6*, 466. [CrossRef]
40. Pirotta, V.; Smith, A.; Ostrowski, M.; Russell, D.; Jonsen, I.D.; Grech, A.; Harcourt, R. An Economical Custom-Built Drone for Assessing Whale Health. *Front. Mar. Sci.* **2017**, *4*, 425. [CrossRef]
41. Dixon, O.F.L.; Gallagher, A.J.; Towner, A.V. Novel aerial observations of a group of killer whales *Orcinus orca* in The Bahamas. *Front. Mar. Sci.* **2023**, *10*, 1265064. [CrossRef]
42. Cilulko, J.; Janiszewski, P.; Bogdaszewski, M.; Szczygielska, E. Infrared thermal imaging in studies of wild animals. *Eur. J. Wildl. Res.* **2013**, *59*, 17–23. [CrossRef]
43. Pabst, D.A.; Rommel, S.A.; McLellan, W.A.; Williams, T.M.; Rowles, T.K. Thermoregulation of the Intra-Abdominal Testes of the Bottlenose Dolphin (*Tursiops Truncatus*) During Exercise. *J. Exp. Biol.* **1995**, *198*, 221–226. [CrossRef] [PubMed]
44. Katsumata, E.; Jaroenporn, S.; Katsumata, H.; Konno, S.; Maeda, Y.; Watanabe, G.; Taya, K. Body Temperature and Circulating Progesterone Levels before and after Parturition in Killer Whales (*Orcinus orca*). *J. Reprod. Dev.* **2006**, *52*, 65–71. [CrossRef]
45. Tersawa, F.; Yokoyama, Y.; Kitamura, M. Rectal temperatures before and after parturition in bottlenose dolphins. *Zoo. Biol.* **1999**, *18*, 153–156. [CrossRef]
46. Shore, R. Ailing Orca Successfully Dosed with Antibiotics in Emergency Bid to Save Her Life. The Vancouver Sun. Available online: https://vancouversun.com/news/local-news/international-team-gives-medication-to-sick-killer-whale-at-sea (accessed on 10 August 2018).

animals

MDPI

Article

Characterization of Thermal Patterns Using Infrared Thermography and Thermolytic Responses of Cattle Reared in Three Different Systems during the Transition Period in the Eastern Amazon, Brazil

Welligton Conceição da Silva [1,*], Jamile Andréa Rodrigues da Silva [2], Éder Bruno Rebelo da Silva [1], Antônio Vinicius Correa Barbosa [3], Carlos Eduardo Lima Sousa [4], Katarina Cardoso de Carvalho [4], Maria Roseane Pereira dos Santos [5], Kedson Alexandri Lobo Neves [6], Lucieta Guerreiro Martorano [7], Raimundo Nonato Colares Camargo Júnior [1] and José de Brito Lourenço-Júnior [1]

1 Postgraduate Program in Animal Science (PPGCAN), Institute of Veterinary Medicine, Federal University of Para (UFPA), Federal Rural University of the Amazônia (UFRA), Brazilian Agricultural Research Corporation (EMBRAPA), Castanhal 68746-360, Brazil; eder.b.rebelo@gmail.com (É.B.R.d.S.); camargojunior@gmail.com (R.N.C.C.J.); joselourencojr@yahoo.com.br (J.d.B.L.-J.)
2 Institute of Animal Health and Production, Federal Rural University of the Amazônia (UFRA), Belem 66077-830, Brazil; jamileandrea@yahoo.com.br
3 Cyberspace Institute, Federal Rural University of the Amazon, Belem 66077-830, Brazil; profvinibarbo@gmail.com
4 Department of Veterinary Medicine, University Center of the Amazon (UNAMA), Santarem 68010-200, Brazil; cadu34.medvet@gmail.com (C.E.L.S.); katarinacc4@gmail.com (K.C.d.C.)
5 Institute of Engineering and Geosciences, Federal University of Western Pará (UFOPA), Santarem 68040-255, Brazil; roseanemaria022@gmail.com
6 Institute of Animal Science, Federal University of Western Pará (UFOPA), Santarem 68040-255, Brazil; kedson_neves@hotmail.com
7 Embrapa Eastern Amazon, Santarem 68010-180, Brazil; lucieta.martorano@embrapa.br
* Correspondence: welligton.medvet@gmail.com; Tel.: +55-(93)-988070692

Citation: Silva, W.C.d.; Silva, J.A.R.d.; Silva, É.B.R.d.; Barbosa, A.V.C.; Sousa, C.E.L.; Carvalho, K.C.d.; Santos, M.R.P.d.; Neves, K.A.L.; Martorano, L.G.; Camargo Júnior, R.N.C.; et al. Characterization of Thermal Patterns Using Infrared Thermography and Thermolytic Responses of Cattle Reared in Three Different Systems during the Transition Period in the Eastern Amazon, Brazil. *Animals* **2023**, *13*, 2735. https://doi.org/10.3390/ani13172735

Academic Editors: Daniel Mota-Rojas and Peter Henzi

Received: 5 August 2023
Revised: 25 August 2023
Accepted: 26 August 2023
Published: 28 August 2023

Simple Summary: This study aimed to characterize the thermal patterns of cattle raised in three different production systems during the transition period in the eastern Amazon, Brazil, using infrared thermography and by evaluating thermolytic responses. The evaluated systems were silvopastoral (with shade and no access to the bathing area), traditional (without shade and with no access to the bathing area), and integrated (with access to the shade and the bathing area). The thermographic images showed that the silvopastoral system, with shading, provided more favorable thermal conditions compared to the other systems. Physiological analyses revealed that the cattle maintained a rectal temperature within normal ranges in all systems. However, the respiratory rate was higher in the traditional system, indicating possible thermal stress in this environment. The thermal comfort indices indicated moderate stress conditions at times of greater solar intensity in all systems. In general, the results suggest that the silvopastoral system offers better thermal conditions for cattle, providing shade and reducing thermal stress. Thus, management practices aimed at animal welfare and productivity in production systems in the eastern Amazon region can be adopted by ranchers.

Abstract: In the Lower Amazon mesoregion, there are basically three types of production systems: the traditional (without shade and no bathing area), the silvopastoral (with shade and no bathing area), and the integrated (with shade and bathing area). It is considered that the type of production system influences the thermal comfort and productivity of cattle, so this research aims to evaluate the influence of these three types of production systems on the thermoregulation of Nellore cattle. The experiment was carried out on a rural property for raising cattle, located in Mojuí dos Campos, Pará, Brazil, during the transition period (June/July). Thirty bovine males (not castrated, aged between 18 and 20 months, average weight of 250 ± 36 kg, body condition score of 3.5, clinically healthy) were randomly divided into three groups: Silvopastoral System—SS (n = 10), Traditional System—TS (n = 10), and Integrated System—IS (n = 10). Climate variables were collected (air temperature (AT °C), relative humidity (RH %), wind speed (WS, m/s), solar radiation (SR), black globe temperature (BGT °C), and physiological parameters, such as respiratory rate (RR) and rectal temperature (RT)) at 6 a.m., 12 p.m., 6 p.m., and 12 a.m. to determine the thermal comfort situation of the animals. Thermographic images of the environment and animals were captured in order to obtain the body surface temperature (BST) through infrared thermography. The Benezra Thermal Comfort Index (BTCI), Environmental Stress Index (ESI), Equivalent Temperature Index (ETI), and Iberian Heat Tolerance Index (Iberian HTI) were used. The results showed that the silvopastoral system, with shading by chestnut trees and an ample vegetative area, presented better thermal conditions, with an average of 28.98 °C, in comparison with the traditional system (35.93 °C) and the integrated one (34.11 °C). It was observed that the body surface temperature of cattle did not differ significantly between the anatomical regions of the body and the studied systems ($p > 0.05$). As for the respiratory rate, the traditional system registered higher values, with an average of 41 movements per minute, indicating possible thermal stress ($p < 0.05$). The thermal comfort indices revealed that all systems presented moderate stress conditions during times of higher solar intensity. It is concluded that the silvopastoral system proved to be more favorable for cattle, providing shade and reducing thermal stress, which may have a positive impact on animal welfare and productivity in this region.

Keywords: heat; ambient climate; thermal stress; thermoregulation

1. Introduction

Cattle production in the eastern Amazon is an activity of great economic and social importance; with vast areas of natural pastures and rich biodiversity, the Amazon has significant potential for the development of cattle raising [1,2]. Thus, in the Amazon region, three models of ruminant rearing systems are mainly used: the traditional system, in which the animals do not have access to shade or water; the silvopastoral system, which provides shade for the animals, but without access to water for bathing; and the integrated system, which allows the animals to enjoy both shade and water for bathing.

The effects of these different production systems can lead to significant changes in the physiology and performance of these animals. Thus, it is possible to deduce that several weaknesses persist in the management of cattle. Furthermore, there is little information about the influences that these different breeding environments can have on beef cattle produced in the region. Thus, it can be inferred that there are no studies about the occasional influences on beef cattle resulting from these breeding environments widespread in the region, which justifies this study.

In this context, some indices help in determining the resulting influences of these production systems, such as the Temperature and Humidity Index (THI), the Black Globe Temperature and Humidity Index (BGTI), the Benezra Thermal Comfort Index (BTCI), the Environmental Stress Index (ESI), the Temperature Equivalent Index (ETI), and the Iberian Heat Tolerance Index (Iberian HTI), calculated according to meteorological variables and physiological parameters [3–7]. Thus, in livestock, infrared thermography has proven to be a valuable tool for monitoring animal health and well-being, allowing for the detection of thermal anomalies, the identification of injuries, the evaluation of the thermal performance of installations, and support for the early diagnosis of health, thus providing a more efficient and precise approach to the management and care of herds [8–11].

Thus, infrared thermography (ITR) helps in the determination of thermal patterns, being a passive, remote, and non-invasive thermal visualization technology that helps in the determination of high temperature zones through images obtained, which directly demonstrate the distribution temperature on the surface of the animals and in the studied production systems [9,12,13].

Cattle are animals sensitive to temperature variations, and when exposed to conditions of excessive heat, they can face serious health and well-being problems. During periods of high temperatures, cattle have difficulties in dissipating body heat, which leads to an increase in respiratory rate and sweating as thermoregulation mechanisms; this suffering is directly influenced by the rearing system [14].

Furthermore, the increase in body temperature is directly related to the increase in physiological parameters, such as respiratory rate, heart rate, and panting score; all this can be used as a sign of thermal stress [15,16]. According to McDowell et al. [17], an increase in rectal temperature of 1 °C is enough to reduce the productivity and performance of different animal species, which results in difficulties in weight gain and adaptive processes.

In this way, the present study presented relevant data about the raising of cattle in systems that include shaded areas and access to water for bathing. This information offers valuable insights for the adoption of appropriate management practices in the production of beef cattle in the eastern Amazon region. Therefore, the objective of this study was to characterize the thermal patterns using infrared thermography and thermolytic responses of cattle reared in three different systems during the transition period in the eastern Amazon, Brazil.

2. Materials and Methods

2.1. Ethical Aspects

This study was submitted to the Committee for Ethics in the Use of Animals (CEUA) and obtained Approved status, under protocol CEUA-UNAMA 0001-87/2023, in May 2023.

2.2. Location

This study was carried out on a rural property for raising cattle, located in Mojuí dos Campos, Pará, Brazil (Figure 1), in the transition period of the year (rainier to less rainy—June/July).

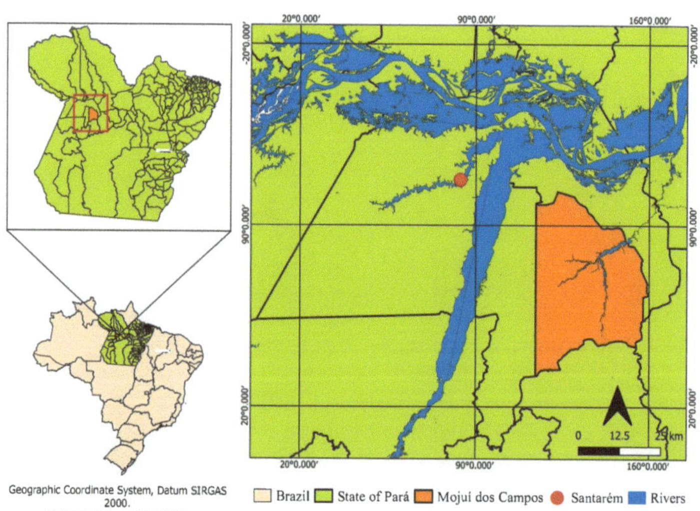

Figure 1. Location map of the study area.

The climate of the mesoregion is hot and humid (Am4); it is characterized by total rainfall lower than 60 mm in the least rainy month and by annual rainfall between 1900 and 2100 mm. It is also characterized by an annual average air temperature of 25.6 °C and relative humidity ranging from 84 to 86% [18]. The rainiest quarter occurs between the months of February and April, and the least rainy quarter occurs between the months of August and October [19]. The forage plant *Brachiaria humidicola* was implanted in the systems.

2.3. Experimental Animals, Management, and Characterization of the Production System

Thirty Nellore bovines (male, not castrated, with similar coloring, aged between 18 and 20 months, average weight of 250 ± 36 kg, body condition score of 3.5 (scale from 1 to 5), clinically healthy) were randomly divided into three groups: Silvopastoral System—SS (n = 10), Traditional System—TS (n = 10), and Integrated System—IS (n = 10) (Figure 2). The SS group remained in a paddock of the same size, with around 20% shade from chestnut trees (*Bertholletia excelsa*), with access to drinking water and mineral salt *ad libitum*. The TS group was led to a paddock without shade from trees, with a *Brachiaria humidicola* pasture, and access to drinking water and mineral salt ad libitum. The IS group remained in a paddock of the same size and pasture, but with access to 20% shade from Brazil nut trees (*Bertholletia excelsa*) and access to water for drinking and bathing, as well as mineral salt, ad libitum. The total experimental area was 10.2 ha of *Brachiaria humidicola*, divided into six paddocks of 1.7 ha, with two per treatment.

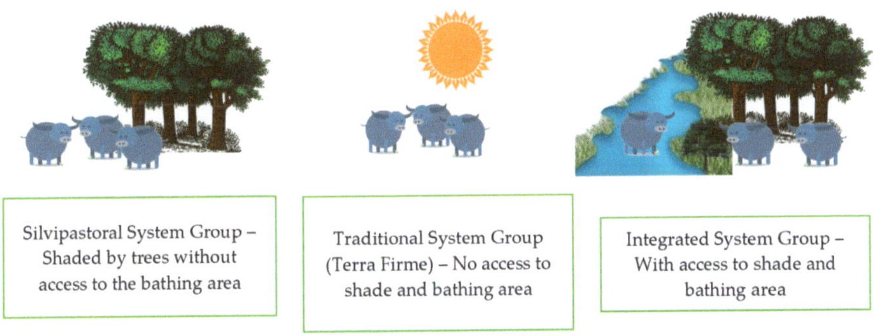

Figure 2. Infographic of the experimental design by evaluated group.

The period of adaptation of the animals to the handling was seven consecutive days, where the animals were taken to the chute so that data of the physiological variables could be collected during the experiment. The production systems used are characterized as follows:

I. Traditional System—no shade and no access to the bathing area. In this system, the animals were subjected to pasture without the presence of trees or other elements that could provide shade; they also did not have access to the bathing area.
II. Silvopastoral System—with shade and no access to the bathing area. In this system, the animals were subjected to pasture with the presence of trees and other elements that could provide shade; they also did not have access to the bathing area.
III. Integrated System—with access to shade and bathing area. In this system, the animals were subjected to pasture with trees and other elements that can provide shade, as well as access to the bathing area.

2.4. Meteorological Variables

The climatic variables evaluated were air temperature (AT °C), relative air humidity (RH %), wind speed (WS, m s^{-1}), dew point temperature (DPT °C), wet bulb temperature (WBT °C), and black globe temperature (BGT °C) obtained using a conventional station. The microclimate conditions, specifically temperature and humidity during this study, are shown in Figure 3.

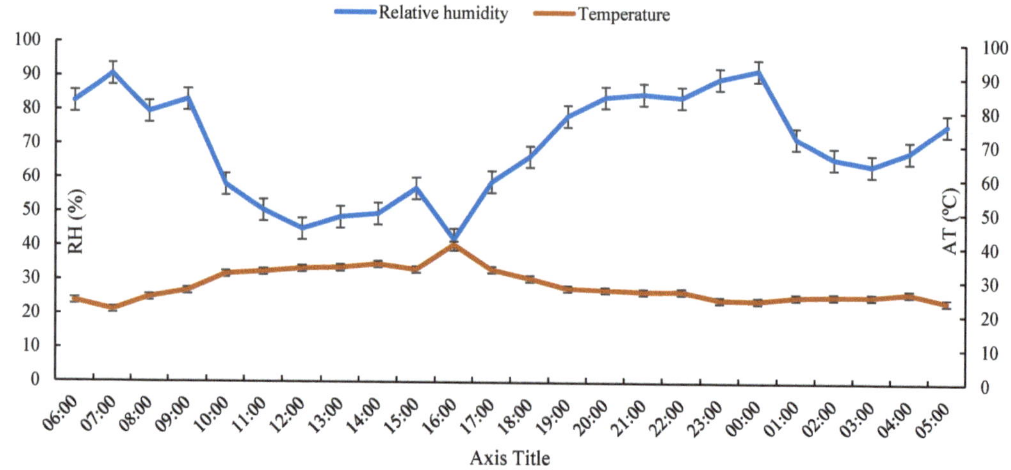

Figure 3. Average agrometeorological data during the period of the experiment.

2.5. *Physiological Variables*

All physiological variables were collected at 6:00 a.m., 12:00 p.m., 6:00 p.m., and 12:00 a.m. during the transition period (June and July) during two days of performance in the months of June and July. On the days of collection, the animals were taken to the management corral, remaining on hold for 30 min before the start of activities to avoid the influence of management on the physiological variables. For this, the animals were managed to the corral on foot and contained in batch type trunks in environments protected from direct sunlight and rain. The animals were handled in batches of ten animals to the corral, avoiding a fixed order of animal entry, and preventing animal data from being influenced by advancing time, within the stipulated time window.

2.6. *Respiratory Rate (RR)*

The RR was obtained by inspecting and counting the thoracic-abdominal movements for one minute [20] with the help of a digital stopwatch. Data were collected in the transition period (June/July) during two days of the experiment at 6:00 a.m., 12:00 p.m., 6:00 p.m., and 12:00 a.m. This assessment was carried out by a single, previously trained, observer.

2.7. *Rectal Temperature (RT)*

The RT was obtained using a veterinary clinical thermometer (Model-5198.10, Incoterm®, São Paulo, Brazil), with a maximum scale of up to 44 °C inserted transrectally into the animals, with the results expressed in degrees Celsius, as per described by Dirksen et al. [21]. This evaluation was carried out by a single observer, during the transition period (June/July) during two days of the experimental period, at 6:00 a.m., 12:00 p.m., 6:00 p.m., and 12:00 a.m.

2.8. *Infrared Thermography*

The diagnosis of the thermal patterns of the targets in the environment of the three production systems was carried out using infrared thermography, and this collection was carried out on 1 July 2023 in an area that totals 1.7 ha per production system. Thus, thermographic image captures were carried out between 12:00 and 15:00, a period with an intense effect of solar radiation on the targets diagnosed in the field research. Thermographic images were acquired on the right side of the animals to reflect the real fluctuation of the IRT, thus preventing interference from ruminal movements.

We monitored the targets by taking images with a near-infrared thermograph (FLIR T650sc), considering an emissivity of 0.95 because it is an open system composed of different materials. Thermograms were analyzed using the Flir Tools program, 6.3 v [22], with the Rainbow HC palette chosen.

The images were acquired from four areas: the head region, the armpit, the flank, and the rump, according to the description of thermal windows described in Mota-Rojas et al. [10].

The camera had high precision with a fixed 25 mm lens, a temperature range from −40 to 150 °C, and thermal sensitivity of 50 mK (>0.05 °C at an ambient temperature of 30 °C). The spectral range ranged from 0.7 to 100 μm, but the imaged targets presented a response between 0.7 and 3.0 μm and an optical resolution of 640 × 480 pixels.

2.9. Body Surface Temperature (BST)

The BST was obtained with the aid of an infrared thermometer using the scientific thermographic camera in the transition period (June/July); it was specifically carried out on 1 July 2023, from 12:00 to 15:00, in the anatomical regions of the head, the armpit, the flank, and the rump, with ten points being imaged (Figure 4). The thermograms of the animals were acquired in the field, while in the pastures, and with the animal in a standing position. In the systems, thermograms were acquired at an approximate orthogonal distance of 5 m, outside the flying distance of the bull [23], so that the animals can remain calm and in their natural behavior.

Figure 4. Points imaged using infrared thermography. Anatomical region—SP1 and SP2—head; SP3 and SP4—armpit; SP5 and SP6—flank; and SP7 and SP8—rump.

2.10. Heat Storage Calculation

The heat storage calculation, in kg of kelvin^{-1} h^{-1}, was obtained using the formula [24]:

$$\Delta RT = (3600 \times Harm \times A)/(Mv \times Hb)$$

where ΔRT—difference in RT; Harm—stored heat (W·m^{-2}); A—animal's surface (m^2) calculated by the equation: $A = 0.13 \cdot M^{0.556v}$; Mv–body mass (kg); and Hb—animal's specific heat (3400 kJ kg^{-1} K^{-1}).

2.11. Index to Assess Thermal Comfort

2.11.1. Temperature and Humidity Index (THI)

During the experiment, readings of environmental variables were collected simultaneously with measurements of physiological variables throughout the day. From the values recorded for the environmental conditions, the Temperature and Humidity Index (THI), proposed by Thom [25], was calculated using the formula:

$$THI = DBT + 0.36 \times DPT + 41.5$$

where DBT = Dry bulb temperature (°C) and DPT = Dew point temperature (°C).

THI = temperature and humidity index, indicating <72 (green color); >72 to 78 (yellow color); and 79 to 88 (orange color) [25].

2.11.2. Black Globe Temperature and Humidity Index (BGHI)

From obtaining these variables, the index of the temperature of the black globe and humidity (BGHI) will be calculated, according to what was proposed by Buffington et al. [26], using the equation:

$$BGHI = BGT + 0.36\ (DPT) + 41.5$$

where BGHI = Black Globe Temperature and Humidity Index, BGT = black globe temperature (°C), and DPT = dew point temperature (°C).

BGHI = black globe temperature and humidity index, indicating >72 comfort situation (green color) and 74 to 78 warning (yellow color) [26].

2.11.3. Benezra Comfort Index (BTCI)

The Benezra Thermal Comfort Index (BTCI) [27] was calculated by the formula:

$$BCI = RT/38.33 + RR/23$$

where RT = rectal temperature (°C) and RR is the respiratory rate.

A BTCI with values greater than 2.0 indicates difficulty in the degree of adaptation of the animals to the environment. Values below 2.0 indicate greater ease of adaptation [27].

2.11.4. Environmental Stress Index (ESI)

The Environmental Stress Index (ESI) was calculated using the formula [28]:

$$ESI = 0.63DBT - 0.03RH + 0.002SR + 0.0054(DBT \times RH) - 0.073(0.1 \times SR)^{-1}$$

where DBT = dry bulb temperature, in °C; RH = relative humidity, in %; and SR = solar radiation, in $W \cdot m^{-2}$.

ESI = environmental stress index, indicating <25 low risk (green color) and ≤ 25 to ≤ 33 moderate to high risk (yellow color) [28].

The Equivalent Temperature Index (ETI) is a metric developed by Baeta et al. [29], and it takes into account temperature, humidity, and air velocity, being especially applicable for semi-housed animals. This index was calculated using the following formula:

$$ETI = 29.83628 - 0.11519 \times AT + 0.00059 \times RH - 0.30525 \times WS$$

where ETI = Equivalent Temperature Index; AT = Air temperature in °C (dry bulb); RH = relative air humidity, in %; and WS = wind speed, in m/s.

ETI = equivalent temperature index, indicating no problems from 18 to <27 (green color) and caution ≥ 27 to ≥ 32 (yellow color) [29].

2.11.5. Iberian Heat Tolerance Index (Iberian HTI)

The Iberian Heat Tolerance Index (Iberian HTI) [30] was calculated using the formula:

$$HTI = 100 - 18(RT - 38.33)$$

where HTI = Heat Tolerance Index and RT = rectal temperature.

An Iberian HTI index close to 100 indicates greater animal comfort [30].

2.12. Data Analysis

The relationships between the thermal comfort indices and the physiological stress indices were investigated. The climatological data were subjected to descriptive statistics analysis. Statistical analyses were performed in Software R version 4.3.1 (RStudio 2023.03.0 Build 386) using parametric ANOVA with Tukey's post hoc test and non-parametric with chi-square test for equal expected proportions when comparing the times between themselves. All analyses considered a significance level of 5%. Statistical analyses for the TR and FR variables were performed considering repeated measurements over time and with interaction between the rearing systems and schedules, with a parametric analysis being performed after meeting the assumptions of normality of the residues by the Shapiro–Wilk test (p-value > 0.05) and the assumption of equality of variances by Mauchly's sphericity test at 5% significance. After ANOVA, the student's t-test was performed.

3. Results

In Figure 5, thermographic images can be seen in the three different production systems, in which the silvopastoral system (Figure 5A) expresses the thermal conditions in the trees in the URT containing an area shaded by chestnut trees (*Bertholletia excelsa*), with a large vegetative area, which favors the provision of shade and reduces the temperature when compared with the thermal images of the traditional system (Figure 5B) and the integrated system (Figure 5C). In the traditional system, it is noted that the temperatures were higher than the other systems because there is no presence of trees. The integrated system with the supply of water for bathing combined with shading with chestnut trees (*Bertholletia excelsa*) presented values higher than the silvopastoral system and lower than the traditional one.

In Table 1, the values of the thermal conditions in each animal production system are noted, and a statistical difference between the silvopastoral system for the traditional and integrated ($p < 0.05$) systems is observed, with no difference between the traditional and integrated ($p > 0.05$).

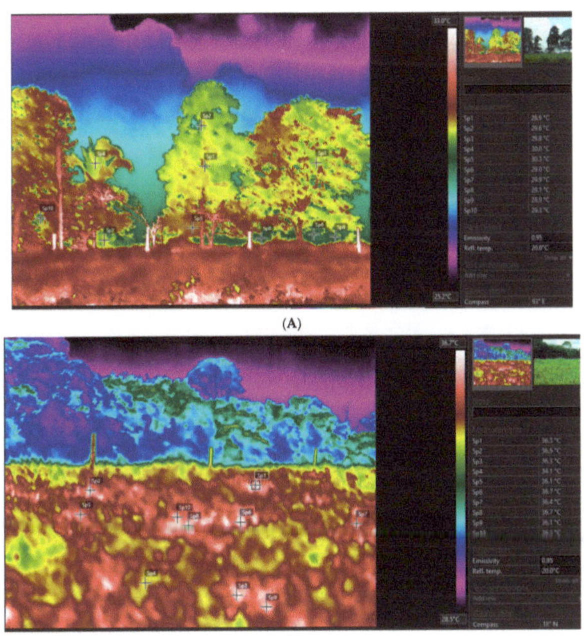

(A)

(B)

Figure 5. *Cont.*

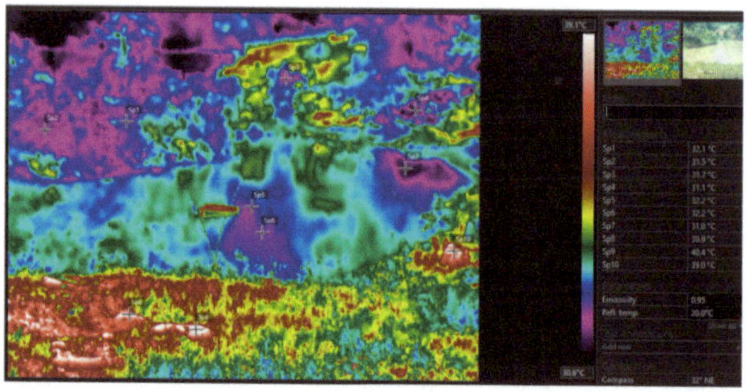

(C)

Figure 5. Thermographic images of the silvopastoral (**A**), traditional (**B**), and integrated (**C**) systems of cattle raising observed between 13:00 and 15:00 during the experimental period in the eastern Amazon. Description of the systems: Silvopastoral—with shade and no access to the bathing area. In this system, the animals were submitted to pasture with the presence of trees and other elements that could provide shade; they did not have access to the bathing area. Traditional—no shade and no access to the bathing area. In this system, the animals were subjected to grazing without the presence of trees or other elements that can provide shade; they did not have access to the bathing area. Integrated—with access to shade and bathing area. In this system, the animals were subjected to pasture with trees and other elements that can provide shade, as well as access to the bathing area.

Table 1. Descriptive analysis of the results of the analysis of thermal data of the targets analyzed in the technological reference unit (TRU) in the different production systems in the eastern Amazon.

Variables	Systems		
	Silvopastoral	**Traditional**	**Integrated**
Sample size	10	10	10
Minimum	23.90	34.10	31.00
Maximum	30.30	36.70	40.40
Total Amplitude	6.40	2.60	9.40
Median	29.45	36.30	32.15
First Quartile (25%)	29.0250	36.1500	31.5500
Third Quartile (75%)	29.8750	36.3750	37.3000
Interquartile deviation	0.8500	0.2250	5.7500
Arithmetic average	28.98 a	35.93 b	34.11 b
Variance	3.40	0.96	15.52
Standard deviation	1.84	0.98	3.94
Standard Error	0.58	0.31	1.25
Coefficient of variation	6.36%	2.73%	11.55%
Asymmetry	−2.79	−1.62	1.02
Kurtosis	8.32	1.12	−1.11

Note: Arithmetic average: a,b different letters in the line indicate statistical differences ($p < 0.05$). Description of the systems: Silvopastoral—with shade and no access to the bathing area. In this system, the animals were submitted to pasture with the presence of trees and other elements that could provide shade; they did not have access to the bathing area. Traditional—no shade and no access to the bathing area. In this system, the animals were subjected to pasture without the presence of trees or other elements that could provide shade; they did not have access to the bathing area. Integrated—with access to shade and bathing area. In this system, the animals were subjected to pasture with trees and other elements that can provide shade, as well as access to the bathing area.

Regarding the body surface temperature (BST), it was noted that the cattle did not show any difference between the anatomical regions of the body and the systems studied ($p > 0.05$) (Table 2).

In the regression analysis of rectal temperature as a function of collection times (06:00, 12:00, 18:00, and 00:00), it was possible to observe that in the silvopastoral, traditional, and integrated systems, there was a higher index of this parameter at 18:00, reaching values of 39.45 °C, 39.5 °C, and 39.5 °C, respectively (Figure 6A–C). Note that there was a gradual increase in the mean TR at 6 a.m., 12 a.m., and 6 p.m., with an exponential reduction at 12 a.m. in all evaluated systems. There was no statistical difference between times per treatment ($p > 0.05$). However, there was a difference between the hours ($p < 0.05$). There was no difference in the treatment and time interaction ($p > 0.05$). There was a difference between 00:00 and 18:00 within the silvopastoral treatment and between 6:00 and 18:00 in the integrated system. In the grouping of times, a difference was observed between all times, except between 00:00 and 12:00. The TR is considered normal for cattle when it is in the range of 38 to 39.5 °C [20]. In this study, cattle reared in silvopastoral, traditional, and integrated systems were within normal limits, with no thermal stress.

Table 2. Surface temperature indices according to the anatomical region of the body and the systems studied in the eastern Amazon.

Variables	Number of Data Points	Silvopastoral	Traditional	Integrated	Average	Standard Deviation
Head	2	35.05	36.7	34.9	35.05	0.815475322
Armpit	2	35.6	36.25	36.1	36.1	0.277888867
Flank	2	38.15	35.95	36.35	36.35	0.956846673
Rump	2	36.55	36.4	34.45	36.4	0.956556323
Average	2	35.6	36.25	36.1	36.1	0.277888867
Standard deviation	-	1.175996918	0.270416346	0.796084166	0.546008242	0.279560169

Note: Description of systems: Silvopastoral—with shade and no access to the bathing area. In this system, the animals were submitted to pasture with the presence of trees and other elements that could provide shade; they did not have access to the bathing area. Traditional—no shade and no access to the bathing area. In this system, the animals were subjected to pasture without the presence of trees or other elements that can provide shade; they did not have access to the bathing area. Integrated—with access to shade and bathing area. In this system, the animals were subjected to pasture with trees and other elements that can provide shade, as well as access to the bathing area.

The regression analysis in relation to the respiratory rate (RR) in the three rearing systems revealed a quadratic effect in relation to the studied time, as observed in Figure 7A–C. When deriving the equation, it was identified that the maximum RR reached 32, 41, and 31 mov min^{-1} in the silvopastoral, traditional, and integrated systems at 6:00 p.m., 12:00 a.m., and 6:00 p.m., respectively ($p < 0.05$) (Figure 7A–C). Added to this, throughout the day, there was a gradual increase in RR in the three rearing systems, with the highest increase observed in animals raised in the traditional system; that is, without the presence of shade, especially at 12:00 a.m., with an average RR of 41 mov min^{-1}. This upward trend of RR throughout the morning and early afternoon indicates a possible physiological response of the study subjects to environmental changes or daily activities. The RR was within the reference values described by Feitosa [20] (within the range of 10 to 30 movements per minute for the silvopastoral system at 6:00 a.m., 12:00 a.m., and 00:00 p.m., with thermal stress at 6:00 p.m. (32 mov min^{-1})), with the same dynamics observed for the animals in the integrated system. However, for the traditional system, stress was noted at all times (6:00 a.m., 12:00 p.m., 6:00 p.m., and 12:00 a.m.), with RR greater than 30 mov min^{-1} [20]. There was a difference for treatment, time, and treatment and time interaction ($p < 0.05$). There was a difference between the silvopastoral and traditional systems and between the traditional and integrated systems ($p < 0.05$); however, there was no difference between the traditional and integrated systems ($p > 0.05$).

There was a greater accumulation of heat during the period of the experiment at 12:00 and 18:00 in the traditional and integrated system and at 18:00 and 00:00 in the silvopastoral system, all associated with the highest rectal temperature index (Figure 8).

In the THI assessment, it was possible to observe the absence of thermal stress at 7:00 a.m. (<72). On the other hand, lower rates were identified at 6:00 a.m., 8:00 a.m., and 9:00 a.m. and after 7:00 a.m., characterizing mild stress (>72 to 78). Moderate stress was evidenced at times of greater solar intensity (from 10:00 a.m. to 6:00 p.m.). The BGHI demonstrated comfort indices in the early hours (6:00 a.m., 7:00 a.m., and 8:00 a.m.) of the day and from 11:00 p.m. to 6:00 a.m.. However, moderate stress was presented between

9:00 a.m. and 10:00 a.m. and from 1:00 p.m. to 10:00 p.m. (Table 3). The environmental stress index (ESI) pointed to a moderate risk at 10:00 p.m., corroborating data from the BGHI, while the other times showed a low risk of stress. The equivalent temperature index (ETI) signaled a state of caution (≥27 to ≥32) at 7 a.m., with the other times showing no stress (18 to <27).

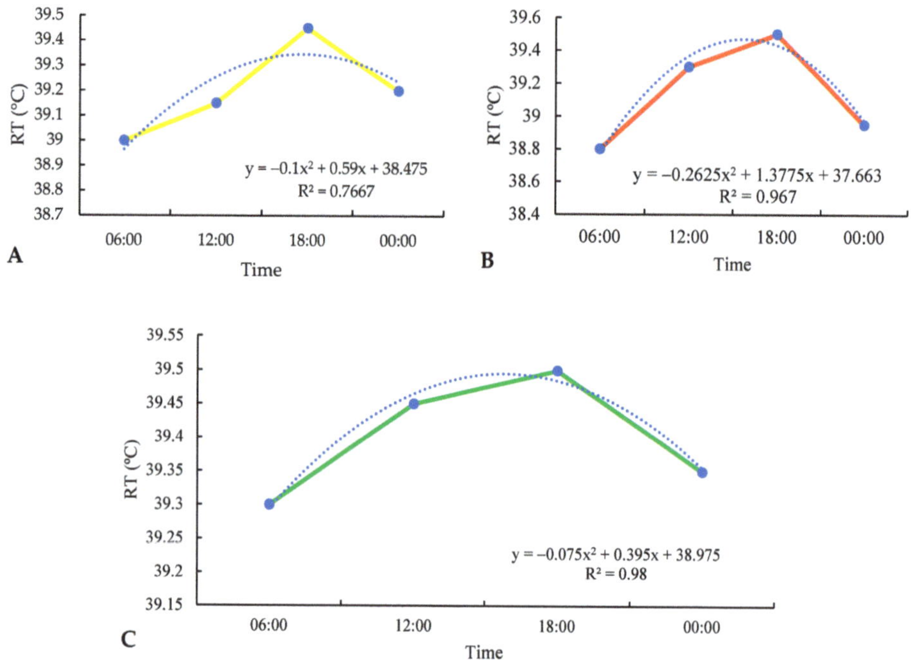

Figure 6. Rectal temperature (RT) in the silvopastoral (**A**), traditional (**B**), and integrated (**C**) systems of cattle observed at 6:00 a.m., 12:00 p.m., 6:00 p.m., and 12:00 a.m. during the experimental period in the eastern Amazon. Description of the systems: Silvopastoral—with shade and no access to the bathing area. In this system, the animals were submitted to pasture with the presence of trees and other elements that could provide shade; they did not have access to the bathing area. Traditional—no shade and no access to the bathing area. In this system, the animals were subjected to pasture without the presence of trees or other elements that could provide shade; they did not have access to the bathing area. Integrated—with access to shade and bathing area. In this system, the animals were subjected to pasture with trees and other elements that can provide shade, as well as access to the bathing area. Dotted line indicates trend line.

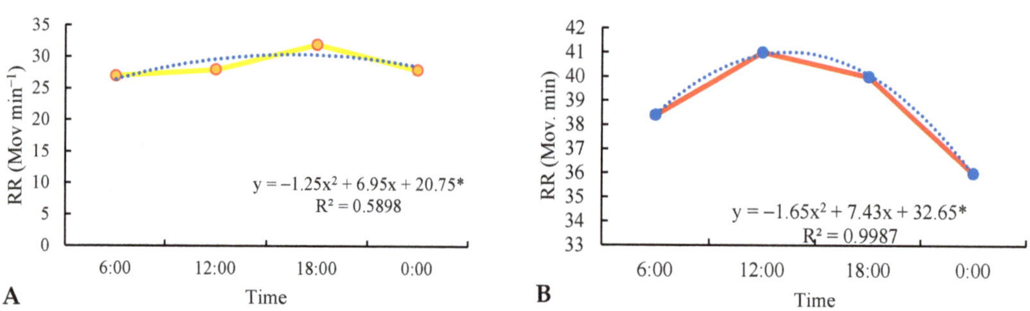

Figure 7. *Cont.*

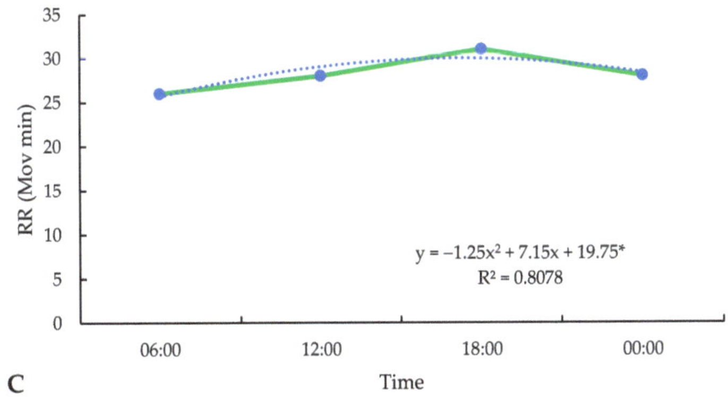

Figure 7. Respiratory rate (RR) in the silvopastoral (**A**), traditional (**B**), and integrated (**C**) systems of cattle observed at 6:00 a.m., 12:00 p.m., 6:00 p.m., and 12:00 a.m. during the experimental period in the eastern Amazon. * $p < 0.05$. Description of systems: Silvopastoral—with shade and no access to the bathing area. In this system, the animals were submitted to pasture with the presence of trees and other elements that could provide shade; they did not have access to the bathing area. Traditional—no shade and no access to the bathing area. In this system, the animals were subjected to pasture without the presence of trees or other elements that can provide shade; they did not have access to the bathing area. Integrated—with access to shade and bathing area. In this system, the animals were subjected to pasture with trees and other elements that can provide shade, as well as access to the bathing area. Dotted line indicates trend line.

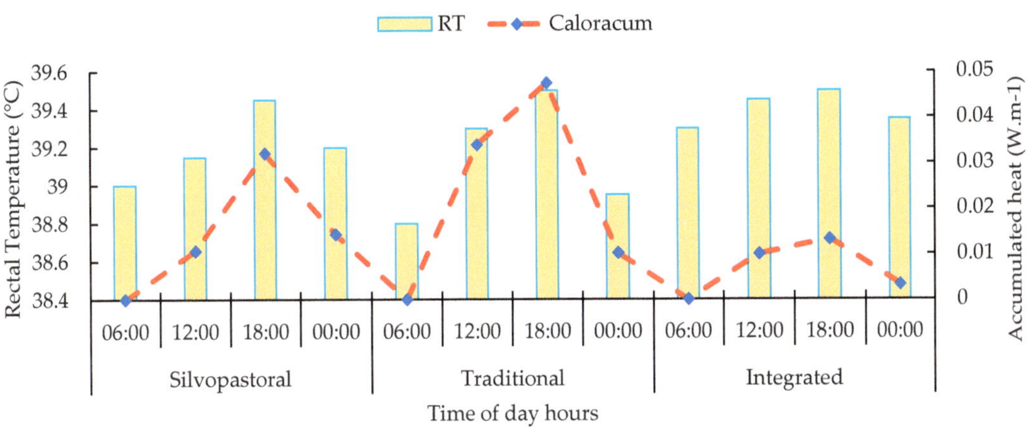

Figure 8. Thermal balance of the interaction between RT and heat accumulation in silvopastoral, traditional, and integrated systems at 06:00, 12:00, 18:00, and 00:00 during the experiment in the eastern Amazon. Description of the systems: Silvopastoral—with shade and no access to the bathing area. In this system, the animals were submitted to pasture with the presence of trees and other elements that could provide shade; they did not have access to the bathing area. Traditional—no shade and no access to the bathing area. In this system, the animals were subjected to pasture without the presence of trees or other elements that can provide shade; they did not have access to the bathing area. Integrated—with access to shade and bathing area. In this system, the animals were subjected to pasture with trees and other elements that can provide shade, as well as access to the bathing area.

The Iberian HTI showed no difference between the studied production systems ($p > 0.05$). The BTCI had a higher index in the traditional system ($p < 0.05$) when compared with the silvopastoral and integrated systems. For the times of day, no difference was observed for either the Iberian HTI or the BTCI. However, the morning shift (87.34) was the most comfortable for the welfare of the cattle, as it had an Iberian HTI index close to 100. In the case of the BTCI, all values were greater than 2.0, signaling difficulty in the degree of adaptability of the animals to the environment (Table 4).

Table 3. Indices used to evaluate the heat stress of cattle raised in silvopastoral (A), traditional (B), and integrated (C) systems from 6 a.m. to 6 a.m. during the experimental period in the eastern Amazon.

Time	THI	BGHI	ESI	ETI
06:00	72.716	70.426	23.08268	26.82964
07:00	69.62	68.43	20.85035	27.14537
08:00	73.996	71.256	23.95235	26.70116
09:00	76.832	74.512	26.36654	26.48436
10:00	81.35	74.82	28.16738	25.89942
11:00	81.438	73.478	27.7384	25.81436
12:00	81.978	72.828	27.72533	25.70745
13:00	82.818	74.308	28.51822	25.67495
14:00	84.336	75.876	29.61209	25.5546
15:00	83.11	76.25	29.33885	25.73756
16:00	83.686	76.826	29.33885	25.73756
17:00	83.29	76.43	29.63687	25.73874
18:00	80.546	75.546	28.18279	26.04266
19:00	77.646	74.646	26.74612	26.38379
20:00	77.498	75.308	26.96888	26.43312
21:00	76.84	74.81	26.48708	26.49706
22:00	76.976	74.796	26.48126	26.48495
23:00	74.036	72.666	24.4403	26.76465
00:00	73.908	72.878	24.46934	26.78917
01:00	73.9604	70.1604	23.57642	26.65066
02:00	73.7248	69.0548	23.13398	26.62408
03:00	73.804	68.844	22.91858	26.6229
04:00	75.2396	70.7796	24.34658	26.51007
05:00	72.5236	69.3436	22.65038	26.80276

Note: THI = temperature and humidity index, indicating <72 (green color); >72 to 78 (yellow color); and 79 to 88 (orange color) [31,32]. BGHI = black globe temperature and humidity index, indicating >72 comfort situation (green color) and 74 to 78 warning (yellow color) [26]. ESI = environmental stress index, indicating <25 low risk (green color) and ≤25 to ≤33 moderate to high risk (yellow color) [28]. ETI = equivalent temperature index, indicating no problems from 18 to <27 (green color) and caution ≥27 to ≥32 (yellow color) [29].

Table 4. Iberian Heat Tolerance Index (Iberian HTI) and Benezra Thermal Comfort Index (BTCI) during the experimental period in the eastern Amazon.

Times	Iberian HTI	BTCI
06H	87.34	2.330651
12H	82.54	2.418684
18H	79.24	2.510365
00H	84.94	2.342784

4. Discussion

The meteorological variables showed higher rates of AT at the hottest times of the day, specifically from 11:00 a.m. to 5:00 p.m., this being the period of highest solar intensity and the lowest RH of the air. Therefore, these are characterized as stress factors due to high temperatures, high humidity, and exposure to intense solar radiation at the times mentioned above [26,33–36]. In a study carried out by Hooper et al. [37], it was observed that air and surface temperatures are higher when there is a decrease in RH during the afternoon compared to the morning, influencing the physiological characteristics of Nellore animals.

The use of infrared thermography proved to be an efficient tool to evaluate and compare the three different cattle production systems in the eastern Amazon. The captured thermographic images clearly revealed the differences in the thermal conditions of the silvopastoral, traditional, and integrated systems. The study by Cândido et al. [38] sought to evaluate the silvopastoral system through infrared thermography, thus reinforcing the effectiveness of the technique.

Figure 6A, which represents the silvopastoral system, shows areas shaded by trees, mainly by the presence of chestnut trees (*Bertholletia excelsa*), which provide a cooler and more comfortable environment for the animals. This large vegetative area acts as a barrier against direct solar radiation, reducing the absorption of heat by the animals and, consequently, decreasing the local temperature [39–41], as shaded areas can reduce the heat load by up to 30% [42].

In contrast, Figure 6B, which depicts the traditional system without tree shading, shows higher temperatures in relation to the silvopastoral system. Without the protection of the trees, the animals are exposed to direct solar radiation and the heating of the soil, resulting in a significant increase in body temperature. Figure 6C, representing the integrated system with shading provided by the trees and the supply of water for bathing, shows intermediate temperatures between the silvopastoral and traditional systems. Tree shading contributes to thermal relief [23,43–50], while the availability of bathing water allows animals to regulate their body temperature through water evaporation, providing a more pleasant environment.

Another justification for the integrated system exhibiting higher temperatures compared to the silvopastoral system is due to the lower leaf density and wider spacing within the tree canopies. The sparser foliage in the integrated system allows for greater penetration of direct solar radiation through the canopy, resulting in increased heating of the underlying environment. In contrast, the silvopastoral system, characterized by more densely distributed leaves within the tree canopies, tends to filter a significant portion of direct solar radiation, contributing to a milder temperature within the system.

On the other hand, the absence of trees in the traditional system can lead to greater exposure of animals to solar heat, resulting in higher temperatures. Studies show that direct solar radiation can considerably increase the temperature of the soil surface and, consequently, the body temperature of animals, providing indices of thermal stress [51].

The BST showed no difference ($p > 0.05$) between the systems and for the anatomical regions studied; this can be explained by the combination of environmental, management, and adaptive factors that may have leveled the thermal responses of the animals in study.

The analysis of rectal temperature regression at different times (6:00 a.m., 12:00 a.m., 6:00 p.m., and 12:00 a.m.) in silvopastoral, traditional, and integrated systems is important to understand how animals respond to thermal variations throughout the day and how these production systems can influence the heat stress in cattle. Thus, the results show that there was a gradual increase in RT throughout the day, reaching its maximum peak at 6:00 p.m. in all evaluated systems. This increase in RT is related to the increase in ambient temperature during the day, especially in the hours of greater solar intensity. Direct solar radiation and ambient temperature can lead to an increase in thermal stress in animals, leading to an increase in RT [52,53].

However, despite the increase in RT throughout the day, the values observed in the silvopastoral, traditional, and integrated systems, with temperatures of 39.45 °C, 39.5 °C, and 39.5 °C, respectively, still remained within the range considered normal for cattle (38 to 39.5 °C), indicating the absence of thermal stress in the animals.

These results are in line with the scientific literature that highlights the importance of shading in silvopastoral systems to reduce heat stress in cattle. The presence of trees in silvopastoral systems provides shading, which reduces direct exposure to solar radiation and helps to reduce the accumulation of heat in the environment [51,54–59]. In addition, the presence of bathing areas, as in the integrated system, can also contribute to heat dissipation by the animals.

The regression analysis of RR in the three cattle breeding systems revealed a quadratic effect in relation to collection times, which indicates a significant variation of this parameter throughout the day. This variation may be related to environmental changes, such as fluctuations in temperature and solar radiation, which directly affect thermal stress in animals. The traditional system showed higher RR values at all times, indicating that the animals in this system are under greater thermal stress, probably due to the lack of shading and adequate spaces for bathing. The silvopastoral and integrated systems showed RR within the reference values, with heat stress evident only at 6 p.m. [20].

The physiological response observed in animals reared in the traditional system can be attributed to direct exposure to solar radiation and the lack of mechanisms to dissipate excessive heat. When cattle face high temperatures, their thermoregulatory system kicks in, increasing the RR to facilitate heat dissipation through breathing. This increase in RR is a way for animals to lose heat by evaporating water from the respiratory surface, but this physiological response can also indicate discomfort and heat stress [41,60].

On the other hand, the silvopastoral and integrated systems provide shaded areas and spaces for bathing, allowing cattle to protect themselves from excessive heat and maintain a more balanced body temperature [61,62]. The shade provided by the trees in the silvopastoral system and the availability of water for bathing in the integrated system help to reduce the ambient temperature and, consequently, the need to dissipate heat through respiratory rate. This results in lower RR values in these systems, keeping cattle within normal heat stress limits [63,64].

The RR regression analysis in the three cattle breeding systems revealed a quadratic effect in relation to collection times. This means that the RR showed significant variation throughout the day, with a curvilinear pattern, reaching maximum values at 6:00 p.m., 12:00 p.m., and 6:00 p.m. in the silvopastoral, traditional, and integrated systems, respectively. This variation indicates that cattle are physiologically responding to environmental changes, especially in relation to temperature and heat stress. Thus, in situations of thermal stress, sweating and peripheral vasodilation initially occur, resulting in a reduction in blood pressure, compensated by an increase in heart and respiratory rate [65–67]. These changes generate a greater demand for energy [68] and can lead to reduced productivity [69]; as a result, the need to maintain homeothermy leads to changes in basic physiological patterns, such as respiratory rate and rectal temperature. Thus, the monitoring of these variables becomes useful to assess the thermal balance in animals [70].

Heat stress in cattle can lead to an increase in RR as an adaptive response to reduce body heat. RR is an important indicator of thermal well-being in animals, and its elevation can indicate discomfort and thermal stress [71–75]. Hahn et al. [76] reported an increase in breathing rate of about 4.3 breaths per minute above the basal rate, i.e., 60 breaths per minute, for every degree Celsius above the threshold temperature of 21.3 °C.

The RR regression analysis in cattle-raising systems showed the importance of shading and providing bathing areas as effective strategies to reduce heat stress in animals. These thermal comfort measures can minimize the physiological responses of cattle to heat, promoting greater well-being and better production conditions in agricultural systems [77,78].

The accumulation of heat at 12:00 and 18:00 in the traditional and integrated systems, as well as at 18:00 and 00:00 in the silvopastoral system, is associated with the highest rectal temperature index observed in cattle during the period of the experiment. This relationship between the accumulation of heat and the increase in rectal temperature reflects the direct influence of the thermal conditions of the environment in the animals, a fact evidenced in the analysis using infrared thermography, which showed higher temperature indices in the traditional system. Rectal temperature is an important indicator of the physiological response of cattle to heat stress, and its elevation can be attributed to the body's effort to dissipate excessive heat to maintain homeostasis [79–86].

During peak solar hours, such as 12:00 and 18:00, solar radiation falls directly on the animals, resulting in a significant increase in ambient temperature. In traditional and integrated systems, where there is not adequate shading, cattle are exposed to intense heat, which can lead to an excessive accumulation of heat in their bodies. Lack of shade can also limit the ability of cattle to shelter and protect themselves from the heat, aggravating heat stress and contributing to increased RT [87,88].

In the silvopastoral system, the greater accumulation of heat observed at 18:00 p.m. and 00:00 p.m. can be explained by the effect of solar radiation accumulated during the day. Trees provide shade throughout the day, which can reduce ambient temperatures compared to traditional and integrated systems. However, as solar radiation decreases after sunset, the heat stored on tree and soil surfaces begins to be released, causing the ambient temperature to rise at these times. This phenomenon is known as the "heat island effect", which can be observed in shaded areas at night [51,89,90].

The results of the THI evaluation indicated no thermal stress at 7:00 a.m., with values below 72, which suggests that this time is more favorable in terms of thermal conditions for cattle. However, at 6:00 a.m., 8:00 a.m., 9:00 a.m., and from 7:00 p.m. onwards, lower THI indices were observed, characterizing a mild stress in the animals in these periods. This result may be related to the beginning of the day, when the ambient temperature has not yet reached its peak, and to the afternoon and night, when the temperature begins to decrease. These times of day can offer some thermal relief for the animals, especially if they are in an environment with adequate shading, as thermal comfort affects the performance of beef cattle [91].

On the other hand, the hours from 10:00 a.m. to 6:00 p.m. showed higher THI indices, indicating moderate thermal stress. These are the periods of greater solar intensity when environmental temperatures are higher and cattle are more susceptible to thermal stress. During these hours, direct exposure to solar radiation can lead to excessive accumulation of heat in animals, increasing body temperature and making thermoregulation difficult [32,92].

The BGHI showed that the hours of 6:00 a.m., 7:00 a.m., and 8:00 a.m. are more comfortable for cattle, with a low risk of thermal stress. This may be associated with nocturnal cooling and lower solar intensity during this period. In addition, from 11:00 p.m. to 6:00 a.m., the BGHI indicated comfort, reflecting the drop in temperatures during the night and early morning [39].

However, between 1:00 p.m. and 10:00 p.m., the BGHI presented moderate thermal stress (75–78—warning on a scale of ≤74: situation of thermal comfort; 75–78: warning; 79–84: danger; and ≥85: emergency) [93]. This time range corresponds to the hottest moments of the day, when cattle are more exposed to direct solar radiation and the accumulation of heat in the environment. The lack of adequate shade and poor air circulation during this period can contribute to the thermal stress of the animals, impairing their well-being and performance [46,50,94,95].

The ESI corroborated the BGHI data, indicating moderate thermal stress at 1:00 p.m. and mild stress from 7:00 p.m. This coherence between the indices reinforces the importance of these times in the occurrence of thermal stress in animal production systems. The afternoon period presented a moderate risk of stress, probably due to the accumulation of heat in the environment during the day and the absence of efficient cooling mechanisms for the animals [39].

The ETI signaled a state of caution (≥27 to ≥32) at 7 a.m., which suggests that at this time, cattle need attention regarding the thermal conditions of the environment. The other times showed no problems (18 to <27), which may be related to the early morning and the night shift, when temperatures tend to be milder. These results reinforce the importance of monitoring thermal conditions throughout the day to ensure the comfort and well-being of cattle, especially in environments with greater exposure to solar radiation and without adequate protection against heat. Corroborating with the results of this research, Campos et al. [6], in the rural area of Bela Vista de Goiás, showed ETI averages equal to 26 for both treatments (sun exposure and shade), with ETI values greater than 27 being considered as a situation of discomfort.

With regard to the BTCI, all values were greater than 2.0, indicating difficulties in the animals' adaptability to the environment. This result reinforces the importance of providing adequate thermal comfort conditions for cattle, especially in production systems with less shading, as in the traditional system. The lack of thermal comfort can lead to a series of negative impacts on the health and performance of animals, affecting their growth, production, and reproduction [32,39,96].

On the other hand, the BTCI showed a significant difference between the production systems, being higher in the traditional system compared to the silvopastoral and integrated systems. This indicates that cattle in the traditional system had greater difficulty in adapting to the environment in relation to the other systems, as values close to two are considered to provide greater animal comfort; that is, the cattle would be presenting RT and RR considered ideal [27]. This difference can be attributed to the lack of adequate shading in the traditional system, which exposes animals directly to solar radiation, increasing thermal stress and compromising their comfort and well-being [41,97].

The morning shift had the highest index of Iberian HTI, approaching 100 (the closer to 100, the more adapted the animal [98]), indicating a greater tolerance to heat for part of the cattle during this period. This result can be explained by the drop in temperatures during the night and the cooling at night, providing a milder environment for the animals in the early hours of the day [39]. Similar results were described by Marins et al. [4] in an experiment on a farm located in the tropical savanna from February to November, which obtained higher Iberian HTI results during the morning period (104.9 to 116.0); in the afternoon, they were 85.4 to 107.0 during all months.

Among the study limitations are the acquisition distance of the thermograms or the potential need to use an alternative lens and the absence of repeated measurements over a specific time interval.

5. Conclusions

The results of the thermographic images indicated that the silvopastoral system, with shading by chestnut trees and an ample vegetative area, provided more favorable thermal conditions compared to the traditional and integrated systems. Physiological analyses showed that the cattle maintained rectal temperatures within normal ranges in all systems. The respiratory rate was higher in the traditional system, indicating possible thermal stress. The thermal comfort indices pointed to moderate stress conditions at times of higher solar intensity. Considering the thermal stress indices evaluated, the silvopastoral system proved to be more suitable for the well-being of the animals, with lower thermal stress levels throughout most of the day. On the other hand, the traditional system showed higher levels of thermal stress, especially during hours of peak solar intensity. The integrated system occupied an intermediate position between the other two systems, with thermal stress values varying depending on the time of day.

It is important to emphasize that studies involving a larger number of animals, conducted across different seasons of the year, and especially with a greater number of repetitions, will provide a more definitive insight and confirmation as to whether a silvopastoral system, with shading from chestnut trees and ample vegetative cover, indeed offers more favorable thermal conditions compared to traditional and integrated systems.

Author Contributions: Conceptualization, W.C.d.S., J.d.B.L.-J., and J.A.R.d.S.; methodology, W.C.d.S., J.A.R.d.S., É.B.R.d.S., A.V.C.B., C.E.L.S., K.C.d.C., M.R.P.d.S., K.A.L.N., R.N.C.C.J., J.d.B.L.-J., and L.G.M.; thermography monitoring and data analysis support, L.G.M.; software, W.C.d.S. and J.d.B.L.-J.; validation, W.C.d.S., J.d.B.L.-J., and J.A.R.d.S.; formal analysis, W.C.d.S. and A.V.C.B.; investigation, W.C.d.S., J.A.R.d.S., É.B.R.d.S., A.V.C.B., C.E.L.S., K.C.d.C., M.R.P.d.S., KALN, L.G.M., R.N.C.C.J., and J.d.B.L.-J.; software, W.C.d.S. and J.d.B.L.-J.; resources, W.C.d.S. and J.d.B.L.-J.; data curation, W.C.d.S. and J.d.B.L.-J.; writing—original draft preparation, W.C.d.S. and J.d.B.L.-J.; writing—review and editing, all authors; visualization, W.C.d.S.; supervision, J.d.B.L.-J. and J.A.R.d.S.; project administration, W.C.d.S. and J.d.B.L.-J.; funding acquisition, J.d.B.L.-J. All authors have read and agreed to the published version of the manuscript.

Funding: This study was funded in part by the Federal University of Pará and Coordenação de Aperfeiçoamento de Pessoal de Nível Superior (CAPES) Brazil. This study also received financial support for the publication fee from the Pró-Reitoria de Pesquisa e Pós-Graduação (PROPESP/UFPA—Announcement—02/2023).

Institutional Review Board Statement: The experimental procedures were carried out in accordance with Brazilian laws and were previously approved by the Animal Ethics Committee (CEUA/UNAMA), CEUA-UNAMA 0001-87/2023, approved on 31 March 2023.

Informed Consent Statement: Not applicable.

Data Availability Statement: The data presented in this study are available upon reasonable request from the corresponding author.

Acknowledgments: We would like to thank the Coordination for the Improvement of Higher Education Personnel (CAPES)—Brazil. This study also received financial support for the publication fee from the Dean of Research and Graduate Studies (PROPESP/UFPA). We also extend our gratitutde to the Programa de Pós-graduação em Ciência Animal (PPGCAN).

Conflicts of Interest: The authors declare that the research was conducted in the absence of any commercial or financial relation-ships that could be construed as a potential conflict of interest.

References

1. Santos, M.A.S.; Júnior, J.D.B.L.; Santana, A.C.; Homma, A.K.O.; Martins, C.M.; Rebello, F.K.; Soares, B.C.; Silva, A.G.M. Comportamento da produção e dos preços de bovinos de corte na Amazônia Brasileira. *Semin. Ciências Agrárias* **2019**, *40*, 1639–1652. [CrossRef]
2. Silva, W.C.D.; Camargo, R.N.C.; Silva, É.B.R.D.; Silva, J.A.R.D.; Picanço, M.L.R.; Santos, M.R.P.D.; Araújo, C.V.; Barbosa, A.V.C.; Bonin, M.N.; Oliveira, A.S.; et al. Perspectives of economic losses due to condemnation of cattle and buffalo carcasses in the northern region of Brazil. *PLoS ONE* **2023**, *18*, e0285224. [CrossRef]
3. Rosanova, C.; Rebouças, G.F.; Silva, M.D.M.P.; Rezende, D.M.L.C.; Rocha, A.S.; Junior, A.P.; Ismar, M.G.; Ferreira, C.C.B.; Silva, E.W. Determinação do ITU–índice de temperatura e umidade da região de Araguaína-TO para avaliação do conforto térmico de bovinos leiteiros. *Braz. J. Dev.* **2020**, *6*, 69254–69258. [CrossRef]
4. Marins, T.N.; Almeida, I.G.B.; Lôbo, B.V.; Pessoa, C.M.B.; Teixeira, R.C.; Alves, B.G.; Gambarini, M.L. Índices de estresse e conforto térmico associados aos parâmetros fisiológicos e perfil energético em vacas Girolando criadas a pasto na savana tropical. *Res. Soc. Dev.* **2020**, *9*, e111973672. [CrossRef]
5. Kemer, A.; Glienke, C.L.; Bosco, L.C. Índices de conforto térmico para bovinos de leite em Santa Catarina Sul do Brasil. *Braz. J. Dev.* **2020**, *6*, 29655–29672. [CrossRef]
6. Campos, J.C.D.; Passini, R.; Sousa, L.J.M.P.; Amaral, A.D.G. Alterações comportamentais de bezerras criadas em bezerreiros tropicais com cobertura alternativa. *Ciência Anim. Bras.* **2023**, *24*, 1–9.
7. Lemos, M.J.; Braga, T.V.S.; Leite, D.K.V.H.; Galindo, E.L.O. Efeito climático sobre a produção de leite de vacas Guzerá criadas em dois ambientes diferentes: Climate effect on milk production of Guzera cows reared in two diferente environments. *Braz. J. Anim. Environ. Res.* **2023**, *6*, 1721–1726. [CrossRef]
8. Silva, W.C.; Martorano, L.G.; Silva, L.K.X.; da Costa Guedes, J.C.; Fernandes, G.B.; dos Santos, A.X.; da Silva, É.B.R. Use of infrared thermography in the assessment of mammary glands and eyeballs in dairy cattle in Western Pará. *Res. Soc. Dev.* **2020**, *9*, e46191211421. [CrossRef]
9. Mota-Rojas, D.; Wang, D.; Titto, C.G.; Gómez-Prado, J.; Fuente, V.C.; Ghezzi, M.; Boscato-Funes, L.; Barrios-Garcia, H.; Torres-Bernal, F.; Casas-Alvarado, A.; et al. Pathophysiology of fever and application of infrared thermography (IRT) in the detection of sick domestic animals: Recent advances. *Animals* **2021**, *11*, 2316. [CrossRef]
10. Mota-Rojas, D.; Pereira, A.M.; Wang, D.; Martínez-Burnes, J.; Ghezzi, M.; Hernández-Avalos, I.; Lendez, P.; Mora-Medina, P.; Casas, A.; Olmos-Hernández, A.; et al. Clinical applications and factors involved in validating thermal windows used in infrared thermography in cattle and river buffalo to assess health and productivity. *Animals* **2021**, *11*, 2247. [CrossRef]
11. Oliveira, A.V.D.; Reis, E.M.B.; Ferraz, P.F.P.; Barbari, M.; Santos, G.S.; Cruz, M.V.R.; Silva, G.F.; Silva, A.O.L. Infrared thermography as a technique for detecting subclinical bovine mastitis. *Arq. Bras. Med. Veterinária Zootec.* **2023**, *74*, 992–998. [CrossRef]
12. Knizkova, I.; Petr, K.U.N.C.; Gürdil, G.; Pinar, Y.; Selvi, K.Ç. Applications of infrared thermography in animal production. *Anadolu Tarım Bilim. Derg.* **2007**, *22*, 329–336.
13. Mota-Rojas, D.; Pereira, A. Infrared Thermography in Large Ruminants. *J. Encycl.—Peer-Rev. Content* **2021**, *1*, 1–15.
14. Almeida, J.V.N.; Marques, L.R.; Marques, T.C.; Guimarães, K.C.; Leão, K.M. Influência do estresse térmico sobre os aspectos produtivos e reprodutivos de bovinos–Revisão. *Res. Soc. Dev.* **2020**, *9*, e230973837. [CrossRef]
15. McManus, C.; Tanure, C.B.; Peripolli, V.; Seixas, L.; Fischer, V.; Gabbi, A.M.; Menegassi, S.R.O.; Stumpf, M.T.; Kolling, G.J.; Dias, E.; et al. Infrared thermography in animal production: An overview. *Comput. Electron. Agric.* **2016**, *123*, 10–16. [CrossRef]

16. Rohleder, L.A.S.; Querino, C.A.S.; Alves, P.V.; Querino, J.K.A.D.S.; Junior, A.L.P.; Vaz, M.A.B. Avaliação de parâmetros ambientais em uma microrregião no sul do estado do Amazonas e suas relações com estresse térmico de bovinos leiteiros. *Ciência Anim. Bras.* **2022**, *23*, 1–7.
17. McDowell, R.E.; Hooven, N.W.; Camoens, J.K. Effects of climate on performance of Holsteins in first lactation. *J. Dairy Sci.* **1976**, *59*, 965–973. [CrossRef]
18. Martorano, L.G.; Nechet, D.; Pereira, L.C. Tipologia climática do Estado do Pará: Adaptação do método de Köppen. *Bol. Geogr. Teorética* **1993**, *23*, 45–46.
19. Martorano, L.G.; Vitorino, M.I.; Silva, B.P.P.C.; Lisboa, L.S.; Sotta, E.D.; Reichardt, K. Climate conditions in the eastern amazon: Rainfall variability in Belem and indicative of soil water deficit. *Afr. J. Agric. Res.* **2017**, *12*, 1801–1810.
20. Feitosa, F.L.F. Exame físico geral ou de rotina. In *Semiologia Veterinária: A Arte do Diagnóstico*; Roca: Barcelona, Spain, 2014; Volume 3, pp. 51–68.
21. Dirksen, G.; Stöber, M.; Gründer, H.D. Rosenberger—Exame clínico dos bovinos. In *Sistema Digestivo*; Dirksen, G., Ed.; Guanabara Koogan: Rio De Janeiro, Brazil, 1993; pp. 166–228.
22. FLIR T650sc. Manual da FLIR T650sc Systems 6.3v. Wilsonville, OR, EUA. 2015. Available online: https://www.flir.com.br/support/products/t650sc/#Overview (accessed on 1 March 2023).
23. Giro, A.; Pezzopane, J.R.M.; Junior, W.B.; Pedroso, A.F.; Lemes, A.P.; Botta, D.; Romanello, N.; Barreto, A.N.; Garcia, A.R. Behavior and body surface temperature of beef cattle in integrated crop-livestock systems with or without tree shading. *Sci. Total Environ.* **2019**, *684*, 587–596. [CrossRef] [PubMed]
24. McGovern, R.E.; Bruce, J.M. A model of the thermal balance for cattle in hot conditions. *J. Agric. Eng. Res.* **2000**, *77*, 81–92. [CrossRef]
25. Thom, E.C. The discomfort index. *Weatherwise* **1959**, *12*, 57–61. [CrossRef]
26. Buffington, D.E.; Collazo-Arocho, A.; Canton, G.H.; Pitt, D.; Thatcher, W.W.; Collier, R.J. Black globe-humidity index (BGHI) as comfort equation for dairy cows. *Trans. ASAE* **1981**, *24*, 711–714. [CrossRef]
27. Benezra, M.V. A new index measuring the adaptability of cattle to tropical conditions. *J. Anim. Sci.* **1954**, *13*, 1015.
28. Moran, D.S.; Pandolf, K.B.; Shapiro, Y.; Heled, Y.; Shani, Y.; Mathew, W.T.; Gonzalez, R.R. An environmental stress index (ESI) as a substitute for the wet bulb globe temperature (WBGT). *J. Therm. Biol.* **2001**, *26*, 427–431. [CrossRef]
29. Baêta, F.C.; Souza, C.F. *Ambiência em Edificações Rurais: Conforto Animal*; UFV: Viçosa, Brazil, 1997; p. 246.
30. Rhoad, A.O. The iberia heat tolerance test for cattle. *Trop. Agric.* **1944**, *21*, 162–164.
31. Armstrong, D. Heat stress interaction with shade and cooling. *J. Dairy Sci.* **1994**, *77*, 2044–2050. [CrossRef] [PubMed]
32. Silva, W.C.D.; Printes, O.V.N.; Lima, D.O.; Silva, É.B.R.D.; Santos, M.R.P.D.; Camargo-Júnior, R.N.C.; Barbosa, A.V.C.; Silva, J.A.R.; Silva, A.G.M.; Silva, L.K.X.; et al. Evaluation of the Temperature and Humidity Index (THI) to support the implementation of a rearing system for ruminants in the Western Amazon. *Front. Vet. Sci.* **2023**, *10*, 1198678. [CrossRef]
33. Shioya, S.; Terada, F.; Iwama, Y. Physiological responses of lactating dairy cows under hot environments. *Eiyoseirikenkyukaiho* **1997**, *41*, 61–68.
34. Herbut, P.; Angrecka, S.; Godyń, D. Effect of the duration of high air temperature on cow's milking performance in moderate climate conditions. *Ann. Anim. Sci.* **2018**, *18*, 195. [CrossRef]
35. McManus, C.M.; Faria, D.A.; Bem, A.D.; Maranhão, A.Q.; Paiva, S.R. Physiology and genetics of heat stress in cattle. *CABI Rev.* **2020**, *53*, 1–8.
36. Lima, S.B.; Stafuzza, N.B.; Pires, B.V.; Bonilha, S.F.; Cyrillo, J.N.; Negrão, J.A.; Paz, C.C. Effect of high temperature on physiological parameters of Nelore (*Bos taurus* indicus) and Caracu (*Bos taurus* taurus) cattle breeds. *Trop. Anim. Health Prod.* **2020**, *52*, 2233–2241. [CrossRef]
37. Hooper, H.B.; Titto, C.G.; Gonella-Diaza, A.M.; Henrique, F.L.; Pulido-Rodríguez, L.F.; Longo, A.L.S.; Leme-dos-Santos, T.M.D.C.; Geraldo, A.C.A.P.D.M.; Pereira, A.M.F.; Binelli, M.; et al. Heat loss efficiency and HSPs gene expression of Nellore cows in tropical climate conditions. *Int. J. Biometeorol.* **2019**, *63*, 1475–1486. [CrossRef]
38. Cândido, A.C.T.F.; Martorano, L.G.; Cândido, B.U.F.; Nascimento, W.; Dias, C.T.D.S.; Lisboa, L.S.S.; Beldini, T.P. Infrared Thermal Profiles in Silvopastoral and Full-Sun Pastures in the Eastern Amazon, Brazil. *Forests* **2023**, *14*, 1463. [CrossRef]
39. Silva, W.C.D.; Silva, É.B.R.D.; Santos, M.R.P.D.; Camargo-Junior, R.N.C.; Barbosa, A.V.C.; Silva, J.A.R.D.; Vinhote, J.A.; Sousa, E.D.V.D.; Lourenço-Júnior, J.D.B. Behavior and thermal comfort of light and dark coat dairy cows in the Eastern Amazon. *Front. Vet. Sci.* **2022**, *9*, 1006093. [CrossRef]
40. Romanello, N.; Barreto, A.N.; Sousa, M.A.P.; Balieiro, J.C.C.; Brandão, F.Z.; Tonato, F.; Bernardi, A.C.C.; Pezzopane, J.R.M.; Portugal, J.A.B.; Garcia, A.R. Thermal comfort of Nelore (*Bos indicus*) and Canchim (*Bos taurus x Bos indicus*) bulls kept in an integrated crop-livestock-forestry system in a tropical climate. *Agric. Syst.* **2023**, *209*, 103687. [CrossRef]
41. Silva, W.C.D.; Silva, J.A.R.D.; Camargo-Júnior, R.N.C.; Silva, É.B.R.D.; Santos, M.R.P.D.; Viana, R.B.; Lourenço-Júnior, J.D.B. Animal welfare and effects of per-female stress on male and cattle reproduction—A review. *Front. Vet. Sci.* **2023**, *10*, 1083469. [CrossRef] [PubMed]
42. Salman, A.K.D.; Giustina, C.D.; Martínez, G.B.; Carnevalli, R.A. *Agroforestry Systems for Dairy Production in Capítulo em Livro Técnico*; INFOTECA-E, Embrapa: Brazilia, Brazil, 2020; Volume 16, pp. 371–390.
43. Nardone, A.; Ronchi, B.; Lacetera, N.; Ranieri, M.S.; Bernabucci, U. Effects of climate changes on animal production and sustainability of livestock systems. *Livest. Sci.* **2010**, *130*, 57–69. [CrossRef]

44. Gil, J.; Siebold, M.; Berger, T. Adoption and development of integrated crop–livestock–forestry systems in Mato Grosso, Brazil. *Agric. Ecosyst. Environ.* **2015**, *199*, 394–406. [CrossRef]
45. Hagenmaier, J.A.; Reinhardt, C.D.; Bartle, S.J.; Thomson, D.U. Effect of shade on animal welfare, growth performance, and carcass characteristics in large pens of beef cattle fed a beta agonist in a commercial feedlot. *J. Anim. Sci.* **2016**, *94*, 5064–5076. [CrossRef]
46. Karvatte, N.; Klosowski, E.S.; Almeida, R.G.; Mesquita, E.E.; Oliveira, C.C.; Alves, F.V. Shading effect on microclimate and thermal comfort indexes in integrated crop-livestock-forest systems in the Brazilian Midwest. *Int. J. Biometeorol.* **2016**, *60*, 1933–1941. [CrossRef]
47. Oliveira, C.C.; Alves, F.V.; Almeida, R.G.; Gamarra, E.L.; Villela, S.D.J.; Martins, P.G.M.D.A. Thermal comfort indices assessed in integrated production systems in the Brazilian savannah. *Agrofor. Syst.* **2018**, *92*, 1659–1672. [CrossRef]
48. Pezzopane, J.R.M.; Nicodemo, M.L.F.; Bosi, C.; Garcia, A.R.; Lulu, J. Animal thermal comfort indexes in silvopastoral systems with different tree arrangements. *J. Therm. Biol.* **2019**, *79*, 103–111. [CrossRef]
49. Deniz, M.; Filho, A.L.S.; Hötzel, M.J.; Sousa, K.T.; Filho, L.C.P.M.; Sinisgalli, P.A. Microclimate and pasture area preferences by dairy cows under high biodiversity silvopastoral system in Southern Brazil. *Int. J. Biometeorol.* **2020**, *64*, 1877–1887. [CrossRef]
50. Barreto, A.N.; Junior, W.B.; Pezzopane, J.R.M.; Bernardi, A.C.C.; Pedroso, A.F.; Marcondes, C.R.; Jacintho, M.A.C.; Romanello, N.; Sousa, M.A.P.; Costa, L.N.; et al. Thermal comfort and behavior of beef cattle in pasture-based systems monitored by visual observation and electronic device. *Appl. Anim. Behav. Sci.* **2022**, *253*, 105687. [CrossRef]
51. Lemes, A.P.; Garcia, A.R.; Pezzopane, J.R.M.; Brandão, F.Z.; Watanabe, Y.F.; Cooke, R.F.; Sponchiado, M.; Paz, C.C.P.; Camplesi, A.C.; Binelli, M.; et al. Silvopastoral system is an alternative to improve animal welfare and productive performance in meat production systems. *Sci. Rep.* **2021**, *11*, 14092. [CrossRef]
52. Santos, M.M.; Souza-Junior, J.B.F.; Dantas, M.R.T.; Costa, L.L.M. An updated review on cattle thermoregulation: Physiological responses, biophysical mechanisms, and heat stress alleviation pathways. *Environ. Sci. Pollut. Res.* **2021**, *28*, 30471–30485. [CrossRef] [PubMed]
53. Jasinski, F.; Evangelista, C.; Basiricò, L.; Bernabucci, U. Responses of Dairy Buffalo to Heat Stress Conditions and Mitigation Strategies: A Review. *Animals* **2023**, *13*, 1260. [CrossRef] [PubMed]
54. Junior, N.A.V.; Silva, M.A.D.A.; Caramori, P.H.; Nitsche, P.R.; Corrêa, K.A.B.; Alves, D.S. Temperature, thermal comfort, and animal ingestion behavior in a silvopastoral system. *Semin. Ciências Agrárias* **2019**, *40*, 403–416. [CrossRef]
55. Huertas, S.M.; Bobadilla, P.E.; Alcántara, I.; Akkermans, E.; Eerdenburg, F.J.V. Benefits of silvopastoral systems for keeping beef cattle. *Animals* **2021**, *11*, 992. [CrossRef] [PubMed]
56. Deniz, M.; Sousa, K.T.; Moro, M.F.; Vale, M.M.; Dittrich, J.R.; Machado-Filho, L.C.P.; Hötzel, M.J. Social hierarchy influences dairy cows' use of shade in a silvopastoral system under intensive rotational grazing. *Appl. Anim. Behav. Sci.* **2021**, *244*, 105467. [CrossRef]
57. Pérez-Lombardini, F.; Mancera, K.F.; Suzán, G.; Campo, J.; Solorio, J.; Galindo, F. Assessing sustainability in cattle silvopastoral systems in the Mexican tropics using the SAFA framework. *Animals* **2021**, *11*, 109. [CrossRef] [PubMed]
58. Oyelami, B.A.; Osikabor, B. Adoption of silvopastoral agroforestry system for a sustainable cattle production in Nigeria. *J. Appl. Sci. Environ. Manag.* **2022**, *26*, 1397–1402. [CrossRef]
59. Deniz, M.; De-Sousa, K.T.; Vieira, F.M.C.; Vale, M.M.D.; Dittrich, J.R.; Daros, R.R.; Hötzel, M.J. A systematic review of the effects of silvopastoral system on thermal environment and dairy cows' behavioral and physiological responses. *Int. J. Biometeorol.* **2023**, *67*, 409–422. [CrossRef] [PubMed]
60. Islam, M.A.; Lomax, S.; Doughty, A.; Islam, M.R.; Jay, O.; Thomson, P.; Clark, C. Automated monitoring of cattle heat stress and its mitigation. *Front. Anim. Sci.* **2021**, *2*, 60. [CrossRef]
61. Mota-Rojas, D.; Napolitano, F.; Braghieri, A.; Guerrero-Legarreta, I.; Bertoni, A.; Martínez-Burnes, J.; Cruz-Monterrosa, R.; Gómez, J.; Ramírez-Bribiesca, E.; Barrios-García, H.; et al. Thermal biology in river buffalo in the humid tropics: Neurophysiological and behavioral responses assessed by infrared thermography. *J. Anim. Behav. Biometeorol.* **2020**, *9*, 2103. [CrossRef]
62. Thornton, P.; Nelson, G.; Mayberry, D.; Herrero, M. Increases in extreme heat stress in domesticated livestock species during the twenty-first century. *Glob. Change Biol.* **2021**, *27*, 5762–5772. [CrossRef] [PubMed]
63. Brcko, C.C.; Silva, J.A.R.D.; Martorano, L.G.; Vilela, R.A.; Nahúm, B.D.S.; Silva, A.G.M.; Barbosa, A.V.C.; Bezerra, A.S.; Lourenço Júnior, J.D.B. Infrared thermography to assess thermoregulatory reactions of female buffaloes in a humid tropical environment. *Front. Vet. Sci.* **2020**, *7*, 180. [CrossRef]
64. Hernandez, A.; Galina, C.S.; Geffroy, M.; Jung, J.; Westin, R.; Berg, C. Cattle welfare aspects of production systems in the tropics. *Anim. Prod. Sci.* **2022**, *62*, 1203–1218. [CrossRef]
65. Marai, I.F.M.; Haeeb, A.A.M. Buffaloes reproductive and productive traits as affected by heat stress. *Trop. Subtrop. Agroecosyst.* **2010**, *12*, 193–217.
66. Detweiler, N.D.; Vigil, K.G.; Yan, S.; Resta, T.C.; Walker, B.R.; Jernigan, N.L. Contribution of Acid-Sensing Ion Channels to Hypoxia-and Hypercapnia-Induced Ventilatory Drive in Conscious Unrestrained Mice. *FASEB J.* **2017**, *31*, 1072–1078.
67. Godyń, D.; Herbut, P.; Angrecka, S. Measurements of peripheral and deep body temperature in cattle–A review. *J. Therm. Biol.* **2019**, *79*, 42–49. [CrossRef]
68. Garcia, A.R.; Matos, L.B.; Junior, J.D.B.L.; Nahúm, B.D.S.; Costa, N.A.D. Increased reproductive efficiency of dairy buffaloes due to silvopastoral system adoption in the Eastern Amazon. *Rev. Vet.* **2010**, *21*, 914–915.

69. Wang, F.X.; Shao, D.F.; Li, S.L.; Wang, Y.J.; Azarfar, A.; Cao, Z.J. Effects of stocking density on behavior, productivity, and comfort indices of lactating dairy cows. *J. Dairy Sci.* **2016**, *99*, 3709–3717. [CrossRef]
70. Alves, J.R.A.; Andrade, T.A.A.; Assis, D.M.; Gurjão, T.A.; Melo, L.R.B.; Souza, B.B. Productive and reproductive performance, behavior and physiology of cattle under heat stress conditions. *J. Anim. Behav. Biometeorol.* **2020**, *5*, 91–96. [CrossRef]
71. Brown-Brandl, T.M.; Nienaber, J.A.; Eigenberg, R.A.; Mader, T.L.; Morrow, J.L.; Dailey, J.W. Comparison of heat tolerance of feedlot heifers of different breeds. *Livest. Sci.* **2006**, *105*, 19–26. [CrossRef]
72. Mader, T.L.; Davis, M.S.; Brown-Brandl, T. Environmental factors influencing heat stress in feedlot cattle. *J. Anim. Sci.* **2006**, *84*, 712–719. [CrossRef] [PubMed]
73. Nienaber, J.A.; Hahn, G.L. Livestock production system management responses to thermal challenges. *Int. J. Biometeorol.* **2007**, *52*, 149–157. [CrossRef]
74. Idris, M.; Uddin, J.; Sullivan, M.; McNeill, D.M.; Phillips, C.J. Non-invasive physiological indicators of heat stress in cattle. *Animals* **2021**, *11*, 71. [CrossRef]
75. Arias, R.A.; Mader, T.L. Evaluation of Four Thermal Comfort Indices and Their Relationship with Physiological Variables in Feedlot Cattle. *Animals* **2023**, *13*, 1169. [CrossRef]
76. Hahn, G.; Parkhurst, A.; Gaughan, J. Cattle respiration rate as a function of ambient temperature. *Am. Soc. Agric. Biol. Eng.* **1997**, *121*, 1–8.
77. Mitlöhner, F.M.; Morrow, J.L.; Dailey, J.W.; Wilson, S.C.; Galyean, M.L.; Miller, M.F.; McGlone, J.J. Shade and water misting effects on behavior, physiology, performance, and carcass traits of heat-stressed feedlot cattle. *J. Anim. Sci.* **2001**, *79*, 2327–2335. [CrossRef]
78. Becker, C.A.; Stone, A.E. Graduate Student Literature Review: Heat abatement strategies used to reduce negative effects of heat stress in dairy cows. *J. Dairy Sci.* **2020**, *103*, 9667–9675. [CrossRef] [PubMed]
79. Koga, A.; Kuhara, T.; Kanai, Y. Comparison of body water retention during water deprivation between swamp buffaloes and Friesian cattle. *J. Agric. Sci.* **2004**, *138*, 435–440. [CrossRef]
80. Morais, D.A.E.F.; Maia, A.S.C.; Silva, R.G.D.; Vasconcelos, A.M.D.; Lima, P.D.O.; Guilhermino, M.M. Annual thyroid hormone variation and thermo regulators traits of milk cows in hot environment. *Rev. Bras. Zootec.* **2008**, *37*, 538–545. [CrossRef]
81. Rhoads, M.L.; Rhoads, R.P.; VanBaale, M.J.; Collier, R.J.; Sanders, S.R.; Weber, W.J.; Crooker, B.A.; Baumgard, L.H. Effects of heat stress and plane of nutrition on lactating Holstein cows: I. Production, metabolism, and aspects of circulating somatotropin. *J. Dairy Sci.* **2009**, *92*, 1986–1997. [CrossRef] [PubMed]
82. Taylor, N.A.; Tipton, M.J.; Kenny, G.P. Considerations for the measurement of core, skin and mean body temperatures. *J. Therm. Biol.* **2014**, *46*, 72–101. [CrossRef]
83. Falkenberg, S.M.; Ridpath, J.; Ley, B.V.; Bauermann, F.V.; Sanchez, N.B.; Carroll, J.A. Comparison of temperature fluctuations at multiple anatomical locations in cattle during exposure to bovine viral diarrhea virus. *Livest. Sci.* **2014**, *164*, 159–167. [CrossRef]
84. Bharati, J.; Dangi, S.S.; Mishra, S.R.; Chouhan, V.S.; Verma, V.; Shankar, O.; Bharti, M.K.; Paul, A.; Mahato, D.K.; Rajesh, G.; et al. Expression analysis of toll like receptors and interleukins in Tharparkar cattle during acclimation to heat stress exposure. *J. Therm. Biol.* **2017**, *65*, 48–56. [CrossRef]
85. Das, S. Impact of climate change on livestock, various adaptive and mitigative measures for sustainable livestock production. *Approaches Poult. Dairy Vet. Sci.* **2017**, *33*, 64–70. [CrossRef]
86. Mishra, S.R. Behavioural, physiological, neuro-endocrine and molecular responses of cattle against heat stress: An updated review. *Trop. Anim. Health Prod.* **2021**, *53*, 400. [CrossRef]
87. Grandin, T. Evaluation of the welfare of cattle housed in outdoor feedlot pens. *Vet. Anim. Sci.* **2016**, *1*, 23–28. [CrossRef]
88. Edwards-Callaway, L.N.; Cramer, M.C.; Cadaret, C.N.; Bigler, E.J.; Engle, T.E.; Wagner, J.J.; Clark, D.L. Impacts of shade on cattle well-being in the beef supply chain. *J. Anim. Sci.* **2021**, *99*, skaa375. [CrossRef]
89. Castro, A.C.; Lourenço-Júnior, J.D.B.; Santos, N.D.F.A.D.; Monteiro, E.M.M.; Aviz, M.A.B.D.; Garcia, A.R. Sistema silvipastoril na Amazônia: Ferramenta para elevar o desempenho produtivo de búfalos. *Ciência Rural.* **2008**, *38*, 2395–2402. [CrossRef]
90. Silva, J.A.R.D.; Araújo, A.A.D.; Lourenço-Júnior, J.D.B.; Santos, N.D.F.A.D.; Garcia, A.R.; Nahúm, B.D.S. Conforto térmico de búfalas em sistema silvipastoril na Amazônia Oriental. *Pesqui. Agropecuária Bras.* **2011**, *46*, 1364–1371. [CrossRef]
91. Lees, A.M.; Sejian, V.; Wallage, A.L.; Steel, C.C.; Mader, T.L.; Lees, J.C.; Gaughan, J.B. The impact of heat load on cattle. *Animals* **2019**, *9*, 322. [CrossRef]
92. Finch, V.A. Body temperature in beef cattle: Its control and relevance to production in the tropics. *J. Anim. Sci.* **1986**, *62*, 531–542. [CrossRef]
93. Baeta, F.D.C.; Souza, C. *Ambiência em Edificações Rurais: Conforto Ambiental*; Editora UFV: Viçosa, Brasil, 2010; p. 269.
94. Abreu, L.V.; Labaki, L.C. Conforto térmico propiciado por algumas espécies arbóreas: Avaliação do raio de influência através de diferentes índices de conforto. *Ambiente Construído* **2010**, *10*, 103–117. [CrossRef]
95. Magalhães, C.A.; Zolin, C.A.; Lulu, J.; Lopes, L.B.; Furtini, I.V.; Vendrusculo, L.G.; Zaiatz, A.P.S.R.; Pedreira, B.C.; Pezzopane, J.R.M. Improvement of thermal comfort indices in agroforestry systems in the southern Brazilian Amazon. *J. Therm. Biol.* **2020**, *91*, 102636. [CrossRef]
96. Yan, G.; Liu, K.; Hao, Z.; Shi, Z.; Li, H. The effects of cow-related factors on rectal temperature, respiration rate, and temperature-humidity index thresholds for lactating cows exposed to heat stress. *J. Therm. Biol.* **2021**, *100*, 103041. [CrossRef]

97. Blackshaw, J.K.; Blackshaw, A.W. Heat stress in cattle and the effect of shade on production and behaviour: A review. *Aust. J. Exp. Agric.* **1994**, *34*, 285–295. [CrossRef]
98. Silva, T.P.D.; Júnior, S.C.S.; Santos, K.R.; Marques, C.A.T.; Torreão, J.N.C. Características termorreguladoras e ganho de peso de cordeiros Santa Inês no sul do estado do Piauí no período de transição seca/águas. *Agrar. Acad. J.* **2013**, *6*, 198–204.

animals

MDPI

Article

Events with Different Emotional Valence Affect the Eye's Lacrimal Caruncle Temperature Changes in Sheep

Marta Comin [1,†], Elie Atallah [1,†], Matteo Chincarini [2], Silvia Michela Mazzola [1], Elisabetta Canali [1], Michela Minero [1], Bruno Cozzi [3], Emanuela Rossi [4], Giorgio Vignola [2] and Emanuela Dalla Costa [1,*]

1 Dipartimento di Medicina Veterinaria e Scienze Animali, Università degli Studi di Milano, 26900 Lodi, Italy; marta.comin@unimi.it (M.C.); elie.atallah@unimi.it (E.A.); silvia.mazzola@unimi.it (S.M.M.); elisabetta.canali@unimi.it (E.C.); michela.minero@unimi.it (M.M.)
2 Facoltà di Medicina Veterinaria, Università degli Studi di Teramo, 64100 Teramo, Italy; m.chincarini@unite.it (M.C.); g.vignola@unite.it (G.V.)
3 Dipartimento di Biomedicina Comparata e Alimentazione, Università degli Studi di Padova, 35131 Padova, Italy; bruno.cozzi@unipd.it
4 Istituto Zooprofilattico Sperimentale dell'Abruzzo e del Molise G. Caporale, 64100 Teramo, Italy; e.rossi@izs.it
* Correspondence: emanuela.dallacosta@unimi.it
† These authors contributed equally to this work.

Simple Summary: We explored the application of infrared thermography (IRT) as a non-invasive method to measure the eye's lacrimal caruncle temperature in sheep, specifically focusing on its potential to detect negative emotions such as fear. Fourteen ewes, divided into two study groups according to the exposed emotional (positive $n = 7$ and negative $n = 7$), were subjected to classical conditioning associating a neutral auditory stimulus with either a positive event (food reward) or a negative event (opening an umbrella). Then, lacrimal caruncle temperatures using IRT, behavioral observation of ear postures, and saliva samples were collected to assess cortisol levels at rest and after treatment. The results reveal a significant increase in lacrimal caruncle temperature post-treatment only in the negative group ($p = 0.017$), indicating a potential association between IRT-measured eye temperature and negative emotional states in sheep. Additionally, behavioral observations align with the learned association between the neutral stimulus and events of different emotional valence. As for cortisol, despite not being significant, its levels increased for both groups post-treatment. The study concludes that IRT proves valuable in non-invasively evaluating the physiological impact of positive and negative events on sheep.

Abstract: Infrared thermography (IRT) has been recently applied to measure lacrimal caruncle temperature non-invasively since this region is related to the sympathetic response, and it seems a promising technique that is able to infer negative emotions in sheep (e.g., fear). However, the scientific literature so far is limited in understanding whether a caruncle's temperature changes also in response to positive emotional states in sheep. Through classical conditioning, we aimed to assess how a positive or a negative event affects the physiological (lacrimal caruncle temperature measured with IRT and cortisol levels) and behavioral responses of sheep (ear position). Fourteen ewes from the same flock were randomly assigned to two treatment groups: positive ($n = 7$) and negative ($n = 7$). Each group was then trained through classical conditioning to associate a neutral auditory (ring bell) stimulus to an oncoming event: for the positive group, the presence of a food reward (maize grains), while for the negative one, the opening of an umbrella. After three weeks of training, before (at rest) and after (post-treatment), lacrimal caruncle temperature was non-invasively measured via IRT, and saliva samples were gently collected to measure cortisol levels. During treatment, sheep behavior was videorecorded and then analyzed using a focal animal sampling technique. At rest, the eye's lacrimal caruncle temperature was similar in both groups, while post-treatment, a significant increase was shown only in the negative group (t-test; $p = 0.017$). In the anticipation phase, sheep in the positive group kept their ears forward longer compared to those in the negative one (Mann–Whitney; $p < 0.014$), 8.3 ± 2.1 s and 5.2 ± 4.2 s, respectively. The behavioral response observed reflects a learnt association between a neutral stimulus and events with different emotional valence. Cortisol

Citation: Comin, M.; Atallah, E.; Chincarini, M.; Mazzola, S.M.; Canali, E.; Minero, M.; Cozzi, B.; Rossi, E.; Vignola, G.; Dalla Costa, E. Events with Different Emotional Valence Affect the Eye's Lacrimal Caruncle Temperature Changes in Sheep. *Animals* **2024**, *14*, 50. https://doi.org/10.3390/ani14010050

Academic Editor: Daniel Mota-Rojas

Received: 1 December 2023
Revised: 19 December 2023
Accepted: 19 December 2023
Published: 22 December 2023

concentration slightly increased in both groups post-treatment. Our results confirm that IRT is a non-invasive technique that can be useful when applied to assess how positive and negative events may affect the physiological response in sheep.

Keywords: classical conditioning; emotional state; fear; infrared thermography; non-invasive measure; sheep welfare

1. Introduction

The recognition that animals are sentient beings, capable of feeling emotions, is a driving force to finding non-invasive indicators for recognizing animals' emotional state, which is an important aspect of their welfare [1,2]. Emotions can be defined as a short-lived and intense affective response to an event characterized by universal facial expressions, phylogenetic continuity of expressed behaviors, physiological activation patterns, and unconscious stimuli evaluation mechanisms [3,4]. From an evolutionary standpoint, emotions are processes that might have evolved from basic mechanisms aimed at either directing other physiological programs or solving adaptive problems faced by a species over time [5], with the final goal of giving the animals the ability to avoid harm/punishment or to seek valuable resources/reward [6,7]. The basic emotions are found throughout all mammals and birds [8,9], and may be categorized into two broad groups, based on their valence, as negative or positive emotions. Negative emotions, like fear and anger, hold a pivotal role in ensuring survival by triggering specific responses aimed at avoiding threats or dealing with stressors through the activation of the sympathetic nervous system [10–12]. On the other hand, positive emotions promote social bonding by activating the parasympathetic nervous system [13,14].

Determining the precise feelings of any individual farm animal is challenging due to their inability to verbally express emotions [15]. Additionally, a single emotional expression measurement, whether derived from facial features or physiological functions, is insufficient to accurately show the farm animal's emotional changes [16]. The study of animal emotions is still a challenge as there is no gold standard measure that can be reliably used to infer the affective state of animals, meaning that it is not possible to determine the emotion the animal is feeling by using a single indicator or measure, but a multimodal approach is still preferred [16–18]. The ability to non-invasively measure emotions plays a central role in animal welfare assessment as it provides valuable information regarding the outcomes of interventions aimed at improving the welfare of the animals [19].

It is well-known that emotions are linked to specific physiological changes within the body [20]: for example, plasma glucocorticoid levels can rise in reaction to acute negative stimuli [21] or in anticipation of a positive situation [22]. The measurement of cortisol may yield confusing results; its detection in high levels in farm animals may enhance performance by affecting immune and endocrine pathways [23], and may have positive effects on traits related to robustness and adaptation, such as newborn survival, resistance to bacteria and parasites, and tolerance to heat stress [24]. Moreover, the limitations of cortisol as an invasive approach [25] suggest exploring less invasive alternatives. Furthermore, emotions may influence body temperature, as they are reactions to events, whether internal or external, which hold significant importance for the organism, and this can manifest as emotional fever or stress-induced hyperthermia [26]. Hence, identifying these changes leads to the possibility to infer negative or positive emotions in animals. Travain and Valsecchi [26] in their review reported the latest studies in which infrared thermography (IRT) has been applied to measure the emotional reactions of domestic animals with a focus on stress, fear, and negative situations [26]. Moreover, they reported that the eye is a particularly suitable zone due to its abundant capillary networks, which rapidly respond to variations in blood flow present on the small areas around the rear edge of the eyelid and the *caruncula lacrimalis* [27–30]. In sheep, Cannas and colleagues [19] showed that IRT seems to be a promising technique for inferring stress and fear in sheep. The physiological reaction (change in eye temperature) was measured in ewes exposed to handling and to a human–animal test, and the findings indicated that the stressful situation increased the lacrimal caruncle temperature [19]. The possible explanation for this is that the perception of stressors such as fear or anxiety can induce an increase in heart rate and blood pressure due to increased sympathetic tone [31]. This increases body temperature, referred to as stress hyperthermia [31]. Other studies have related the surface thermal response in companion animals exposed to social isolation [32–34]. However, it has not been possible to establish a clear relationship between this thermal response and its meaning as an emotional state, something that could be complemented with behavioral indicators.

Indeed, considering the behavioral components of the emotion changes in ear posture might be a reliable and non-invasive tool for assessing emotional valence in sheep [20,35–38]. While most research has shown an increase in ocular temperature after a negative stimulus, some studies have shown an immediate and transient reduction [30]. Reefmann and colleagues [35] found that a negative experience (e.g., social isolation) leads to the highest number of ear posture changes and forward and asymmetric ear postures. While Boissy and colleagues [37] found a correlation between ear postures in sheep and distinct emotional responses. The backward posture is associated with fear in unfamiliar and uncontrollable unpleasant situations, the raised ear posture indicates anger in response to controllable unpleasant situations, and the asymmetric ear posture expresses surprise in sudden situations and startles responses. These identified ear postures could serve as specific emotional indicators in sheep. Furthermore, Vogeli and colleagues [39] highlighted that the duration of the ear postures is also instructive, not only the frequency with which they were shown, because different ear postures are likely to occur for a different amount of time each time they are held.

To date, there is no current scientific literature available to explore whether eye temperature undergoes changes in response to positive emotional states in sheep, such as the presence of a food reward. We hypothesize that eye temperature may vary in response to positive events, and ear posture may be correlated with physiological parameters such as cortisol and IRT, providing valuable indicators for investigating the emotional states of the sheep. We hypothesize that eye temperature may vary in response to different valence emotional events, and ear posture may be correlated with physiological measures such as cortisol and IRT, providing valuable indicators for investigating the emotional states of the sheep. The aim of this study was to evaluate, through classical conditioning, how positive or negative events influence the physiological responses (lacrimal caruncle temperature measured with IRT and cortisol levels) and behavioral reaction (i.e., ear position) in sheep.

2. Materials and Methods

2.1. Ethical Statement

Data collection on animals was carried out in accordance with relevant guidelines and regulations, and all experimental protocols were approved by the national ethical commission (Ministero della Salute, authorization n° 457/2016-PR, 919/2017-PR). If any sheep was deemed to be in greater-than-mild stress (assessed live by an independent veterinarian), then it was immediately removed from the study.

2.2. Animals, Housing, and Husbandry

This study was conducted at the experimental farm of Istituto Zooprofilattico Sperimentale dell'Abruzzo e del Molise (Italy) during the month of June in 2019. Fourteen 1-year-old Sarda ewes, not gestating nor lactating and clinically healthy, were selected and housed together in a 45 m^2 resting pen with straw provided for bedding. In the resting pen, ewes had free access to water, and they were fed with hay twice a day (8 a.m. and 6 p.m.). Their diet was supplemented with a commercial concentrate (Mangimi Ariston Srl, Teramo, Italy; 250–300 g/sheep). The resting pen was connected to the experimental pen by a corridor 2 m long and delimited with gates specially placed to direct the path of the animals from one pen to the other (Figure 1).

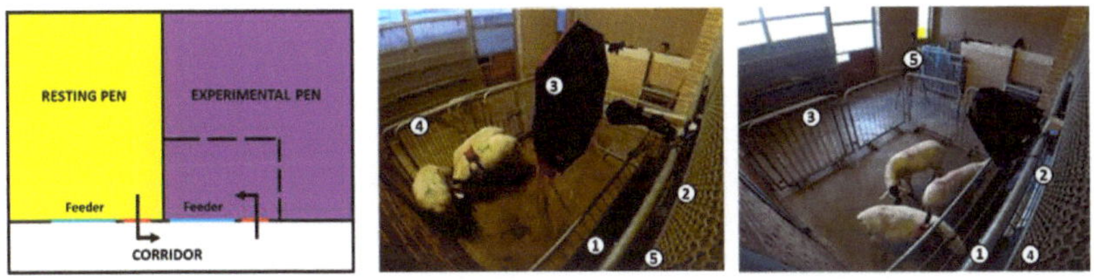

(a) (b) (c)

Figure 1. (**a**) Schematic representation of the resting and experimental pen; (**b**) experimental pen for negative group (1: feeder; 2: hidden car horn and wall covered with black plastic sheets to prevent the operator from being seen by the animals; 3: umbrella; 4: fence; 5: camera); (**c**) experimental pen for positive group (1: feeder; 2: hidden doorbell and wall covered with black plastic sheets to prevent the operator from being seen by the animals; 3: fence; 4 and 5: cameras).

The experimental pen was equipped with a 2 m long feeder, which was fixed in such a way as to remain constantly accessible by both the animals inside the box and the operators from outside. The door was covered on its outer side with black plastic sheets to prevent the operators from being seen by the sheep.

2.3. *Animal and Study Design*

Ewes were randomly assigned to two treatment groups: positive (n = 7) and negative (n = 7). Each group was trained through classical conditioning to associate a neutral auditory stimulus (doorbell and car horn for the positive and negative group, respectively) to an oncoming event: for the positive group, the presence of a food reward (maize grains), while for the negative one, the automatic opening of an umbrella covered with colored wires to make it as scary as possible for the animals.

Training Strategy Workflow

The training of ewes, lasting three weeks, comprised two different aspects: habituation to the experimental set up and classical conditioning of a neutral sound. To prevent animals from hearing the neutral sound when they were not in the experimental pen, the rest of the flock, not involved in the training, was kept outside in a pasture far from the experimental pen. Sheep were gradually habituated to be split from their flock in small groups of three and moved to the experimental pen (2×2 m^2), where they were socially and visually isolated from the rest of the flock. To avoid social isolation stress, within each group, two "accompanying sheep" were selected. The accompanying sheep (always the same two individuals for each group) performed the training together with each individual experimental sheep, but they were not included in the data collection. The training of ewes took place daily in the afternoon at the same time between 14 and 18 and it was run by two operators. During the habituation phase (5 consecutive days), ewes entered the experimental pen and remained there for 30 min, receiving a standard food reward (i.e., 100 g of maize grains) in a bucket. The conditioning phase started after the habituation, and it was divided into two different steps. In the first step (8 consecutive days), sheep learned to associate a neutral sound lasting five seconds to an oncoming event. Through classic Pavlovian conditioning, animals of the positive group associated the sound, performed by an operator located in the corridor, with the release of maize grains in the feeder, while sheep of the negative group associated the sound with a sudden opening of the umbrella by an operator located in the corridor. Each training day, sheep underwent 10 conditioning trials. As a second step (8 consecutive days), the time between the sound and the oncoming event was gradually increased to 10 s. Ten seconds was selected as a good time interval to evaluate anticipatory behavior in sheep. All sheep were conditioned at the end of 3 weeks by ensuring that after the sound, they moved toward the feeder (positive group) or ran to a corner (negative group).

2.4. Data Collection and Analysis

The following table reports the different phases for each treatment group (Table 1).

Table 1. Summary representation of the test phases for both of the treatment groups.

Positive Group	Negative Group
Baseline (10 s)	Baseline (10 s)
Sound (5 s)	Sound (5 s)
Anticipation (10 s)	Anticipation (10 s)
Feed distribution (5 s)	Umbrella opening (5 s)
Consumption phase (10 s)	Sheep reaction (10 s)

2.4.1. Lacrimal Caruncle Infrared Thermography

Data collection took place on the last day of the conditioning phase. The temperature (°C) of the lacrimal caruncle of each sheep was measured in the resting and in the experimental pen after completing the 10 training trials of the day after the end of the last trial consumption phase (positive group) and sheep reaction (negative group) by using an infrared camera (NEC Avio TVS500; Nippon Avionics Co., Ltd., Tokyo, Japan) with a standard optic system (accuracy $\pm$ 2 °C and resolution 320 $\times$ 240 px). Room temperature and humidity in both resting and experimental pens were automatically controlled and stable (minimum = 19.30 °C, maximum = 20.43 °C; and mean = 19.92 °C). To optimize the accuracy of the thermographic image, the same image of a Lambert surface was taken to define the radiance emission and to nullify the effect of surface reflections on the tested animals before every work session [40]. Lacrimal caruncle was selected as the target area because it has a temperature that is unaffected by the presence of hair [41,42]. Pictures (3 for each eye, left and right, taken with approximately 10 s between them) were collected from the same location, on a side with an approximate angle of 90° and from a distance range of 0.5 m from the animal. Grayess IRT Analyzer 6.0 software (Bradenton, FL, USA) was used to determine the temperature of the lacrimal caruncle, and the maximum temperature (°C) inside a circular region traced around the area was measured [43]. The emissivity values employed in the analyses were those measured for high emissivity objects (0.97 and 0.98). An example of a picture is shown in Figure 3a.

2.4.2. Ear Posture

Videos were then analyzed with a focal animal continuous recording method using the software Solomon Coder beta 17.03.22. The durations of different ear postures in the frontal plane (pointing forward, backward, or asymmetrical), as described by Boissy and colleagues [37] (Figure 2), were evaluated during the baseline (10 s) and anticipation phase (10 s) of the oncoming event by an independent observer not aware of the aim of the study. If the ears' posture is oriented forward, it means that the top of the ear is in front of the frontal plane, and, on the contrary, if pointing backward, the top of the ear is behind the frontal plane. Lastly, if the two ears differ in their position with respect to the frontal plane, they are evaluated as asymmetrical.

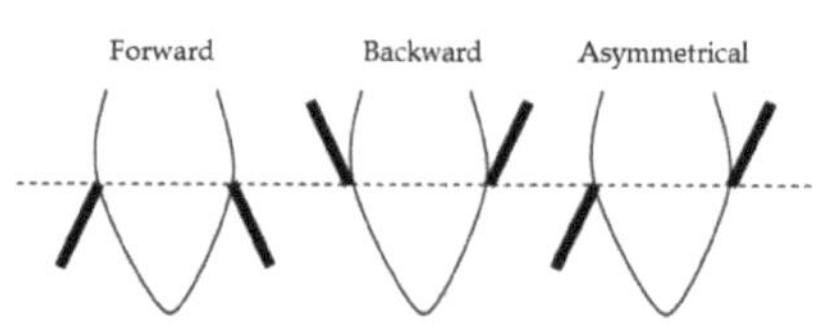

Figure 2. Position of the ears in relation to the frontal plane of the head (modified from Boissy et al., 2011 [37]).

2.4.3. Cortisol

For the assessment of cortisol concentrations, saliva specimens were obtained from each sheep with a commitment to ensuring minimal distress during the procedure (sheep were previously habituated to this manipulation). SalivaBio Children's Swabs (Salimetrics®, Carlsbad, CA, USA) were utilized to collect samples in both the resting pen and post-treatment, immediately following the acquisition of thermographic images. The swab, introduced into the sheep's oral cavity for approximately 1 min, was administered with gentle restraint, allowing the animal to remain in proximity to the flock without isolation. Subsequent to sampling, the swab was deposited in the designated device tube, sealed with a plastic stopper to avert evaporation, and promptly placed in an ice container. Within an hour, the samples were frozen at −20 °C, maintaining this temperature until analysis. During analysis, the samples were thawed at room temperature, followed by centrifugation (3500 rpm for 15 min at 4 °C) in accordance with the salivary sample protocol provided by the kit manufacturer. A commercially available multispecies cortisol enzyme-linked immunosorbent assay (ELISA) kit (Enzo Life Sciences, Farmingdale, NY, USA), validated through established protocols [44], was employed for analysis. Duplicate aliquots (100 µL) of the samples were placed in wells, and absorbance was measured at a wavelength of 405 nm using a microplate reader (Multiskan EX, LabSystem, Thermo Fisher Scientific, Milan, Italy). To assess the accuracy of our samples, a recovery test was implemented, yielding a mean recovery of 109.1% ± 8.4. The intra- and inter-assay coefficients of variation averaged 3.9 and 7.8%, respectively. The assay sensitivity was determined to be 56.72 pg/mL (range 156–10,000 pg/mL). Importantly, the laboratory researcher conducting the analysis remained blinded to the study hypotheses and conditions.

2.5. Statistical Analysis

Data were entered into Microsoft Excel (Microsoft Corporation, 2010, Washington, DC, USA) before being analyzed with SPSS statistical package (SPSS Statistic 28, IBM, Armonk, NY, USA). The descriptive statistics (mean and standard deviations) were calculated for each of the studied variables. The data were tested for normality and homogeneity of variance using Kolmogorov–Smirnov and Levene's tests, respectively. For thermographic data, t-tests were used to compare data at rest and post-treatment in both groups. Cortisol levels and ear posture data were not normally distributed; therefore, a non-parametric Mann–Whitney U test was used to identify differences between treatment groups and the Wilcoxon test was used to identify possible differences in cortisol levels at rest and post-treatment in both groups. Differences were considered statistically significant if $p \leq 0.05$.

3. Results

The lacrimal caruncle's temperature at rest was 38.31 ± 0.41 °C in the negative group and 38.35 ± 0.45 °C in the positive group with no significant differences (independent t test; p = 0.802) between the groups. After treatment, the sheep in the negative group showed a significant increase (paired samples t-test; p = 0.017) in lacrimal caruncle temperature (38.77 ± 0.72 °C), while in the positive group eye temperature remained unchanged (38.37 ± 0.33 °C), as shown in Figure 3b.

In the baseline phase, there are no significant differences in ear positions between both the positive and negative treatment groups (Figure 4a), while during the anticipation phase, there were statistically significant differences in ear posture between the two treatment groups (Figure 4b). Ewes in the positive group kept their ears forward longer (Mann–Whitney U test; $p < 0.014$) and backward for less seconds (Mann–Whitney U test; $p < 0.008$).

The cortisol concentration at rest was 0.43 ± 0.07 pg/mL in the negative group and 0.41 ± 0.12 pg/mL in the positive group with no significant differences (Mann–Whitney U test; $p > 0.05$) between the groups. After treatment, the sheep in both groups showed an increase in cortisol levels, although not significant (Wilcoxon test; $p > 0.05$), and the negative group's level was 0.67 ± 0.39 pg/mL, while for the positive group, it was 0.59 ± 0.39 pg/mL.

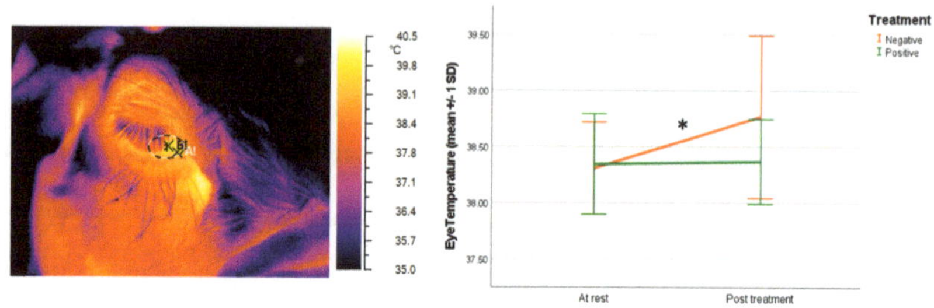

(a) (b)

Figure 3. (**a**) Thermographic infrared image of the sheep's eye with the lacrimal caruncle as the target area; (**b**) graph showing mean (±1 SD) lacrimal caruncle temperature (°C) at rest and after the treatment for both negative and positive groups (paired samples *t*-test; * $p < 0.05$).

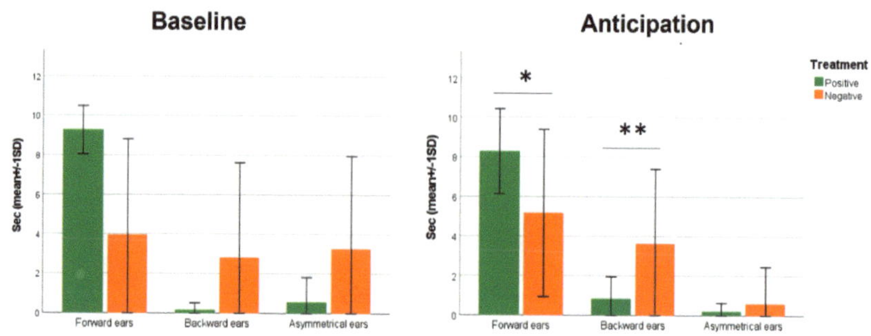

(a) (b)

Figure 4. Graphs reporting the mean (±SD) duration in seconds of different ear positions of the negative of positive treatment groups during (**a**) baseline and (**b**) anticipation phases (Mann–Whitney U test; * $p < 0.05$; ** $p < 0.01$).

4. Discussion

The study aimed at evaluating, through a classical conditioning, how positive or negative events influence the physiological responses and behavioral reaction in sheep. This was achieved by evaluating ear position and measuring eye temperature using infrared thermography and salivary cortisol levels in two groups of sheep conditioned to two different valence stimuli. The findings indicated that, post-treatment, there was a significant increase in the temperature of the lacrimal caruncle only in the negative treatment group, as opposed to prior to its commencement with the sheep in the resting box. Previous studies have provided evidence indicating that the temperature of the lacrimal caruncle in sheep is elevated during short stressful events such as restraint [19] and shearing [45]. Furthermore, several studies have also demonstrated that castration [46], transport [47], and forced lateralization test in cattle [48], trimming [49] and noseband tightening [50] in horses, and veterinary visits in dogs [33] resulted in elevations in ocular temperature. The higher temperature of the eye caruncle herein found could be related to ocular vasodilatation causing the greater heat dissipation identified here by the thermographic measurements. Vasodilation can be due to fear-induced activation of the parasympathetic nervous system, in this case by the facial nerve (VII pair) through the pterygopalatine ganglion and the subsequent postganglionic fibers [51]. The eye and its surrounding skin tissue provide an image that may reflect the sympathetic/parasympathetic balance of the animal [52]. Furthermore, when a "fight/flight" reaction is triggered [42], there is an increase in core body temperature [53], primarily observed as a transient acute response, lasting less than one hour [31]. Given that cortisol induces various thermogenic processes in tissue metabolism, such as gluconeogenesis, glycolysis, proteolysis, and lipolysis [54,55], it is possible to suggest that alterations in eye temperature could be influenced, to some extent, by the activation of the hypothalamic–pituitary–adrenal (HPA) axis and the visceral nervous system in response to stimulation [55]. On the contrary, other studies showed that the application of a painful stimulus did not result in any observable alterations in the ocular temperature of sheep [56] or in some instances observed a rapid decrease in eye temperature following exposure to a stressor, followed by a subsequent increase back to baseline values [28,30,57]. This decrease is believed to be followed by thermogenic reactions in tissue metabolism, which are induced by glucocorticoids and may be responsible for the subsequent increase in temperature [54]. The reasons behind the fluctuation of ocular temperature in response to stress remain uncertain. It is possible that variations in data collection methods, such as the timing of sampling, could contribute to the observed discrepancies among studies [45]. Additionally, it would be interesting to explore whether the level of the stimulus could be a factor in the absence of an emotional response.

In terms of behavioral reaction, our findings showed a significant increase in the time spent with ears forward in the positive group, awaiting a food reward, while sheep in the negative group spent more time with their ears backwards waiting for the opening of the umbrella. The presence of the "accompanying sheep" reduced the potential for stress induced by separation from the flock, and the gradual habituation to the experimental set up allowed for the assessment of behavioral reactions without the influence of environmental novelty. Several studies in different animal species [37,58–61] have investigated the possible role of ear postures in social communication and the expression of the internal state [37,58–61]. For example, in dairy cows, a relaxed ear posture is indicative of a positive, low-arousal emotional state [62], and, in horses and pigs, a negative emotional experience might be associated with backwards ears [63,64]. Ruminants possess well-developed muscles around their ears, allowing them a high degree of independent movement in both the forward and backward directions [36]; however, ear postures are highly species-dependent [62], and, for instance, both backwards and forwards ears might be associated either with positive or negative situations [35,36,62,65]. In sheep, Reefmann and colleagues [35,36] found that regarding the number of ear posture changes, forward and asymmetric ear postures were highest during negative experiences of social isolation, while a "passive" ear posture were observed during positive ones. Boissy et al. (2011) [37] found similar results; in their study, sheep were exposed to situations of varying degrees of suddenness and familiarity, negative contrast, and controllability [37]. Asymmetrical ear posture was mainly displayed in response to sudden situations in relation with a startle response (surprise). During the anticipation phase, this is reduced by the effects of the conditioning. In our study, sheep in the positive group kept their ears forward more often, since the experience was of a pleasant nature. However, these results are in contrast with the findings of the mentioned studies [35–37]. Reefmann and colleagues [35,36] focused their conclusions mainly on the frequency of ear posture changes, rather than on the duration, in positive–negative contrast paradigms [35,36]. Positive–negative contrasts pertain to the phenomenon observed when an animal undergoes conditioning to anticipate a specific reward or event, and, subsequently, that reward or event is altered to be of either diminished perceived value or heightened perceived value [66]. Moreover, there is a strong interrelation of attention and emotion [67], suggesting that the scrutiny of ear postures remains unaffected by attention, emphasizing that attention forms an integral component of an emotional response. The level of attention is increased during the anticipation of feeding as a positive event [68], enabling prompt detection of the occurrence or absence of a reward. The certainty of anticipation of positive events might be perceived as a controllable situation, as displayed by the raised ear posture reported by Boissy and colleagues (2011) [37]. Moreover, the increased level of activity when anticipating the upcoming event was reported as a result of the learnt association between the occurrence of a neutral sound and the oncoming food reward [69]. This could explain the increased time spent in the forward position during the anticipation phase in the positive group, whereas the successful acquisition of these rewards induces positive affective states [70]; however, further studies are needed to investigate and clarify how the positive and negative value of an experience affects active behavior [69].

In our study, cortisol levels increased, although not significantly, in both positive and negative groups post-treatment. The anticipation of positive events could potentially induce transitioning stress and frustration if the reward is bestowed belatedly [36]. We can also hypothesize that the certain anticipation of a positive experience, such as waiting for a food reward, can also make the context stimulating where the sheep actively direct their attention to the surroundings, as observed for dairy cows [61].

Combining lacrimal caruncle temperature, cortisol levels, and ear position in sheep with positive and negative expectations can be considered an innovative approach for studying the emotional reactions of sheep. Additional research is required to further examine the relationship between these parameters in order to gain more comprehensive insights into the optimal timing for sample collection. It is important to consider that the present study included a small sample size and was run in an experimental situation; therefore, some attention should be paid when we want to generalize the results to on-farm situations.

5. Conclusions

Our results confirmed that infrared thermography could be applied as a non-invasive tool to identify negative and positive events that may elicit emotional states affecting physiological response in sheep since the thermal, physiological, and behavioral facial response were influenced by negative events in sheep. Therefore, such evidence shows that the use of IRT might be able to assess emotional response in this species. Further research is needed to investigate, in more depth, the correlation between eye temperature, cortisol-level variations, and behavioral reactions to gather information about the best timing for collecting samples, thus improving the reliability and applicability of this technology in on-farm situations.

Author Contributions: Conceptualization, M.C. (Matteo Chincarini), S.M.M., M.M., B.C., G.V. and E.D.C.; methodology, M.C. (Matteo Chincarini), S.M.M., M.M., B.C., G.V. and E.D.C.; formal analysis, E.D.C.; investigation, M.C. (Matteo Chincarini), E.C., E.R., M.M., B.C., G.V. and E.D.C.; resources, B.C.; data curation, M.C. (Matteo Chincarini), M.M. and E.D.C.; writing—original draft preparation, M.C. (Marta Comin) and E.A.; writing—review and editing, M.C. (Marta Comin), E.A., S.M.M., M.C. (Matteo Chincarini), B.C., E.R. and E.D.C.; supervision, M.M., G.V., B.C. and E.D.C.; funding acquisition, B.C. All authors have read and agreed to the published version of the manuscript.

Funding: This research was funded by MIUR-PRIN2015 (grant number 2015Y5W9YP).

Institutional Review Board Statement: The animal study protocol was approved by the national ethical commission for studies involving animals (Ministero della Salute, authorization n° 457/2016-PR, 919/2017-PR).

Informed Consent Statement: Not applicable.

Data Availability Statement: The data presented in this study are available upon request from the corresponding author E.D.C. The data are not publicly available due to privacy restrictions.

Acknowledgments: The authors would like to thank all the staff of the experimental farm of Istituto Zooprofilattico Sperimentale dell'Abruzzo e del Molise that helped with the management of the animals during the experiments.

Conflicts of Interest: The authors declare no conflict of interest.

References

1. Chincarini, M.; Dalla Costa, E.; Qiu, L.; Spinelli, L.; Cannas, S.; Palestrini, C.; Canali, E.; Minero, M.; Cozzi, B.; Ferri, N.; et al. Reliability of FNIRS for Noninvasive Monitoring of Brain Function and Emotion in Sheep. *Sci. Rep.* **2020**, *10*, 14726. [CrossRef] [PubMed]
2. Hemsworth, P.; Mellor, D.; Cronin, G.; Tilbrook, A. Scientific Assessment of Animal Welfare. *N. Z. Vet. J.* **2015**, *63*, 24–30. [CrossRef] [PubMed]

3. Ekman, P. *Emotion in the Human Face*; Cambridge University Press: New York, NY, USA, 1982.
4. Fox, E. *Emotion Science Cognitive and Neuroscientific Approaches to Understanding Human Emotions*; Palgrave Macmillan: London, UK, 2008; ISBN 978-0-230-00517-4.
5. Boissy, A.; Manteuffel, G.; Jensen, M.B.; Moe, R.O.; Spruijt, B.; Keeling, L.J.; Winckler, C.; Forkman, B.; Dimitrov, I.; Langbein, J.; et al. Assessment of Positive Emotions in Animals to Improve Their Welfare. *Physiol. Behav.* **2007**, *92*, 375–397. [CrossRef] [PubMed]
6. Panksepp, J. Evolution Constructed the Potential for Subjective Experience within the Neurodynamics of the Mammalian Brain. In *The Nature of Emotion: Fundamental Questions*; Eckman, P., Davidson, R.J., Eds.; Oxford University Press: Oxford, UK, 1994; pp. 336–399.
7. Rolls, E. Precis of the Brain and Emotion. *Behav. Brain Sci.* **2000**, *23*, 177–191. [CrossRef] [PubMed]
8. Jarvis, E.D.; Güntürkün, O.; Bruce, L.; Csillag, A.; Karten, H.; Kuenzel, W.; Medina, L.; Paxinos, G.; Perkel, D.J.; Shimizu, T.; et al. Avian Brains and a New Understanding of Vertebrate Brain Evolution. *Nat. Rev. Neurosci.* **2005**, *6*, 151–159. [CrossRef] [PubMed]
9. Reiner, A.; Perkel, D.J.; Bruce, L.L.; Butler, A.B.; Csillag, A.; Kuenzel, W.; Medina, L.; Paxinos, G.; Shimizu, T.; Striedter, G.; et al. Revised Nomenclature for Avian Telencephalon and Some Related Brainstem Nuclei. *J. Comp. Neurol.* **2004**, *473*, 377–414. [CrossRef] [PubMed]
10. Tooby, J.; Cosmides, L. The Past Explains the Present. *Ethol. Sociobiol.* **1990**, *11*, 375–424. [CrossRef]
11. Levenson, R.W. Human Emotion: A Functionalist View. In *The Nature of Emotion: Fundamental Questions*; Ekman, P., Davidson, R.J., Eds.; Oxford University Press: New York, NY, USA, 1994; pp. 123–126.
12. Lazarus, R.S. Cognition and Motivation in Emotion. *Am. Psychol.* **1991**, *46*, 352–367. [CrossRef]
13. Fredrickson, B.L. Cultivated Emotions: Parental Socialization of Positive Emotions and Self-Conscious Emotions. *Psychol. Inq.* **1998**, *9*, 279–281. [CrossRef]
14. Fredrickson, B.L. The Role of Positive Emotions in Positive Psychology: The Broaden-and-Build Theory of Positive Emotions. *Am. Psychol.* **2001**, *56*, 218–226. [CrossRef]
15. Broom, D. Welfare, Stress, and Evolution of Feelings. *Adv. Study Behav.* **1998**, *27*, 371–403.
16. Neethirajan, S.; Reimert, I.; Kemp, B. Measuring Farm Animal Emotions—Sensor-Based Approaches. *Sensors* **2021**, *21*, 553. [CrossRef] [PubMed]
17. Lezama-García, K.; Orihuela, A.; Olmos-Hernández, A.; Reyes-Long, S.; Mota-Rojas, D. Facial Expressions and Emotions in Domestic Animals. *CABI Rev.* **2019**, *14*, 1–12. [CrossRef]
18. Whittaker, A.L.; Marsh, L.E. The Role of Behavioural Assessment in Determining "positive" Affective States in Animals. *CABI Rev.* **2019**, 1–13. [CrossRef]
19. Cannas, S.; Palestrini, C.; Canali, E.; Cozzi, B.; Ferri, N.; Heinzl, E.; Minero, M.; Chincarini, M.; Vignola, G.; Dalla Costa, E.; et al. Thermography as a Non-Invasive Measure of Stress and Fear of Humans in Sheep. *Animals* **2018**, *8*, 146. [CrossRef]
20. Dantzer, R. *Les Émotions*; Presses Universitaires de France: Paris, France, 1988.
21. Mason, J.W. A Re-Evaluation of the Concept of "Non-Specificity" in Stress Theory. In *Principles, Practices, and Positions in Neuropsychiatric Research*; Elsevier: Amsterdam, The Netherlands, 1972; pp. 323–333.
22. Colborn, D.R.; Thompson, D.L.; Roth, T.L.; Capehart, J.S.; White, K.L. Responses of Cortisol and Prolactin to Sexual Excitement and Stress in Stallions and Geldings. *J. Anim. Sci.* **1991**, *69*, 2556. [CrossRef]
23. Borgs, P.; Mallard, B.A. Immune-Endocrine Interactions in Agricultural Species: Chromium and Its Effect on Health and Performance. *Domest. Anim. Endocrinol.* **1998**, *15*, 431–438. [CrossRef]
24. Mormède, P.; Foury, A.; Terenina, E.; Knap, P.W. Breeding for Robustness: The Role of Cortisol. *Animal* **2011**, *5*, 651–657. [CrossRef]
25. Kammersgaard, T.S.; Malmkvist, J.; Pedersen, L.J. Infrared Thermography—A Non-Invasive Tool to Evaluate Thermal Status of Neonatal Pigs Based on Surface Temperature. *Animal* **2013**, *7*, 2026–2034. [CrossRef]
26. Travain, T.; Valsecchi, P. Infrared Thermography in the Study of Animals' Emotional Responses: A Critical Review. *Animals* **2021**, *11*, 2510. [CrossRef]
27. Stewart, M.; Schaefer, A.L.; Haley, D.B.; Colyn, J.; Cook, N.J.; Stafford, K.J.; Webster, J.R. Infrared Thermography as a Non-Invasive Method for Detecting Fear-Related Responses of Cattle to Handling Procedures. *Anim. Welf.* **2008**, *17*, 387–393. [CrossRef]
28. Stewart, M.; Webster, J.R.; Verkerk, G.A.; Schaefer, A.L.; Colyn, J.J.; Stafford, K.J. Non-Invasive Measurement of Stress in Dairy Cows Using Infrared Thermography. *Physiol. Behav.* **2007**, *92*, 520–525. [CrossRef] [PubMed]
29. Stewart, M.; Wilson, M.T.; Schaefer, A.L.; Huddart, F.; Sutherland, M.A. The Use of Infrared Thermography and Accelerometers for Remote Monitoring of Dairy Cow Health and Welfare. *J. Dairy Sci.* **2017**, *100*, 3893–3901. [CrossRef] [PubMed]
30. Stewart, M.; Stafford, K.J.J.; Dowling, S.K.K.; Schaefer, A.L.L.; Webster, J.R.R. Eye Temperature and Heart Rate Variability of Calves Disbudded with or without Local Anaesthetic. *Physiol. Behav.* **2008**, *93*, 789–797. [CrossRef] [PubMed]
31. Adriaan Bouwknecht, J.; Olivier, B.; Paylor, R.E. The Stress-Induced Hyperthermia Paradigm as a Physiological Animal Model for Anxiety: A Review of Pharmacological and Genetic Studies in the Mouse. *Neurosci. Biobehav. Rev.* **2007**, *31*, 41–59. [CrossRef] [PubMed]
32. Hernández-Avalos, I.; Mota-Rojas, D.; Mendoza-Flores, J.E.; Casas-Alvarado, A.; Flores-Padilla, K.; Miranda-Cortes, A.E.; Torres-Bernal, F.; Gómez-Prado, J.; Mora-Medina, P. Nociceptive Pain and Anxiety in Equines: Physiological and Behavioral Alterations. *Vet. World* **2021**, *14*, 2984–2995. [CrossRef] [PubMed]

33. Travain, T.; Colombo, E.S.; Heinzl, E.; Bellucci, D.; Prato Previde, E.; Valsecchi, P. Hot Dogs: Thermography in the Assessment of Stress in Dogs (*Canis familiaris*)—A Pilot Study. *J. Vet. Behav.* **2015**, *10*, 17–23. [CrossRef]
34. Hall, C.; Kay, R.; Yarnell, K. Assessing Ridden Horse Behavior: Professional Judgment and Physiological Measures. *J. Vet. Behav.* **2014**, *9*, 22–29. [CrossRef]
35. Reefmann, N.; Wechsler, B.; Gygax, L. Behavioural and Physiological Assessment of Positive and Negative Emotion in Sheep. *Anim. Behav.* **2009**, *78*, 651–659. [CrossRef]
36. Reefmann, N.; Bütikofer Kaszàs, F.; Wechsler, B.; Gygax, L. Ear and Tail Postures as Indicators of Emotional Valence in Sheep. *Appl. Anim. Behav. Sci.* **2009**, *118*, 199–207. [CrossRef]
37. Boissy, A.; Aubert, A.; Désiré, L.; Greiveldinger, L.; Delval, E.; Veissier, I.; Desire, L. Cognitive Sciences to Relate Ear Postures to Emotions in Sheep. *Anim. Welf.* **2011**, *20*, 47–56. [CrossRef]
38. Chapagain, D.; Uvnäs-Moberg, K.; Lidfors, L.M. Investigating the Motivation to Play in Lambs. *Appl. Anim. Behav. Sci.* **2014**, *160*, 64–74. [CrossRef]
39. Vögeli, S.; Wechsler, B.; Gygax, L. Welfare by the Ear: Comparing Relative Durations and Frequencies of Ear Postures by Using an Automated Tracking System in Sheep. *Anim. Welf.* **2014**, *23*, 267–274. [CrossRef]
40. Mallick, S.P.; Zickler, T.E.; Kriegman, D.J.; Belhumeur, P.N. Beyond Lambert: Reconstructing Specular Surfaces Using Color. *Proc. IEEE Comput. Soc. Conf. Comput. Vis. Pattern Recognit.* **2005**, *2*, 619–626. [CrossRef]
41. Bartolomé, E.; Sánchez, M.J.J.; Molina, A.; Schaefer, A.L.L.; Cervantes, I.; Valera, M. Using Eye Temperature and Heart Rate for Stress Assessment in Young Horses Competing in Jumping Competitions and Its Possible Influence on Sport Performance. *Animal* **2013**, *7*, 2044–2053. [CrossRef] [PubMed]
42. Dai, F.; Cogi, N.H.; Heinzl, E.U.L.; Dalla Costa, E.; Canali, E.; Minero, M. Validation of a Fear Test in Sport Horses Using Infrared Thermography. *J. Vet. Behav. Clin. Appl. Res.* **2015**, *10*, 128–136. [CrossRef]
43. Grayess. *IRT Analyser Users' Manual*; GRAYESS, Inc.: Bradenton, FL, USA, 2007.
44. Bonelli, F.; Rota, A.; Aurich, C.; Ille, N.; Camillo, F.; Panzani, D.; Sgorbini, M. Determination of Salivary Cortisol in Donkey Stallions. *J. Equine Vet. Sci.* **2019**, *77*, 68–71. [CrossRef]
45. Arfuso, F.; Acri, G.; Piccione, G.; Sansotta, C.; Fazio, F.; Giudice, E.; Giannetto, C. Eye Surface Infrared Thermography Usefulness as a Noninvasive Method of Measuring Stress Response in Sheep during Shearing: Correlations with Serum Cortisol and Rectal Temperature Values. *Physiol. Behav.* **2022**, *250*, 113781. [CrossRef]
46. Stewart, M.; Verkerk, G.A.; Stafford, K.J.; Schaefer, A.L.; Webster, J.R. Noninvasive Assessment of Autonomic Activity for Evaluation of Pain in Calves, Using Surgical Castration as a Model. *J. Dairy Sci.* **2010**, *93*, 3602–3609. [CrossRef]
47. Lei, M.C.; Félix, L.; Cardoso, R.; Monteiro, S.M.; Silva, S.; Venâncio, C. Non-Invasive Biomarkers in Saliva and Eye Infrared Thermography to Assess the Stress Response of Calves during Transport. *Animals* **2023**, *13*, 2311. [CrossRef]
48. Uddin, J.; McNeill, D.M.; Phillips, C.J.C. Measuring Emotions in Dairy Cows: Relationships between Infrared Temperature of Key Body Parts, Lateralised Behaviour and Milk Production. *Appl. Anim. Behav. Sci.* **2023**, *269*, 106115. [CrossRef]
49. Yarnell, K.; Hall, C.; Billett, E. An Assessment of the Aversive Nature of an Animal Management Procedure (Clipping) Using Behavioral and Physiological Measures. *Physiol. Behav.* **2013**, *118*, 32–39. [CrossRef] [PubMed]
50. Fenner, K.; Yoon, S.; White, P.; Starling, M.; McGreevy, P. The Effect of Noseband Tightening on Horses' Behavior, Eye Temperature, and Cardiac Responses. *PLoS ONE* **2016**, *11*, e0154179. [CrossRef] [PubMed]
51. Cozzi, B.; Granato, A.; Merighi, A. *Neuroanatomia Dell'uomo*, 2nd ed.; Antonio Delfino Editore srl: Roma, Italy, 2017.
52. Sutherland, M.A.; Worth, G.M.; Dowling, S.K.; Lowe, G.L.; Cave, V.M.; Stewart, M. Evaluation of Infrared Thermography as a Non-Invasive Method of Measuring the Autonomic Nervous Response in Sheep. *PLoS ONE* **2020**, *15*, e0233558. [CrossRef] [PubMed]
53. Oka, T.; Oka, K.; Hori, T. Mechanisms and Mediators of Psychological Stress-Induced Rise in Core Temperature. *Psychosom. Med.* **2001**, *63*, 476–486. [CrossRef] [PubMed]
54. Sapolsky, R.M.; Romero, L.M.; Munck, A.U. How Do Glucocorticoids Influence Stress Responses? Integrating Permissive, Suppressive, Stimulatory, and Preparative Actions. *Endocr. Rev.* **2000**, *21*, 55–89. [CrossRef] [PubMed]
55. Valera, M.; Bartolomé, E.; Sánchez, M.J.; Molina, A.; Cook, N.; Schaefer, A. Changes in Eye Temperature and Stress Assessment in Horses During Show Jumping Competitions. *J. Equine Vet. Sci.* **2012**, *32*, 827–830. [CrossRef]
56. Stubsjøen, S.M.; Flø, A.S.; Moe, R.O.; Janczak, A.M.; Skjerve, E.; Valle, P.S.; Zanella, A.J. Exploring Non-Invasive Methods to Assess Pain in Sheep. *Physiol. Behav.* **2009**, *98*, 640–648. [CrossRef]
57. Church, J.S.; Hegadoren, P.R.; Paetkau, M.J.; Miller, C.C.; Regev-Shoshani, G.; Schaefer, A.L.; Schwartzkopf-Genswein, K.S. Influence of Environmental Factors on Infrared Eye Temperature Measurements in Cattle. *Res. Vet. Sci.* **2014**, *96*, 220–226. [CrossRef]
58. Schmied, C.; Waiblinger, S.; Scharl, T.; Leisch, F.; Boivin, X. Stroking of Different Body Regions by a Human: Effects on Behaviour and Heart Rate of Dairy Cows. *Appl. Anim. Behav. Sci.* **2008**, *109*, 25–38. [CrossRef]
59. Wathan, J.; McComb, K. The Eyes and Ears Are Visual Indicators of Attention in Domestic Horses. *Curr. Biol.* **2014**, *24*, R677–R679. [CrossRef] [PubMed]
60. Bellegarde, L.G.A.; Haskell, M.J.; Duvaux-Ponter, C.; Weiss, A.; Boissy, A.; Erhard, H.W. Face-Based Perception of Emotions in Dairy Goats. *Appl. Anim. Behav. Sci.* **2017**, *193*, 51–59. [CrossRef]

61. Battini, M.; Agostini, A.; Mattiello, S. Understanding Cows' Emotions on Farm: Are Eye White and Ear Posture Reliable Indicators? *Animals* **2019**, *9*, 477. [CrossRef] [PubMed]
62. Proctor, H.S.; Carder, G. Can Ear Postures Reliably Measure the Positive Emotional State of Cows? *Appl. Anim. Behav. Sci.* **2014**, *161*, 20–27. [CrossRef]
63. Reimert, I.; Bolhuis, J.E.; Kemp, B.; Rodenburg, T.B. Indicators of Positive and Negative Emotions and Emotional Contagion in Pigs. *Physiol. Behav.* **2013**, *109*, 42–50. [CrossRef]
64. Dalla Costa, E.; Pascuzzo, R.; Leach, M.C.; Dai, F.; Lebelt, D.; Vantini, S.; Minero, M. Can Grimace Scales Estimate the Pain Status in Horses and Mice? A Statistical Approach to Identify a Classifier. *PLoS ONE* **2018**, *13*, e0200339. [CrossRef]
65. Edgar, J.L.; Nicol, C.J.; Pugh, C.A.; Paul, E.S. Surface Temperature Changes in Response to Handling in Domestic Chickens. *Physiol. Behav.* **2013**, *119*, 195–200. [CrossRef]
66. Flaherty, C.F.; Rowan, G.A. Successive, Simultaneous, and Anticipatory Contrast in the Consumption of Saccharin Solutions. *J. Exp. Psychol. Anim. Behav. Process.* **1986**, *12*, 381–393. [CrossRef]
67. Carretié, L.; Mercado, F.; Tapia, M.; Hinojosa, J.A. Emotion, Attention, and the 'Negativity Bias', Studied through Event-Related Potentials. *Int. J. Psychophysiol.* **2001**, *41*, 75–85. [CrossRef]
68. Melges, F.T.; Poppen, R.L. Expectation of Rewards and Emotional Behavior in Monkeys. *J. Psychiatr. Res.* **1976**, *13*, 11–21. [CrossRef]
69. Chincarini, M.; Qiu, L.; Spinelli, L.; Torricelli, A.; Minero, M.; Dalla Costa, E.; Mariscoli, M.; Ferri, N.; Giammarco, M.; Vignola, G.; et al. Evaluation of Sheep Anticipatory Response to a Food Reward by Means of Functional Near-Infrared Spectroscopy. *Animals* **2018**, *9*, 11. [CrossRef] [PubMed]
70. Burman, O.; McGowan, R.; Mendl, M.; Norling, Y.; Paul, E.; Rehn, T.; Keeling, L. Using Judgement Bias to Measure Positive Affective State in Dogs. *Appl. Anim. Behav. Sci.* **2011**, *132*, 160–168. [CrossRef]

animals

MDPI

Article

Is the Weight of the Newborn Puppy Related to Its Thermal Balance?

Karina Lezama-García [1], Julio Martínez-Burnes [2], Míriam Marcet-Rius [3], Angelo Gazzano [4], Adriana Olmos-Hernández [5], Patricia Mora-Medina [6], Adriana Domínguez-Oliva [7], Alfredo M. F. Pereira [8], Ismael Hernández-Ávalos [6], Uri Baqueiro-Espinosa [9], Ana de Mira Geraldo [8], Alejandro Casas-Alvarado [7] and Daniel Mota-Rojas [7,*]

1 PhD Program in Biological and Health Sciences, [Doctorado en Ciencias Biológicas y de la Salud], Universidad Autónoma Metropolitana, Mexico City 04960, Mexico
2 Animal Health Group, Facultad de Medicina Veterinaria y Zootecnia, Universidad Autónoma de Tamaulipas, Victoria City 87000, Mexico
3 Animal Behaviour and Welfare Department, IRSEA (Research Institute in Semiochemistry and Applied Ethology), Quartier Salignan, 84400 Apt, France
4 Department of Veterinary Sciences, University of Pisa, 56124 Pisa, Italy
5 Division of Biotechnology—Bioterio and Experimental Surgery, Instituto Nacional de Rehabilitación-Luis Guillermo Ibarra Ibarra (INR-LGII), Mexico City 14389, Mexico
6 Facultad de Estudios Superiores Cuautitlán, Universidad Nacional Autónoma de México (UNAM), Cuautitlán Izcalli 54714, Mexico
7 Neurophysiology, Behavior and Animal Welfare Assessment, DPAA, Universidad Autónoma Metropolitana (UAM), Unidad Xochimilco, Calzada del Hueso 1100, Coapa, Col Villa Quietud, Coyoacán, Mexico City 04960, Mexico
8 Mediterranean Institute for Agriculture, Environment and Development (MED), Institute for Advanced Studies and Research, Universidade de Évora, Pólo da Mitra, Ap. 94, 7006-554 Évora, Portugal
9 School of Biological Sciences, Queen's University Belfast, Belfast BT9 5DL, UK
* Correspondence: dmota100@yahoo.com.mx

Citation: Lezama-García, K.; Martínez-Burnes, J.; Marcet-Rius, M.; Gazzano, A.; Olmos-Hernández, A.; Mora-Medina, P.; Domínguez-Oliva, A.; Pereira, A.M.F.; Hernández-Ávalos, I.; Baqueiro-Espinosa, U.; et al. Is the Weight of the Newborn Puppy Related to Its Thermal Balance?*Animals* **2022**, *12*, 3536. https://doi.org/10.3390/ani12243536

Academic Editor: Maria Cristina Veronesi

Received: 20 November 2022
Accepted: 9 December 2022
Published: 14 December 2022

Simple Summary: Newborns experience a significant thermal change at birth, leading their bodies to adjust and reduce their metabolism to survive. In this study, newborn puppies' weights and their temperatures in different body areas and at different measure times were evaluated to determine if there is a relationship between weight and their ability to reach thermostability. It was observed that there is a positive relationship between the weight of the puppies and their thermoregulatory capacity.

Abstract: Hypothermia, a factor associated with neonatal mortality, can occur immediately after birth as a protective mechanism to prevent hypoxic damage in neonates, or to reduce the metabolic rate to improve the chances of survival in the first hours of life. The heat interchange through the superficial temperature of animals can be evaluated with infrared thermography (IRT). However, to date, there is no information on thermal windows in puppies. This study aimed to evaluate, with the use of IRT, the microcirculatory alterations in 8 different thermal windows identified at 7 different times in 289 newborn puppies assigned to different groups. Three thermograms were taken from four zones of each puppy: the facial, frontal, right lateral, and left lateral regions. Newborn puppies were grouped in 4 quartiles according to their weight: Q_1 (126–226 g) n = 73, Q_2 (227–330 g) n = 72, Q_3 (331–387 g) n = 74, and Q_4 (388–452 g) n = 70. A total of 8 thermal windows were considered at 7 evaluation times from Wet at birth until 24 h after birth (AB). Two-way mixed ANOVA within and between subjects' design for each thermal window (eight models) was performed. Results revealed a positive correlation between the puppy's weight and its ability to achieve thermostability in all the evaluated thermal windows. Statistically significant differences ($p < 0.0001$) between the 4 quartiles (Q_1, Q_2, Q_3, and Q_4) were found. The lowest temperatures were recorded when the pups were still wet and the highest at 24 h AB. Thermal windows with the highest temperatures were abdominal (34.234 ± 0.056 °C), thoracic (33.705 ± 0.049 °C), nasal (30.671 ± 0.110 °C), and upper left palpebral (34.066 ± 0.052 °C), while the lowest were thoracic limb brachial biceps (27.534 ± 0.051 °C), thoracic limb elbow (27.141 ± 0.049 °C), thoracic limb metacarpal (27.024 ± 0.062 °C), and femoral pelvic limb

(27.654 ± 0.055 °C). Assessing the thermal response in newborn puppies can help identify drastic temperature reductions or deficient thermoregulatory compensation during the first hours of life, preventing the consequences of hypothermia.

Keywords: dogs; puppy welfare; animal perinatology; newborn puppy; thermoregulation; infrared thermography

1. Introduction

Endothermic animals regulate their body temperature through heat interchange between the amount produced by metabolism and the environment [1]. The control of heat exchange between the body surface and external environment plays a significant role in the regulation of body temperature during different physiological phases and/or activities throughout the life of homeotherms. Thermoregulatory adjustments can be induced not only by changes in environmental temperature but also by a variety of physiological situations including age, fasting and food intake, stress circumstances, and inflammation status, inducing changes in internal temperature which are followed by changes in body surface temperature. Under stressful conditions, animal's reaction is mainly centered on the activation of the sympathetic system and the hypothalamic–hypophysis–adrenal axis (HPA) through the release of effector hormones, namely, catecholamine and glucocorticoid, respectively [2]. One of the known results of the sympathetic system's activation is stress-induced hyperthermia, consisting in increased core body temperature with consequent changes in involucre temperature. Therefore, that allows obtaining distance images of a specific body region—represents a valuable tool to monitor the physiologic status, welfare, and stress responses of animals [3–5].

Thermoregulation plays an important role in newborns, especially in altricial species [6]. Through its extensive dermal vascularization, the body avoids losing temperature by producing changes in the blood flow of specific body regions [7], which are called biological thermal windows [8].

These changes can be indirectly quantified by evaluating tissue heat radiation through infrared thermography (IRT) [9]. Using a tool that can record these changes could help to identify states of hypothermia, a factor related to mortality and one of the most important causes of death in neonates [10,11].

Perinatal mortality in some species, such as pigs [12] and dogs [13,14], is one of the biggest concerns for producers and breeders since it can represent an average of one puppy out of ten live births dying before two months of age [15]. In addition, the immaturity of newborn puppies makes them highly vulnerable. Neonate thermoregulation involves biochemical, anatomical, physiological [16,17], and endocrine mechanisms to trigger respiratory and vascular changes and activate metabolism to produce energy [18]. It has been suggested that hypothermia occurs immediately after birth as a protective mechanism to prevent hypoxic damage in the neonate and reduce the metabolic rate to improve the survival of the newborn in the first hours [19–21]. Another important risk factor for neonatal mortality in diverse species, such as humans, pigs, and cattle, is low birth weight [22,23].

Similarly, in dogs, low birth weight neonates have a higher risk of death. More precisely, this risk is 12 times greater than in animals with normal birth weight [22]. In this species, studies have been carried out on the effect of the dam's weight on the puppy's weight at birth, litter size, vitality, and survival [13,24,25]. Likewise, the effect of the dam´s weight on the presentation of asphyxia and newborn hematological values have been evaluated [26]. According to Plavec et al. [27], puppies that experienced severe distress postpartum had significantly worse survival rates in the first week of their lives (p = 0.0113 and p = 0.0231, respectively). Attempts have been made to establish the most appropriate thermal regions or windows that provide better information on thermal changes in veterinary medicine [28]. Thus, thermal utility windows in dogs are the lacrimal caruncle, eye, ear, thorax flanks, appendicular area, and face [29–31]. However, there still needs to be more information on thermal windows in puppies. This could aid in our understanding of the process of thermoregulation in newborn dogs and, thus, help reduce the high mortality rates in this species.

This study aims to evaluate the microcirculatory changes in the different thermal windows of puppies assigned to different groups according to weight. The repercussions of their presentation of hypothermia were also analyzed. Our hypothesis assumed that (1) bigger newborns will have fewer thermoregulation problems than those of small sizes; (2) the highest temperatures will be obtained in thermal windows closest to the brain and vital organs (palpebral, thoracic, abdominal, or nasal); and (3) the lowest temperatures will be the ones recorded in thermal windows furthest from vital structures (thoracic limbs and pelvic limbs).

2. Materials and Methods

2.1. Facilities

The study was carried out in 10 veterinary clinics in the central zone of the municipality of Campeche, Campeche state, Mexico, a zone with a tropical climate and a temperature range between 36 and 40 °C. The participation of owners of pregnant bitches was requested. Bitches were monitored and offered medical attention to carry out prenatal stage control from day 25 of pregnancy to 48 h of the puppies' life.

2.2. Study Population

In total, 289 puppies from 60 parturient bitches were recruited and divided into 4 groups using size-specific quartiles, following Mugnier et al. [32] and Tessi et al.'s [33] method. The first quartile (Q_1) represents the lowest 25% of registered values, the second quartile (Q_2) represents 25–50%, the third quartile (Q_3) represents 50–75%, and the fourth quartile represents 75–100% (Q_4). Animals in the group Q_1 were considered low-weight, while those belonging to Q_4, the highest 25% of registered values, were considered high-weight puppies. This classification is due to the great variety of sizes between the breeds of dogs—the greatest morphological variability within any land mammal species. In this sense, in dogs, we can find adult body weight ranges from 500 g in miniature breeds, such as Chihuahuas, to more than 100 kg in giant breeds, such as mastiffs [34]. Among the breeds included in this study were Chihuahua, German shepherd, Labrador, golden retriever, Great Dane, standard schnauzer, cocker spaniel, poodle, Scottish terrier, and Belgian shepherd. As a result of this considerable variation in body weights, birth weight should be analyzed according to breed size. Quartiles were calculated at the puppy level with this formula: Qa = Li ((aN/4 + Fi-1)/Fi) Ai, where Li is the lower limit of the class where the quartile is located, N is the sum of the absolute frequencies, Fi-1 is the accumulated frequency of the previous class, and Ai is the amplitude of the class, that is, the number of values contained in the interval. The groups were Q_1 (126–226 g) $n = 73$ puppies, Q_2 (227–330 g) $n = 72$ puppies, Q_3 (331–387 g) $n = 74$ puppies, and Q_4 (388–452 g) $n = 70$ puppies.

The weight of the puppies was obtained using a digital scale from Salter Weight Tronix Ltd., West Bromwich, UK, as soon as the dam stopped licking and cleaning the amnionic fluids and placental membranes from them.

2.3. Infrared Thermography

In the present study, in each puppy, 8 thermal windows were identified at 7 different times, according to the methodology used by Lezama-García et al. [35]. The assessment times were: Wet (after the mother stopped licking the amniotic fluids from the pup), Dry (after drying the puppies with a towel for one minute), Colostrum (immediately after colostrum intake and after the puppy's release of the teat), 30 min after birth (AB), 1 h AB, 4 h AB, and 24 h AB. The thermal windows are explained in detail in Figure 1. A total of 16,240 records of the puppies' minimum, maximum, and average surface temperature were obtained. These data resulted from 289 puppies, from which 3 thermograms were taken from 4 zones: facial, frontal, right lateral, and left lateral regions. The facial images included the following regions: the upper left palpebral (ULP) and nasal (N). Frontal included thoracic limb metacarpal (TLM). Right lateral recorded femoral pelvic limb (FPL), thoracic limb biceps brachial (TLBB), and thoracic limb elbow (TLE), while on the left side, thoracic (T) and abdominal temperatures were assessed. Thermographic images were obtained with an infrared camera, FLIR® model Thermal TM E80, FLIR Systems, Wilsonville, OR, USA, with the following specifications: IR resolution 320 × 240 pixels, thermal sensitivity <0.045 °C, accuracy ±2 °C or 2% of reading in the ambient temperature of 10 °C to 35 °C, and image frequency of 60 Hz. All images were collected with an emissivity of 0.95 at a uniform distance of 30 cm. The thermographic images were saved in JPEG format to analyze them later using specialized FLIR Tools software® (FLIR Systems, Wilsonville, OR USA). The different evaluation times registered the minimum, average, and maximum temperatures for every image.

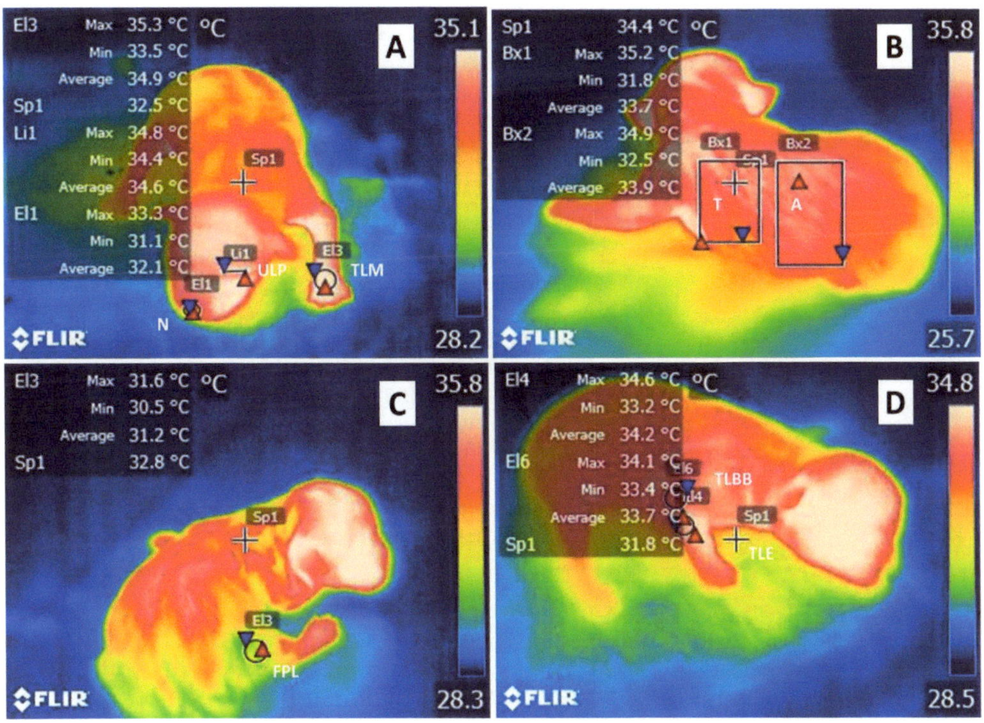

Figure 1. Thermal windows in newborn puppies. (**A**) Nasal (N) (El1), upper left palpebral (ULP) (Li1), and thoracic limb metacarpal (TLM) (El3) windows were made with circular figures delimited by the edges of the nasal mucosa, at the area of the edge of the left upper eyelid, and in the joint formed by the metacarpals covering the area from medial to lateral end, respectively. (**B**) Thoracic (T) window (Bx1) was made with rectangular figures delimited by the axillary area, the area of the last rib, and from the region of the spinal vertebrae to the ventral part of the abdomen. (**A**) Abdominal window (Bx2) was delimited by two millimeters after the last rib, to the inguinal area, and from the region of the spinal vertebrae to the ventral part of the abdomen. (**C**) Femoral pelvic limb (FPL) window (El3) was delimited by the space bounded by the edge of the pelvic limb in the biceps femoris region. (**D**) Thoracic limb brachial biceps (TLBB) (El6) and thoracic limb elbow (TLE) windows (El4) were obtained by placing circular figures from the area where the armpit begins to half the width of the thoracic limb and placing circular figures in the elbow area covering the vertex formed by the humerus-radio-ulnar joint, respectively.

2.4. Statistical Analysis

Two-way mixed ANOVA within and between subjects' design for each thermal window (eight models) was performed. This analysis has 4 independent categories (the groups according to weight of subjects) and 7 factors that were related or dependent (within subjects), also known as repeated measures.

Analysis components:

a. Within subjects' factor: 7 levels or times: (1) Wet; (2) Dry; (3) Colostrum; (4) 30 min; (5) 1 h; (6) 4 h; (7) 24 h.
b. Between subjects' factor: quartiles (4 levels): (1) Quartile 1 (Q_1): 126–226 g n = 73; (2) Quartile 2 (Q_2): 227–330 g n = 72; (3) Quartile 3 (Q_3): 331–387 g n = 74; (4) Quartile 4 (Q_4): 388–452 g n = 70.
c. Interaction between factors: groups.

2.5. Ethical Statement

All the animal owners were asked to grant informed consent to carry out the research. This project was approved by the Ph.D. Program in Biological and Health Science Academic Committee, with approval number CBS.114.19. The animals included in the present study were treated gently and were not touched or stressed, since infrared thermography is a non-invasive technique. All work was performed under lineaments of Mexico's Official Norm NOM-062-ZOO-1999 on technical specifications for animal production, care, and ethical use in applied ethological studies [36].

3. Results

The means and standard errors of the eight thermal windows were analyzed between the four quartiles. Statistically significant differences ($p < 0.0001$) between the 4 quartiles (Q_1, Q_2, Q_3, and Q_4) were found. However, in general and in all thermal windows, quartiles Q_2 and Q_3 were similar. In the same way, the evidence indicates significant differences between the temperature means of thermal windows and the different puppies' evaluation times (Wet, Dry, Colostrum, 30 min AB, 1 h AB, 4 h AB, and 24 h AB).

The highest temperatures were recorded in the Q4 group puppies and the lowest in the Q1 puppies in all evaluated thermal windows (Tables 1–8).

Table 1. Average of the temperature of upper left palpebral (ULP) window at different times in newborn puppies classified into 4 groups according to weight.

Time	Q_1 n = 73	Q_2 n = 72	Q_3 n = 74	Q_4 n = 70
Wet	$30.800 \pm 0.065^{b,5}$	$30.830 \pm 0.066^{b,5}$	$30.836 \pm 0.065^{a,b,5}$	$31.083 \pm 0.067^{a,5}$
Dry	$31.621 \pm 0.083^{b,4}$	$31.637 \pm 0.084^{b,4}$	$31.740 \pm 0.082^{a,b,4}$	$32.015 \pm 0.085^{a,4}$
Colostrum	$31.964 \pm 0.072^{b,3}$	$32.073 \pm 0.073^{b,3}$	$32.186 \pm 0.072^{a,b,3}$	$32.401 \pm 0.074^{a,3}$
30 min AB	$32.929 \pm 0.052^{b,2}$	$32.905 \pm 0.053^{a,b,2}$	$33.158 \pm 0.052^{a,2}$	$33.264 \pm 0.053^{a,2}$
1 h AB	$32.901 \pm 0.039^{c,2}$	$32.958 \pm 0.039^{b,c,2}$	$33.092 \pm 0.038^{b,2}$	$33.322 \pm 0.039^{a,2}$
4 h AB	$33.025 \pm 0.078^{b,2}$	$33.065 \pm 0.079^{a,b,2}$	$33.330 \pm 0.078^{a,2}$	$33.436 \pm 0.080^{a,2}$
24 h AB	$33.600 \pm 0.051^{b,1}$	$33.723 \pm 0.051^{b,1}$	$33.923 \pm 0.050^{a,1}$	$34.066 \pm 0.052^{a,1}$

(Two-way mixed ANOVA.) Notes: n, number of puppies; weight of puppies according to category (Q_1, 126–226 g; Q_2, 227–330 g; Q_3, 331–387 g; Q_4, 388–452 g); AB, after birth; least-squares mean ± standard error. ^{a,b,c}: different superscripts among columns indicate significant temperature differences between puppy's weight groups at the same time. ^{1,2,3,4,5}: different numbers among rows indicate significant differences between times in the same puppy's weight group.

Table 2. Average of the temperature of nasal (N) window at different times in newborn puppies classified into 4 groups according to its weight.

Time	Q_1 $n = 73$	Q_2 $n = 72$	Q_3 $n = 74$	Q_4 $n = 70$
Wet	28.187 ± 0.064 [c,4]	28.649 ± 0.064 [b,4]	28.711 ± 0.063 [b,3]	29.127 ± 0.065 [a,3]
Dry	29.757 ± 0.059 [c,1,2,3]	30.002 ± 0.060 [b,1,3]	30.331 ± 0.059 [a,1]	30.431 ± 0.060 [a,1,2]
Colostrum	29.445 ± 0.076 [c,2]	29.860 ± 0.077 [a,b,2,3]	30.036 ± 0.076 [a,2]	30.293 ± 0.078 [a,1,2]
30 min AB	29.385 ± 0.084 [b,2]	29.645 ± 0.085 [a,b,2]	29.924 ± 0.084 [a,2]	30.167 ± 0.086 [a,1,2]
1 h AB	29.494 ± 0.144 [b,1,2]	29.619 ± 0.145 [a,b,2]	29.994 ± 0.143 [a,1,2]	30.224 ± 0.148 [a,1]
4 h AB	29.366 ± 0.146 [b,2]	29.697 ± 0.147 [a,b,3]	30.080 ± 0.145 [a,1,2]	30.377 ± 0.149 [a,1]
24 h AB	29.977 ± 0.107 [b,c,1]	30.281 ± 0.108 [a,1,3]	30.453 ± 0.107 [a,1]	30.671 ± 0.110 [a,1]

(Two-way mixed ANOVA.) Notes: n, number of puppies; weight of puppies according to category (Q_1, 126–226 g; Q_2, 227–330 g; Q_3, 331–387 g; Q_4, 388–452 g); AB, after birth; least-squares mean ± standard error. [a,b,c]: different superscripts among columns indicate significant differences of temperature between puppy's weight groups at the same time. [1,2,3,4]: different numbers among rows indicate significant differences between times in the same puppy's weight group.

Table 3. Average of the temperature of thoracic limb metacarpal (TLM) window at different times in newborn puppies classified into 4 groups according to weight.

Time	Q_1 $n = 73$	Q_2 $n = 72$	Q_3 $n = 74$	Q_4 $n = 70$
Wet	27.024 ± 0.062 [c,6]	27.222 ± 0.062 [c,5]	27.482 ± 0.061 [b,6]	27.743 ± 0.063 [a,5]
Dry	29.858 ± 0.066 [c,3,5]	30.155 ± 0.066 [b,2,4]	30.281 ± 0.065 [a,b,2,5]	30.446 ± 0.067 [a,3]
Colostrum	29.028 ± 0.038 [c,4]	29.255 ± 0.038 [b,3]	29.344 ± 0.037 [b,4]	29.742 ± 0.038 [a,4]
30 min AB	29.582 ± 0.081 [b,3]	29.536 ± 0.082 [b,3]	29.886 ± 0.081 [a,3]	30.186 ± 0.083 [a,3]
1 h AB	30.294 ± 0.057 [b,c,2]	30.452 ± 0.057 [b,2]	30.644 ± 0.056 [b,2]	30.940 ± 0.058 [a,2]
4 h AB	30.239 ± 0.196 [b,1,2]	30.430 ± 0.197 [b,1,2]	30.664 ± 0.194 [a,b,1,2]	31.177 ± 0.200 [a,1,2]
24 h AB	30.778 ± 0.086 [b,1]	30.971 ± 0.086 [a,b,1]	31.271 ± 0.085 [a,1]	31.543 ± 0.088 [a,1]

(Two-way mixed ANOVA.) Notes: n, number of puppies; weight of puppies according to category (Q_1, 126–226 g; Q_2, 227–330 g; Q_3, 331–387 g; Q_4, 388–452 g); AB, after birth; least-squares mean ± standard error. [a,b,c]: different superscripts among columns indicate significant temperature differences between puppy's weight groups at the same time. [1,2,3,4,5,6]: different numbers among rows indicate significant differences between times in the same puppy's weight group.

Table 4. Average of the temperature of thoracic (T) window at different times in newborn puppies classified into 4 groups according to weight.

Time	Q_1 $n = 73$	Q_2 $n = 72$	Q_3 $n = 74$	Q_4 $n = 70$
Wet	28.884 ± 0.160 [a,6]	28.861 ± 0.161 [a,7]	29.089 ± 0.159 [a,5]	29.194 ± 0.163 [a,6]
Dry	31.525 ± 0.042 [c,5]	31.657 ± 0.042 [b,c,6]	31.806 ± 0.042 [a,b,4]	31.902 ± 0.043 [a,5]
Colostrum	31.718 ± 0.056 [c,4]	31.904 ± 0.057 [b,c,5]	31.982 ± 0.056 [a,b,4]	32.119 ± 0.057 [a,4]
30 min AB	32.067 ± 0.040 [c,3]	32.176 ± 0.040 [c,4]	32.358 ± 0.040 [b,3]	32.520 ± 0.041 [a,3]
1 h AB	32.218 ± 0.050 [c,3]	32.380 ± 0.050 [b,c,3]	32.453 ± 0.049 [a,b,3]	32.601 ± 0.051 [a,3]
4 h AB	32.504 ± 0.058 [c,2]	32.666 ± 0.059 [b,c,2]	32.734 ± 0.058 [a,b.2]	32.910 ± 0.059 [a,2]
24 h AB	33.251 ± 0.048 [c,1]	33.451 ± 0.048 [b,1]	33.596 ± 0.048 [a,b,1]	33.705 ± 0.049 [a,1]

(Two-way mixed ANOVA.) Notes: n, number of puppies; weight of puppies according to category (Q_1, 126–226 g; Q_2, 227–330 g; Q_3, 331–387 g; Q_4, 388–452 g); AB, after birth; least-squares mean ± standard error. [a,b,c]: different superscripts among columns indicate significant temperature differences between puppy's weight groups at the same time. [1,2,3,4,5,6,7]: different numbers among rows indicate significant differences between times in the same puppy's weight group.

Table 5. Average of the temperature of abdominal (A) window at different times in newborn puppies classified into 4 groups according to weight.

Time	Q_1 n = 73	Q_2 n = 72	Q_3 n = 74	Q_4 n = 70
Wet	29.664 ± 0.100 ^{a,5}	29.739 ± 0.101 ^{a,5}	29.800 ± 0.099 ^{a,5}	29.939 ± 0.102 ^{a,5}
Dry	32.353 ± 0.039 ^{c,4}	32.494 ± 0.039 ^{b,c,4}	32.608 ± 0.038 ^{b,4}	32.764 ± 0.039 ^{a,4}
Colostrum	32.501 ± 0.063 ^{b,4}	32.696 ± 0.063 ^{a,b,4}	32.708 ± 0.062 ^{a,b,4}	32.862 ± 0.064 ^{a,4}
30 min AB	32.836 ± 0.062 ^{b,3}	33.007 ± 0.063 ^{a,b,3}	33.059 ± 0.062 ^{a,b,3}	33.191 ± 0.064 ^{a,3}
1 h AB	32.941 ± 0.069 ^{b,3}	33.094 ± 0.069 ^{a,b,3}	33.112 ± 0.068 ^{a,b,3}	33.337 ± 0.070 ^{a,3}
4 h AB	33.322 ± 0.071 ^{b,2}	33.432 ± 0.072 ^{a,b,2}	33.438 ± 0.071 ^{a,b,2}	33.642 ± 0.073 ^{a,2}
24 h AB	33.788 ± 0.055 ^{c,1}	33.995 ± 0.055 ^{b,1}	34.087 ± 0.055 ^{a,b,1}	34.234 ± 0.056 ^{a,1}

(Two-way mixed ANOVA.) Notes: n, number of puppies; weight of puppies according to category (Q_1, 126–226 g; Q_2, 227–330 g; Q_3, 331–387 g; Q_4, 388–452 g); AB, after birth; least-squares mean ± standard error. ^{a,b,c}: different superscripts among columns indicate significant temperature differences between puppy's weight groups at the same time. ^{1,2,3,4,5}: different numbers among rows indicate significant differences between times in the same puppy's weight group.

Table 6. The average temperature of the femoral pelvic limb (FPL) window at different times in newborn puppies classified into 4 groups according to weight.

Time	Q_1 n = 73	Q_2 n = 72	Q_3 n = 74	Q_4 n = 70
Wet	27.654 ± 0.055 ^{c,5}	27.979 ± 0.055 ^{b,5}	28.001 ± 0.055 ^{b,5}	28.348 ± 0.056 ^{a,5}
Dry	28.670 ± 0.111 ^{b,3}	28.826 ± 0.112 ^{a,b,3,4}	29.072 ± 0.111 ^{a,b,3,4}	29.116 ± 0.114 ^{a,3}
Colostrum	28.020 ± 0.062 ^{c,4}	28.280 ± 0.063 ^{b,4}	28.515 ± 0.062 ^{a,4}	28.714 ± 0.064 ^{a,4}
30 min AB	28.363 ± 0.097 ^{b,3,4}	28.535 ± 0.098 ^{a,b,4}	28.611 ± 0.097 ^{a,b,4}	28.846 ± 0.099 ^{a,3,4}
1 h AB	28.732 ± 0.081 ^{b,3}	29.072 ± 0.081 ^{a,3}	29.137 ± 0.080 ^{a,3}	29.366 ± 0.082 ^{a,3}
4 h AB	29.815 ± 0.070 ^{b,2}	30.068 ± 0.071 ^{a,b,2}	30.226 ± 0.070 ^{a,2}	30.434 ± 0.072 ^{a,2}
24 h AB	30.825 ± 0.089 ^{b,1}	31.163 ± 0.090 ^{a,1}	31.186 ± 0.088 ^{a,1}	31.483 ± 0.091 ^{a,1}

(Two-way mixed ANOVA.) Notes: n, number of puppies; weight of puppies according to category (Q_1, 126–226 g; Q_2, 227–330 g; Q_3, 331–387 g; Q_4, 388–452 g); AB, after birth; least-squares mean ± standard error. ^{a,b,c}: different superscripts among columns indicate significant temperature differences between puppy's weight groups at the same time. ^{1,2,3,4,5}: different numbers among rows indicate significant differences between times in the same puppy's weight group.

Table 7. Average of the temperature of thoracic limb biceps brachial (TLBB) window at different times in newborn puppies classified into 4 groups according to weight.

Time	Q_1 n = 73	Q_2 n = 72	Q_3 n = 74	Q_4 n = 70
Wet	27.534 ± 0.051 ^{c,6}	27.840 ± 0.051 ^{b,5}	27.883 ± 0.050 ^{b,5}	28.344 ± 0.052 ^{a,5}
Dry	28.439 ± 0.114 ^{b,3,4}	28.615 ± 0.115 ^{a,b,3}	28.948 ± 0.113 ^{a,3}	29.043 ± 0.116 ^{a,3}
Colostrum	27.884 ± 0.053 ^{d,5}	28.127 ± 0.053 ^{c,4}	28.440 ± 0.052 ^{b,4}	28.693 ± 0.054 ^{a,3,4}
30 min AB	28.206 ± 0.088 ^{c,4}	28.399 ± 0.088 ^{b,c,3,4}	28.543 ± 0.087 ^{a,b,4}	28.839 ± 0.089 ^{a,3,4}
1 h AB	28.630 ± 0.078 ^{c,3}	28.988 ± 0.079 ^{b,3}	29.111 ± 0.078 ^{b,3}	29.421 ± 0.080 ^{a,3}
4 h AB	29.567 ± 0.090 ^{c,2}	29.876 ± 0.091 ^{b,c,2}	30.149 ± 0.090 ^{a,b,2}	30.329 ± 0.092 ^{a,2}
24 h AB	30.712 ± 0.080 ^{c,1}	31.023 ± 0.081 ^{b,1}	31.024 ± 0.080 ^{b,1}	31.500 ± 0.082 ^{a,1}

(Two-way mixed ANOVA.) Notes: n, number of puppies; weight of puppies according to category (Q_1, 126–226 g; Q_2, 227–330 g; Q_3, 331–387 g; Q_4, 388–452 g); AB, after birth; least-squares mean ± standard error. ^{a,b,c}: different superscripts among columns indicate significant temperature differences between puppy's weight groups at the same time. ^{1,2,3,4,5,6}: different numbers among rows indicate significant differences between times in the same puppy's weight group.

Table 8. Average of the thoracic limb elbow (TLE) window temperature at different times in newborn puppies classified into 4 groups according to weight.

Time	Q_1 $n = 73$	Q_2 $n = 72$	Q_3 $n = 74$	Q_4 $n = 70$
Wet	27.141 ± 0.049 ^{d,7}	27.355 ± 0.049 ^{c,7}	27.675 ± 0.049 ^{b,5}	28.005 ± 0.050 ^{a,5}
Dry	30.798 ± 0.062 ^{b,6}	30.993 ± 0.062 ^{b,6}	31.341 ± 0.062 ^{a,4}	31.351 ± 0.063 ^{a,3}
Colostrum	30.365 ± 0.066 ^{b,5}	30.535 ± 0.067 ^{b,5}	30.787 ± 0.066 ^{a,3}	30.851 ± 0.068 ^{a,4}
30 min AB	31.095 ± 0.069 ^{c,4}	31.272 ± 0.069 ^{b,c,4}	31.450 ± 0.068 ^{a,b,4}	31.565 ± 0.070 ^{a,2,3}
1 h AB	30.093 ± 0.061 ^{c,3}	30.239 ± 0.062 ^{c,3}	30.540 ± 0.061 ^{b,3}	30.895 ± 0.062 ^{a,2,4}
4 h AB	32.149 ± 0.069 ^{b,2}	32.415 ± 0.070 ^{a,2}	32.615 ± 0.069 ^{a,2}	32.485 ± 0.071 ^{a,1}
24 h AB	32.848 ± 0.081 ^{a,b,1}	33.094 ± 0.082 ^{b,1}	33.012 ± 0.081 ^{b,1}	32.594 ± 0.083 ^{a,1}

(Two-way mixed ANOVA.) Notes: n, number of puppies; weight of puppies according to category (Q_1, 126–226 g; Q_2, 227–330 g; Q_3, 331–387 g; Q_4, 388–452 g); AB, after birth; least-squares mean ± standard error. ^{a,b,c,d}: different superscripts among columns indicate significant temperature differences between puppy's weight groups at the same time. ^{1,2,3,4,5,6,7}: different numbers among rows indicate significant differences between times in the same puppy's weight group.

Tables 1–8 show the estimated marginal means ± standard error. "Bonferroni corrections were used to adjust p values for multiple comparisons" to avoid a type-I statistical error, and the normal distribution of data was assessed using visual inspection of histogram and Q-Q plots.

Table 2 shows the average temperatures of the nasal window (N); a significant statistical difference ($p < 0.0001$) between group Q_1 and Q_4, with a difference of 2.48 °C between Wet time in Q_1 and 24 h AB on Q_4, can be observed.

Table 3 shows the range between different times on the TLM window in newborns with different weights classified into four quartiles. Moreover, like in previous tables, the average temperature in this thermal window shows that puppies with lower weights (Q_1) recorded low temperatures, and puppies with higher weights show the highest temperatures. It is important to note that the lowest temperature of all windows at Q_1 Wet time was recorded in this thermal window.

In Table 5, it can be observed that there is a significant difference ($p < 0.0001$) between the four groups at different times. However, there are similarities between groups Q_2 and Q_3, a pattern repeated in almost all thermal windows evaluated in this study. Similarly, it can be observed that the highest temperatures recorded in this study occurred in this thermal window at 24 h AB.

Table 6 shows statistically significant differences between the lowest and highest quartiles; however, at 24 h AB, the 4 quartiles practically remain without significant differences, except for Q_1.

Table 7 shows that when comparing the Q_1 group with Q_4 at 24 h AB, the difference between these groups was 4.78 °C, and when comparing the temperatures of wet puppies of the Q_1 group with temperatures recorded in the Q_4 group at 24 h AB, a difference of 3.96 °C was recorded, but similitudes were observed again between group Q_2 y Q_3.

Table 8 shows that statistically significant differences between Q_1 and Q_4 at different times were more evident than in the rest of the thermal windows, and these differences were observed between groups Q_2 and Q_3, unlike what was observed in the remainder of the tables.

Subsequently, the correlations between the puppy's weight and their temperatures at different times and in the different thermal windows were evaluated using Spearman's rank correlations. Tables 9–16 report the existing correlations between the puppies' weight and their temperature, where the weights of the four groups (Q_1, Q_2, Q_3, and Q_4) and their respective temperatures by time and thermal window indicate how the temperature tends to vary at different times and in different thermal windows. In almost all cases, there is a positive correlation with statistically significant differences ($p < 0.0001$) (Tables 9–15). These results mean that the higher the weight of the puppies, the higher temperatures recorded in them. Only one case was observed where the correlation was negative (Table 16) in the thermal window TLE at 24 h AB. In this case, as the puppies' weight increased, their temperature decreased in the TLE window. The *r* values in all tables were between 0.146 and 0.651, except in Table 16 of the TLE thermal window, where $r = -0.024$ ($p = 0.679$) at 24 h AB. The highest correlations were observed in Table 15 TLBB window (Colostrum $r = 0.624$) and Table 11 TLM window (Colostrum $r = 0.632$).

Table 9. Significant correlations between puppies' weight and their superficial temperature in UPL thermal window.

Variables	Correlation Coefficient (*r*)	*p*-Value
Wet	0.190	0.001
Dry	0.189	0.001
Colostrum	0.248	<0.001
30 min AB	0.379	<0.001
1 h AB	0.444	<0.001
4 h AB	0.292	<0.001
24 h AB	0.364	<0.001

Spearman's rank correlation coefficients and their statistical significance between puppies' weight and their temperature at different times. AB: after birth.

Table 10. Significant correlations between puppy's weight and their superficial temperature at N thermal window.

Variables	Correlation Coefficient (*r*)	*p*-Value
Wet	0.542	<0.001
Dry	0.510	<0.001
Colostrum	0.423	<0.001
30 min AB	0.365	<0.001
1 h AB	0.226	<0.001
4 h AB	0.353	<0.001
24 h AB	0.331	<0.001

Spearman's rank correlation coefficients and their statistical significance between puppies' weight and their temperature at different times. AB: after birth.

Table 11. Significant correlations between puppies' weight and their superficial temperature at TLM thermal window.

Variables	Correlation Coefficient (*r*)	*p*-Value
Wet	0.551	<0.001
Dry	0.372	<0.001
Colostrum	0.632	<0.001
30 min AB	0.315	<0.001
1 h AB	0.469	<0.001
4 h AB	0.452	<0.001
24 h AB	0.353	<0.001

Spearman's rank correlation coefficients and their statistical significance between puppies' weight and their temperature at different times. AB: after birth.

Table 12. Significant correlations between puppies' weight and their superficial temperature at T thermal window.

Variables	Correlation Coefficient (*r*)	*p*-Value
Wet	0.182	0.002
Dry	0.409	<0.001
Colostrum	0.307	<0.001
30 min AB	0.449	<0.001
1 h AB	0.337	<0.001
4 h AB	0.328	<0.001
24 h AB	0.441	<0.001

Spearman's rank correlation coefficients and their statistical significance between puppies' weight and their temperature at different times. AB: after birth.

Table 13. Significant correlations between puppies' weight and their superficial temperature at A thermal window.

Variables	Correlation Coefficient (*r*)	*p*-Value
Wet	0.146	0.013
Dry	0.419	<0.001
Colostrum	0.258	<0.001
30 min AB	0.252	<0.001
1 h AB	0.273	<0.001
4 h AB	0.205	<0.001
24 h AB	0.376	<0.001

Spearman's rank correlation coefficients and their statistical significance between puppies' weight and their temperature at different times. AB: after birth.

Table 14. Significant correlations between puppies' weight and their superficial temperature at FPL thermal window.

Variables	Correlation Coefficient (*r*)	*p*-Value
Wet	0.429	<0.001
Dry	0.415	<0.001
Colostrum	0.494	<0.001
30 min AB	0.224	<0.001
1 h AB	0.338	<0.001
4 h AB	0.365	<0.001
24 h AB	0.292	<0.001

Spearman's rank correlation coefficients and their statistical significance between puppies' weight and their temperature at different times. AB: after birth.

Table 15. Significant correlations between puppies' weight and their superficial temperature at TLBB thermal window.

Variables	Correlation Coefficient (*r*)	*p*-Value
Wet	0.520	<0.001
Dry	0.444	<0.001
Colostrum	0.624	<0.001
30 min AB	0.311	<0.001
1 h AB	0.397	<0.001
4 h AB	0.406	<0.001
24 h AB	0.360	<0.001

Spearman's rank correlation coefficients and their statistical significance between puppies' weight and their temperature at different times. AB: after birth.

Table 16. Significant correlations between puppies' weight and their superficial temperature at TLE thermal window.

Variables	Correlation Coefficient (*r*)	*p*-Value
Wet	0.651	<0.001
Dry	0.391	<0.001
Colostrum	0.429	<0.001
30 min AB	0.400	<0.001
1 h AB	0.525	<0.001
4 h AB	0.259	<0.001
24 h AB	−0.024	0.679

Spearman's rank correlation coefficients and their statistical significance between puppies' weight and their temperature at different times. AB: after birth

4. Discussion

The present study assessed the superficial temperature in 289 newborn puppies born from 60 bitches with different weights at 8 anatomical regions or thermal windows. The puppies were evaluated at different times and assigned to four groups using size-specific quartiles (Q_1, Q_2, Q_3, and Q_4). According to the results, there was an association between puppies' thermal response and their weight, with the highest temperature recorded in Q_4 in all thermal windows at all evaluation times. Additionally, although the temperatures at Wet and Dry times differed between the four groups (Q_1, Q_2, Q_3, and Q_4), both (Wet and Dry times) had the lowest values in all groups. Interestingly, at 24 h AB, the puppies' temperature in every thermal window was similar between the groups, showing a similar thermoregulatory capacity independent of their weight; however, it was still the lowest in group Q_1 and highest in Q_4. These results differed from those of another study carried out by this workgroup [35], where the groups were classified according to the dam's weight, and it was found that in the groups with greater dams' weight, the thermoregulation of the puppies stabilized at 24 h AB, showing positive correlations with higher weight of the mother, due to better thermoregulation of her pups.

The present results show the effect of puppies' weight on their superficial temperature. Newborns with low weight at birth had lower temperature values for all thermal windows at every measured time, compared with the heaviest puppies. According to Harri et al. [37], birth weight is important for the survival of neonates, and it is one of the most important factors that influence temperature loss, since the body mass index (BMI) determines the capacity of thermoregulation in newborns, as with foxes. However, the litter size also significantly influences ($p = 0.003$) the weight of the newborns. In this sense, it is common to have a higher number of low-birth-weight puppies in large litters compared to small litters [25]. This effect can be seen in piglets, where large litters can have smaller piglets with higher mortality rates, due to hypothermia or trauma generated by the mother when they are too weak to move [38].

This thermoregulatory dependence on weight and time has also been reported in dog puppies that cannot stabilize their temperature rhythm for several days after birth, but their rectal temperatures become stable at six weeks AB [39]. It is important to consider that the temperature of the environment in which puppies live could have a favorable or unfavorable effect on their welfare [40]. However, in this study, the temperature of the environment was not controlled, but similar conditions (34–40 °C) and foam mats were maintained in the place of parturition to avoid temperature loss by convection. Schrank et al. [41] reported that birth weight could impact newborns' vitality and the health. It also has been reported that dogs of small size tend to lose heat quickly because they have a greater surface-to-area volume ratios [42]; therefore, they require a greater heat production to maintain their thermostability [43]. This can also be attributed to the wide variability of canine breeds [39].

Significant differences have been observed throughout the times evaluated, from wet to 24 h AB, in the evaluated regions. When the fetuses are still in the uterus of the bitch, they maintain a temperature of 0.3 to 0.5 °C higher than the dam's body temperature. However, at birth, the puppies experience a significant thermal change and loss of temperature through evaporation occurs, because they are born wet, due to the amniotic fluid [44]. This happens because the insulation and protection function that the coat provides is not efficient, since the amniotic fluid has a high thermal conductivity which generates greater hypothermia. In addition to this, other contributing factors include decreased body fat, age, and lack of acclimatization to the environment [45,46]. Therefore, increased body temperatures when passing from Wet to Dry time may be explained either by the decrease in heat loss through evaporation, or the decrease in convection favored by air currents. Another critical factor is rubbing puppies, which stimulates dermal vascular microcirculation of peripheral blood vessels (peripheral vasodilation), increasing microvascular hyperemia at the dermal level.

Hypothermia was observed in newborn puppies of the present study at the first three evaluation times (Wet, Dry, and Colostrum), where the lowest superficial temperatures were recorded.

The thermostability observed at 24 h AB in puppies from all groups could be attributed to the efficacy of the thermoregulatory response of newborns regardless of their initial weight, as shown in the values of IRT in all thermal windows, where the four groups, when compared to the temperature at wet, had a similar average increase. Several factors can influence this response, such as glycogen reserves, colostrum intake, increased digestion, the presence of the dam, and the members of the litter. The thoracic windows demonstrate an important fact: as other studies carried out with other species showed, the temperature tends to rise after the consumption of colostrum; in this study, we can see that immediately after the consumption of colostrum, the temperature tends to rise in the four quartiles in the T window.

As observed in the results presented, at 24 h AB all puppies reached thermostability after their coat was dried and following colostrum intake and heat transfer of the mother's temperature by convection. According to Münnich and Küchenmeister [21], these thermoregulatory behaviors are important during newborns' first days since puppies cannot maintain their body temperature when exposed to cold environments up to six days after birth.

According to the results reported in this study, the most notable thermoregulatory changes in newborn puppies were low temperatures in the distal regions, that is, in the thermal windows of TLM and FPL, and high temperatures in thermal windows A, T, and ULP. This could be because the most important structures related to metabolism and vital functions are found in the thoracic, abdominal, and cranial regions [47,48]. The appendicular windows presented the lowest temperatures, which can be explained, in the case of newborn puppies, because they still do not present locomotion, as they are an altricial species, and their degree of neurological development, responsible for the mechanism of thermal modulation, is immature. Neonates rely on peripheral vasoconstriction and arteriovenous shunts at the skin level to reduce heat loss, but also use non-shivering thermogenesis by lipolysis of brown adipose tissue and the oxidation of fatty acids in the mitochondria through adenosine triphosphate (ATP) synthesis, because it is reported that neonatal skeletal muscles are immature at birth. It is precisely this point that provides the explanation for why thermogenesis by shivering in newborn puppies is not so effective [44]. The time the altricial species develop the ability to obtain thermostability also influences this. According to Pineda and Dooley [49], thermostability is not reached until day 18 of life, and autonomic thermoregulation is known to be wholly developed at the end of the fourth week [19].

Additional research that could be performed based on the present findings includes study of the influence of factors such as environmental temperature, size of litter, and quality of maternal care on the thermostability of newborn puppies during the first hours of life. It is important to mention that one of the main limitations of this study was that, since births did not always occur in the same place, the environmental temperatures could not be controlled and standardized for all cases. Likewise, applying IRT to study the association with the percentage of male puppies would be interesting, since it was found that the percentage of male puppies was positively correlated with lateral nursing, a position that provides puppies the easiest access to the milk. Looking at this data from an evolutionary point of view, the mothers spend more attention on male puppies because it is more convenient in terms of fitness [50] and could alter the peripheral thermal response of neonates.

5. Conclusions

Results suggest that, during the first hours of the life of newborn puppies, their thermoregulation mechanism is deficient in anatomical and metabolic factors that can be reduced by colostrum intake, maintaining an adequate body temperature, and their mechanisms to conserve or dissipate heat in the evaluated thermal windows.

These results also suggest a positive correlation between the puppy's weight and its ability to achieve thermostability, which is continuously observed in all the evaluated thermal windows. The puppies born with lower weights (Q_1, 126–226 g) presented the lowest temperatures, while puppies with higher weights (Q_4, 388–452 g) presented the highest temperatures.

Likewise, the time effect of the temperature recording showed relevant changes in them; the lowest temperatures were obtained when the puppies were still wet and the highest at 24 h AB.

Regarding the body regions where temperatures were evaluated, the findings suggest that the thermal windows with the highest temperatures were A, T, N, and ULP, and those with the lowest temperatures were TLBB, TLE, TLM, and FPL. Assessing the thermal response in newborn puppies can help identify drastic temperature reductions or deficient thermoregulatory compensation during the first hours of life, preventing the consequences of hypothermia.

Author Contributions: Conceptualization, D.M.-R.; Methodology, K.L.-G., U.B.-E. and D.M.-R., Formal analysis, U.B.-E., Writing—original draft, K.L.-G. and A.D.-O.; Writing—review & editing, J.M.-B., M.M.-R., A.G., A.O.-H., P.M.-M., A.M.F.P., I.H.-Á., A.d.M.G., A.C.-A. and D.M.-R.; Supervision, J.M.-B., M.M.-R., A.G., A.O.-H., P.M.-M., A.D.-O., A.M.F.P., I.H.-Á., A.d.M.G., A.C.-A. and D.M.-R. All authors have read and agreed to the published version of the manuscript.

Funding: This research received no external funding.

Institutional Review Board Statement: The study was conducted according to the guidelines of the Declaration of Helsinki. The Doctoral Program in Biological and Health Science Academic Committee approved this project with approval number CBS.114.19. The animals included in the present study were treated gently and were not touched or stressed, since infrared thermography is a non-invasive technique.

Informed Consent Statement: All the owners of the study animals were asked to grant informed consent to carry out the procedures.

Data Availability Statement: Not applicable.

Conflicts of Interest: The authors declare that the research was conducted in the absence of any commercial or financial relationships that could be construed as a potential conflict of interest.

References

1. Tortora, G.; Derrickson, B. Principios de anatomía y fisiología. In *Principios de Anatomía y Fisiología*; Tortora, G., Derrickson, B., Eds.; Médica Panamericana: Madrid, Spain, 2013; pp. 1048–1051.
2. Arfuso, F.; Acri, G.; Piccione, G.; Sansotta, C.; Fazio, F.; Giudice, E.; Giannetto, C. Eye surface infrared thermography usefulness as a noninvasive method of measuring stress response in sheep during shearing: Correlations with serum cortisol and rectal temperature values. *Physiol. Behav.* **2022**, *250*, 113781. [CrossRef] [PubMed]
3. Gianesella, M.; Arfuso, F.; Fiore, E.; Giambelluca, S.; Giudice, E.; Armato, L.; Piccione, G. Infrared thermography as a rapid and non-invasive diagnostic tool to detect inflammatory foot diseases in dairy cows. *Pol. J. Vet. Sci.* **2018**, *21*, 299–305. [CrossRef] [PubMed]
4. Arfuso, F.; Rizzo, M.; Giannetto, C.; Giudice, E.; Fazio, F.; Piccione, G. Age-related changes of serum mitochondrial uncoupling 1, rumen and rectal temperature in goats. *J. Therm. Biol.* **2016**, *59*, 47–51. [CrossRef] [PubMed]
5. Rizzo, M.; Arfuso, F.; Alberghina, D.; Giudice, E.; Gianesella, M.; Piccione, G. Monitoring changes in body surface temperature associated with treadmill exercise in dogs by use of infrared methodology. *J. Therm. Biol.* **2017**, *69*, 64–68. [CrossRef]
6. Lezama-García, K.; Mota-Rojas, D.; Pereira, A.M.F.; Martínez-Burnes, J.; Ghezzi, M.; Domínguez, A.; Gómez, J.; de Mira Geraldo, A.; Lendez, P.; Hernández-Ávalos, I.; et al. Transient receptor potential (TRP) and thermoregulation in animals: Structural biology and neurophysiological aspects. *Animals* **2022**, *12*, 106. [CrossRef]
7. Bertoni, A.; Mota-Rojas, D.; Álvarez-Macias, A.; Mora-Medina, P.; Guerrero-Legarreta, I.; Morales-Canela, A.; Gómez-Prado, J.; José-Pérez, N.; Martínez-Burnes, J. Scientific findings related to changes in vascular microcirculation using infrared thermography in the river buffalo. *J. Anim. Behav. Biometeorol.* **2020**, *8*, 288–297. [CrossRef]
8. Andrade, D.V. Thermal windows and heat exchange. *Temperature* **2015**, *2*, 451. [CrossRef]
9. Villanueva-García, D.; Mota-Rojas, D.; Martínez-Burnes, J.; Olmos-Hernández, A.; Mora-Medina, P.; Salmerón, C.; Gómez, J.; Boscato, L.; Gutiérrez-Pérez, O.; Cruz, V.; et al. Hypothermia in newly born piglets: Mechanisms of thermoregulation and pathophysiology of death. *J. Anim. Behav. Biometeorol.* **2021**, *9*, 2101. [CrossRef]

10. Mota-Rojas, D.; López, A.; Martínez-Burnes, J.; Muns, R.; Villanueva-García, D.; Mora-Medina, P.; González-Lozano, M.; Olmos-Hernández, A.; Ramírez-Necoechea, R. Is vitality assessment important in neonatal animals? *CAB Rev. Perspect. Agric. Vet. Sci. Nutr. Nat. Resour.* **2018**, *13*, 1–13. [CrossRef]
11. Uchanska, O.; Ochota, M.; Eberhardt, M.; Nizanski, W. Dead or Alive? A review of perinatal factors that determine canine neonatal viability. *Animals* **2022**, *12*, 1402. [CrossRef]
12. Houška, L.; Wolfová, M.; Nagy, I.; Csörnyei, Z.; Komlósi, I. Economic values for traits of pigs in Hungary. *Czech J. Anim. Sci.* **2010**, *55*, 139–148. [CrossRef]
13. Veronesi, M.C. Assessment of canine neonatal viability-the Apgar score. *Reprod. Domest. Anim.* **2016**, *51*, 46–50. [CrossRef] [PubMed]
14. Veronesi, M.C.; Panzani, S.; Faustini, M.; Rota, A. An Apgar scoring system for routine assessment of newborn puppy viability and short-term survival prognosis. *Theriogenology* **2009**, *72*, 401–407. [CrossRef] [PubMed]
15. Chastant-Maillard, S.; Guillemot, C.; Feugier, A.; Mariani, C.; Grellet, A.; Mila, H. Reproductive performance and pre-weaning mortality: Preliminary analysis of 27,221 purebred female dogs and 204,537 puppies in France. *Reprod. Domest. Anim.* **2017**, *52*, 158–162. [CrossRef] [PubMed]
16. Mota-Rojas, D.; Pereira, A.M.F.; Wang, D.; Martínez-Burnes, J.; Ghezzi, M.; Hernández-Avalos, I.; Lendez, P.; Mora-Medina, P.; Casas, A.; Olmos-Hernández, A.; et al. Clinical applications and factors involved in validating thermal windows used in infrared thermography in cattle and river buffalo to assess health and productivity. *Animals* **2021**, *11*, 2247. [CrossRef]
17. Martínez-Burnes, J.; Muns, R.; Barrios-García, H.; Villanueva-García, D.; Domínguez-Oliva, A.; Mota-Rojas, D. Parturition in mammals: Animal models, pain and distress. *Animals* **2021**, *11*, 2960. [CrossRef]
18. Hillman, N.H.; Kallapur, S.G.; Jobe, A.H. Physiology of transition from intrauterine to extrauterine life. *Clin. Perinatol.* **2012**, *39*, 769–783. [CrossRef]
19. Lawler, D.F. Neonatal and pediatric care of the puppy and kitten. *Theriogenology* **2008**, *70*, 384–392. [CrossRef]
20. Lezama-García, K.; Mota-Rojas, D.; Martínez-Burnes, J.; Villanueva-García, D.; Domínguez-Oliva, A.; Gómez-Prado, J.; Mora-Medina, P.; Casas-Alvarado, A.; Olmos-Hernández, A.; Soto, P.; et al. Strategies for hypothermia compensation in altricial and precocial newborn mammals and their monitoring by infrared thermography. *Vet. Sci.* **2022**, *9*, 246. [CrossRef]
21. Münnich, A.; Küchenmeister, U. Causes, diagnosis and therapy of common diseases in neonatal puppies in the first days of life: Cornerstones of practical approach. *Reprod. Domest. Anim.* **2014**, *49*, 64–74. [CrossRef]
22. Fix, J.S. Relationship of Piglet Birth Weight with Growth, Efficiency, Composition, and Mortality. Ph.D. Thesis, North Carolina State University, Raleigh, NC, USA, 2010.
23. Mota-Rojas, D.; Wang, D.; Titto, C.G.; Martínez-Burnes, J.; Villanueva-García, D.; Lezama-García, K.; Domínguez, A.; Hernández-Avalos, I.; Mora-Medina, P.; Verduzco, A.; et al. Neonatal infrared thermography images in the hypothermic ruminant model: Anatomical-morphological physiological aspects and mechanisms for thermoregulation. *Front. Vet. Sci.* **2022**, *9*, 963205. [CrossRef] [PubMed]
24. Groppetti, D.; Ravasio, G.; Bronzo, V.; Pecile, A. The role of birth weight on litter size and mortality within 24 h of life in purebred dogs: What aspects are involved? *Anim. Reprod. Sci.* **2015**, *163*, 112–119. [CrossRef] [PubMed]
25. Mila, H.; Grellet, A.; Feugier, A.; Chastant-Maillard, S. Differential impact of birth weight and early growth on neonatal mortality in puppies. *J. Anim. Sci.* **2015**, *93*, 4436–4442. [CrossRef] [PubMed]
26. Reyes-Sotelo, B.; Mota-Rojas, D.; Mora-Medina, P.; Ogi, A.; Mariti, C.; Olmos-Hernández, A.; Martínez-Burnes, J.; Hernández-Ávalos, I.; Sánchez-Millán, J.; Gazzano, A. Blood biomarker profile alterations in newborn canines: Effect of the mother's weight. *Animals* **2021**, *11*, 2307. [CrossRef] [PubMed]
27. Plavec, T.; Knific, T.; Slapšak, A.; Raspor, S.; Lukanc, B.; Pipan, M.Z. Canine neonatal assessment by vitality score, amniotic fluid, urine, and umbilical cord blood analysis of glucose, lactate, and cortisol: Possible influence of parturition type? *Animals* **2022**, *12*, 1247. [CrossRef] [PubMed]
28. Mota-Rojas, D.; Titto, C.G.; Orihuela, A.; Martínez-Burnes, J.; Gómez-Prado, J.; Torres-Bernal, F.; Flores-Padilla, K.; Carvajal-de la Fuente, V.; Wang, D. Physiological and behavioral mechanisms of thermoregulation in mammals. *Animals* **2021**, *11*, 1733. [CrossRef]
29. Casas-Alvarado, A.; Martínez-Burnes, J.; Mora-Medina, P.; Hernández-Avalos, I.; Domínguez-Oliva, A.; Lezama-García, K.; Gómez-Prado, J.; Mota-Rojas, D. Thermal and circulatory changes in diverse body regions in dogs and cats evaluated by infrared thermography. *Animals* **2022**, *12*, 789. [CrossRef]
30. Reyes-Sotelo, B.; Mota-Rojas, D.; Martínez-Burnes, J.; Olmos-Hernández, A.; Hernández-Ávalos, I.; José, N.; Casas-Alvarado, A.; Gómez, J.; Mora-Medina, P. Thermal homeostasis in the newborn puppy: Behavioral and physiological responses. *J. Anim. Behav. Biometeorol.* **2021**, *9*, 2112. [CrossRef]
31. Travain, T.; Colombo, E.S.; Grandi, L.C.; Heinzl, E.; Pelosi, A.; Prato Previde, E.; Valsecchi, P. How good is this food? A study on dogs' emotional responses to a potentially pleasant event using infrared thermography. *Physiol. Behav.* **2016**, *159*, 80–87. [CrossRef]
32. Mugnier, A.; Mila, H.; Guiraud, F.; Brévaux, J.; Lecarpentier, M.; Martinez, C.; Mariani, C.; Adib-Lesaux, A.; Chastant-Maillard, S.; Saegerman, C.; et al. Birth weight as a risk factor for neonatal mortality: Breed-specific approach to identify at-risk puppies. *Prev. Vet. Med.* **2019**, *171*, 104746. [CrossRef]

33. Tesi, M.; Miragliotta, V.; Scala, L.; Aronica, E.; Lazzarini, G.; Fanelli, D.; Abramo, F.; Rota, A. Relationship between placental characteristics and puppies' birth weight in toy and small sized dog breeds. *Theriogenology* **2020**, *141*, 1–8. [CrossRef] [PubMed]
34. Boyko, A.R.; Quignon, P.; Li, L.; Schoenebeck, J.J.; Degenhardt, J.D.; Lohmueller, K.E.; Zhao, K.; Brisbin, A.; Parker, H.G.; VonHoldt, B.M.; et al. A Simple genetic architecture underlies morphological variation in dogs. *PLoS Biol.* **2012**, *8*, e1000451. [CrossRef] [PubMed]
35. Lezama-García, K.; Martínez-Burnes, J.; Villanueva-García, D.; Pérez-Jiménez, J.C.; Domínguez-Oliva, A.; Mora-Medina, P.; Hernández-Ávalos, I.; Olmos-Hernández, A.; Mota-Rojas, D. Relation between the dam's weight on vascular microcirculation and in the temperature of her puppies at different stages post-partum. *Vet. Sci.* **2022**, *9*, 673. [CrossRef]
36. Sherwin, C.M.; Christiansen, S.B.; Duncan, I.J.; Erhard, H.W.; Lay, D.C.; Mench, J.A.; O'Connor, C.E.; Carol Petherick, J. Guidelines for the ethical use of animals in applied ethology studies. *Appl. Anim. Behav. Sci.* **2003**, *81*, 291–305. [CrossRef]
37. Harri, M.; Mononen, J.; Haapanen, K.; Korhonen, H. Postnatal changes in hypothermic response in farm born blue foxes and raccoon dogs. *J. Therm. Biol.* **1991**, *16*, 71–76. [CrossRef]
38. Gómez-Prado, J.; Pereira, A.M.F.; Villanueva-García, D.; Wang, D.; Domínguez-Oliva, A.; Mora-Medina, P.; Hernández-Avalos, I.; Martínez-Burnes, J.; Casas-Alvarado, A.; Olmos-Hernández, A.; et al. Thermoregulation mechanisms and perspectives for validating thermal windows in pigs with hypothermia and hyperthermia: An overview. *Front. Vet. Sci.* **2022**, *in press*. [CrossRef]
39. Piccione, G.; Fazio, F.; Giudice, E.; Refinetti, R. Body size and the daily rhythm of body temperature in dogs. *J. Therm. Biol.* **2009**, *34*, 171–175. [CrossRef]
40. Jordan, M.; Bauer, A.; Stella, J.; Croney, C. Temperature Requirements for Dogs. 2022. Available online: https://www.extension.purdue.edu/extmedia/va/va-16-w.pdf (accessed on 1 May 2022).
41. Schrank, M.; Mollo, A.; Contiero, B.; Romagnoli, S. Bodyweight at birth and growth rate during the neonatal period in three canine breeds. *Animals* **2019**, *10*, 8. [CrossRef]
42. Rigotti, C.F.; Jolliffe, C.T.; Leece, E.A. Effect of prewarming on the body temperature of small dogs undergoing inhalation anesthesia. *J. Am. Vet. Med. Assoc.* **2015**, *247*, 765–770. [CrossRef]
43. Ewart, S. Homeostasis. In *Fisiología Veterinaria*; Klein, B.G., Ed.; Elsevier: Barcelona, Spain, 2009; pp. 596–607.
44. Asakura, H. Fetal and neonatal thermoregulation. *J. Nippon Med. Sch.* **2004**, *71*, 360–370. [CrossRef]
45. Mallet, M.L. Pathophysiology of accidental hypothermia. *QJM* **2002**, *95*, 775–785. [CrossRef] [PubMed]
46. Oncken, A.; Kirby, R.; Rudloff, E. Hypotherma in critically ill dogs and cats. *Compend. Contin. Educ. Pract. Vet.* **2001**, *23*, 506–520.
47. Casas-Alvarado, A.; Mota-Rojas, D.; Hernández-Ávalos, I.; Mora-Medina, P.; Olmos-Hernández, A.; Verduzco-Mendoza, A.; Reyes-Sotelo, B.; Martínez-Burnes, J. Advances in infrared thermography: Surgical aspects, vascular changes, and pain monitoring in veterinary medicine. *J. Therm. Biol.* **2020**, *92*, 102664. [CrossRef] [PubMed]
48. Mota-Rojas, D.; Martínez-Burnes, J.; Casas-Alvarado, A.; Gómez-Prado, J.; Hernández-Avalos, I.; Domínguez-Oliva, A.; Lezama-García, K.; Jacome-Romero, J.; Rodríguez-González, D.; Pereira, A.M.F. Clinical usefulness of infrared thermography to detect sick animals: Frequent and current cases. *CABI Rev.* **2022**, *22*, 1–27. [CrossRef]
49. Pineda, M.; Dooley, M. *McDonald's Veterinary Endocrinology and Reproduction*; Wiley-Blackwell: Hoboken, NJ, USA, 2008; p. 618.
50. Ogi, A.; Mariti, C.; Pirrone, F.; Baragli, P.; Gazzano, A. The influence of oxytocin on maternal care in lactating dogs. *Animals* **2021**, *11*, 1130. [CrossRef]

animals

MDPI

Commentary

Thermoregulation during Field Exercise in Horses Using Skin Temperature Monitoring

Elisabeth-Lidwien J. M. M. Verdegaal [1,2,3,*], Gordon S. Howarth [1,2], Todd J. McWhorter [1,2] and Catherine J. G. Delesalle [1,3]

1 Thermoregulation Research Group, Equine Health and Performance Centre, University of Adelaide, Roseworthy, SA 5116, Australia; gordon.howarth@adelaide.edu.au (G.S.H.); todd.mcwhorter@adelaide.edu.au (T.J.M.); catherine.delesalle@ugent.be (C.J.G.D.)
2 School of Animal and Veterinary Sciences, Roseworthy Campus, University of Adelaide, Roseworthy, SA 5371, Australia
3 Research Group of Comparative Physiology, Department of Translational Physiology, Infectiology and Public Health, Faculty of Veterinary Medicine, Ghent University, 9820 Merelbeke, Belgium
* Correspondence: lidwien.verdegaal@adelaide.edu.au

Simple Summary: Heat stress is an important performance and welfare issue for exercising horses. The process of thermoregulation is crucial for equine athletes. Excessive metabolic heat generation during exercise, combined with inefficient heat dissipation, can lead to hyperthermia if not detected in time and not effectively managed. Accurately monitoring heat generation during exercise allows for early preventative intervention as body temperature rises. Skin temperature monitoring is commonly used as a non-invasive method to assess body temperature responses pre- and post-exercise. To date, few studies have used infrared thermographic techniques to monitor body temperature continuously during exercise under laboratory conditions and in the field. In reviewing these results, the accuracy of measuring skin temperature as a reliable indication of overall body temperature is discussed. This commentary summarizes the results of studies measuring surface skin temperature in horses, particularly using infrared thermography for exercise-focused monitoring in the field.

Abstract: Hyperthermia and exertional heat illness (EHI) are performance and welfare issues for all exercising horses. Monitoring the thermoregulatory response allows for early recognition of metabolic heat accumulation during exercise and the possibility of taking prompt and effective preventative measures to avoid a further increase in core body temperature (T_c) leading to hyperthermia. Skin temperature (T_{sk}) monitoring is most used as a non-invasive tool to assess the thermoregulatory response pre- and post-exercise, particularly employing infrared thermographic equipment. However, only a few studies have used thermography to monitor skin temperature continuously during exercise. This commentary provides an overview of studies investigating surface skin temperature mainly by infrared thermography (IRT) during exercise. The scientific evidence, including methodologies, applications, and challenges associated with (continuous) skin temperature monitoring in horses during field exercise, is discussed. The commentary highlights that, while monitoring T_{sk} is straightforward, continuous T_{sk} alone does not always reliably estimate T_c evolvement during field exercise. In addition, inter-individual differences in thermoregulation need to be recognized and accounted for to optimize individual wellbeing. With the ongoing development and application of advanced wearable monitoring technology, there may be future advances in equipment and modeling for timely intervention with horses at hyperthermic risk to improve their welfare. However, at this point, infrared thermographic assessment of T_{sk} should always be used in conjunction with other clinical assessments and veterinary examinations for a reliable monitoring of the welfare of the horse.

Keywords: thermoregulation; body temperature monitoring; skin temperature (T_{sk}); infrared thermography (IRT); sports medicine, exercise; horse; equine; core temperature (T_c); continuous monitoring

Citation: Verdegaal, E.-L.J.M.M.; Howarth, G.S.; McWhorter, T.J.; Delesalle, C.J.G. Thermoregulation during Field Exercise in Horses Using Skin Temperature Monitoring. *Animals* **2024**, *14*, 136. https://doi.org/10.3390/ani14010136

Academic Editor: Daniel Mota-Rojas

Received: 29 November 2023
Revised: 24 December 2023
Accepted: 26 December 2023
Published: 30 December 2023

1. Introduction

Field exercise is an integral aspect of equine sports and competition where horses face a myriad of physiological and environmental challenges. Efficient thermoregulation is paramount to ensure optimal performance and prevent heat-related stress in equine athletes. The incidence of heat stress and exertional heat illness (EHI) is projected to rise further due to the ongoing effects of global climate change. Climate change is a generally accepted phenomenon, and the scientific community is warning the world about its long-term consequences both for human and animal health [1–3]. Particularly, more acute changes in environmental conditions such as heat waves, entail that athletes may need to perform under circumstances to which they are not acclimatized. These sudden increased ambient temperatures (T_a) are the dominant external risk factor for heat stress for human and equine athletes [1–3].

EHI is associated with various kinds of exercise and impacts the overall well-being of all mammals, including humans, and may lead to exertional heat stroke (EHS) and ultimately death. In exercising human athletes, EHS is among the top three causes of sudden death, and in summer, it is the number one cause of athlete death in the USA [4]. For example, EHI is common in American football, with an average incidence rate ranging from 0.01 to 0.4%. Heat monitoring studies being developed to investigate the EHI problem in human athletes have used many monitoring approaches such as ingestible thermo-monitoring pills or a rectal thermistor, skin temperature (T_{sk}) sensors, and measures of subjective perception of heat load [5].

Both EHI and EHS are also problematic conditions in equine athletes and can occur during a wide variety of different types of exercise [6]. However, heat exhaustion triggered by thermoregulatory-induced physiological feedback failure is most commonly reported when horses need to perform submaximal exercise during extended time intervals such as endurance rides. The prevalence of metabolic disorders in endurance horses ranges from 2% to 15% [7–14]. In comparison, for maximal intensity exercise, EHI screening studies have recently reported a prevalence between 0.1% and 1.1% in racehorses in Australia, the UK, and Japan [15–19]. A study in eastern Australia focused on selected EHI cases post-exercise at the racetrack and suggested an EHI incidence of up to 9.5% during hot summer months [17]. The latter study used four severity levels of EHI (1 = least severe) reported by Brownlow et al. [20] and concluded that 96% of horses could be categorized as level 1. This suggests that many EHI cases of low-level severity with vague clinical signs may have been overlooked in the past. At present, the association between EHI and exercise injuries leading to fatalities is not well understood, and hence the roles that heat exhaustion and EHI play are unknown.

All equestrian sports are facing growing scrutiny from both the general public and animal welfare organizations. When unfortunate incidents result in fatalities, spectators are swift to criticize sporting and racing events and increasingly call for safeguarding the well-being of sports horses in order to maintain the "social license" to operate in equestrian sports [21]. Consequently, from an animal welfare perspective, it is paramount and timely to prevent occurrence of heat stress on all occasions. The aforementioned welfare issues are a concerning reality that is motivating research groups worldwide to develop practical methods for reliable monitoring during field exercise to ensure the thermoregulatory welfare of horses [21–26]. To achieve this, there is a need for minimally invasive body temperature monitoring systems, also termed 'heat monitoring'. The ultimate goal of monitoring is to accurately detect a potentially dangerous rise in core body heat in real-time such that effective preventative actions can then be taken.

In an ideal world, all temperature monitoring equipment including wearable technological tools should be non-invasive and enable horse owners and veterinarians to have a continuous real-time view of the evolution of body temperature during exercise in the field [22–25]. An approach that is gaining popularity for monitoring the thermoregulatory response is the follow-up of the T_{sk} in horses performing in the field exercise using wearable T_{sk} equipment [24]. While this equipment provides output of physiological data, the question remains whether these data have physiological relevance for the evolution of the core body temperature, which is the core point of interest to safeguard welfare in a reliable way. This article reviews and comments on the benefits and challenges of monitoring thermoregulation in exercising horses using non-invasive surface T_{sk}, particularly through infrared thermography.

2. Thermoregulation: Core vs. Shell Temperature

Thermoregulation is the internal physiological process which balances metabolic heat load and the exchange of this heat with the environment. Overall body temperature in mammals is divided into an inner core body temperature (T_c) and an outer shell temperature. T_c in horses is regulated within a narrow range (37.4–38.0 °C), while shell temperature varies more widely in response to ongoing central thermoregulatory processes [24,26]. The inner core temperature refers to deep-body temperatures, while shell temperature includes intramuscular, subcutaneous, and surface skin temperatures (T_{sk}). Heat is moved between core and shell areas to equilibrate heat production and heat loss and thus regulate the overall body temperature. Consequently, temperature is not uniform at different locations in and on the body. Thermoregulatory processes are controlled via neurophysiologic control centers in the hypothalamus. Both peripheral and central thermoreceptors are nerve endings that respond to changes in T_c and T_a, all in communication with the hypothalamus (Figure 1). The peripheral thermoreceptors in the skin detect the peripheral shell temperature and respond to a range of external temperatures from cold to hot (5–60 °C), with cold thermoreceptors being the most numerous.

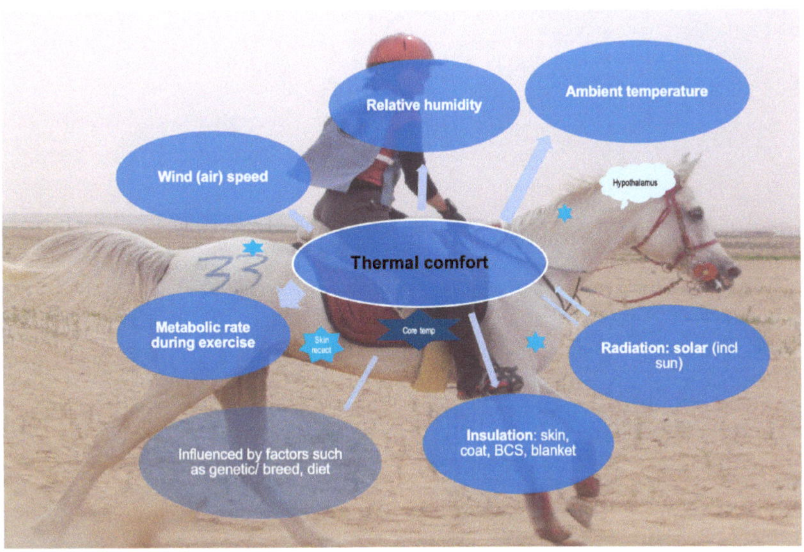

Figure 1. Factors affecting thermoregulation in horses and its feedback control mechanisms including thermoreceptors in the skin (spread over body, examples displayed as bright-blue 6-point stars), and the core temperature receptors (dark-blue 6-point star) which provide information to the hypothalamus (displayed as light-blue call-out), which adjusts the responses either to produce or to lose heat to ensure thermal comfort; BCS: body condition score.

3. Mechanism of Metabolic Heat Loss

During exercise, heat originating from exercising muscle groups will be dissipated to the surrounding environment through different tissues and by using transportation via the bloodstream towards the skin surface. *H* will then be released into the surrounding environment through four mechanisms: radiation, conduction, convection, and evaporation

(Figure 1). Therefore, the surface area (cm^2) of the body is a critical factor in determining rates of heat exchange. Because heat exchange with the environment is proportional to the relative size of the body surface area (BSA), the rate of heat exchange per unit of body mass (BM) is largest in the smallest animal (if other variables are equal). For example, the BSA-to-volume ratio is 50% less in a 500 kg horse when compared to humans [26,27]. This lower BSA-to-BM ratio of horses results in greater demands being imposed on their thermoregulatory system during exercise [28–30]. Only if T_{sk} is higher than the surroundings can the body lose heat by radiation, conduction, and evaporation.

Among these heat-loss pathways, the most critical and pivotal for horses is the activation of sweat glands to evaporate sweat from the skin surface [27,29,31–34]. The efficacy of heat dissipation via this evaporative process relies on the thermal contrast between locally perfusion-based T_{sk} and its immediate surroundings, with consideration of variables such as vapor pressure, airflow, and solar radiation, especially during outdoor exercise [17,28,35] (Figure 1). Consequently, when considering using T_{sk} as a method for monitoring temperature, it is essential to anticipate that all pathways for dissipating H to the environment can influence the T_{sk} data output.

Additionally, horses are hindgut fermenters, which entails that important fermentation processes take place within the gastro-intestinal (GI) system [26]. These processes are recognized for their propensity to produce an enormous amount of heat which in turn represents an additional challenge for the thermoregulatory system. It is quite plausible that the type of dietary load (composition of intestinal content) inside the GI system of a horse performing exercise has its impact on T_c evolution at that time point [26]. How quickly the latter is translated towards the outer shell temperature is currently unknown, but again, requires notice as a factor that may influence the T_{sk} data output. From a comparative perspective, in humans it is known that food containing high amounts of fats, complex carbohydrates, and proteins stimulate the T_c increase during digestion [26].

4. Field-Based Thermoregulation Research—Non-Invasive Monitoring

Until recently, little was known about how thermoregulation functions in real-life circumstances during field exercise and recovery in different equine sports disciplines. An important reason for this was the complete lack of effective, reliable temperature monitoring equipment that can be safely and comfortably applied during field exercise. In the past, most equine thermoregulatory research has been performed either under laboratory conditions, for example on a treadmill [36–38], or under field conditions, only focussed on the pre- and post-exercise period. A few studies have aimed at continuous monitoring of the T_c during exercise and recovery in real-time field training and competitions. One of those studies used intra-uterine temperature loggers as a nonsurgical, minimally invasive method [39], an approach which only allows for monitoring in female horses. Recently, a reliable novel method is the oral administration of a small telemetric GI temperature pill to monitor T_c on a continuous basis [40–43].

The aforementioned field studies reveal that in-the-field exercise monitoring is essential to obtain a holistic view on how the thermoregulatory system actually works. This is not only challenged by the performed exercise itself, but also by the environmental conditions in which environmental temperature and humidity are pivotal factors [43]. Additionally, the terrain itself influences exercise intensity. For example, the presence of slopes and descents in a competition course as well as varying properties of the surface over the course, such as sand or mud, are challenging for optimal exercise capacity [43]. These factors must all be accounted for when monitoring T_c during exercise, and for that purpose, monitoring thermoregulation during field exercise has been shown to be more reliable than lab-based testing [43]. In addition to thermoregulation monitoring, it is important to consider thermal comfort when assessing the welfare of horses (Figure 1). Thermal comfort refers to a state in which horses can still successfully thermoregulate, but may already feel thermally uncomfortable. This is similar to evaluating a perception of heat exertion during heat monitoring in human athletes. Many factors that impact thermal comfort, such as solar radiation, skin insulation, and wind speed, can also affect T_{sk} values.

5. Skin Temperature Measurement

Non-invasive methods such as T_{sk} measurement are of significant interest and have found various applications. Generally, four types of temperature sensor equipment are employed in thermoregulation research: thermometers, thermistors, thermocouples, and infrared thermography (IRT) devices. Of these, IRT has received the most extensive attention in recent equine studies and is commonly used remotely (non-contact temperature sensor) [22,44–56]. Digital temperature sensors may include technology that comprises a logger consisting of a temperature sensor in a stainless-steel computer chip with an enclosed battery [57]. Thermistors are temperature-sensitive resistors that change resistance with temperature and have been used in several equine studies [34,58–60]. Both digital temperature sensors and thermistors have a small size, fast response time, and relatively low cost. However, they are invasive, are limited to point measurements, and may alter the local skin conditions. Thermocouples are temperature sensors based on the Seebeck effect, where a voltage is generated in response to a temperature difference (thermo-electric sensors) and have been used in some equine studies [57,61,62]. In addition, fiber optic sensors can be used to record temperature by measuring the change in light transmission properties through an optical fiber to measure temperature [63]. In human thermoregulatory research, conductive textiles have received increasing attention recently. These are textiles with embedded conductive fibers that can be used to measure T_{sk}. However, this technique still encounters many calibration challenges, and the accuracy of data output is influenced by textile placement [63]. In addition, the dermal patch technology is a small adhesive patch with integrated sensors applied to the skin surface [63]. Lastly, IRT measures the infrared radiation emitted by an object to create a temperature map [22] and has multiple applications in equine medicine, including detecting inflammation and musculoskeletal injuries such as lameness, tendonitis, and laminitis [51,64,65]. In several studies, IRT has been used as an indicator of stress, but those studies investigated only a limited number of horses (five horses) [44,46,65,66] and one study investigated the correlation of increased T_{sk} with blood lactate concentration post exercise in 30 horses [67]. Lastly, infrared contact sensors (IRCSs) are contact-based sensors using infrared technology to measure T_{sk} [24,60].

6. Skin Temperature at Point(s)-in-Time or Continuous Monitoring during Exercise

Changes in T_{sk} during field exercise are complex. As exercise continues, T_{sk} varies by body region and may be transient and rapidly changing in relation to local heat exchange from the muscles to the environment, an increased sweat rate, and forced convection (wind), all highly influenced by T_a.

6.1. T_{sk} Measurement at a Point-in-Time (Pre-Exercise and Post-Exercise)

The most common method to record T_c in field competition is serial measurement of the rectal temperature (T_{re}) pre- and, more often, only post-exercise. These are "snapshots" (points-in-time) of temperature evolvement. Importantly, research has shown that T_{re} lags behind or is always lower than T_c [28,42]; consequently, serial T_{re} measurement does not allow for early-stage remedial intervention. Importantly, there is also a clear lack of correlation between T_{sk} and T_c [24]. To date, several methods are employed for monitoring T_{sk} such as infrared thermography using cameras. While handheld IRT cameras are becoming increasingly popular in equine sports medicine due to their non-invasiveness, most exercise studies involve point-in-time T_{sk} measurements only post-exercise and without pre-exercise and intra-exercise periods [22,30,45–48,52,53,58,68–70]. A non-exhaustive overview of T_{sk} studies associated with exercising horses, either monitoring non-continuously or continuously in horses, is presented in Table 1.

Table 1. Non-exhaustive overview of equine thermoregulatory monitoring studies measuring T_{sk} (°C) evolvement associated with exercise (E): non-continuous data points (N/C, at point-in-time) or continuous monitoring (C). Aim: aim of study; IRT: infrared thermography; N: number of horses; Type: Thermo-sensor type; ROIs: regions of interest (number of locations of IRT measurement); SH: shoulder; number ICS: the location at an intercostal space; Delta T_{sk}: T_{sk} change pre- to post-exercise. Only essential data are presented. T_{sk}: skin temperature; T_{CV}: central venous blood temperature; T_{re}: rectal temperature; T_{RA}: right atrial blood temperature; T_{PA}: pulmonary artery blood temperature; SET: standardized exercise testing treadmill; T_a: ambient temperature; max.: maximum; min: minutes; s: seconds; field: field studies; BSA: body surface area; BM: body mass; TX: therapy; RR: respiratory rate; EHI: exertional heat illness; HH, HD, CD: hot-humid, hot-dry, cool-dry T_a (table adjusted from Verdegaal et al. [24]).

Reference	Aim	N	Type	C or N/C	Compared T_c Y/N	ROIs	Delta T_{sk}	Conclusion
Verdegaal et al., 2022 [24]	E-field, C T_{sk} compared to C T_c	12	IRT device	C	Y, T_c GI	1: girth at ventral thorax	11–13 °C	T_{sk} is not proxy for T_c in field
Janczarek et al., 2022 [59]	E-field, post- cooling effect	19	IRT camera	N/C	Y, T_{re}	6: 4 legs, neck, hip	2.1–3.7 °C	T_{sk} and delta T_{sk} varied with body location
Brownlow and Smith, 2020 [48]	E-field, post exercise T_{sk}	260	IRT camera	N/C: post-exercise	N	3: neck, shoulder, thorax	-	T_{sk} > 39 °C direct post-race = higher risk for EHI development
Klous et al., 2020 [57]	E-field, pre-cooling effect	10	i-Button® & glue	C	Y: C T_{re}	2: Shoulder, rump	−3 °C	Pre-cooling (rinsing cold water for 8 min) -> T_{re} median 0.3 °C & T_{sk} mean −3 °C
Takahashi et al., 2020 [52]	E-SET—to 42 °C T_{PA} post-cooling methods until T_{PA} < 39 °C	5	IRT camera	N/C	Y: T_{PA}	1: left thorax	N/A	When T_{PA} 42 °C, best T_{sk} at 17[e] ICS > 40 °C Shower for 30 min with tap water most affective to lower T_c
Wilk et al., 2020 [53]	E–field, ridden–compare BM riders	12	IRT camera	N/C	Y, T_{re} (N/C)	7	~6 °C	Rider > 20% of horse BW -> higher T_{sk}
Witkowska-Pilaszewicz et al., 2020 [67]	E–field, correlate with blood lactate	30	IRT camera	N/C	N	11	~max. 4 °C	>T_{sk} at 30 min post-exercise correlated with > blood lactate (10.4 mmol/L)

Table 1. *Cont.*

Reference	Aim	N	Type	C or N/C	Compared T_c Y/N	ROIs	Delta T_{sk}	Conclusion
Redaelli et al., 2019 [45]	E–field, intensity with T_{sk} & cortisol	8	IRT camera	N/C	N	7	~max. 17 °C	T_{sk} at crown of head correlated with E intensity & increased serum cortisol
Soroko et al., 2019 [46]	E–SET pre & post T_{sk}: ridden & blood values	9	IRT camera	N/C	N	7 muscle regions	~1–2 °C	Higher T_{sk} in ridden horses compared to non-ridden
Soroko, 2018 [44]	E–SET dynamic IRT camara	5	Dynamic use IRT camara q15s	C	N	4: SH, neck, croup, chest	~max. 4 °C	Neck was hottest ROI (34.7 °C)
Rizzo et al., 2018 [66]	E–field, Acupuncture & transport	5	IRT camera	N/C	N	6	~4 °C	Both flank T_{sk} and T_{re} higher post acupuncture, transport, exercise
Yarnell et al., 2014 [58]	E–SET Dry/water treadmill	7	IRT camera	N/C	N	1	~6 °C	T_{sk} during dry treadmill assess changes in blood flow muscles
Wallsten et al., 2012 [30]	E–field, clipping & blanket in cold T_a	3	Thermistor probes with sensors on skin	N/C	N	2: neck, biceps, tail	~7 °C	Unclipped & covered with blanket → higher RR & T_{re} (38.2 °C)
Jodkowska et al., 2011 [70]	E–field, T_{max} whole BSA, jumping	35	IRT camera	N/C	Y, N/C T_{re}	36 ROIs and 25 ROIs post- exercise	~5 °C	T_{sk} rise highest at head, neck, trunk. T_{re} WNL
Simon et al., 2006 [69]	E–SET time return base T_{sk}	6	IRT camera	N/C	N	2: FL & HL	~5–6 °C	All return to base T_{sk} in 45 min
Morgan et al., 2002 [68]	E–SET coat clipping on T_{CV}, T_{sk}, T_{sk}	6	IRT thermometer	N/C	N	1	-	Clipping results in better heat loss
Geor et al., 2000 [62]	E–field, acclimation HH and CD, HD	6	Thermocouples with sticky tape	C	Y: T_{PA}	1- Shaved at thorax	~2.5 °C	No difference in delta T_{sk} (no T_{RA} corr. ~3 °C difference)

Table 1. *Cont.*

Reference	Aim	N	Type	C or N/C	Compared T_c Y/N	ROIs	Delta T_{sk}	Conclusion
Marlin et al., 1999 [38]	E–SET acclimation & sweating	5	Thermistor probe	C	Y: T_{RA} N	1: tail skin (T_{PA}, T_{re})	~7 °C	Onset of sweating occurred at lower T_{sk} following acclimation
Marlin et al., 1998 [37]	E–SET cooling	5	Thermistor probe	C	Y: T_{PA} N	2: tail skin, coat	~5 °C overall	Both T_{sk} & T_{PA} decreased 0.8 °C in response to cooling (6 × 30 s cold water) over 6.5 min
Marlin et al., 1996 [36]	E–SET Cool and hot T_a	4	Thermistor probe	C	Y: T_{RA} Y	1: tail skin (T_{RA}, T_{re})	~6 °C	T_{sk} follows T_{RA} pattern
Geor et al., 1995 [61]	E–field, HH and CD, HD	5	Thermocouples with sticky tape	C	Y: T_{RA} N	1- Shaved thorax		T_{sk} different from T_c

A systematic review of 45 T_{sk} studies of human endurance athletes reported that only a few studies involved continuous temperature monitoring [23]. From a physiological viewpoint, it is more than plausible that not only the actual temperature increases, but also the rate of increase and the time profile which are important parameters to reliably estimate how an individual athlete copes with heat load. Those parameters remain a "black box" when only non-continuous temperature monitoring is applied. Some human studies have applied highly frequent monitoring intervals. Two studies monitored T_{sk} every 30 s, one study every 180 s, and four studies every 5 min [23]. These human studies highlighted essential inter-individual differences concerning the T_{sk} response time profile [23]. Similarly, significant inter-individual T_{sk} differences have been reported in all mammals [33], including a recent equine exercise study [24]. In addition, T_{sk} in 21 horses monitored at rest also revealed significant inter-individual differences, which entails that defining "thermo-comfort" reference ranges actually needs a customized approach [50].

In horses, several studies have revealed possible correlations between recorded point-in-time T_{sk} values and various other physiological parameters [24,45]. For example, a study involving eight endurance horses analyzed the association between endurance training intensity and T_{sk} evolvement using IRT cameras focused at different locations on the body. This revealed that T_{sk} at the coronary band increased with training intensity, but maximum T_{sk} did not [45]. Another study monitoring endurance horses revealed that a higher T_{sk} at the end-of-exercise period was associated with a lower T_c at the end-of-recovery period (60 min) [24]. In addition, there was no significant correlation between T_{sk} at the end of exercise and the duration of HR recovery to 60 bpm [24]. Several hypotheses could be considered to explain the marked difference between T_{sk} and T_c, both with respect to temperature values and evolvement over time. For example, it could be hypothesized that the raised T_{sk} functions as an "early whistle blower" for the launch of an active thermoregulatory response to anticipate the increased T_c, and once the metabolic heat is successfully dissipated, the T_c decreases. This argument could be coupled with the effect of cooling-down measures being applied post-exercise, which may be more prominent when T_{sk} is greater and ultimately result in higher dissipation of metabolic heat and swifter reduction of T_c. However, the important fact that both temperature values and time evolvement patterns importantly differ between T_{sk} and T_c shows that the physiological link between both is much more complicated than initially anticipated in studies investigating the many monitoring wearables on the equine market.

Interestingly, an immediate post-race, high point-in-time IRT T_{sk} measurement > 39 °C in the neck region of racehorses exercising in a warm-hot T_a has been advocated as an early indicator for increased risk for development of post-exercise EHI [48]. Possibly in the case of the manifestation of a narrow temperature gradient between T_c and T_{sk}, the capacity of the equine body to transfer heat to the skin is greatly reduced and thus most probably hinders heat dissipation through evaporation [24,48]. Therefore, additional research has tested the effectiveness of different cooling-down techniques with the obvious goal of increasing the temperature gradient between the T_{sk} and the T_c post racing. A recent study evaluated five cooling methods for racehorses—walking with and without air fans, intermittent application of cold water with or without scraping, and continuous pouring of tap water over the body [52]. The study revealed a mean T_{sk} of 40 °C at the end-of-race exercise in a warm environment (mean T_a 31.8 °C) [52] and a T_{sk} lower than 34.5 °C at the end of moderate intensity exercise in a moderate environment (mean T_a 24 °C) [59]. Of considerable concern is the occurrence of a markedly increased T_{sk} post-exercise (>39 °C) which could indicate the horse is potentially already manifesting mild to severe manifestations of EHI.

Significantly, all of the aforementioned studies were performed under a wide range of T_a, from hot to cool conditions, and tested large disparities in exercise intensity ranging from racing to endurance exercise [24,45,48]. The pattern of T_{sk} and its variability can partly be attributed to acute fluctuations in blood flow associated with differing exercise intensities [32,60]. In racehorses executing high-intensity, short-duration exercise, heat dissipation will typically occur post-exercise, while endurance horses manage heat dissipation throughout their submaximal prolonged exercise and their mandatory monitored rest periods [43,48,71,72].

6.2. Continuous T_{sk} Monitoring during Exercise

To date, study results have clearly shown that the relationship between T_{sk} and T_c is very complex [32,33,72], and those studies that involve continuous temperature monitoring are pivotal to better understand this complexity. Only eight studies have applied continuous T_{sk} monitoring during exercise, of which four studies involved a controlled laboratory environment using treadmills [36–38,44] (Table 1). These laboratory-based investigations cannot address the authentic environmental conditions encountered by sports horses during open-air competitions. By comparison, different important variables can be deduced from the analysis of the IRT data time profiles during exercise in the field such as the average, maximum and minimum T_{sk}, standard deviation, and delta T_{sk} (T_{sk} variation during the exercise) [24].

Currently, only a few studies have tried to correlate both continuous T_c (through either arterial, venous blood, or GI pill temperature) and T_{sk} in an attempt to analyze if T_{sk} could function as a proxy for the core body thermoregulatory response [24,36–38,62] (Table 1). Importantly, the four laboratory-based treadmill investigations consistently affirmed the absence of a significant correlation between T_{sk} and T_c. For instance, two submaximal exercise studies used arterial blood temperature as a reference T_c to compare the impact of different environmental conditions varying from cool, dry (room T_a and 45% relative humidity (RH)) to hot, humid (34 °C, 85% RH) on the thoracic surface T_{sk} using thermocouples and revealed dissimilarities between T_{sk} and T_c [61,62]. Two high-intensity exercise studies investigated tail surface T_{sk} evolvement response to cooling-down techniques (6 times cold water shower for 30 s) and acclimation (30 °C and 80% RH over 15 days), respectively, and both demonstrated clear distinctions between T_{sk} and T_c recordings [37,38]. One other study did not use T_{sk} monitoring but recorded muscle temperature using intra-muscular microchips under laboratory conditions [73]. In that study, the muscle temperature continuously increased and lagged behind the central venous temperature during the recovery phase. This non-alignment is in accordance with previous studies reporting on the lack of direct correlation between muscle temperature and T_c [29,74,75].

Only four studies have applied continuous T_{sk} monitoring during field exercise and thus avoided laboratory conditions [24,57,61,62] (Table 1). None of these studies has identified a correlation between T_{sk} and T_c. In a recent study, we investigated the possible correlation of T_{sk} with T_c involving 13 endurance horses in a real-life competition [24]. The T_c was continuously measured using an ingested telemetric GI pill, and T_{sk} was continuously recorded every 15 s using an infrared thermistor sensor integrated into a modified belt during an endurance ride. Our study revealed no significant correlation between the profiles of T_{sk} and T_c during exercise and recovery phases, which demonstrated that monitoring T_{sk} does not reliably serve as a proxy for assessing the important core body thermoregulatory response during field exercise. Additionally, our study highlighted significant inter-individual variations in T_{sk} and T_c profiles, emphasizing the importance of implementing an individualized temperature monitoring model for all sport horses. Our previous field study [43] confirmed a substantial inter-individual variability in the T_{sk} time profiles despite the same exercise protocol. Similar findings have been reported in human athlete studies [76–79]. Consequently, researchers have developed integrative models that utilize T_{sk} to assess heat balance during exercise. This approach has been explored in both human studies and one equine study [80–82]. However, in a field study involving 17 human marathon runners, no regression model could effectively predict physiological heat stress load using single-point IRT T_{sk} measurements [83]. It is important to consider that this lack of reliable prediction of heat load can also apply to exercising horses in the field.

7. Overview of Factors Influencing the T_{sk} Measurement

While efforts are being made to incorporate continuous T_{sk} monitoring into field-based thermoregulation assessment, researchers need to be fully aware of several factors that add to the complexity of the physiological relationship between T_{sk} and T_c. Skin temperature evolvement depends on several individual equine variables such as breed and genetic factors, sex, age, body composition, coat color, and physical fitness level. These factors are coupled with environmental parameters (such as T_a, RH, and wind speed), attributes of the exercise (duration and intensity, additionally influenced by terrain conditions), and the possible additional application of cooling-down interventions. Notably, T_{sk} represents the outer shell temperature and thus is much more easily influenced by additional factors when compared to the T_c. From a physiological viewpoint, the T_c should be guarded with much greater scrutiny and precision when compared to the outer shell temperature. It is critical to remember that every monitoring device relies on a specific type of technology, each having advantages and shortcomings. Therefore, the choices of the type of temperature sensor equipment as well as the specific anatomical location of the T_{sk} measurement need to be taken into account throughout the studies (Table 1).

7.1. Environmental Factors

Evidently, the added value of performing field studies is the capacity to involve the effect of environmental factors on thermoregulation in exercising horses. Due to climate change, some environmental conditions can worsen quickly, pressing highly trained horses to perform under sudden changed climate conditions (such as heat waves) without adaptation and acclimatization.

Evaporation is the endothermic process in humans and horses during which liquid (sweat) transforms into vapor. The capacity to sweat relies on an active vasodilatory and vasoconstrictor mechanism controlled by the autonomic nervous system in horses during exercise [26,32,84]. Evaporation of sweat also depends on the skin-to-ambient-vapor pressure difference as well as on the T_{sk} to T_a difference. Because of the dependence of the evaporation process on the temperature and the vapor gradient between the skin and the immediate surroundings, this process will have a reduced capacity in hot and humid weather. Environmental factors play a substantial role in shaping the dynamics of T_{sk} and T_c despite different exercise intensities and their interplay, as these external factors have the capacity to swiftly modify T_{sk} without directly influencing T_c [33,68,85–88]. These factors include T_a, solar radiation, soil radiation, RH, shade, and wind speed (vs. enclosed areas without wind). For instance, fluctuations in T_a within the range of 20 °C to 30 °C have been associated with the initiation of skin vasodilation and sweat evaporation [89]. Solar radiation, by heating the skin, may contribute up to 15% of the total heat gain in an exercising horse depending on local air velocity (wind speed). In one study, T_{sk} and T_{re} reached their peaks during maximum solar radiation in daytime hours [87,90].

Exercise studies conducted in warmer environments have documented post-exercise T_{sk} measurements surpassing 39 °C. For example, a recent study found that 28 out of 38 horses exercising in a hot, dry setting (mean T_a 38.8 °C) and 6 out of 37 horses exercising in a warm, humid environment (mean T_a 31.1 °C) exhibited post-exercise T_{sk} values exceeding 39 °C. The researchers suggested that horses with T_{sk} recordings exceeding 39 °C are at risk of heat stress and EHI, emphasizing that this T_{sk} response may identify racehorses in need of cooling down [48]. Verdegaal et al. [24] found that the mean T_{sk} for endurance horses after exercise in a cool environment (mean T_a 15.3 °C) was 28 °C, and none of the horses had a T_{sk} exceeding 39 °C. It is important to note that in the latter study, the T_{sk} sensor was placed on the chest and protected from solar radiation by a belt and girth which prevented it from being influenced by environmental factors like solar radiation.

7.2. Individual Equine Factors and Thermoregulation

Horses have several individual factors that may impact their thermoregulation, including breed, body condition score, age, and temperament (including nervousness). Various skin-related factors such as sweating rate, skin thickness, blood vessel density, hair coat characteristics, presence of hair, clipping practices, coat color, and anhidrosis, all influence the efficacy of their sweat evaporation and the T_{sk} measurement [27,30,33,53,54,56,68,80,91,92]. The influence of breed on T_{sk} relates to the BSA-to-BM ratio, where a higher ratio leads to increased heat dissipation [89]. Clipping of the hair coat resulted in reduced T_{sk} and T_{re} [30,54,68]. Additionally, coat color may hold some relevance [80], although other studies did not reveal a significant influence of coat color on T_{sk} [24,85]. Individual variations in the temperament, such as nervousness, can trigger sympathetic nerve activity resulting in vasoconstriction of skin blood vessels. This neurophysiological response may contribute to disparities in local T_{sk}, reduced heat dissipation, and hyperthermia [93].

7.3. Location of T_{sk} Equipment

T_{sk} recording at specific locations on the skin body surface shows there is a delicate equilibrium between heat influx from arterial blood, metabolic processes, local skin activities, and heat exchange with the surroundings through conduction, radiation, and evaporation. Any factors disrupting this equilibrium can induce alterations in T_{sk}. In addition, numerous elements that influence T_c during exercise concurrently impact T_{sk}, including various performance-related factors and environmental conditions. Importantly, the precise anatomical positioning of sensors on horses for T_{sk} measurement has been demonstrated to impact T_{sk} outcomes [22,50,59,70,86]. For example, highest T_{sk} readings using IRT were measured at the neck, chest, and shoulders [50,59,70]. Variations in T_{sk} across different bodily regions (ROIs: regions of interest) such as neck, shoulder, and hip can be attributed to their varying network of blood vessels to facilitate thermal heat exchange with the surrounding environment after the competition [33,70,71]. In addition, exercise intensity (submaximal vs. maximal) and duration (prolonged vs. short) can influence T_{sk} [22,24]. Consequently, monitoring the physiological response of the local T_{sk} to alterations in T_c during exercise over time, lacks correlation between T_c and T_{sk} [24].

7.4. Choice of T_{sk} Measurement Equipment

Studies with variations in surface T_{sk} measurements may depend on the equipment type and may not be clinically relevant for horses exercising in the field [35]. For instance, a study comparing IRT and thermocouples in 12 human athletes at various time points (pre-exercise, during exercise, and post-exercise) revealed poor correlation and low reliability between the two methods [94].

8. Challenges and Limitations of IRT and Its Studies

It is essential to recognize that IRT sensors rely on distinct physical processes to collect data, potentially resulting in disparities in the generated data due to those variations such as type (including emissivity and the thermal window of the camera) and location of the T_{sk} equipment (ROIs). A study of human athletes conducted in a hot environment showed close agreement between a telemetric thermistor system and a standard hard-wired thermistor system but poor agreement with a thermal camera [95]. In addition, the calibration and validation of IRT sensors before a study, particularly through comparisons with certified thermocouples in a controlled water bath, are infrequently performed.

Another limitation is differences in the methodology of IRT equipment use. For example, studies creating T_{sk} images using a remote IRT camera positioned approximately 30 cm away from the skin surface at different body parts or of the entire body yielded diverse results [22,50,96]. The advantage of placing the IRT camera remotely away from the skin is that the local T_{sk} will not be affected by direct contact with the equipment. However, it is important to note that the surrounding environment, such as T_a, solar radiation, and coat characteristics, can influence remote T_{sk} measurements [22,35]. In contrast, when temperature sensors are in direct contact with the skin and secured by a belt, a reliable sensor-to-skin contact is ensured which is crucial for accurate readings and minimizing the potential influence of local skin alterations and environmental factors [22,24,35]. Interestingly, one study comparing IRT T_{sk} measurements with T_{re} in 40 adult horses concluded that T_{sk} was not a reliable method for determining T_c [96].

In summary, IRT techniques exhibit wide variability of results in both human and equine medicine, arising from aspects such as thermographic camera placement and environmental control measures [22–24,35]. While consensus guidelines have been established for multiple data collection methods of human T_{sk} using IRT [97], specific protocols for equine IRT use are still emerging [22]. Considering that additional factors such as air movement and sweating can easily and rapidly change T_{sk} without directly affecting T_c, further research is needed into reliable monitoring methods to enhance sport horse wellbeing.

9. Future Implications

Although monitoring T_{sk} is generally considered straightforward, especially with the promotion of all kinds of wearable technology in the market, it is essential to accept that that there is no method of T_{sk} monitoring that has been proven to be a reliable technique to safeguard thermo-wellbeing in horses at this point in equine research [24].

Recent advancements in human exercise research have shifted toward the application of models and algorithms that incorporate variables such as heart rate (HR) and HR variability to investigate if the T_c could be reliably estimated [60,81,86,98,99]. Physiologically, HR reflects both the blood flow rate to the muscles, which is related to metabolic heat production, and the blood flow to the skin, which is related to heat loss. For example, recent studies combined continuous insulated T_{sk} monitoring with HR monitoring in 13 and 8 human athletes exposed to hot (35 °C) and warm (25 °C) environments, respectively, and suggested a potential predictive model for T_{re} or T_c (using GI temperature pills) [81,99]. However, contrary to these findings, the HR recovery in the endurance horses in an equine study did not exhibit a direct relationship with T_{sk} [24]. Further investigation is required to explore the potential association between T_{sk} and HR before promoting HR and HR variability for the accurate predictive modeling of T_c in equine athletes. With the ongoing development of wearable technology, future advances in temperature monitoring equipment especially targeting T_c and in temperature modeling, can be expected to emerge to enable timely intervention in order to improve the welfare of the horse. Another critical motivator to advance this research field is aimed at improving the performance and health of all exercising horses.

10. Conclusions

The recent increased awareness of heat-related illness associated with exercise highlights the need to monitor equine core body temperature response accurately and continuously. Utilization of IRT in the realm of exercise is a relatively recent practice, and numerous fundamental different benefits and challenges warrant ongoing discussion, particularly in relation to various methodological aspects. Importantly, based on current evidence as presented, monitoring T_{sk} during field exercise is not a valid proxy for T_c monitoring. Further research could include the application of machine learning algorithms integrating T_{sk} data in a more "holistic physiological model". However, for that purpose, more standardized studies of equine thermoregulation specifically in the field are required.

Author Contributions: Conceptualization, E.-L.J.M.M.V. and C.J.G.D.; writing—original draft preparation, E.-L.J.M.M.V. and C.J.G.D.; writing—review and editing, E.-L.J.M.M.V., G.S.H., T.J.M. and C.J.G.D. All authors have read and agreed to the published version of the manuscript.

Funding: This research received no external funding.

Institutional Review Board Statement: Not applicable.

Informed Consent Statement: Not applicable.

Data Availability Statement: The data presented in this study are available in article.

Acknowledgments: The authors are grateful to colleagues for the editorial assistance.

Conflicts of Interest: The authors declare no conflicts of interest.

References

1. Bouchama, A.; Abuyassin, B.; Lehe, C.; Laitano, O.; Jay, O.; O'Connor, F.G.; Leon, L.R. Classic and exertional heatstroke. *Nat. Rev. Dis. Primers* **2022**, *8*, 8. [CrossRef] [PubMed]
2. Luber, G.; McGeehin, M. Climate change and extreme heat events. *Am. J. Prev. Med.* **2008**, *35*, 429–435. [CrossRef] [PubMed]
3. Solymosi, N.; Torma, C.; Kern, A.; Maroti-Agots, A.; Barcza, Z.; Konyves, L.; Berke, O.; Reiczigel, J. Changing climate in Hungary and trends in the annual number of heat stress days. *Int. J. Biometeorol.* **2010**, *54*, 423–431. [CrossRef] [PubMed]
4. Casa, D.J.; DeMartini, J.K.; Bergeron, M.F.; Csillan, D.; Eichner, E.R.; Lopez, R.M.; Ferrara, M.S.; Miller, K.C.; O'Connor, F.; Sawka, M.N.; et al. National athletic trainers' association position statement: Exertional Heat Illnesses. *J. Athl. Train.* **2015**, *50*, 986–1000. [CrossRef] [PubMed]
5. Nocera, A.; Sbrollini, A.; Romagnoli, S.; Morettini, M.; Gambi, E.; Burattini, L. Physiological and biomechanical monitoring in American football players: A scoping review. *Sensors* **2023**, *23*, 3538. [CrossRef] [PubMed]
6. Geor, R.J.; McCutcheon, L.J. Hydration effects on physiological strain of horses during exercise-heat stress. *J. Appl. Physiol.* **1998**, *84*, 2042–2051. [CrossRef]
7. Barnes, A.; Kingston, J.; Beetson, S.; Kuiper, C. Endurance veterinarians detect physiologically compromised horses in a 160 km ride. *Equine Vet. J.* **2010**, *42*, 6–11. [CrossRef]
8. Nagy, A.; Murray, J.K.; Dyson, S. Elimination from elite endurance rides in nine countries: A preliminary study. *Equine Vet. J.* **2010**, *42*, 637–643. [CrossRef]
9. Fielding, C.L.; Meier, C.A.; Balch, O.K.; Kass, P.H. Risk factors for the elimination of endurance horses from competition. *J. Am. Vet. Med. Assoc.* **2011**, *239*, 493–498. [CrossRef]
10. Nagy, A.; Murray, J.K.; Dyson, S.J. Descriptive epidemiology and risk factors for eliminations from Federation Equestre Internationale endurance rides due to lameness and metabolic reasons (2008–2011). *Equine Vet. J.* **2014**, *46*, 38–44. [CrossRef]
11. Younes, M.; Robert, C.; Cottin, F.; Barrey, E. Speed and cardiac recovery variables predict the probability of elimination in equine endurance events. *PLoS ONE* **2015**, *10*, e0137013. [CrossRef] [PubMed]
12. Fielding, C.L.; Meier, C.A.; Fellers, G.K.; Magdesian, K.G. Ability of clinicopathologic variables and clinical examination findings to predict race elimination in endurance horses. *Am. J. Vet. Res.* **2017**, *78*, 50–56. [CrossRef] [PubMed]
13. Bennet, E.D.; Parkin, T.D.H. Fédération Equestre Internationale endurance events: Risk factors for failure to qualify outcomes at the level of the horse, ride and rider (2010–2015). *Vet. J.* **2018**, *236*, 44–48. [CrossRef] [PubMed]
14. Legg, K.A.; Weston, J.F.; Gee, E.K.; Bolwell, C.F.; Bridges, J.P.; Rogers, C.W. Characteristics of endurance competitions and risk factors for elimination in New Zealand during six seasons of competition (2010/11–2015/16). *Animals* **2019**, *9*, 611. [CrossRef] [PubMed]
15. Nomura, M.; Shiose, T.; Ishikawa, Y.; Mizobe, F.; Sakai, S.; Kusano, K. Prevalence of post-race exertional heat illness in Thoroughbred racehorses and climate conditions at racecourses in Japan. *J. Equine Sci.* **2019**, *30*, 17–23. [CrossRef] [PubMed]
16. Takahashi, Y.; Takahashi, T. Risk factors for exertional heat illness in *Thoroughbred racehorses* in flat races in Japan (2005–2016). *Equine Vet. J.* **2020**, *52*, 364–368. [CrossRef]
17. Brownlow, M.A.; Brotherhood, J.R. An investigation into environmental variables influencing post-race exertional heat illness in Thoroughbred racehorses in temperate eastern Australia. *Aust. Vet. J.* **2021**, *99*, 473–481. [CrossRef]
18. Trigg, L.E.; Lyons, S.; Mullan, S. Risk factors for, and prediction of, exertional heat illness in Thoroughbred racehorses at British racecourses. *Sci. Rep.* **2023**, *13*, 3063. [CrossRef]
19. Verdegaal, E.-L.J.M.M.; McWhorter, T.J.; Halstead, F.; Prior, C.; Kennedy, H.K.; Delesalle, C.J.G. Retrospective study of the prevalence and associated risk factors of exertional heat illness in racing Thoroughbreds in Australia. *Sch. Anim. Vet. Sci. Univ. Adel. Aust.* **2023**. *manuscript in preparation; to be submitted.*
20. Brownlow, M.A.; Dart, A.J.; Jeffcott, L.B. Exertional heat illness: A review of the syndrome affecting racing Thoroughbreds in hot and humid climates. *Aus. Vet. J.* **2016**, *94*, 240–247. [CrossRef]
21. Heleski, C.; Stowe, C.J.; Fiedler, J.; Peterson, M.L.; Brady, C.; Wickens, C.; Macleod, J.N. Thoroughbred racehorse welfare through the lens of 'social license to operate—With an emphasis on a U.S. perspective. *Sustainability* **2020**, *12*, 1706. [CrossRef]
22. Soroko, M.H.K. Infrared thermography: Current applications in equine medicine. *J. Equine Vet. Sci.* **2018**, *60*, 90–96. [CrossRef]
23. Rojas-Valverde, D.; Tomás-Carús, P.; Timón, R.; Batalha, N.; Sánchez-Ureña, B.; Gutiérrez-Vargas, R.; Olcina, G. Short-term skin temperature responses to endurance exercise: A systematic review of methods and future challenges in the use of infrared thermography. *Life* **2021**, *11*, 1286. [CrossRef] [PubMed]
24. Verdegaal, E.J.M.M.; Howarth, G.S.; McWhorter, T.J.; Delesalle, C.J.G. Is continuous monitoring of skin surface temperature a reliable proxy to assess the thermoregulatory response in endurance horses during field exercise? *Front. Vet. Sci.* **2022**, *9*, 894146. [CrossRef] [PubMed]
25. Kee, P.; Anderson, N.; Gargiulo, G.D.; Velie, B.D. A synopsis of wearable commercially available biometric-monitoring devices and their potential applications during gallop racing. *Equine Vet. Ed.* **2023**, *35*, 551–560. [CrossRef]
26. Ewart, S.L. Thermoregulation. In *Cunningham's Textbook of Veterinary Physiology*, 6th ed.; Bradley, G.K., Ed.; Elsevier: St. Louise, MO, USA, 2020; pp. 596–607.

27. Hodgson, D.R. Thermoregulation. In *The Athletic Horse, Principles and Practice of Equine Sport Medicine*, 2nd ed.; Hodgson, D.R., McGowan, C.M., McKeever, K.H., Eds.; Saunders/Elsevier: St. Louis, MO, USA, 2014; pp. 108–124.
28. Hodgson, D.R.; McCutcheon, L.J.; Byrd, S.K.; Brown, W.S.; Bayly, W.M.; Brengelmann, G.L.; Gollnick, P.D. Dissipation of metabolic heat in the horse during exercise. *J. Appl. Physiol.* **1993**, *74*, 1161–1170. [CrossRef]
29. Hodgson, D.R.; Davis, R.E.; McConaghy, F.F. Thermoregulation in the horse in response to exercise. *Br. Vet. J.* **1994**, *150*, 219–235. [CrossRef]
30. Wallsten, H.; Olsson, K.; Dahlborn, K. Temperature regulation in horses during exercise and recovery in a cool environment. *Acta Vet. Scan.* **2012**, *54*, 42. [CrossRef]
31. McCutcheon, L.J.; Geor, R.J. Thermoregulation and exercise-associated heat stress. In *Equine Exercise Physiology: The Science of Exercise in the Athletic Horse*; Hinchcliff, K.W., Geor, R.J., Kaneps, A.J., Eds.; Saunder/Elsevier: St Louis, MO, USA, 2008; pp. 382–396.
32. Mota-Rojas, D.; Titto, C.G.; Orihuela, A.; Martínez-Burnes, J.; Gómez-Prado, J.; Torres-Bernal, F.; Flores-Padilla, K.; Carvajal-De La Fuente, V.; Wang, D. Physiological and behavioral mechanisms of thermoregulation in mammals. *Animals* **2021**, *11*, 1733. [CrossRef]
33. Mota-Rojas, D.; Wang, D.; Titto, C.G.; Gómez-Prado, J.; Carvajal-De La Fuente, V.; Ghezzi, M.; Boscato-Funes, L.; Barrios-García, H.; Torres-Bernal, F.; Casas-Alvarado, A.; et al. Pathophysiology of fever and application of infrared thermography (IRT) in the detection of sick domestic animals: Recent advances. *Animals* **2021**, *11*, 2316. [CrossRef]
34. Brownlow, M.A.; Mizzi, J.X. Thermoregulatory capacity of the Thoroughbred racehorse and its relationship to the pathogenesis of exertional heat illness. *Equine Vet. Ed.* **2022**, *34*, 214–221. [CrossRef]
35. MacRae, B.A.; Annaheim, S.; Spengler, C.M.; Rossi, R.M. Skin temperature measurement using contact thermometry: A systematic review of setup variables and their effects on measured values. *Front. Physiol.* **2018**, *9*, 29. [CrossRef] [PubMed]
36. Marlin, D.J.; Scott, C.M.; Schroter, R.C.; Mills, P.C.; Harris, R.C.; Harris, P.A.; Orme, C.E.; Roberts, C.A.; Marr, C.M.; Dyson, S.J.; et al. Physiological responses in nonheat acclimated horses performing treadmill exercise in cool (20 °C/40% RH), hot dry (30 °C/40% RH) and hot humid (30 °C/80% RH) conditions. *Equine Vet. J.* **1996**, *28* (Suppl. 22), 70–84. [CrossRef]
37. Marlin, D.J.; Scott, C.M.; Roberts, C.A.; Casas, I.; Holah, G.; Schroter, R.C. Post exercise changes in compartmental body temperature accompanying intermittent cold water cooling in the hyperthermic horse. *Equine Vet. J.* **1998**, *30*, 28–34. [CrossRef] [PubMed]
38. Marlin, D.J.; Scott, C.M.; Schroter, R.C.; Harris, R.C.; Harris, P.A.; Roberts, C.A.; Mills, P.C. Physiological responses of horses to a treadmill simulated speed and endurance test in high heat and humidity before and after humid heat acclimation. *Equine Vet. J.* **1999**, *31*, 31–42. [CrossRef] [PubMed]
39. Smith, J.E.; Barnes, A.L.; Maloney, S.K. A nonsurgical method allowing continuous core temperature monitoring in mares for extended periods, including during endurance exercise. *Equine Vet. J.* **2006**, *38* (Suppl. 36), 65–69. [CrossRef]
40. Green, A.R.; Gates, S.G.; Lawrence, L.M. Measurement of horse core body temperature. *J. Therm. Biol.* **2005**, *30*, 370–377. [CrossRef]
41. Green, A.R.; Gates, R.S.; Lawrence, L.M.; Wheeler, E.F. Continuous recording reliability analysis of three monitoring systems for horse core body temperature. *Comp. Electr. Agric.* **2008**, *61*, 88–95. [CrossRef]
42. Verdegaal, E.J.M.M.; Delesalle, C.; Caraguel, C.G.B.; Folwell, L.E.; McWhorter, T.J.; Howarth, G.S.; Franklin, S.H. Evaluation of a telemetric gastrointestinal pill for continuous monitoring of gastrointestinal temperature in horses at rest and during exercise. *Am. J. Vet. Res.* **2017**, *78*, 778–784. [CrossRef]
43. Verdegaal, E.J.M.M.; Howarth, G.S.; McWhorter, T.J.; Boshuizen, B.; Franklin, S.H.; Vidal Moreno de Vega, C.; Jonas, S.E.; Folwell, L.E.; Delesalle, C.J.G. Continuous monitoring of the thermoregulatory response in endurance horses and trotter horses during field exercise: Baselining for future hot weather studies. *Front. Physiol.* **2021**, *12*, 708737. [CrossRef]
44. Soroko, M.; Howell, K.; Dudek, K.; Wilk, I.; Zastrzeżyńska, M.; Janczarek, I. A pilot study into the utility of dynamic infrared thermography for measuring body surface temperature changes during treadmill exercise in horses. *J. Equine Vet. Sci.* **2018**, *62*, 44–46. [CrossRef]
45. Redaelli, V.; Luzi, F.; Mazzola, S.; Bariffi, G.D.; Zappaterra, M.; Nanni Costa, L.; Padalino, B. The use of infrared thermography (IRT) as stress indicator in horses trained for endurance: A pilot study. *Animals* **2019**, *9*, 84. [CrossRef] [PubMed]
46. Soroko, M.; Spitalniak-Bajerska, K.; Zaborski, D.; Pozniak, B.; Dudek, K.; Janczarek, I. Exercise-induced changes in skin temperature and blood parameters in horses. *Arch. Anim. Breed* **2019**, *62*, 205–213. [CrossRef] [PubMed]
47. Soroko, M.; Zaborski, D.; Dudek, K.; Yarnell, K.; Gorniak, W.; Vardasca, R. Evaluation of thermal pattern distributions in racehorse saddles using infrared thermography. *PLoS ONE* **2019**, *14*, e0221622. [CrossRef] [PubMed]
48. Brownlow, M.; Smith, T. The use of the hand-held infrared thermometer as an early detection tool for exertional heat illness in Thoroughbred racehorses: A study at racetracks in eastern Australia. *Equine Vet. Ed.* **2020**, *33*, 296–305. [CrossRef]
49. Giannetto, C.; Arfuso, F.; Giudice, E.; Gianesella, M.; Fazio, F.; Panzera, M.; Piccione, G. Infrared methodologies for the assessment of skin temperature daily rhythm in two domestic mammalian species. *J. Therm. Biol.* **2020**, *92*, 102677. [CrossRef] [PubMed]
50. Meisfjord Jorgensen, G.H.; Mejdell, C.M.; Boe, K.E. Effects of hair coat characteristics on radiant surface temperature in horses. *J. Therm. Biol.* **2020**, *87*, 102474. [CrossRef]
51. Prochno, H.C.; Barussi, F.M.; Bastos, F.Z.; Weber, S.H.; Bechara, G.H.; Rehan, I.F.; Michelotto, P.V. Infrared thermography applied to monitoring musculoskeletal adaptation to training in Thoroughbred race horses. *J. Equine Vet. Sci.* **2020**, *87*, 102935. [CrossRef]

52. Takahashi, Y.; Ohmura, H.; Mukai, K.; Shiose, T.; Takahashi, T. A comparison of five cooling methods in hot and humid environments in Thoroughbred horses. *J. Equine Vet. Sci.* **2020**, *91*, 103130. [CrossRef]
53. Wilk, I.; Wnuk-Pawlak, E.; Janczarek, I.; Kaczmarek, B.; Dybczynska, M.; Przetacznik, M. Distribution of superficial body temperature in horses ridden by two riders with varied body weights. *Animals* **2020**, *10*, 340. [CrossRef]
54. Maśko, M.; Witkowska-Piłaszewicz, O.; Jasiński, T.; Domino, M. Thermal features, ambient temperature and hair coat lengths: Limitations of infrared imaging in pregnant primitive breed mares within a year. *Reprod. Dom. Anim.* **2021**, *56*, 1315–1328. [CrossRef] [PubMed]
55. Zielińska, P.; Soroko, M.; Howell, K.; Godlewska, M.; Hildebrand, W.; Dudek, K. Comparison of the effect of high-intensity laser therapy (HILT) on skin surface temperature and vein diameter in pigmented and non-pigmented skin in healthy racehorses. *Animals* **2021**, *11*, 1965. [CrossRef] [PubMed]
56. Domino, M.; Borowska, M.; Trojakowska, A.; Kozłowska, N.; Zdrojkowski, Ł.; Jasiński, T.; Smyth, G.; Maśko, M. The Effect of rider:horse bodyweight ratio on the superficial body temperature of horse's thoracolumbar region evaluated by advanced thermal image processing. *Animals* **2022**, *12*, 195. [CrossRef] [PubMed]
57. Klous, L.; Siegers, E.; van den Broek, J.; Folkerts, M.; Gerrett, N.; van Oldruitenborgh-Oosterbaan, M.S.; Munsters, C. Effects of pre-cooling on thermophysiological responses in elite eventing horses. *Animals* **2020**, *10*, 1664. [CrossRef] [PubMed]
58. Yarnell, K.; Fleming, J.; Stratton, T.D.; Brassington, R. Monitoring changes in skin temperature associated with exercise in horses on a water treadmill by use of infrared thermography. *J. Therm. Biol.* **2014**, *45*, 110–116. [CrossRef] [PubMed]
59. Janczarek, I.; Wiśniewska, A.; Tkaczyk, E.; Wnuk-Pawlak, E.; Kaczmarek, B.; Liss-Szczepanek, M.; Kędzierski, W. Effect of different water cooling treatments on changes in rectal and surface body temperature in leisure horses after medium-intensity effort. *Animals* **2022**, *12*, 525. [CrossRef]
60. Hillen, B.; Pfirrmann, D.; Nägele, M.; Simon, P. Infrared thermography in exercise physiology: The dawning of exercise radiomics. *Sports Med.* **2019**, *50*, 263–282. [CrossRef]
61. Geor, R.J.; McCutcheon, L.J.; Ecker, G.L.; Lindinger, M.I. Thermal and cardiorespiratory responses of horses to submaximal exercise under hot and humid conditions. *Equine Vet. J.* **1995**, *27* (Suppl. 20), 125–132. [CrossRef]
62. Geor, R.J.; McCutcheon, L.J.; Ecker, G.L.; Lindinger, M.I. Heat storage in horses during submaximal exercise before and after humid heat acclimation. *J. Appl. Physiol.* **2000**, *89*, 2283–2293. [CrossRef]
63. Li, Q.; Zhang, L.N.; Tao, X.M.; Ding, X. Review of flexible temperature sensing networks for wearable physiological monitoring. *Adv. Healthc. Mat.* **2017**, *6*, 1601371. [CrossRef]
64. Turner, T.A. Diagnostic thermography. *Vet. Clin. N. Am. Equine Pr.* **2001**, *17*, 95–113. [CrossRef] [PubMed]
65. Soroko, M.; Howell, K.; Dudek, K. The effect of ambient temperature on infrared thermographic images of joints in the distal forelimbs of healthy racehorses. *J. Therm. Biol.* **2017**, *66*, 63–67. [CrossRef] [PubMed]
66. Rizzo, L.; Thompson, M.W. Cardiovascular adjustments to heat stress during prolonged exercise. *J. Sports Med. Phys. Fit.* **2018**, *58*, 727–743. [CrossRef] [PubMed]
67. Witkowska-Pilaszewicz, O.; Masko, M.; Domino, M.; Winnicka, A. Infrared thermography correlates with lactate concentration in blood during race training in horses. *Animals* **2020**, *10*, 2072. [CrossRef] [PubMed]
68. Morgan, K.; Funkquist, P.; Nyman, G. The effect of coat clipping on thermoregulation during intense exercise in trotters. *Equine Vet. J.* **2002**, *34* (Suppl. 34), 564–567. [CrossRef] [PubMed]
69. Simon, E.L.; Gaughan, E.M.; Epp, T.; Spire, M. Influence of exercise on thermographically determined surface temperatures of thoracic and pelvic limbs in horses. *J. Am. Vet. Med. Assoc.* **2006**, *229*, 1940–1944. [CrossRef]
70. Jodkowska, M.; Oblacinska, A.; Tabak, I.; Radiukiewicz, K. Differences in dietary patterns between overweight and normal-weight adolescents. *Med. Wieku Rozw.* **2011**, *15*, 266–273.
71. Brownlow, M.; Mizzi, J. Exertional heat illness in Thoroughbred racehorses–Pathophysiology, case definition and treatment rationale. *Equine Vet. Ed.* **2021**, *34*, 259–271. [CrossRef]
72. Brownlow, M.; Mizzi, J.X. Epidemiology of exertional heat illness in Thoroughbred racehorses in temperate eastern Australia: The role of extrinsic (environmental) factors in disease causation. *Equine Vet. Ed.* **2022**, *34*, 660–672. [CrossRef]
73. Kang, H.; Zsoldos, R.R.; Woldeyohannes, S.M.; Gaughan, J.B.; Sole Guitart, A. The Use of Percutaneous Thermal Sensing Microchips for Body Temperature Measurements in Horses Prior to, during and after Treadmill Exercise. *Animals* **2020**, *10*, 2274. [CrossRef]
74. Byrd, S.K.; McCutcheon, L.J.; Hodgson, D.R.; Gollnick, P.D. Altered sarcoplasmic reticulum function after high-intensity exercise. *J. Appl. Physiol.* **1989**, *67*, 2072–2077. [CrossRef] [PubMed]
75. McCutcheon, L.J.; Byrd, S.K.; Hodgson, D.R. Ultrastructural changes in skeletal muscle after fatiguing exercise. *J. Appl. Physiol.* **1992**, *72*, 1111–1117. [CrossRef] [PubMed]
76. Fenemor, S.P.; Gill, N.D.; Sims, S.T.; Beaven, C.M.; Driller, M.W. Validity of a tympanic thermometer and thermal Imaging camera for measuring core and skin temperature during exercise in the heat. *Meas. Phys. Ed Exerc. Sci.* **2020**, *24*, 49–55. [CrossRef]
77. Foster, J.; Hodder, S.G.; Lloyd, A.B.; Havenith, G. Individual responses to heat stress: Implications for hyperthermia and physical work capacity. *Front. Physiol.* **2020**, *11*, 1483. [CrossRef] [PubMed]
78. De Korte, J.Q.; Bongers, C.C.W.G.; Hopman, M.T.E.; Eijsvogels, T.M.H. Exercise performance and thermoregulatory responses of elite athletes exercising in the heat: Outcomes of the thermo Tokyo study. *Sports Med.* **2021**, *51*, 2423–2436. [CrossRef] [PubMed]

79. Westwood, C.S.; Fallowfield, J.L.; Delves, S.K.; Nunns, M.; Ogden, H.B.; Layden, J.D. Individual risk factors associated with exertional heat illness: A systematic review. *Exp. Physiol.* **2021**, *106*, 191–199. [CrossRef]
80. Mostert, H.J.; Lund, R.J.; Guthrie, A.J.; Cilliers, P.J. Integrative model for predicting thermal balance in exercising horses. *Equine Vet. J.* **1996**, *28* (Suppl. 22), 7–15. [CrossRef]
81. Eggenberger, P.; MacRae, B.A.; Kemp, S.; Burgisser, M.; Rossi, R.M.; Annaheim, S. Prediction of core body temperature based on skin temperature, heat flux, and heart rate under different exercise and clothing conditions in the heat in young adult males. *Front. Physiol.* **2018**, *9*, 1780. [CrossRef]
82. Tanda, G. A simplified approach to describe the mean skin temperature variations during prolonged running exercise. *J. Therm. Biol.* **2021**, *99*, 103005. [CrossRef]
83. Pérez-Guarner, A.; Priego-Quesada, J.I.; Oficial-Casado, F.; Cibrián Ortiz de Anda, R.M.; Carpes, F.P.; Palmer, R.S. Association between physiological stress and skin temperature response after a half marathon. *Physiol. Meas.* **2019**, *40*, 034009. [CrossRef]
84. McConaghy, F.F.; Hodgson, D.R.; Rose, R.J.; Hales, J.R. Redistribution of cardiac output in response to heat exposure in the pony. *Equine Vet. J.* **1996**, *28* (Suppl. 22), 42–46. [CrossRef] [PubMed]
85. Robinson, T.R.; Hussey, S.B.; Hill, A.E.; Heckendorf, C.C.; Stricklin, J.B.; Traub-Dargatz, J.L. Comparison of temperature readings from a percutaneous thermal sensing microchip with temperature readings from a digital rectal thermometer in equids. *J. Am. Vet. Med.* **2008**, *233*, 613–617. [CrossRef] [PubMed]
86. Wartzek, T.; Muhlsteff, J.; Imhoff, M. Temperature measurement. *Biomed. Tech.* **2011**, *56*, 241–257. [CrossRef] [PubMed]
87. Holcomb, K.E.; Tucker, C.B.; Stull, C.L. Physiological, behavioral, and serological responses of horses to shaded or unshaded pens in a hot, sunny environment1. *J. Sci.* **2013**, *91*, 5926–5936. [CrossRef] [PubMed]
88. Raymond, C.; Matthews, T.; Horton, R.M. The emergence of heat and humidity too severe for human tolerance. *Sci Adv.* **2020**, *6*, eaaw1838. [CrossRef] [PubMed]
89. Morgan, E.K.M. Climatic energy demand of horses. *Equine Vet. J.* **1995**, *27*, 396–399. [CrossRef]
90. Guthrie, A.J.; Lund, R.J. Thermoregulation. Base mechanisms and hyperthermia. *Vet. Clin. N. Am. Equine Pr.* **1998**, *14*, 45–59. [CrossRef]
91. Mayhew, I.G.; Ferguson, H.O., 2nd. Clinical, clinicopathologic, and epidemiologic features of anhidrosis in central Florida Thoroughbred horses. *J. Vet. Intern. Med.* **1987**, *1*, 136–141. [CrossRef]
92. Johnson, S.R.; Rao, S.; Hussey, S.B.; Morley, P.S.; Traub-Dargatz, J.L. Thermographic eye temperature as an index to body temperature in ponies. *J. Equine Vet. Sci.* **2011**, *31*, 63–66. [CrossRef]
93. Hetem, R.S.; Mitchell, D.; de Witt, B.A.; Fick, L.G.; Meyer, L.C.; Maloney, S.K.; Fuller, A. Cheetah do not abandon hunts because they overheat. *Biol. Lett.* **2013**, *9*, 20130472. [CrossRef]
94. Fernandes, A.d.A.; Amorim, P.R.d.S.; Brito, C.J.; de Moura, A.G.; Moreira, D.G.; Costa, C.M.A.; Sillero-Quintana, M.; Marins, J.C.B. Measuring skin temperature before, during and after exercise: A comparison of thermocouples and infrared thermography. *Physiol. Meas.* **2014**, *35*, 189–203. [CrossRef] [PubMed]
95. James, C.A.; Richardson, A.J.; Watt, P.W.; Maxwell, N.S. Reliability and validity of skin temperature measurement by telemetry thermistors and a thermal camera during exercise in the heat. *J. Therm. Biol.* **2014**, *45*, 141–149. [CrossRef] [PubMed]
96. Ramey, D.; Bachmann, K.; Lee, M.L. A comparative study of non-contact infrared and digital rectal thermometer measurements of body temperature in the horse. *J. Equine Vet. Sci.* **2011**, *31*, 191–193. [CrossRef]
97. Moreira, D.G.; Costello, J.T.; Brito, C.J.; Adamczyk, J.G.; Ammer, K.; Bach, A.J.E.; Costa, C.M.A.; Eglin, C.; Fernandes, A.A.; Fernández-Cuevas, I.; et al. Thermographic imaging in sports and exercise medicine: A Delphi study and consensus statement on the measurement of human skin temperature. *J. Therm. Biol.* **2017**, *69*, 155–162. [CrossRef]
98. Cuddy, J.S.; Buller, M.; Hailes, W.S.; Ruby, B.C. Skin temperature and heart rate can be used to estimate physiological strain during exercise in the heat in a cohort of fit and unfit males. *Mil. Med.* **2013**, *178*, e841–e847. [CrossRef]
99. Welles, A.P.; Buller, M.J.; Looney, D.P.; Rumpler, W.V.; Gribok, A.V.; Hoyt, R.W. Estimation of metabolic energy expenditure from core temperature using a human thermoregulatory model. *J. Therm. Biol.* **2018**, *72*, 44–52. [CrossRef]

animals

MDPI

Article

Thermographic Profiles in Livestock Systems under Full Sun and Shaded Pastures during an Extreme Climate Event in the Eastern Amazon, Brazil: *El Niño* of 2023

Welligton Conceição da Silva [1,*], Jamile Andréa Rodrigues da Silva [2], Lucieta Guerreiro Martorano [3], Éder Bruno Rebelo da Silva [1], Carlos Eduardo Lima Sousa [4], Kedson Alessandri Lobo Neves [5], Cláudio Vieira de Araújo [6], Leonel António Joaquim [1], Thomaz Cyro Guimarães de Carvalho Rodrigues [1], Tatiane Silva Belo [4], Raimundo Nonato Colares Camargo-Júnior [1] and José de Brito Lourenço-Júnior [1]

1 Postgraduate Program in Animal Science (PPGCAN), Institute of Veterinary Medicine, Federal University of Para (UFPA), Castanhal 68746-360, PA, Brazil; eder.b.rebelo@gmail.com (É.B.R.d.S.); jleokim.a@hotmail.com (L.A.J.); thomazguimaraes@yahoo.com.br (T.C.G.d.C.R.); camargojunior@gmail.com (R.N.C.C.-J.); joselourencojr@yahoo.com.br (J.d.B.L.-J.)

2 Institute of Animal Health and Production, Federal Rural University of the Amazon (UFRA), Belem 66077-580, PA, Brazil; jamileandrea@yahoo.com.br

3 Embrapa Eastern Amazon, Santarem 68010-180, PA, Brazil; lucieta.martorano@embrapa.br

4 Department of Veterinary Medicine, University Center of the Amazon (UNAMA), Santarem 68010-200, PA, Brazil; cadu34.medvet@gmail.com (C.E.L.S.); tatianebelovet@gmail.com (T.S.B.)

5 Institute of Animal Science, Federal University of Western Pará (UFOPA), Santarem 68040-255, PA, Brazil; kedson_neves@hotmail.com

6 Department of Agricultural and Environmental Sciences, Federal University of Mato Grosso (UFMT), Sinop 78550-728, MT, Brazil; cvaufmt@gmail.com

* Correspondence: welligton.medvet@gmail.com; Tel.: +55-(93)-988070692

Citation: Silva, W.C.d.; Silva, J.A.R.d.; Martorano, L.G.; Silva, É.B.R.d.; Sousa, C.E.L.; Neves, K.A.L.; Araújo, C.V.d.; Joaquim, L.A.; Rodrigues, T.C.G.d.C.; Belo, T.S.; et al. Thermographic Profiles in Livestock Systems under Full Sun and Shaded Pastures during an Extreme Climate Event in the Eastern Amazon, Brazil: *El Niño* of 2023. *Animals* **2024**, *14*, 855. https://doi.org/10.3390/ani14060855

Academic Editor: Daniel Mota-Rojas

Received: 2 February 2024
Revised: 29 February 2024
Accepted: 29 February 2024
Published: 11 March 2024

Simple Summary: The *El Niño* presents itself as a serious problem in the pastures of the northern region of Brazil, as it compromises the availability and quality of forage and water. Therefore, the objective of this study was to characterize the thermographic profile of three production systems in the Eastern Amazon, Brazil. The results show significant differences between areas with and without chestnut tree shade. Between August and November, the highest temperatures were recorded in full sun pastures, contrasting with lower temperatures in shaded areas. The interaction between the systems revealed significant thermal variations, highlighting the positive impact of native trees on thermal regulation and indicating possible strategies to mitigate the adverse effects of *El Niño*.

Abstract: The *El Niño* represents a substantial threat to pastures, affecting the availability of water, forage and compromising the sustainability of grazing areas, especially in the northern region of Brazil. Therefore, the objective of this study was to characterize the thermographic profile of three production systems in the Eastern Amazon, Brazil. The study was conducted on a rural cattle farm in Mojuí dos Campos, Pará, Brazil, between August and November 2023. The experiment involved livestock production systems, including traditional, silvopastoral and integrated, with different conditions of shade and access to the bathing area. An infrared thermographic (IRT) camera was used, recording temperatures in different zones, such as areas with trees, pastures with forage and exposed pastures. The highest mean temperatures ($p = 0.02$) were observed in pastures with full sun from August to November. On the other hand, the lowest average temperatures were recorded in areas shaded by chestnut trees (*Bertholletia excelsa*). The highest temperature ranges were found in sunny areas and the lowest were recorded in shaded areas. The highest temperatures were observed in the pasture in full sun, while the lowest were recorded in areas shaded by chestnut trees ($p < 0.0001$). The interaction between the systems and treatments revealed significant temperature differences ($p < 0.0001$), with the native trees showing an average temperature of 35.9 °C, lower than the grasses and soil, which reached 61.2 °C. This research concludes that, under *El Niño* in the Eastern Amazon, areas shaded by Brazil nut trees had lower temperatures, demonstrating the effectiveness of shade. Native trees, compared to grasses and soil, showed the ability to create cooler environments, highlighting the positive influence on different species such as sheep, goats and cattle.

Keywords: grazing areas; thermal images; shade cover; Eastern Amazon region

1. Introduction

Three systems of cattle farming can be found in the Eastern Amazon: traditional, which is the one most cattle farmers use, with no trees or bathing places for the animals; integrated, with shade and water for bathing; and silvopastoral, with shade from trees but no water for bathing. Thus, systems where the animals have access to water from existing rivers and lakes to cool off during periods of intense solar radiation and areas with pastures shaded by native tree species are important to ensure the animals' thermal comfort during the period of highest solar radiation. However, there are still systems where pastures are predominantly in full sun, with access to water only at drinking fountains [1–3]. A major scientific concern is reconciling animal welfare with productive performance in Brazilian livestock farming [4–7].

Exposure to direct sunlight makes animals susceptible to heat stress. Heat stress has a negative impact on the animal's physiology and productivity, as the animal's heat dissipation capacity can be exceeded [8,9]. Therefore, animal welfare (AW) advocates the combination of environment and productivity, in which behavioral reactions can be observed in animals, depending on where they are, influencing production and economic impacts [6,10–14].

Even considering that cattle have thermoregulatory mechanisms that help dissipate accumulated internal heat, such as increased vasodilation, respiratory rate, heart rate, body and rectal temperature, as well as frequent urination and defecation, it is necessary to observe the limiting conditions in terms of extreme thermal values, especially in regions with high temperatures, as is the case in regions in the equatorial belt. Research shows that high or very low temperatures tend to hinder thermoregulation, leading to various physiological reflexes, such as increased heart and respiratory rates, sialorrhea and metabolic or reproductive problems, such as an increase in cortisol, glucose and a reduction in the animals' digestion process, as well as a reduction in the quality of female oocytes and spermatogenesis, causing losses. In this sense, cattle regulate the excess solar radiation that comes into direct contact with their bodies by seeking shade, shelter or water to mitigate heat stress [15–22].

Among the technologies used to investigate animal comfort and environmental conditions quickly, non-invasively and with reliable and replicable results, monitoring using near-infrared thermography has been gaining ground, with applications in different areas of scientific knowledge, such as early diagnosis of diseases, non-invasive assessment of rectal temperature, determination of animal thermal stress and thermal characterization of the environment in which they live [23,24]. The radiometric imaging allow data to be obtained by evaluating targets such as the surface of animals and correlating them with the production systems studied [25–30].

Thus, characterizing the thermographic profile of different animal production systems in the Brazilian Amazon becomes important due to the region's peculiar environmental conditions. The Amazon faces intense climatic challenges, such as high temperatures and humidity, which can have a significant impact on animal welfare and performance. By evaluating the thermal variations of the environment in which the animals live, thermography allows for a more in-depth understanding of the thermal conditions of production systems, which makes it possible to adopt management strategies to improve the productive efficiency and health of the herds in each system [28,31–35].

Some factors are beyond human control, such as years when extreme climatic phenomena occur, for example, the effects of *El Niño*/Southern Oscillation years, which in the Amazon have been associated with an intensification of the period of reduced rainfall in the region, causing periods of water scarcity in the soil and an increase in air and soil temperature. This adverse climate pattern represents a significant threat to pastures and, consequently, to the animals that depend on these grazing areas to guarantee the daily food supply of their livestock [36,37]. Therefore, rising temperatures and decreasing rainfall during *El Niño* events contribute to a water deficit in pastures, reducing the quantity and quality of plants available for animal grazing. In addition, water scarcity and reduced forage availability directly compromise pasture capacity, further increasing animal nutrition and health problems [38–41]. Water stress can also cause soil degradation, making it more susceptible to erosion. In addition, the lack of water and feed in pastures during periods of *El Niño* can force farmers to resort to alternative practices [42–45].

Furthermore, understanding how animal production systems interact with the environment, it is possible to develop more sustainable practices and ensure more efficient production. Thus, integrating thermographic data into cattle management contributes to developing regional production systems. Furthermore, this study is a first in the thermal characterization of pastures during *El Niño* in the Eastern Amazon. The hypothesis in this study was that the environment during the period of the extreme weather event would be challenging and high in temperature and that the trees would be lower in temperature, providing a suitable environment for the animals. Therefore, this study aimed to characterize the thermographic profile of three different production systems in the Eastern Amazon, Brazil.

2. Materials and Methods

2.1. Location of the Study Area and Evaluation Period

The experiment was carried out in the municipality of Mojuí dos Campos, Pará State, Brazil, on a farm, during the rainy season (December–May) and the least rainy period (June–November) and during the transition periods of the year (from the rainiest to the least rainy—June/July and from the least rainy to the rainiest—December/January). The thermal profile presented in this work refers to the period between August and November 2023. It is worth noting that even in the wettest regions of the state of Pará, in *El Niño* years, from August onwards, estimates using the water balance considering the dominant root zone in pastures already indicate water deficits in the soil [46,47]. The months of October to November are considered the months with the highest heat index in different regions, including the north of the country [35].

Figure 1 shows the location map of the study area, which was carried out in an area with livestock in the municipality of Mojuí dos Campos (average altitude of 127 m, latitude 02°10′17″ S and longitude 56°44′42″ W), in the west of the state of Pará, which is part of the Eastern Amazon production hub.

2.2. Climate Information for the Region

The climate in the region is humid, with rainfall totals of less than 60 mm in the less rainy months and rainfall ranging from 1900 to 2100 mm, in conditions governed by the Am4 typology, in which the average annual air temperature is around 25.6 °C and the relative humidity varies from 84 to 86% [46]. In the mesoregion, the wettest quarter occurs between the months of February and April and the least rainy between August and October [47].

2.3. Animals and Production Systems

The study was carried out with 30 cattle weighing 250 ± 36 kg of the white Nelore breed, aged between 18 and 20 months, all uncastrated males, classified as clinically healthy, divided into three farming systems, with an ideal body score of 3.5, on a scale of 1 to 5. On the property with a livestock system, the forage *Brachiaria brizantha* cv. Marandú was planted and managed in rotational grazing with 14 days of occupation and 28 days of rest per paddock. When the pasture was planted, the necessary nutrients were applied for the two periods of the year, corresponding to the rainiest and least rainy, according to the soil analysis.

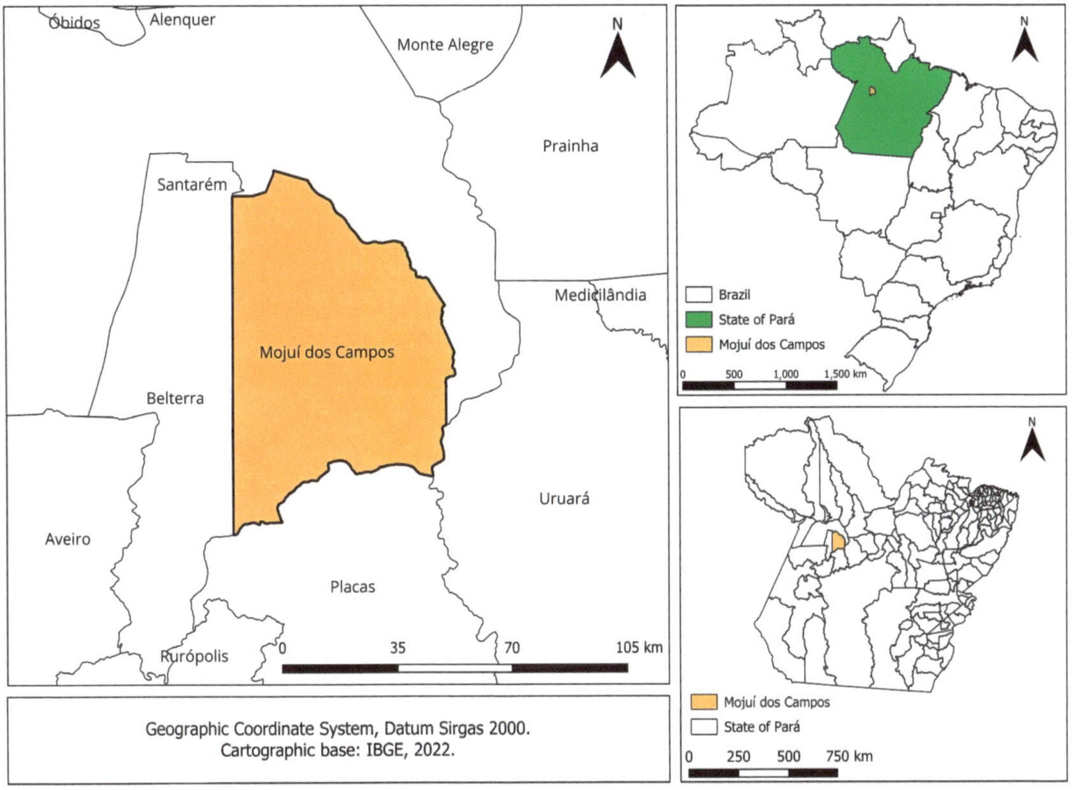

Figure 1. The location of the municipality where the experiment was carried out in the Eastern Amazon, Brazil.

The production systems used in this study were classified as a traditional system (TS) when there were no trees in the paddocks, but they had free access to water for consumption, while in the silvopastoral system (SP) the animals were in an environment with shade from trees and access to water for consumption, while in the integrated system (IS) there was access to shade from trees, water for consumption and also for bathing.

The TS was composed only of *Brachiaria brizantha* cv. Marandú forage grass, with no trees or areas of shade. In the SP and IS, the paddocks also had the same forage plant cover, with the presence of trees native to the Amazon, such as Brazil nut trees (*Bertholletia excelsa*), in both groups. However, the animals in the SP only had water to drink, while in the IS the animals had water for both bathing and drinking. In addition, salt was offered ad libitum in all the TS, SP and IS systems. The total experimental area was 15.3 ha of *Brachiaria brizantha* cv. Marandú is divided into nine 1.7 ha paddocks, three per treatment.

2.4. Infrared Thermography

The data were recorded using an IRT camera [48], with emissivity set at 0.95. The camera used has a fixed lens of 25 mm, temperature range from −40 to 150 °C, sensitivity of 50 mk (>0.05 °C ambient temperature of 30 °C), spectral range from 0.7 to 100 μm, but the photographed targets have a response between 0.7 and 3.0 μm with an optical resolution of 640 × 480 pixels with a maximum emissivity index of 0.9. Temperature and humidity were controlled by positioning against the wind and the observer was positioned against the sun so as not to interfere with the camera. The thermograms were then examined using Flir Tools software, version 6.3, with the Rainbow HC palette chosen. The operator used the camera at eye level, always maintaining consistency and without the aid of a tripod, focusing perpendicular to the target, following the methodological guidelines proposed by Barreto et al. [49] and Silva et al. [50].

For each zone assessed, eight radiometric images were analyzed for soil temperatures in areas with trees, pastures with forage and exposed pastures. For each thermogram and for and each color (T1—white, T2—red, T3—yellow and T4—green), 10 thermographic points were imaged (Figure 2A–D). The images were captured between 12:30 p.m. and 1:30 p.m. in August and November 2023, corresponding to the dry season in the region, which was affected by a strong *El Niño* event. In August and September, soil water deficits can be seen, even in places with higher rainfall (climate typology Af2), such as Belém. In this region, soil water deficits begin in August and persist until November, with October being the month of greatest water scarcity. Because of this radiometric images were taken in this region in September and October to highlight and monitor thermal responses in dry soil in the face of climate changes observed during the *El Niño* climate event.

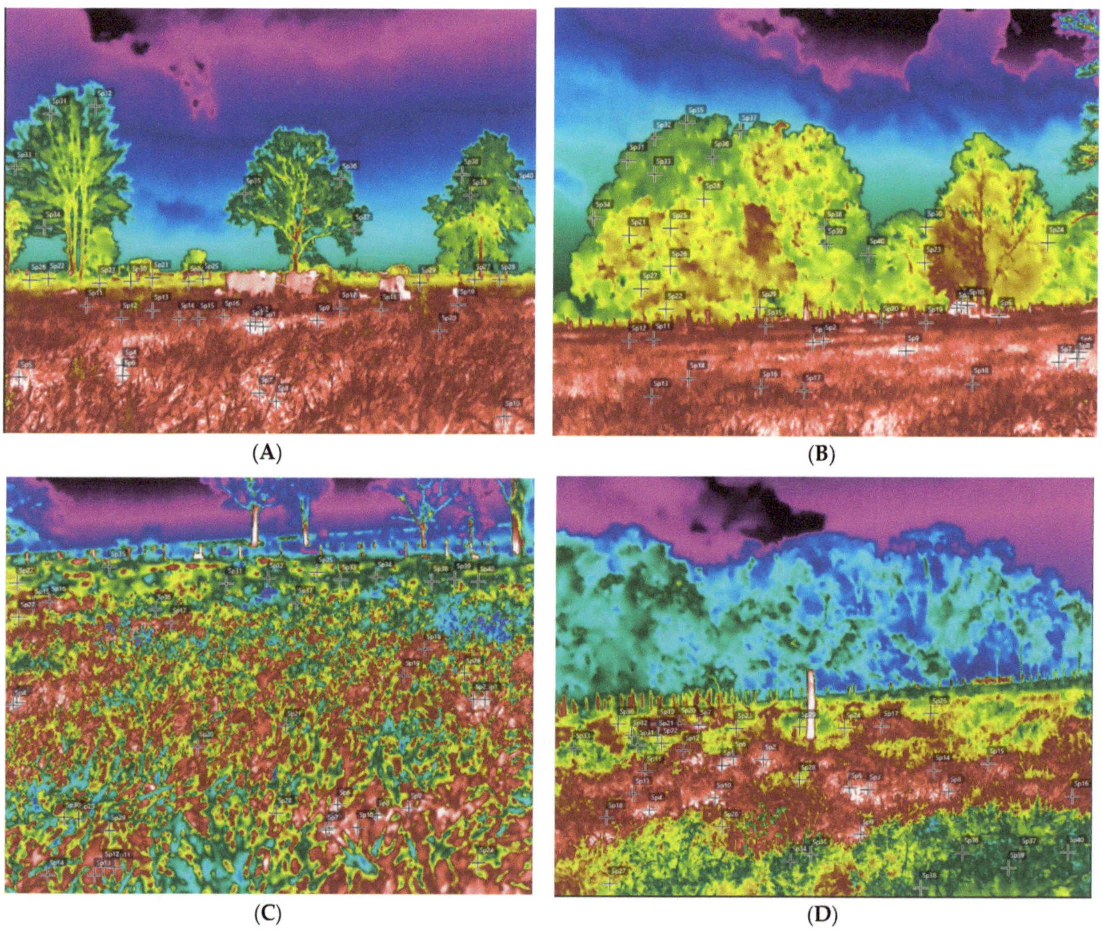

Figure 2. Thermographic images of the shaded paddock. (**A**) Area of ground and trees imaged. (**B**) Full sun areas imaged. (**C**) Soil exposed in a sunny area. (**D**) Exposed soil and forage.

2.5. Statistical Analysis

To verify the homogeneity of the variance of the residuals, the Bartlett's test was used and the Shapiro–Wilk test was used to evaluate the normality of these residuals. In both cases, the hypothesis of nullity was accepted. The temperature was thus considered a continuous random variable with a normal distribution and treatments showed homogeneity of variance and were analyzed in an entirely randomized design and a subdivided plot structure, according to the line-air model:

$$Yijk = \mu + \alpha\_i + D\_((i)k) + \beta\_j + (\alpha\beta)\_ij + \varepsilon\_((ij)k)$$

where Yijk is the vector of temperature observations in the i-th system, in the j-th treatment of the k-th repetition, the term "μ" is a general constant inherent to all observations in Yijk; α_i represents the systems (i = 1 and 2); the term D_((i)k) re-presents the plot error, characterized by the "k" repetitions nested in the "i" levels of systems, with k = 10 animals; β_j represents the effects of the treatments arranged in the subplot, with j = 1, 2, 3 and 4 (T1 = white color; T2 = red color; T3 = yellow color; T4 = green color); the term (αβ)_ij refers to the interaction between the i-th level of systems with the j-th level of treatments and is the error associated with the subplot that received level i of systems and level j of the treatment factor in repetition k, with ~NIID(0,σ_e^2).

The Tukey's mean comparison test was used to compare levels of treatment effects, with a significance level of 0.05. The Package Statistical Analysis System [51] was used for the statistical analysis.

3. Results

The climate parameters assessed were air temperature (TA °C) and relative humidity (RH %), from the agrometeorological mini-stations set up in the experimental area. Figure 3 illustrates the agrometeorological conditions, expressed by air temperature and humidity, in the study area, showing the effect of the *El Niño* severe weather event. During this period, the temperature reached 37.7 °C at the time of highest heat load.

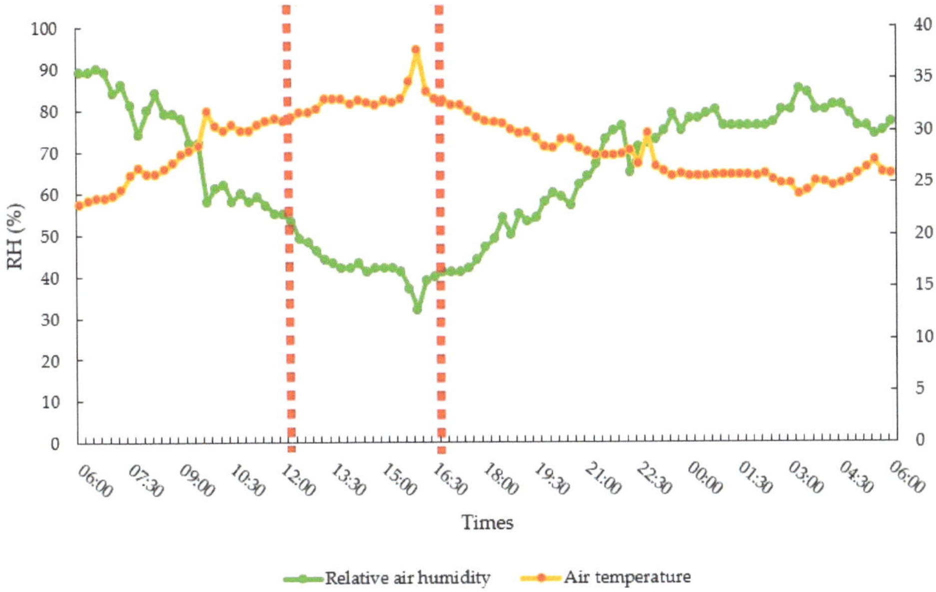

Figure 3. Weather data averaged from August to November 2023. Vertically dotted red bars show the time the thermographic images were taken.

Figure 4A–D shows the evolution of the pasture during the collection of thermographic data over the months. It is possible to see the pasture with a water deficit as the months go by, and the soil becoming exposed due to the death of the forage plants, which is very evident from September 2023 onwards.

(A)

(B)

(C)

(D)

Figure 4. Images of the pasture area according to the months of the year during *El Niño*. (**A**) August. (**B**) September. (**C**) October. (**D**) November.

Table 1 shows the extreme temperature values (minimum and maximum), average temperature and descriptive analysis of the SH and SN environments. From August to November, the highest average temperatures were observed in the full sun pasture (*Brachiaria brizantha* cv. Marandú) (p = 0.02). On the other hand, the lowest average temperatures were recorded in areas shaded by chestnut trees (*Bertholletia excelsa*). The highest temperature ranges were found in sunny areas and the lowest in shaded areas.

Table 1. Descriptive data of the thermal targets analyzed in the different production systems in the Eastern Amazon.

Variables	Systems							
	SH				SN			
	T1	T2	T3	T4	T1	T2	T3	T4
Sample size	10	10	10	10	10	10	10	10
Minimum	36.1	36.2	34.8	32.7	38.9	36.2	34.3	32.7
Maximum	40.9	37.3	34.6	33.7	40.9	37.3	34.6	33.9
Total Amplitude	4.8	1.0	0.3	0.9	2.0	1.09	0.3	0.9
Median	39.29	36.4	34.5	33.3	39.5	36.4	34.5	33.3
First Quartile (25%)	39.1	36.2	34.4	33.1	39.1	36.2	34.4	33.1
Third Quartile (75%)	40.5	36.8	34.6	33.5	40.5	36.8	34.6	33.5
Interquartile deviation	1.4	0.5	0.1	0.3	1.3	0.5	0.1	0.3
Arithmetic average	39.4	36.5	34.5	33.2	39.7	36.5	34.5	33.2
Variance	1.9	0.1	0.01	0.8	0.6	0.1	0.01	0.08
Standard deviation	1.39	0.7	0.3	0.1	0.2	0.36	0.1	0.29
Standard Error	0.4	0.2	0.1	0.03	0.09	0.1	0.03	0.09
Coefficient of variation	3.55	2.0	0.9	0.2	0.8	0.9	0.2	0.8
Asymmetry	−1.32	0.6	0.9	0.1	−0.4	0.9	0.1	−0.4
Kurtosis	2.9	−1.4	0.1	−1.1	−0.4	0.1	−1.1	−0.4

Note: SH = Shadow, SN = Sun; T1 = white, T2 = red, T3 = yellow, T4 = green.

There were significant differences between the treatments in full sun and under the shade of the Brazil nut tree (*Bertholletia excelsa*) for the thermal responses at the imaged points (Figure 5). For the sun and shade treatments where the temperatures were higher, within the lighter-colored thermal patterns (Figure 5A,B), it can be seen that the temperatures in the pasture area in full sun were higher than those in the shade ($p < 0.0001$). However, it was possible to observe that the darker color patterns showed lower temperatures in both treatments (Figure 5C,D) ($p < 0.0001$).

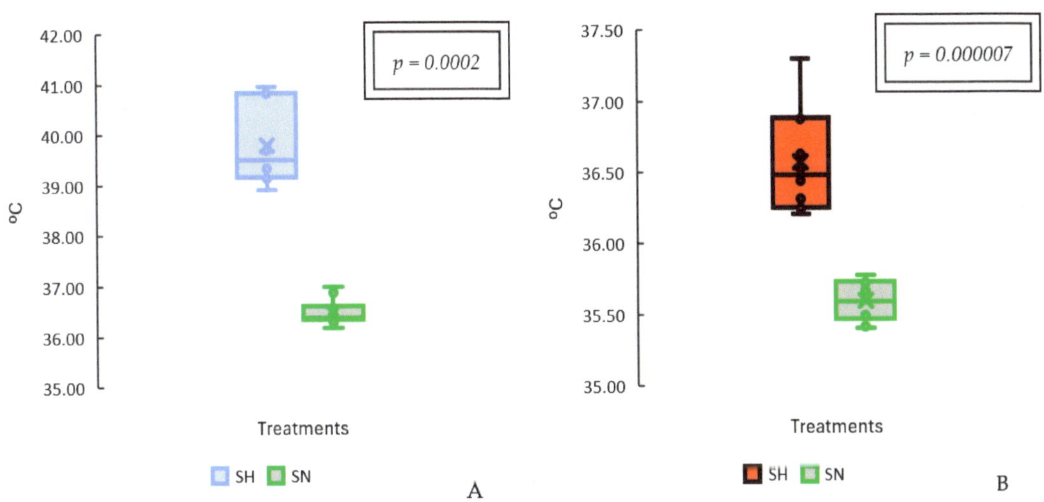

Figure 5. *Cont.*

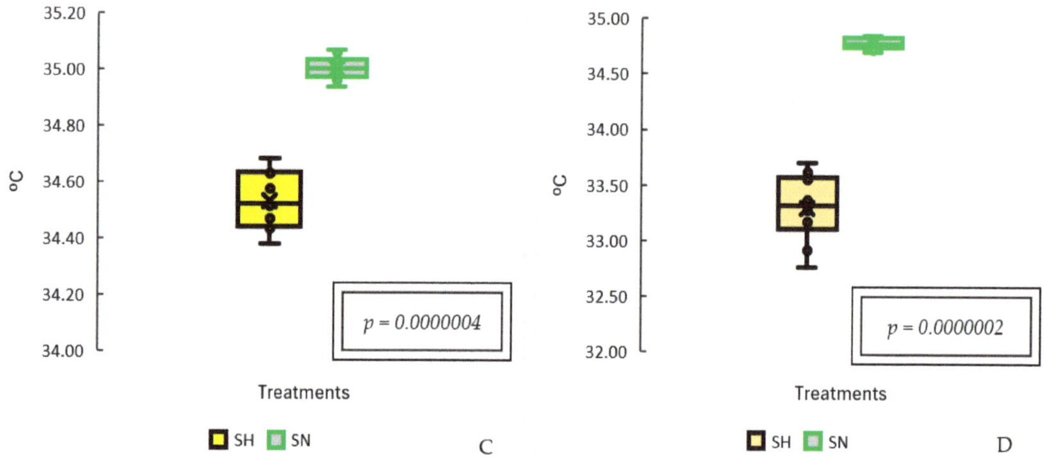

Figure 5. Estimated means and standard deviations for temperature in each combination of system levels and treatments. (**A**) white color standard. (**B**) red color standard. (**C**) yellow color standard. (**D**) green color standard. SH = Shadow, SN = Sun.

There was a significant interaction effect between the effects of systems and treatments ($p < 0.0001$). In the sun system, the highest temperature values were found for treatments T1 and T2, respectively, with treatments T3 and T4 showing similar temperatures to each other and lower than the other treatments (Table 2). In the chestnut tree (*Bertholletia excelsa*) shade system, all the treatments differed in temperature, with higher average values for T1, T2, T2 and T4, in that respective order.

Table 2. Estimated means and standard deviations for temperature in each combination of system levels and treatments.

	SH		SN		Total	
	Average	**SD**	**Average**	**SD**	**Average**	**SD**
T1	36.472 a	0.255	39.796 a	0.798	38.134	1.800
T2	35.592 b	0.141	36.572 b	0.360	36.082	0.569
T3	35.005 c	0.040	34.534 c	0.101	34.769	0.253
T4	34.782 c	0.050	33.296 d	0.297	34.039	0.79
Total	35.463	0.677	36.049	2.530	35.756	1.863

Note: Different letters in the column indicate statistical differences ($p < 0.05$). SH = Shadow, SN = Sun. SD = standard deviations. T1 = white, T2 = red, T3 = yellow, T4 = green.

The average temperature of the native trees in the Amazon region was 35.9 °C, 25.3 °C lower than the temperature of the grasses and soil, which reached 61.2 °C (Figure 6) ($p < 0.05$).

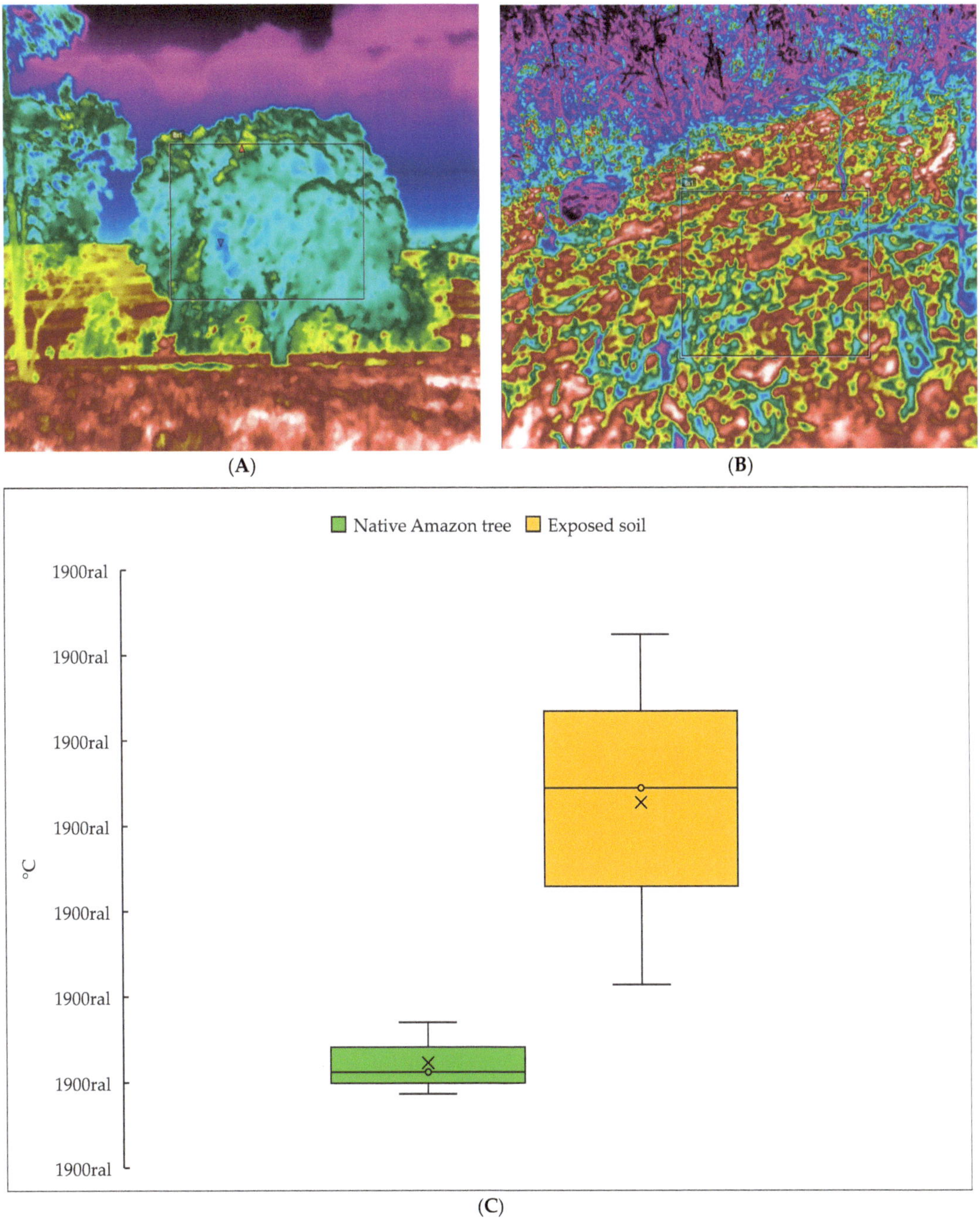

Figure 6. Thermal images and box plots of the minimum, maximum and average temperatures evaluated. (**A**) native Amazon tree. (**B**) exposed soil. (**C**) Averages for exposed soil (yellow) and shaded areas (green).

4. Discussion

The temperatures recorded in the thermal patterns in this study demonstrate a challenging scenario for raising beef cattle in a tropical region, being outside the range considered ideal by Baêta and Souza [52], who established favorable climatic conditions between 10 and 27 °C. The constant increase in temperatures indicates the need for innovative approaches to tackling the thermal challenges associated with livestock farming in this environment.

The study highlights the role of shading provided by trees in silvopastoral systems as an effective strategy to mitigate the adverse impacts of heat, thus improving the thermal comfort of animals, especially during the hottest periods of the day. Thus, trees in pasture areas are important to help with the thermo-regulation of different species of animals, such as cattle and sheep, as the trees block the passage of the sun's rays, making the environment cooler and more comfortable, reducing heat stress and improving the well-being and health of the animals. In addition, the use of infrared thermography in the silvopastoral system in the western region of the state of Pará offers a more detailed perspective of the thermal variations in the environment during *El Niño*.

Numerous studies have highlighted the relationships between microclimate variables and canopy thermal characteristics in pasture environments [35,53–58]. Do not consider these interactions to influence substantial errors in climate measurements, due to temperature variation between the ground surface and the air masses above the canopy [59]. Thus, this study provides direct evidence of climate change, as it presents the effects of *El Niño* in the Amazon region, as well as its repercussions for the region's ecosystem [60–62], especially the low availability of pasture as a source of food for animals.

This study is a pioneer in the characterization of infrared thermal patterns during *El Niño* in the Amazon region. The evolution of the pasture observed during the collection of thermographic data over the months during the *El Niño* climatic event reveals the presence of a water deficit, especially from September onwards, resulting in the exposure of the soil due to the death of the forage plants. This can be explained because in Brazil, *El Niño* has been associated with increased risks of drought and higher temperatures [63–67], especially in the northern regions of the country.

The adverse weather pattern presented by *El Niño* poses a significant threat to pastures and, consequently, to the animals that depend on these areas for food. The high temperatures and decreased rainfall during *El Niño* contribute to a water deficit in pastures, resulting in a reduction in the quantity and quality of plants available for animal grazing [68–70]. Thus, the concomitant reduction in precipitation and increase in atmospheric vapour pressure deficit (VPD) has led to increased tree mortality [71–76].

This condition, coupled with water scarcity and reduced forage availability, directly compromises the ability of pastures to support animals, leading to nutrition and health problems. In addition, water stress can cause soil degradation, making it more susceptible to erosion. In addition, the lack of water and feed in pastures during *El Niño* periods can force farmers to resort to alternative practices, which can have significant economic and social impacts [77,78].

The data presented in Table 1 reveal distinct patterns in thermal variations between sunny and shaded environment from August to November. The highest average temperatures were recorded in the full sun pasture (*Brachiaria brizantha* cv. Marandú). This result is in line with the scientific understanding that areas directly exposed to solar radiation tend to experience higher temperatures due to the direct absorption of solar energy. The increase in temperature results in a greater thermal load on the environment, which consequently translates into an intensification in the emission of infrared radiation [28,79].

On the other hand, areas shaded by chestnut trees (*Bertholletia excelsa*) had the lowest was recorded average temperatures. The presence of trees or shade structures plays a crucial role in local thermal regulation, providing a cooler environment. Shade reduces direct exposure to the sun's rays, minimizing heat absorption and therefore resulting in milder temperatures, as observed in the data by Silva et al. [6].

The results indicate that the highest thermal amplitudes were found in areas with direct exposure to the sun, while the lowest were recorded in shaded areas. This observation is in line with meteorological principles, as environments more exposed to the sun tend to experience a wide variation in temperatures between day and night due to the greater absorption and release of heat.

The thermal oscillations identified between the environments indicate the ability of the area with the presence of vegetation to alter the environment and its thermal conditions, as this has a direct impact on the incidence of solar radiation. The hottest areas were in places with no vegetation cover, indicating that these areas tend to absorb more thermal radiation, which raises the temperature of these areas, affecting the physiology of the grasses, water availability and the ecology of the environment.

The shade areas proved to be effective in reducing solar radiation, corroborating the previous findings by Lopes et al. [80] and Oliveira et al. [81], who observed thermal comfort in integrated crop-livestock-forest (ICLF) systems due to the reduction of solar radiation provided by the presence of trees. The density and spatial arrangement of trees in integrated pasture production systems induce significant microclimatic changes, even at a distance of 50 m from the trees [81–84].

The average temperature of 35.9 °C in the native trees represents a more moderate condition compared to the extremely high temperature of 61.2 °C recorded in the soil and grasses. This difference can be explained by the shading phenomenon provided by the treetops. The presence of trees in the Amazon region creates a cooler environment and attenuates direct exposure to the sun, limiting the amount of solar radiation that reaches the ground [35]. In addition, the transpiration from the trees helps to cool the environment, maintaining milder temperatures in the surroundings.

The temperature reductions identified in this study exceeded those observed by Barreto et al. [49], where the maximum reduction was 3.7% in the treatment with trees on pasture.

According to Silva [85], in pastures with the presence of trees, a direct impact was observed on the energy balance of the system, making it possible to reduce up to 80% of the incident solar radiation and 30% of the thermal load on the animals. Ref. [49], in Mato Grosso do Sul, Brazil, the presence of native trees reduced thermal load by 24.0% in July, 16.3% in August and 29.6% in September. In the same study, an integrated system with eucalyptus shade reduced thermal load by 17.0% in July, 4.6% in August and 7.5% in September. Ref. [86] indicated that at noon the reduction of the thermal load was 17.8% in the silvopastoral system compared to another treatment without the presence of shading.

It should be noted that the behavioral parameters of grazing cattle were not assessed in this study, but it was possible to observe in the field that in the hottest hours these animals seek shaded areas, a fact reported in the study by [23,80,87–89]. In general, animals look for areas with the presence of shadows during times of greater incidence of solar radiation since they are more susceptible to heat stress. The implementation of trees in pastures has become a sustainable option [90], due to improved welfare for livestock in tropical regions [28,50,91].

In addition to the benefit of shade for animals, silvopastoral systems have the potential to contribute to reducing greenhouse gas emissions by removing carbon from the atmosphere and storing biomass in the soil [92]. In addition, trees can become an extra source of income for landowners [93]. In a study involving buffalo, Garcia et al. [94] and Brcko et al. [3] pointed out that the inclusion of trees in tropical pastures promoted greater thermal comfort for the animals.

In addition, the shading of trees used in silvopastoral systems provides numerous benefits for both the animals and the environment, as it reduces greenhouse gas emissions and removes carbon from the atmosphere by storing it in the soil [92]. Thus, trees can become a source of income for landowners in an additional way [93]. Ref. [94] and Brcko et al. [3], in a study with buffaloes, observed that using trees in tropical pastures promoted greater thermal comfort.

The fact that trees promote well-being and, consequently, shelter animals in periods of greater solar radiation intensity was evidenced in this study, as observed in Figure 7A,B.

(A) **(B)**

Figure 7. Cattle in shaded areas (**A**,**B**).

Climate changes resulting from climatic events such as *El Niño* and their impacts on the pasture environment in the Amazon region, as observed in this study, serve as a subsidy for further research on the effects of changes associated with temperature and humidity on pasture productivity, the importance of shading and bathing areas during the hottest periods, as well as sustainable measures to reduce the impacts of climate change on the nutritional quality of forage and consequently on the diet of animals, in addition to developing sustainable strategies for the region of this study.

5. Conclusions

Under the conditions in which temperatures were highest during August and September, it was possible to conclude that the effects of *El Niño* in the Eastern Amazon, Brazil, were as follows:

i. The full sun pasture area showed higher temperatures than the shaded area with the leafiest native species in the pasture, the Amazonian Brazil nut tree (*Bertholletia excelsa*), reinforcing the importance of maintaining the tree component in the region's pastures. The leftover effect is diagnosed by the darker thermal patterns, which indicate lower temperatures, expressing the effectiveness of the shade provided by the trees in reducing heat and, consequently, increasing animal comfort.
ii. The analysis of average temperatures in native trees in the Amazon region, compared to the temperatures recorded in grasses and on the ground, underlines the ability of trees to create environments with higher environmental quality for cattle with light coats. The thermal difference between the native trees and their surroundings suggests a cooling effect provided by the trees, reinforcing their positive influence on thermal comfort in the area studied.
iii. It was possible to observe the numerous advantages of using infrared thermography to define efficient thermal patterns in the Eastern Amazon. These results corroborate the importance of the presence of trees in pasture management, not only to influence

thermal conditions, but also to provide shade for animals, promote animal welfare and contribute to mitigating heat stress.

Author Contributions: Conceptualization, W.C.d.S., J.d.B.L.-J. and J.A.R.d.S.; methodology and investigation, W.C.d.S., J.A.R.d.S., É.B.R.d.S., C.E.L.S., K.A.L.N., R.N.C.C.-J., T.S.B., L.A.J., C.V.d.A., T.C.G.d.C.R., J.d.B.L.-J. and L.G.M.; thermography monitoring and data analysis support, L.G.M.; software, W.C.d.S. and J.d.B.L.-J.; validation, W.C.d.S., J.d.B.L.-J. and J.A.R.d.S.; formal analysis, W.C.d.S. and C.V.d.A.; software, W.C.d.S. and J.d.B.L.-J.; resources, W.C.d.S. and J.d.B.L.-J.; data curation, W.C.d.S. and J.d.B.L.-J.; writing—original draft preparation, W.C.d.S. and J.d.B.L.-J.; writing—review and editing, all authors; visualization, W.C.d.S.; supervision, J.d.B.L.-J. and J.A.R.d.S.; project administration, W.C.d.S. and J.d.B.L.-J.; funding acquisition, J.d.B.L.-J. All authors have read and agreed to the published version of the manuscript.

Funding: This research was funded by the Federal University of Pará (UFPA) and the Coordination for the Improvement of Higher Education Personnel (CAPES), under grant number (02/2023) and the APC was funded by the Pro-Rectory for Research and Postgraduate Studies (PROPESP/UFPA-02/2023).

Institutional Review Board Statement: This study was submitted to the Animal Ethics Committee (CEUA/UNAMA) and was approved on 31 March 2023, under the number: CEUA-UNAMA 0001-87/2023, in accordance with Brazilian law.

Informed Consent Statement: It was not applied in this study.

Data Availability Statement: The datasets presented in this article are not readily available because the data are part of an ongoing study. Requests for access to the datasets should be directed to welligton.medvet@gmail.com.

Acknowledgments: As authors, we would like to express our gratitude to the Coordination for the Improvement of Higher Education Personnel (CAPES)-Brazil. We would also like to thank the Office of the Dean of Research and Graduate Studies (PROPESP/UFPA) and the Graduate Program in Animal Science (PPGCAN). We would like to thank the members of the Bioclimatology, Animal Welfare and Nutrition in the Amazon (BANA) research group.

Conflicts of Interest: The authors declare no conflicts of interest.

References

1. Cezar, I.M.; Queiroz, H.P.D.; Thiago, L.D.S.; Garagorry, F.L.; Costa, F.P. Sistemas de produção de gado de corte no Brasil: Uma descrição com ênfase no regime alimentar e no abate. *Empresa Bras. Pesqui. Agropecuária (Embrapa)* **2005**, *1*, 40.
2. Joele, M.R.; Lourenço, L.F.; Lourenço-Júnior, J.B.; Araújo, G.S.; Budel, J.C.; Garcia, A.R. Meat quality of buffaloes finished in traditional or silvopastoral system in the Brazilian Eastern Amazon. *J. Sci. Food Agric.* **2017**, *97*, 1740–1745. [CrossRef] [PubMed]
3. Brcko, C.C.; Silva, J.A.R.D.; Garcia, A.R.; Silva, A.G.M.E.; Martorano, L.G.; Vilela, R.A.; Nahúm, B.S.; Barbosa, A.V.C.; Silva, W.C.D.; Rodrigues, T.C.G.C.; et al. Effects of Climatic Conditions and Supplementation with Palm Cake on the Thermoregulation of Crossbred Buffaloes Raised in a Rotational Grazing System and with Natural Shade in Humid Tropical Regions. *Animals* **2023**, *14*, 53. [CrossRef]
4. Santos, N.D.D.; Silva, J.A.D.; Araújo, A.A.D.; Garcia, A.R.; Beldini, T.P.; Rodrigues, L.S.; Bezerra, A.S.; Lourenço-Júnior, J.B. Silvopastoral system mitigates the thermal stress and benefits water buffaloesâ™ comfort in the Eastern Amazon, Brazil. *J. Agric. Stud.* **2020**, *8*, 193–202.
5. Silva, J.A.R.D.; Pantoja, M.H.D.A.; Silva, W.C.D.; Almeida, J.C.F.D.; Noronha, R.D.P.P.; Barbosa, A.V.C.; Lourenço-Júnior, J.D.B. Thermoregulatory reactions of female buffaloes raised in the sun and in the shade, in the climatic conditions of the rainy season of the Island of Marajó, Pará, Brazil. *Front. Vet. Sci.* **2022**, *9*, 998544. [CrossRef]
6. Silva, W.C.D.; Silva, É.B.R.D.; Santos, M.R.P.D.; Camargo-Júnior, R.N.C.; Barbosa, A.V.C.; Silva, J.A.R.D.; Vinhote, J.A.; Sousa, E.D.V.D.; Lourenço-Júnior, J.D.B. Behavior and thermal comfort of light and dark coat dairy cows in the Eastern Amazon. *Front. Vet. Sci.* **2022**, *9*, 1006093. [CrossRef]
7. Silva, W.C.D.; Silva, J.A.R.D.; Júnior, A.G.; Alvarenga, A.B.B.D.; Barbosa, A.V.C.; Silva, É.B.R.D.; Santos, M.R.P.D.; Lourenço-Júnior, J.B.; Camargo-Júnior, R.N.C.; Silva, A.G.M.E. A new proposal for the use of the focal animal technique in buffaloes in the Eastern Amazon. *Front. Vet. Sci.* **2023**, *10*, 9. [CrossRef]
8. Mota-Rojas, D.; Napolitano, F.; Braghieri, A.; Guerrero-Legarreta, I.; Bertoni, A.; Martínez-Burnes, J.; Cruz-Monterrosa, R.; Gómez, J.; Ramírez-Bribiesca, E.; Barrios-García, H.; et al. Thermal biology in river buffalo in the humid tropics: Neurophysiological and behavioral responses assessed by infrared thermography. *J. Anim. Behav. Biometeorol.* **2020**, *9*, 2103. [CrossRef]

9. Gómez-Prado, J.; Pereira, A.M.; Wang, D.; Villanueva-García, D.; Domínguez-Oliva, A.; Mora-Medina, P.; Hernández-Avalos, I.; Martínez-Burnes, J.; Casas-Alvarado, A.; Olmos-Hernández, A.; et al. Thermoregulation mechanisms and perspectives for validating thermal windows in pigs with hypothermia and hyperthermia: An overview. *Front. Vet. Sci.* **2022**, *9*, 1023294. [CrossRef]
10. Dawkins, M.S. Farm animal welfare: Beyond "natural" behavior. *Science* **2023**, *379*, 326–328. [CrossRef]
11. Hansen, B.G.; Langseth, E.; Berge, C. Animal welfare and cow-calf contact-farmers' attitudes, experiences and adoption barriers. *J. Rural Stud.* **2023**, *97*, 34–46. [CrossRef]
12. Kumar, P.; Ahmed, M.A.; Abubakar, A.A.; Hayat, M.N.; Kaka, U.; Ajat, M.; Goh, Y.M.; Sazili, A.Q. Improving animal welfare status and meat quality through assessment of stress biomarkers: A critical review. *Meat Sci.* **2023**, *197*, 109048. [CrossRef] [PubMed]
13. Napolitano, F.; Rosa, G.R.; Chay-Canul, A.; Álvarez-Macías, A.; Pereira, A.M.; Bragaglio, A.; Mora-Medina, P.; Rodríguez-González, D.; García-Herrera, R.; Hernández-Ávalos, I.; et al. The Challenge of Global Warming in Water Buffalo Farming: Physiological and Behavioral Aspects and Strategies to Face Heat Stress. *Animals* **2023**, *13*, 3103. [CrossRef]
14. van Eerdenburg, F.J.; Di Giacinto, A.M.; Hulsen, J.; Snel, B.; Stegeman, J.A. A new, practical animal welfare assessment for dairy farmers. *Animals* **2021**, *11*, 881. [CrossRef] [PubMed]
15. Berman, A. Estimates of heat stress relief needs for Holstein dairy cows. *J. Anim. Sci.* **2005**, *83*, 1377–1384. [CrossRef]
16. Farooq, U.; Samad, H.A.; Shehzad, F.; Qayyum, A. Physiological responses of cattle to heat stress. *World Appl. Sci. J.* **2010**, *8*, 38–43.
17. Rashamol, V. Adaptabilidade fisiológica da pecuária ao estresse térmico: Uma revisão atualizada. *J. Anim. Behav. Biometeorol.* **2018**, *6*, 62–71. [CrossRef]
18. Collier, R.J.; Baumgard, L.H.; Zimbelman, R.B.; Xiao, Y. Heat stress: Physiology of acclimation and adaptation. *Anim. Front.* **2019**, *9*, 12–19. [CrossRef]
19. Cartwright, S.L.; Schmied, J.; Karrow, N.; Mallard, B.A. Impact of heat stress on dairy cattle and selection strategies for thermotolerance: A review. *Front. Vet. Sci.* **2023**, *10*, 1198697. [CrossRef]
20. Mateescu, R.G.; Davila, K.M.S.; Hernandez, A.S.; Andrade, A.N.; Zayas, G.A.; Rodriguez, E.E.; Dikmen, S.; Oltenacu, P.A. Impact of Brahman genetics on skin histology characteristics with implications for heat tolerance in cattle. *Front. Genet.* **2023**, *14*, 1107468. [CrossRef]
21. Romanello, N.; Barreto, A.D.N.; Sousa, M.A.P.D.; Balieiro, J.C.C.; Brandão, F.Z.; Tonato, F.; Bernardi, A.C.D.C.; Pezzopane, J.R.M.; Portugal, J.A.B.; Garcia, A.R. Thermal comfort of Nelore (*Bos indicus*) and Canchim (*Bos taurus* × *Bos indicus*) bulls kept in an integrated crop-livestock-forestry system in a tropical climate. *Agric. Syst.* **2023**, *209*, 103687.
22. Silva, W.C.D.; Silva, J.A.R.D.; Camargo-Júnior, R.N.C.; Silva ÉBR, D.; Santos, M.R.P.D.; Viana, R.B.; Silva, A.G.M.; Silva, C.M.G.; Lourenço-Júnior, J.D.B. Animal welfare and effects of per-female stress on male and cattle reproduction—A review. *Front. Vet. Sci.* **2023**, *10*, 1083469. [CrossRef]
23. Giro, A.; Pezzopane, J.R.M.; Junior, W.B.; Pedroso, A.D.F.; Lemes, A.P.; Botta, D.; Romanello, N.; Barreto, A.D.N.; Garcia, A.R. Behavior and body surface temperature of beef cattle in integrated crop-livestock systems with or without tree shading. *Sci. Total Environ.* **2019**, *684*, 587–596. [CrossRef]
24. Mota-Rojas, D.; Wang, D.; Titto, C.G.; Gómez-Prado, J.; Fuente, V.C.D.L.; Ghezzi, M.; Boscato-Funes, L.; Barrios-Garcia, H.; Torres-Bernal, F.; Casas-Alvarado, A.; et al. Pathophysiology of fever and application of infrared thermography (IRT) in the detection of sick domestic animals: Recent advances. *Animals* **2021**, *11*, 2316. [CrossRef] [PubMed]
25. Mota-Rojas, D.; Pereira, A.M.; Wang, D.; Martínez-Burnes, J.; Ghezzi, M.; Hernández-Avalos, I.; Lendez, P.; Mora-Medina, P.; Casas, A.; Olmos-Hernández, A.; et al. Clinical applications and factors involved in validating thermal windows used in infrared thermography in cattle and river buffalo to assess health and productivity. *Animals* **2021**, *11*, 2247.
26. Mota-Rojas, D.; Ogi, A.; Villanueva-García, D.; Hernández-Ávalos, I.; Casas-Alvarado, A.; Domínguez-Oliva, A.; Lendez, P.; Ghezzi, M. Thermal Imaging as a Method to Indirectly Assess Peripheral Vascular Integrity and Tissue Viability in Veterinary Medicine: Animal Models and Clinical Applications. *Animals* **2023**, *14*, 142. [CrossRef]
27. Whittaker, A.L.; Muns, R.; Wang, D.; Martínez-Burnes, J.; Hernández-Ávalos, I.; Casas-Alvarado, A.; Domínguez-Oliva, A.; Mota-Rojas, D. Assessment of Pain and Inflammation in Domestic Animals Using Infrared Thermography: A Narrative Review. *Animals* **2023**, *13*, 2065. [CrossRef] [PubMed]
28. Silva, W.C.D.; Printes, O.V.N.; Lima, D.O.; Silva, É.B.R.D.; Santos, M.R.P.D.; Camargo-Júnior, R.N.C.; Barbosa, A.V.C.; Silva, J.A.R.D.; Silva, A.G.M.E.; Silva, L.K.X.; et al. Evaluation of the temperature and humidity index to support the implementation of a rearing system for ruminants in the Western Amazon. *Front. Vet. Sci.* **2023**, *10*, 10. [CrossRef]
29. Ghezzi, M.D.; Napolitano, F.; Casas-Alvarado, A.; Hernández-Ávalos, I.; Domínguez-Oliva, A.; Olmos-Hernández, A.; Pereira, A.M. Utilization of Infrared Thermography in Assessing Thermal Responses of Farm Animals under Heat Stress. *Animals* **2024**, *14*, 616. [CrossRef]
30. Marquez, H.P.; Schaefer, A.L.; Bench, C.J. Use of fidget and drinking behaviour in combination with facial infrared thermography for diagnosis of bovine respiratory disease in a spontaneous model. *Animal* **2024**, *18*, 101096. [CrossRef]
31. Zhang, C.; Xiao, D.; Yang, Q.; Wen, Z.; Lv, L. Application of Infrared Thermography in Livestock Monitoring. *Trans. ASABE* **2020**, *63*, 389–399. [CrossRef]

32. Martorano, L.G.; Garcia, A.R.; Silva, L.K.X.; Amaral, J.M.D.; Fernandes, C.; Oliveira, P.P.A. Infrared Thermography to Diagnose Bioclimatic Conditions in Livestock Systems in The Amazon Biome. *Minist. Agric. Livest. Food Supply (MAPA)* **2021**, *1*, 104.
33. Travain, T.; Valsecchi, P. Infrared thermography in the study of animals' emotional responses: A critical review. *Animals* **2021**, *11*, 2510. [CrossRef]
34. Wang, F.K.; Shih, J.Y.; Juan, P.H.; Su, Y.C.; Wang, Y.C. Non-invasive cattle body temperature measurement using infrared thermography and auxiliary sensors. *Sensors* **2021**, *21*, 2425. [CrossRef] [PubMed]
35. Cândido, A.C.T.F.; Martorano, L.G.; Cândido, B.U.F.; Nascimento, W.; Dias, C.T.D.S.; Lisboa, L.S.S.; Fernandes, P.C.C.; Silva, A.R.; Dias-Filho, M.B.; Beldini, T.P. Infrared Thermal Profiles in Silvopastoral and Full-Sun Pastures in the Eastern Amazon, Brazil. *Forests* **2023**, *14*, 1463. [CrossRef]
36. Canedo-Rosso, C.; Hochrainer-Stigler, S.; Pflug, G.; Condori, B.; Berndtsson, R. Drought impact in the Bolivian Altiplano agriculture associated with the El Niño–Southern Oscillation using satellite imagery data. *Nat. Hazards Earth Syst. Sci.* **2021**, *21*, 995–1010. [CrossRef]
37. Singh, M.; Zhu, X. Analysis of how the spatial and temporal patterns of fire and their bioclimatic and anthropogenic drivers vary across the Amazon rainforest in El Niño and non-El Niño years. *PeerJ* **2021**, *9*, e12029. [CrossRef]
38. Lyon, B. The strength of El Niño and the spatial extent of tropical drought. *Geophys. Res. Lett.* **2004**, *3*, L21204. [CrossRef]
39. Goddard, L.; Dilley, M. *El Niño*: Catastrophe or opportunity. *J. Clim.* **2005**, *18*, 651–665. [CrossRef]
40. Mera, Y.E.Z.; Vera, J.F.R.; Pérez-Martín, M.Á. Linking El Niño Southern Oscillation for early drought detection in tropical climates: The Ecuadorian coast. *Sci. Total Environ.* **2018**, *643*, 193–207. [CrossRef]
41. Generoso, R.; Couharde, C.; Damette, O.; Mohaddes, K. The growth effects of El Nino and La Nina: Local weather conditions matter. *Ann. Econ. Stat.* **2020**, *140*, 83–126. [CrossRef]
42. Marengo, J.A.; Espinoza, J.C. Extreme seasonal droughts and floods in Amazonia: Causes, trends and impacts. *Int. J. Climatol.* **2016**, *36*, 1033–1050. [CrossRef]
43. Santos, E.B.; Freitas, E.D.; Rafee, S.A.A.; Fujita, T.; Rudke, A.P.; Martins, L.D.; Souza, R.A.F.D.; Martins, J.A. Spatio-temporal variability of wet and drought events in the Paraná River basin—Brazil and its association with the El Niño—Southern oscillation phenomenon. *Int. J. Climatol.* **2021**, *41*, 4879–4897. [CrossRef]
44. Marengo, J.A.; Cunha, A.P.M.; Nobre, C.A.; Neto, G.G.R.; Magalhães, A.R.; Torres, R.R.; Sampaio, G.; Alexandre, F.; Alves, L.M.; Cuartas, L.A.; et al. Assessing drought in the drylands of northeast Brazil under regional warming exceeding 4 °C. *Nat. Hazards* **2020**, *103*, 2589–2611. [CrossRef]
45. Junqueira, R.; Amorim, J.D.S.; Viola, M.R.; Mello, C.R.D.; Uddameri, V.; Prado, L.F. Drought occurrences and impacts on the upper Grande river basin, Brazil. *Meteorol. Atmos. Phys.* **2022**, *134*, 45. [CrossRef]
46. Martorano, L.G.; Nechet, D.; Pereira, L.C. Tipologia climática do Estado do Pará: Adaptação do método de Köppen. *Bol. Geogr. Teorética* **1993**, *23*, 45–46. Available online: https://periodicos.ufpe.br/revistas/rbgf (accessed on 29 January 2024).
47. Martorano, L.G.; Vitorino, M.I.; Silva, B.P.P.C.; Lisboa, L.S.; Sotta, E.D.; Reichardt, K. Climate conditions in the eastern amazon: Rainfall variability in Belem and indicative of soil water deficit. *Afr. J. Agric. Res.* **2017**, *12*, 1801–1810. [CrossRef]
48. FLIR T650sc. *Manual da FLIR T650sc Systems 6.3v*, FLIR Systems, Inc.: Wilsonville, OR, USA, 2015. Available online: https://www.flir.com.br/support/products/t650sc/#Overview(accessed on 1 March 2023).
49. Barreto, C.D.; Alves, F.V.; Ramos, C.E.C.D.O.; Leite, M.C.D.P.; Leite, L.C.; Junior, N.K. Infrared thermography for evaluation of the environmental thermal comfort for livestock. *Int. J. Biometeorol.* **2020**, *64*, 881–888. [CrossRef] [PubMed]
50. Silva, W.C.D.; Silva, J.A.R.D.; Silva, É.B.R.D.; Barbosa, A.V.C.; Sousa, C.E.L.; Carvalho, K.C.D.; Santos, M.R.P.D.; Neves, K.A.L.; Martorano, L.G.; Júnior, R.N.C.C.; et al. Characterization of Thermal Patterns Using Infrared Thermography and Thermolytic Responses of Cattle Reared in Three Different Systems during the Transition Period in the Eastern Amazon, Brazil. *Animals* **2023**, *13*, 2735. [CrossRef]
51. Rodriguez, R.N. *Statistical Analysis System (SAS)*; WIREs Computational Statistics Institute: Cary, NC, USA, 2023.
52. Baêta, F.D.C.; Souza, C.D.F. Ambiência em edificações rurais: Conforto animal. *Viçosa UFV* **2010**, *2*, 269.
53. Baluja, J.; Diago, M.P.; Balda, P.; Zorer, R.; Meggio, F.; Morales, F.; Tardaguila, J. Assessment of vineyard water status variability by thermal and multispectral imagery using an unmanned aerial vehicle (UAV). *Irrig. Sci.* **2012**, *30*, 511–522. [CrossRef]
54. Meier, F.; Scherer, D. Spatial and temporal variability of urban tree canopy temperature during summer 2010 in Berlin, Germany. *Theor. Appl. Climatol.* **2012**, *110*, 373–384. [CrossRef]
55. Maes, W.H.; Baert, A.; Huete, A.R.; Minchin, P.E.; Snelgar, W.P.; Steppe, K. A new wet reference target method for continuous infrared thermography of vegetations. *Agric. For. Meteorol.* **2016**, *226*, 119–131. [CrossRef]
56. Junior, N.K.; Miyagi, E.S.; Oliveira, C.C.D.; Barreto, C.D.; Mastelaro, A.P.; Bungenstab, D.J.; Alves, F.V. Infrared thermography for microclimate assessment in agroforestry systems. *Sci. Total Environ.* **2020**, *731*, 139252.
57. Kharel, G.; Dhakal, M.; Deb, S.K.; Slaughter, L.C.; Simpson, C.; West, C.P. Effect of Long-Term Semiarid Pasture Management on Soil Hydraulic and Thermal Properties. *Plants* **2023**, *12*, 1491. [CrossRef]
58. Deniz, M.; De-Sousa, K.T.; Vieira, F.M.C.; Vale, M.M.D.; Dittrich, J.R.; Daros, R.R.; Hötzel, M.J. A systematic review of the effects of silvopastoral system on thermal environment and dairy cows' behavioral and physiological responses. *Int. J. Biometeorol.* **2023**, *67*, 409–422. [CrossRef] [PubMed]
59. Hammerle, A.; Meier, F.; Heinl, M.; Egger, A.; Leitinger, G. Implications of atmospheric conditions for analysis of surface temperature variability derived from landscape-scale thermography. *Int. J. Biometeorol.* **2017**, *61*, 575–588. [CrossRef]

60. Schöngart, J.; Junk, W.J.; Piedade, M.T.F.; Ayres, J.M.; Hüttermann, A.; Worbes, M. Teleconnection between tree growth in the Amazonian floodplains and the El Nino–Southern Oscillation effect. *Glob. Chang. Biol.* **2004**, *10*, 683–692. [CrossRef]
61. Santos, V.A.H.F.D.; Ferreira, M.J.; Rodrigues, J.V.F.C.; Garcia, M.N.; Ceron, J.V.B.; Nelson, B.W.; Saleska, S.R. Causes of reduced leaf-level photosynthesis during strong El Niño drought in a Central Amazon forest. *Glob. Chang. Biol.* **2018**, *24*, 4266–4279. [CrossRef]
62. Ávila-Jiménez, J.; Gutiérrez, J.D.; Altamiranda-Saavedra, M. The effect of El Niño and La Niña episodes on the existing niche and potential distribution of vector and host species of American Cutaneous Leishmaniasis. *Acta Trop.* **2024**, *249*, 107060. [CrossRef]
63. Armenteras, D.; Barreto, J.S.; Tabor, K.; Molowny-Horas, R.; Retana, J. Changing patterns of fire occurrence in proximity to forest edges, roads and rivers between NW Amazonian countries. *Biogeosciences* **2017**, *14*, 2755–2765. [CrossRef]
64. Mota, P.H.S.; Rocha, S.J.S.S.D.; Castro, N.L.M.; Marcatti, G.E.; França, L.C.D.J.; Schettini, B.L.S.; Villanova, P.H.; Santos, H.T.; Santos, A.R.D. Forest fire hazard zoning in Mato Grosso state, Brazil. *Land Use Policy* **2019**, *88*, 104206. [CrossRef]
65. Júnior, L.A.S.D.S.; Delgado, R.C.; Pereira, M.G.; Teodoro, P.E.; Junior, C.A.D.S. Fire dynamics in extreme climatic events in western amazon. *Environ. Dev.* **2019**, *32*, 100450. [CrossRef]
66. Santos, I. Connection between El Niño Deep Convection and Precipitation in Northeast Brazil. *J. Geogr. Geol.* **2024**, *16*. [CrossRef]
67. Montanher, O.C.; Minaki, C.; de Morais, E.S.; de Paula Silva, J.; Pereira, P. Geosystemic impacts of the extreme rainfall linked to the El Niño 2015/2016 event in Northern Paraná, Brazil. *Appl. Sci.* **2023**, *13*, 9678. [CrossRef]
68. Dai, A. Increasing drought under global warming in observations and models. *Nat. Clim. Chang.* **2013**, *3*, 52–58. [CrossRef]
69. Rifai, S.W.; Li, S.; Malhi, Y. Coupling of El Niño events and long-term warming leads to pervasive climate extremes in the terrestrial tropics. *Environ. Res. Lett.* **2019**, *14*, 105002. [CrossRef]
70. Cai, W.; McPhaden, M.J.; Grimm, A.M.; Rodrigues, R.R.; Taschetto, A.S.; Garreaud, R.D.; Dewitte, B.; Poveda, G.; Ham, Y.G.; Santoso, A.; et al. Climate impacts of the El Niño–southern oscillation on South America. *Nat. Rev. Earth Environ.* **2020**, *1*, 215–231. [CrossRef]
71. Jiménez-Muñoz, J.C.; Mattar, C.; Barichivich, J.; Santamaría-Artigas, A.; Takahashi, K.; Malhi, Y.; Sobrino, J.A.; Schrier, G.V.lD. Record-breaking warming and extreme drought in the Amazon rainforest during the course of El Niño 2015–2016. *Sci. Rep.* **2016**, *6*, 33130. [CrossRef]
72. Esquivel-Hernández, G.; Mosquera, G.M.; Sánchez-Murillo, R.; Quesada-Román, A.; Birkel, C.; Crespo, P.; Célleri, R.; Windhorst, D.; Breuer, L.; Boll, J. Moisture transport and seasonal variations in the stable isotopic composition of rainfall in Central American and Andean Páramo during El Niño conditions (2015–2016). *Hydrol. Process.* **2019**, *33*, 1802–1817. [CrossRef]
73. Fontes, C.G.; Dawson, T.E.; Jardine, K.; McDowell, N.; Gimenez, B.O.; Anderegg, L.; Negrón-Juáres, R.; Higuchi, N.; Fine, P.V.A.; Araújo, A.C.; et al. Dry and hot: The hydraulic consequences of a climate change–type drought for Amazonian trees. *Philos. Trans. R. Soc. B Biol. Sci.* **2018**, *373*, 20180209. [CrossRef]
74. Miyamoto, K.; Aiba, S.I.; Aoyagi, R.; Nilus, R. Effects of El Niño drought on tree mortality and growth across forest types at different elevations in Borneo. *For. Ecol. Manag.* **2021**, *490*, 119096. [CrossRef]
75. Powers, J.; Vargas, G.G.; Brodribb, T.; Schwartz, N.; Perez-Aviles, D.; Smith-Martin, C.; Becknell, J.; Aureli, F.; Blanco, R.; Calderón-Morales, E.; et al. A catastrophic tropical drought kills hydraulically vulnerable tree species. *Methods* **2021**, *2020*, 03–17. [CrossRef]
76. Hänchen, L.; Klein, C.; Maussion, F.; Gurgiser, W.; Calanca, P.; Wohlfahrt, G. Widespread greening suggests increased dry-season plant water availability in the Rio Santa valley, Peruvian Andes. *Earth Syst. Dyn.* **2022**, *13*, 595–611. [CrossRef]
77. Rifai, S.W.; Girardin, C.A.; Berenguer, E.; Aguila-Pasquel, J.D.; Dahlsjö, C.A.; Doughty, C.E.; Jeffery, K.J.; Moore, S.; Oliveras, I.; Riutta, T.; et al. ENSO Drives interannual variation of forest woody growth across the tropics. *Philos. Trans. R. Soc. B Biol. Sci.* **2018**, *373*, 20170410. [CrossRef]
78. Wigneron, J.P.; Fan, L.; Ciais, P.; Bastos, A.; Brandt, M.; Chave, J.; Saatchi, S.; Baccini, A.; Fensholt, R. Tropical forests did not recover from the strong 2015–2016 *El Niño* event. *Sci. Adv.* **2020**, *6*, eaay4603. [CrossRef]
79. Abreu-Harbich, L.V.D.; Labaki, L.C.; Matzarakis, A. Effect of tree planting design and tree species on human thermal comfort in the tropics. *Landsc. Urban Plan.* **2015**, *138*, 99–109. [CrossRef]
80. Lopes, L.B.; Eckstein, C.; Pina, D.S.; Carnevalli, R.A. The influence of trees on the thermal environment and behaviour of grazing heifers in Brazilian Midwest. *Trop. Anim. Health Prod.* **2016**, *48*, 755–761. [CrossRef] [PubMed]
81. Oliveira, C.C.D.; Alves, F.V.; Almeida, R.G.D.; Gamarra, E.L.; Villela, S.D.J.; Martins, P.G.M.D.A. Thermal comfort indices assessed in integrated production systems in the Brazilian savannah. *Agrofor. Syst.* **2018**, *92*, 1659–1672. [CrossRef]
82. Pereira, L.M.R.; Fischer, V.; Moreno, C.B.; Pardo, M.P.; Gomes, J.F.; Monks, P.L. Diurnal ingestive behavior of Jersey heifers fed diferente supplements. *Rev. Bras. Agrociência* **2005**, *11*, 453–459.
83. Abreu, L.V.; Labaki, L.C. Conforto térmico propiciado por algumas espécies arbóreas: Avaliação do raio de influência através de diferentes índices de conforto. *Ambiente Construído* **2010**, *10*, 103–117. [CrossRef]
84. Karvatte, N.; Klosowski, E.S.; Almeida, R.G.D.; Mesquita, E.E.; Oliveira, C.C.D.; Alves, F.V. Shading effect on microclimate and thermal comfort indexes in integrated crop-livestock-forest systems in the Brazilian Midwest. *Int. J. Biometeorol.* **2016**, *60*, 1933–1941. [CrossRef] [PubMed]
85. Silva, R.G. Predição da configuração de sombras de árvores em pastagens para bovinos. *Eng. Agrícola* **2006**, *26*, 36–281. [CrossRef]
86. Baliscei, M.A.; Barbosa, O.R.; Souza, W.D.; Costa, M.A.T.; Krutzmann, A.; Queiroz, E.D.O. Microclimate without shade and silvopastoral system during summer and winter. *Acta Scientiarum. Anim. Sci.* **2013**, *35*, 49–56. [CrossRef]

87. Baliscei, M.A.; Souza, W.D.; Barbosa, O.R.; Cecato, U.; Krutzmann, A.; Queiroz, E.D.O. Behavior of beef cattle and the microclimate with and without shade. *Acta Scientiarum. Anim. Sci.* **2012**, *34*, 409–415. [CrossRef]
88. Lees, A.M.; Lees, J.C.; Sejian, V.; Sullivan, M.L.; Gaughan, J.B. Influence of shade on panting score and behavioural responses of *Bos taurus* and *Bos indicus* feedlot cattle to heat load. *Anim. Prod. Sci.* **2019**, *60*, 305–315. [CrossRef]
89. Souza, E.C.D.; Salman, A.K.D.; Cruz, P.G.D.; Veit, H.M.; Carvalho, G.A.D.; Silva, F.R.F.D.; Schmitt, E. Thermal comfort and grazing behavior of Girolando heifers in integrated crop-livestock (ICL) and crop-livestock-forest (ICLF) systems. *Acta Scientiarum. Anim. Sci.* **2019**, *41*, 10. [CrossRef]
90. Lemaire, G.; Franzluebbers, A.; Carvalho, P.C.D.F.; Dedieu, B. Integrated crop–livestock systems: Strategies to achieve synergy between agricultural production and environmental quality. *Agric. Ecosyst. Environ.* **2014**, *190*, 4–8. [CrossRef]
91. Garcia, A.R. Thermal comfort on reproductive processes of water buffaloes raised under tropical conditions. *Rev. Bras. Reprod. Anim.* **2013**, *37*, 121–130.
92. Figueiredo, E.B.D.; Jayasundara, S.; Bordonal, R.D.O.; Berchielli, T.T.; Reis, R.A.; Wagner-Riddle, C.; Junior, N.L.S. Greenhouse gas balance and carbon footprint of beef cattle in three contrasting pasture-management systems in Brazil. *J. Clean. Prod.* **2017**, *142*, 420–431. [CrossRef]
93. Dube, F.; Couto, L.; Silva, M.D.; Leite, H.G.; Garcia, R.; Araujo, G.A.A. A simulation model for evaluating technical and economic aspects of an industrial eucalyptus-based agroforestry system in Minas Gerais, Brazil. *Agrofor. Syst.* **2002**, *55*, 73–80. [CrossRef]
94. Garcia, A.R.; Silva, L.K.X.; Barros, D.V.; Lourenço-Junior, J.D.B.; Martorano, L.G.; Lisboa, L.S.S.; Silva, J.A.R.D.; Sousa, J.S.D.; Silva, A.O.A.D. Key points for the thermal comfort of water buffaloes in Eastern Amazon. *Ciência Rural* **2022**, *53*, e20210544. [CrossRef]

animals

MDPI

Article

Caffeine Administration in Piglets with Low Birthweight and Low Vitality Scores, and Its Effect on Physiological Blood Profile, Acid–Base Balance, Gas Exchange, and Infrared Thermal Response

Dina Villanueva-García [1], Marcelo Ghezzi [2], Patricia Mora-Medina [3], Ismael Hernández-Ávalos [3], Adriana Olmos-Hernández [4], Alejandro Casas-Alvarado [5], Karina Lezama-García [5], Adriana Domínguez-Oliva [5,*], Daniela Rodríguez-González [5] and Miriam Marcet-Rius [6,*]

1 Division of Neonatology, National Institute of Health, Hospital Infantil de México Federico Gómez, Mexico City 06720, Mexico; dinavg21@yahoo.com
2 Animal Welfare Area, Faculty of Veterinary Sciences (FCV), Universidad Nacional del Centro de la Provincia de Buenos Aires (UNCPBA), University Campus, Tandil 7000, Argentina
3 Facultad de Estudios Superiores Cuautitlán, Universidad Nacional Autónoma de México (UNAM), Cuautitlán Izcalli 54714, Mexico
4 Division of Biotechnology-Bioterio and Experimental Surgery, Instituto Nacional de Rehabilitación Luis Guillermo Ibarra Ibarra (INR-LGII), Mexico City 14389, Mexico
5 Neurophysiology, Behavior and Animal Welfare Assesment, DPAA, Xochimilco Campus, Universidad Autónoma Metropolitana, México City 04960, Mexico; kikislezama@hotmail.com
6 Department of Animal Behaviour and Welfare, Research Institute in Semiochemistry and Applied Ethology (IRSEA), 84400 Apt, France
* Correspondence: mvz.freena@gmail.com (A.D.-O.); m.marcet@group-irsea.com (M.M.-R.)

Citation: Villanueva-García, D.; Ghezzi, M.; Mora-Medina, P.; Hernández-Ávalos, I.; Olmos-Hernández, A.; Casas-Alvarado, A.; Lezama-García, K.; Domínguez-Oliva, A.; Rodríguez-González, D.; Marcet-Rius, M. Caffeine Administration in Piglets with Low Birthweight and Low Vitality Scores, and Its Effect on Physiological Blood Profile, Acid–Base Balance, Gas Exchange, and Infrared Thermal Response. *Animals* **2023**, *13*, 3491. https://doi.org/10.3390/ani13223491

Academic Editor: Jin-Ho Cho

Received: 18 September 2023
Revised: 18 October 2023
Accepted: 8 November 2023
Published: 12 November 2023

Simple Summary: Intrapartum mortality reaches high figures in piglets, and it has been seen that those born with low weight and those with low or failed vitality scores are the most prone. Caffeine is used to reduce cardiorespiratory problems in neonates, both humans and animals. For this reason, the decision was made to use different doses of caffeine in newborn piglets in this study, to define if its oral administration would help reduce this problem and what would be the optimal dose and time of administration. The following blood gas parameters were evaluated: pH, pO_2, pCO_2, HCO_3, Ca++, glucose, and lactate, as well as superficial and rectal temperature at 1 h, 6 h, and 24 h after birth. It was concluded that the optimal dose of caffeine to improve the vitality score and reduce possible cardiorespiratory problems is 30 mg/kg orally.

Abstract: Intrapartum asphyxia, fetal hypoxia, and their consequences (e.g., acidosis, hypercapnia, hypoglycemia, and hypothermia) are the main factors related to physio-metabolic imbalances that increase neonatal mortality in piglets, particularly in piglets with low birthweight and low vitality scores. This study aimed to evaluate the effect of three different doses of caffeine (10, 20, and 30 mg/kg) administered orally to 480 newborn piglets with low birthweight and low vitality scores. Blood gas parameters (pH, pO_2, pCO_2, and HCO_3^-), physio-metabolic profile (Ca^{++}, glucose, and lactate), and the thermal response assessed through infrared thermography in four thermal windows (ocular, auricular, snout, and hindlimb) and rectal temperature were evaluated during the first 24 h of life. Doses of 30 mg/kg resulted in significant differences at 24 h for all evaluated parameters, suggesting that caffeine administration improved the cardiorespiratory function and metabolic activity of piglets by reducing acidosis, restoring glycemia, and increasing surface and rectal temperature. In conclusion, caffeine at 30 mg/kg could be suggested as an appropriate dose to use in piglets with low birthweight and low vitality scores. Future research might need to study the presentation of adverse effects due to higher caffeine concentrations.

Keywords: caffeine; newborn piglet; blood profile; vitality score; acid–base imbalance

1. Introduction

Newborn mortality in piglets ranges from 15 to 70% during the first three days post-farrowing [1], and from 11 to 20% in the pre-weaning period [2]. Several factors influence the high mortality rates. For example, physiological immaturity at birth, intrauterine growth restriction, and dystocic farrowing [3]. The incidence of dystocia (ranging from 12.78 to 75.3%) [4,5] can result in decreased maternal blood supply to the fetus due to inadequate uterine contractions, umbilical cord rupture, or meconium consumption, leading to intrapartum asphyxia and fetal hypoxia [6,7].

As a result of fetal hypoxia, newborn piglets can be born with low vitality scores due to a decreased cardiorespiratory function that will consequently cause physio-metabolic alterations such as respiratory/metabolic acidosis (increase in lactic acid production) and hypercapnia (low O_2 and high CO_2 concentrations) [8,9]. Low vitality during the first 72 h post-farrowing is also associated with 2 to 30% of neonatal deaths because piglets have delayed access to the udder and are not able to ingest colostrum [10,11]. Moreover, the number of piglets is higher than the number of teats of the sow due to genetic selection; therefore, the weakest piglets have few possibilities to consume colostrum. Since piglets are born with limited glycogen reserves that are rapidly depleted in the first 12 h (by 42%) [3,12,13], inadequate colostrum intake within the first day of life predisposes piglets to starvation, hypoglycemia, and hypothermia [3,14,15].

Newborn piglets are susceptible to hypothermia due to thermoregulatory impairments related to low birth weight—limiting the energy reserves of piglets [16,17]—or to anatomical traits of the species (e.g., low hair density, limited deposits of brown adipose tissue and glycogen) [18–20]. Thermogenesis through shivering and brown adipose tissue, or sympathetic responses such as vasoconstriction to reduce heat loss, are mechanisms triggered by piglets to maintain their body temperature [18].

Caffeine administration has been suggested as a method to improve piglet neonatal survival [10]. Caffeine is a stimulant of the central nervous system (CNS) due to its adenosine receptor antagonism [21–23]. In humans, caffeine increases respiratory volume, decreasing hypoxia and possibly increasing newborn vitality [24–26]. In contrast, in veterinary medicine, the relation between the respiratory stimulatory effect and the improvement in vitality when using caffeine has not been completely established, since there is no defined therapeutic dose to improve the vigor of newborn animals with low birthweight [10,27]. Nonetheless, different doses (e.g., 10–27 mg/kg) of caffeine have recently been studied to reduce neonatal cardiorespiratory alterations [28–31]. In a study carried out in piglets by Jarrat el al. [10], they administered three treatments (caffeine 30 mg, glucose 300 mg and caffeine with glucose combined) and observed that said administration did not cause changes in the performance of the piglets evaluated with the exception of low birthweight piglets, whose growth rates improved during the first 3 days of life.

Although different studies have assessed the effect of different doses of caffeine on parameters such as triglycerides, cholesterol, lactate, glucose, and rectal temperature [10,32], or have used different routes of administration (e.g., subcutaneous) [33], there are no studies to date associating the physio-metabolic profile, acid–base balance, and blood gas exchange parameters with the thermal response of piglets using infrared imaging. For this reason, the objective of this study was to evaluate the effect of three different doses of caffeine administered orally in newborn piglets with low birthweight and low vitality scores. The effect of caffeine on the metabolic profile, acid–base balance, blood gas exchange, and thermal response was studied using infrared imaging. We hypothesized that the administration of caffeine at a dose of 30 mg/kg would improve the biomarkers associated with the energy metabolism, gas exchange, and thermal response of newborn piglets.

2. Materials and Methods

2.1. Study Location

The present study was performed on a commercial swine farm located in the central region of Mexico. The swine farm employs a production regimen with approximately 300 farrows per week in an 'all in–all out' system.

2.2. Animals and Housing Conditions

4456 piglets born from 400 multiparous Landrace X Yorkshire sows were monitored before selecting the piglets for the present study, of which 480 were selected. The sow's parity ranged from two to seven. The sows were individually housed in non-plasticized flooring crates five days before the expected farrowing date, where they stayed for 28 days until weaning. The farrowing facilities included electronic ventilation systems, with both natural and artificial lighting available. The average room temperature of the sow house was maintained at 25 °C with a relative humidity of 62%. Lighting intensity was set at 38.7 candela. The design and specifications of the farrowing facilities were in accordance with the Mexican Official Standard NOM-062-ZOO-1999 [34]. After farrowing, the sows were fed 3.2 kg of a commercial concentrated lactating animal feed (12.2 MJ ME/kg and 15% CP).

Farrowing was induced with intramuscular administration of prostaglandin (Lutalyse; Pharmacia & Upjohn, Mexico), injected 36 h before the expected farrowing date (at 114 days of gestation). On the day of farrowing, sows were not provided feed but had unrestricted access to water. No obstetrical manipulation was performed during farrowing, but there was medical intervention and resuscitation of asphyxiated piglets. Sows that experienced dystocia and required oxytocin administration or medical intervention were excluded from the study.

2.3. Inclusion Criteria

Piglets with a low vitality score (≤5 points) and low birthweight (<1 kg) were included in the present study. To select the piglets, within the first minute after birth, vitality was assessed using Zaleski and Hacker's [35] scale, modified by [36]. Five variables were assessed: heart rate, latency to respiration, latency to stand up, skin color of the snout, and meconium staining. Using a 0–10 scale, piglets with 8–10 points were classified as high vitality, 6–7 as medium vitality, and below 5 as low vitality [9].

To obtain the birthweight, piglets were weighted with a digital scale (Bascule Salter Weight-Tronix, West Bromwich, UK) after vitality assessment. A total of 480 piglets were selected for the evaluation of physiological and thermal parameters.

2.4. Treatments

The 480 newborn piglets were randomly assigned to four treatments (n = 120 piglets in each group). To establish the different doses, 0.1 mL (low dose), 0.2 mL (medium dose), and 0.3 mL (high dose) of caffeine (Loeffler Laboratories, Mexico, 100 mL/10 g of caffeine. Each milliliter contains 100 mg of caffeine) were diluted in 6 mL of 9% saline solution (Pisa Laboratories, Mexico City, Mexico). The treatments were divided as follows: placebo (G_1): oral administration (PO) of 6 mL of saline per piglet; low dose of caffeine (G_2), 10 mg/kg PO; medium dose of caffeine (G_3), 20 mg/kg PO; and high dose of caffeine (G_4), 30 mg/kg PO. PO administration was performed using a syringe, briefly restraining the piglet to ensure that the entire dose of the drug was taken.

2.5. Experimental Design

Neonatal monitoring was performed by four trained D.V.M. M.Sc. Immediately after birth and after vitality and birthweight recording, piglets were completely dried with kraft paper, rubbing the piglets for 2 min. After drying the animals, thermal imaging and blood sampling were performed in four evaluation times by two blind evaluators: basal: before the PO administration of caffeine, after piglets were dried; 1 h: one hour after birth; 6 h: six hours after birth; and 24 h: 24 h after birth.

All newborn piglets were managed according to standard farm protocols, including umbilical cord cutting and colostrum intake within the first hour after birth. Piglets were identified using an indelible marker on the dorsal skin. At basal, 1 h, 6 h, and 24 h, thermal imaging was performed before retro-orbital sinus (ROS) bleeding to avoid handling stress due to blood sampling. Likewise, during the 24 h of the experimental phase, none of the piglets was subjected to routine farm practices such as tail docking, iron injection, or ear tagging. These routine procedures were performed at three days old, after the study's completion.

2.6. Assessed Variables

2.6.1. Thermal Imaging

An infrared thermal camera FLIR® E60 (FLIR Systems, Wilsonville, OR, USA) was used. The camera had a thermal sensibility of <0.045 °C, an IR resolution of 320x240 pixels, and a precision of ± 2 °C or 2%. Thermal imaging was performed at a maximum distance of 1 m from the piglets, focusing on four thermal windows: the periocular area (*Regio orbitalis*) (OCU), delimited by an ellipse covering the entire ocular surface, including a few millimeters of the upper and lower eyelids; the auricular region (*Regio auricularis*) (EAR), set by an ellipse in the central cartilage; the nasal region (*Regio nasalis*) (NOSE), drawn as a circle including the snout, using the upper lip as a ventral limit; and the left hindlimb (LIMB), at the edge of the pelvic limb, in the biceps femoris region. Figure 1 shows the delimitation of said thermal windows.

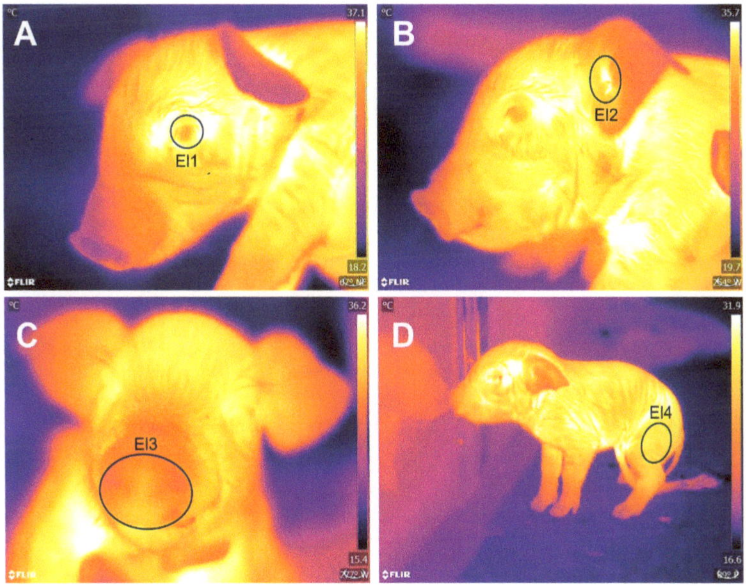

Figure 1. Delimitation of thermal windows. (**A**) OCULAR (El1). (**B**) EAR (El2). (**C**) NOSE (El3). (**D**) LIMB (El4).

During basal, 1 h, 6 h, and 24 h, IRT was performed before handling the piglets for blood sampling. All images were saved in JPG format to analyze them in FLIR Tools (FLIR Systems, USA) software, obtaining the maximum, minimum, and average values.

2.6.2. Blood Gas and Physiological Profile

At all evaluation times (basal, 1 h, 6 h, and 24 h), 150 µL of blood was collected from the ROS with lithium heparin capillary tubes. Immediately after sampling, on the farm, all samples (150 µL) were processed in a GEM Premier blood gas analyzer (Instrumentation Laboratory Diagnostics, Lexington, KY, USA/Milan, Italy) to analyze the following metabolites: glucose (mg/dL), lactate (mg/dL), partial pressure of carbon dioxide (pCO_2) (mmHg), partial pressure of oxygen (pO_2) (mmHg), pH, bicarbonate (HCO_3^-) (mmol/L), and Ca^{++} (mmol/L).

2.7. Statistical Analysis

The analysis of the data was performed using the statistical package GraphPad Prism (Ver. 9.4.0). Descriptive statistics were obtained for each variable of G_1, G_2, G_3, and G_4. A Shapiro–Wilk test was performed to analyze the normality of the data.

The data of blood gases, physio-metabolic blood profile, and surface temperature of the different thermal windows were considered independent variables, while the caffeine dose was considered a dependent variable. To assess the effects of these variables, an analysis of variance (ANOVA) under a mixed linear model was used:

$$Yijk= \mu+\tau i+\tau j + \tau i\tau j + \beta k +eij$$

where Y = response variable gasometry, physio-metabolic blood profile, and superficial thermal response. τi = effect of caffeine dose (G_1, G_2, G_3, and G_4). τj = effect of the evaluation time (basal, 1 h, 6 h, and 24 h). β = random effect (animal). μ = population mean. e = residue.

The correlations between the rectal temperature and the IRT temperature were analyzed using a Pearson correlation. To evaluate the differences between the means, a Tukey post hoc test was used. In all cases, a significance level of $p < 0.05$ was set. Under this same model, the effect of sex on each treatment (G_1, G_2, G_3, and G_4) was evaluated.

2.8. Ethical Statement

The experimental procedure of the current investigation received approval from the Laboratory of Animal Ethology and Neurophysiology (ETO-078-21) and the Postgraduate Veterinary Commission of the Universidad Autónoma Metropolitana in Mexico City, Mexico. The management of the animals was in accordance with FRAME's guidelines for animal research and the ethical considerations for applied ethological studies [37]. Acknowledging that the procedures of the present study might cause mild pain, suffering, and/or distress, the use of cell phones, televisions, and other sources of noise was restricted to minimize stressors. All animals were treated with humane care. The design and specifications of the farrowing facilities were in accordance with the Mexican Official Standard NOM-062-ZOO-1999 [34].

3. Results

Differences in blood gas and physio-metabolic parameters were observed in all newborn piglets according to the caffeine dose. In particular, caffeine at a dose of 30 mg/kg (G_4) registered significant differences and the highest changes from basal to 24 h. Regarding the effect of sex in the piglets, no significant differences were found in the present study. Therefore, results are presented as pooled data from males and females, expressed as mean ± standard error (SE).

3.1. Blood Gas Parameters

Table 1 shows the values of blood gas parameters (pH, pCO_2, pO_2, and HCO_{3-}) and the pH values of the four experimental groups. When comparing G_2 and G_3 with G_4, it can be observed that the pH of G_4 piglets was significantly higher ($p = 0.001$) at 1 h, 6 h, and

24 h by 0.03, 0.01 and 0.09, respectively. Likewise, G_4 significantly differed from basal to 1 h, 6 h, and 24 h ($p = 0.001$), recording progressive increases of 0.01, 0.11, and 0.14, respectively.

Table 1. Value of blood gas parameters (mean ± SE) of newborn piglets with low birthweight and low vitality scores receiving different oral doses of caffeine.

Parameter	Treatment	Basal	1 h	6 h	24 h	*p*-Value
pH	G_1 (Saline)	$7.20 \pm 0.005^{a,3}$	$7.23 \pm 0.005^{c,2}$	$7.22 \pm 0.005^{b,2}$	$7.25 \pm 0.005^{b,c,1}$	0.006
	G_2 (10 mg/kg)	$7.19 \pm 0.008^{a,3}$	$7.26 \pm 0.005^{b,2}$	$7.29 \pm 0.005^{a,1}$	$7.24 \pm 0.008^{b,c,2}$	0.001
	G_3 (20 mg/kg)	$7.21 \pm 0.005^{a,3}$	$7.28 \pm 0.005^{a,2}$	$7.30 \pm 0.005^{a,1}$	$7.27 \pm 0.005^{b,2}$	0.01
	G_4 (30 mg/kg)	$7.17 \pm 0.006^{a,3}$	$7.29 \pm 0.005^{a,2}$	$7.30 \pm 0.005^{a,2}$	$7.33 \pm 0.006^{a,1}$	0.001
	p-value	0.91	0.0001	0.001	0.0001	
pCO_2 (mmHg)	G_1 (Saline)	$85.93 \pm 0.63^{a,1}$	$82.93 \pm 0.63^{a,2}$	$80.93 \pm 0.63^{a,3}$	$79.93 \pm 0.63^{a,4}$	0.001
	G_2 (10 mg/kg)	$85.08 \pm 0.75^{a,1}$	$81.93 \pm 0.63^{b,2}$	$79.93 \pm 0.63^{b,4}$	$83.00 \pm 0.74^{a,3}$	0.005
	G_3 (20 mg/kg)	$85.25 \pm 0.48^{a,1}$	$80.93 \pm 0.63^{b,2}$	$77.93 \pm 0.63^{c,2}$	$77.13 \pm 0.47^{a,2}$	0.002
	G_4 (30 mg/kg)	$87.46 \pm 0.58^{a,1}$	$77.93 \pm 0.63^{c,2}$	$76.95 \pm 0.63^{c,2}$	$60.43 \pm 0.59^{b,3}$	0.001
	p-value	0.87	0.0001	0.001	0.001	
pO_2 (mmHg)	G_1 (Saline)	$16.83 \pm 0.36^{a,2}$	$17.83 \pm 0.36^{c,1}$	$17.84 \pm 0.36^{a,1}$	$17.83 \pm 0.75^{b,1}$	0.0001
	G_2 (10 mg/kg)	$16.98 \pm 0.31^{a,2}$	$19.83 \pm 0.36^{a,1}$	$17.83 \pm 0.36^{a,2}$	$18.00 \pm 0.31^{b,1}$	0.0001
	G_3 (20 mg/kg)	$15.82 \pm 0.21^{a,2}$	$18.83 \pm 0.36^{b,1}$	$18.84 \pm 0.36^{b,1}$	$19.82 \pm 0.21^{a,1}$	0.0001
	G_4 (30 mg/kg)	$16.44 \pm 0.17^{a,3}$	$18.82 \pm 0.36^{b,2}$	$18.83 \pm 0.36^{b,2}$	$21.06 \pm 0.23^{a,1}$	0.0001
	p-value	0.52	0.0001	0.0001	0.0005	
HCO_3^- (mmol/L)	G_1 (Saline)	$18.35 \pm 0.22^{a,1}$	$18.36 \pm 0.22^{b,1}$	$18.34 \pm 0.22^{b,1}$	$18.37 \pm 0.22^{d,1}$	0.99
	G_2 (10 mg/kg)	$18.20 \pm 0.16^{a,2}$	$18.36 \pm 0.22^{b,2}$	$18.35 \pm 0.22^{b,2}$	$19.42 \pm 0.16^{c,1}$	0.001
	G_3 (20 mg/kg)	$18.26 \pm 0.18^{a,2}$	$18.36 \pm 0.22^{b,2}$	$18.44 \pm 0.22^{b,2}$	$20.47 \pm 0.18^{b,1}$	0.001
	G_4 (30 mg/kg)	$18.22 \pm 0.15^{a,3}$	$18.50 \pm 0.22^{a,3}$	$19.50 \pm 0.22^{a,2}$	$22.22 \pm 0.15^{a,1}$	0.001
	p-value	0.89	0.002	0.0001	0.0001	

[a,b,c,d] Different letters indicate significant differences between treatments in the same evaluation time. [1,2,3,4] Different numbers indicate significant differences between evaluation times in a single treatment. $n = 120$ piglets per each group. G_1: placebo (saline); G_2: low dose (10 mg/kg); G_3: medium dose (20 mg/kg); G_4: high dose (30 mg/kg).

Regarding pCO_2, the G_4 dose of caffeine significantly reduced pCO_2 by 4 mmHg, 3 mmHg, and 22.57 mmHg at 1 h ($p = 0.0001$), 6 h ($p = 0.001$), and 24 h ($p = 0.001$), respectively, compared with G_2 and G_3. The same experimental group (G_4) had significant reductions of up to 27.03 mmHg at 24 h ($p = 0.001$).

In contrast to pCO_2 in G_4, pO_2 increased significantly from 16.44 ± 0.17 mmHg at 1 h after administration of caffeine to 21.06 ± 0.23 mmHg at 24 h ($p = 0.0005$) (Table 1). A comparison between evaluation times showed that, while G_2 piglets increased their pO_2 by 2.85 mmHg from basal to 1 h, they had a significant decrease of 2 mmHg from 1 h to 6 h ($p = 0.0001$). In contrast, a progressive increase was observed in G_3 and G_4, increasing up to 4 mmHg and 4.62 mmHg, respectively ($p = 0.0001$).

When addressing HCO_3^-, similarly to other blood parameters, piglets in G_4 had the highest increases compared to G_2 and G_3, registering increases of 1.06 mmol/L at 6h ($p = 0.0001$) and 2.8 mmol/L at 24 h ($p = 0.0001$). Between evaluation times, all groups receiving caffeine increased their HCO_3^- levels by 1.22 mmol/L in G_2 ($p = 0.001$), 2.21 mmol/L in G_3 ($p = 0.001$), and 4 mmol/L in G_4 from basal to 24 h.

3.2. *Physio-Metabolic Profile*

The differences between treatments and Ca^{++} values are presented in Table 2. In G_2, G_3, and G_4, caffeine significantly reduced Ca^{++} by 8%, 12%, and 14%, respectively, at 24 h ($p = 0.002$). In particular, G_4 registered a significant decrease from basal (2.08 ± 0.01 mmol/L) to 24 h (1.80 ± 0.01 mmol/L) ($p = 0.0001$). For glucose, G_4 showed a statistically significant increase of 1-2 mg/dL at 1 h ($p = 0.0001$) and a similar increase at 6 h ($p = 0.0001$). At 24 h, although all groups showed increasing values, G_4 registered the highest glucose concentration of 97.33 ± 0.70 mg/dL ($p = 0.007$). Similarly, it was observed that only G_4

registered a statistically significant progressive increase of 12.28 mg/dL at 1 h, 15.28 mg/dL at 6 h, and 40.25 mg/dL at 24 h (p = 0.02)

Table 2. Values of physio-metabolic parameters (mean ± SE) of newborn piglets with low birthweight and low vitality receiving different oral doses of caffeine.

Parameter	Treatment	Basal	1 h	6 h	24 h	p-Value
Ca (mmol/L)	G_1 (Saline)	2.21 ± 0.02 ^{a,1}	2.10 ± 0.02 ^{b,2}	2.09 ± 0.02 ^{b,2}	2.08 ± 0.02 ^{d,2}	0.0001
	G_2 (10 mg/kg)	2.16 ± 0.01 ^{a,1}	2.09 ± 0.02 ^{b,1}	2.08 ± 0.02 ^{b,1}	2.02 ± 0.01 ^{c,2}	0.001
	G_3 (20 mg/kg)	2.17 ± 0.01 ^{a,1}	2.07 ± 0.02 ^{a,2}	2.06 ± 0.02 ^{a,2}	1.92 ± 0.01 ^{b,3}	0.03
	G_4 (30 mg/kg)	2.08 ± 0.01 ^{a,1}	2.05 ± 0.02 ^{a,1}	2.05 ± 0.02 ^{a,1}	1.80 ± 0.01 ^{a,2}	0.0001
	p-value	0.92	0.0001	0.006	0.002	
Glucose (mg/dL)	G_1 (Saline)	58.36 ± 0.70 ^{a,3}	65.36 ± 0.70 ^{d,2}	66.36 ± 0.70 ^{d,1}	67.36 ± 0.70 ^{c,1}	0.0001
	G_2 (10 mg/kg)	55.88 ± 0.68 ^{a,3}	67.36 ± 0.70 ^{c,1}	67.36 ± 0.70 ^{c,1}	64.88 ± 0.70 ^{d,2}	0.001
	G_3 (20 mg/kg)	56.65 ± 1.25 ^{a,3}	68.36 ± 0.70 ^{b,2}	70.36 ± 0.70 ^{b,1}	69.65 ± 0.70 ^{b,1}	0.0001
	G_4 (30 mg/kg)	57.08 ± 1.18 ^{b,4}	69.36 ± 0.75 ^{a,3}	72.36 ± 0.70 ^{a,2}	97.33 ± 0.70 ^{a,1}	0.02
	p-value	0.95	0.0001	0.0001	0.007	
Lactate (mg/dL)	G_1 (Saline)	93.97 ± 0.76 ^{a,1}	91.97 ± 0.76 ^{a,2}	91.97 ± 0.76 ^{a,2}	89.97 ± 0.76 ^{a,3}	0.0001
	G_2 (10 mg/kg)	96.48 ± 2.09 ^{a,1}	90.97 ± 0.76 ^{b,1}	89.97 ± 0.76 ^{b,1}	84.44 ± 2.09 ^{b,2}	0.001
	G_3 (20 mg/kg)	92.67 ± 0.79 ^{a,1}	84.97 ± 0.76 ^{d,2}	82.97 ± 0.76 ^{c,3}	75.28 ± 0.79 ^{c,4}	0.001
	G_4 (30 mg/kg)	94.69 ± 0.69 ^{a,1}	82.97 ± 0.76 ^{c,2}	74.97 ± 0.76 ^{d,3}	41.69 ± 0.69 ^{d,4}	0.001
	p-value	0.99	0.001	0.0001	0.006	

^{a,b,c,d} Different letters indicate significant differences between treatments in the same evaluation time. ^{1,2,3} Different numbers indicate significant differences between evaluation times in a single treatment. n = 120 piglets per each group. G_1: placebo (saline); G_2: low dose (10 mg/kg); G_3: medium dose (20 mg/kg); G_4: high dose (30 mg/kg).

The differences found in lactate values (presented in Table 2), show that, when comparing G_2 and G_3, lactate concentration significantly decreased by 15 mg/dL and 8 mg/dL, respectively, at 6 h, and by 42.75 mg/dL and 33.59 mg/dL, respectively, at 24 h. For all groups, during the basal event, the animals were in lacto-acidemia, since it was observed that there was a statistically significant decrease in lactate levels by 7%, 11%, and 17% at 1 h, 6 h, and 24 h, respectively, in G_2 and G_3 (p = 0.001). In G_4, decreases of 13% at 1 h, 20% at 6 h, and 55% at 24 h were observed (p = 0.001).

3.3. Thermal Response

Table 3 shows the temperature differences of the four thermal windows (OCU, EAR, NOSE, LIMB). Regarding OCU, in all groups where caffeine was administered, the surface temperature was 0.15 °C higher compared to G_1 at 6 h (p = 0.04) and 0.98 °C higher compared to G_1 at 24 h (p = 0.0001). Likewise, it was found that G_2 and G_1 presented a general increase of 1.20 °C at 6 h, and of 1.45 °C at 24 h (p = 0.001). In G_4, this increase was 1.25 °C at 6 h and 2.22 °C at 24 h (p = 0.001).

For EAR, it was observed that the G_4 temperature was higher by 0.76 °C and 1 °C compared to G_3 and G_1, respectively (p = 0.0005). Between evaluation times, all treatments presented statistically significant decreases of up to 2 °C at 1 h, and a statistically significant increase of up to 2.7 °C at 6 h. At 24 h, G_4 had the greatest statistically significant increase of 2.41 °C compared to basal (p = 0.001).

At NOSE, it was observed that G_4 was statistically 2.62 °C higher compared to G_3, 3.09 °C compared to G_2, and 5.47 °C compared to G_1 at 24 h (p = 0.0001). Between evaluation times, all groups presented a statistically significant decrease of 2.7 °C at 1 h. Subsequently, there was a statistically significant increase of 2.8 °C at 6 h, while at 24 h, a significant increase of 6.3 °C and 7.12 °C was observed in G_2 and G_3, respectively. Nonetheless, G_4 had the highest temperatures (32.29 ± 0.05 °C) and a statistically significant increase of 9.74 °C (p = 0.001).

Table 3. Temperature values of the four thermal windows (OCU, EAR, NOSE, LIMB) (°C) (mean ± SE) of newborn piglets with low birthweight and low vitality receiving different oral doses of caffeine.

Thermal Window	Treatment	Basal	1 h	6 h	24 h	*p*-Value
OCU	G_1 (Saline)	33.63 ± 0.03 ^{a,2}	32.04 ± 0.03 ^{c,3}	34.65 ± 0.03 ^{b,1}	34.79 ± 0.03 ^{b,1}	0.001
	G_2 (10 mg/kg)	33.84 ± 0.03 ^{a,3}	32.14 ± 0.03 ^{b,4}	34.78 ± 0.03 ^{a,2}	35.00 ± 0.03 ^{a,1}	0.001
	G_3 (20 mg/kg)	33.58 ± 0.02 ^{a,3}	32.27 ± 0.02 ^{a,4}	34.79 ± 0.02 ^{a,2}	35.03 ± 0.02 ^{a,1}	0.001
	G_4 (30 mg/kg)	33.55 ± 0.02 ^{a,3}	32.30 ± 0.02 ^{a,4}	34.80 ± 0.02 ^{a,2}	35.77 ± 0.02 ^{a,1}	0.001
	p-value	0.96	0.001	0.04	0.0001	
EAR	G_1 (Saline)	32.54 ± 0.03 ^{a,3}	30.07 ± 0.03 ^{a,4}	33.66 ± 0.03 ^{a,2}	33.91 ± 0.03 ^{c,1}	0.0001
	G_2 (10 mg/kg)	32.58 ± 0.03 ^{a,3}	30.24 ± 0.03 ^{a,4}	33.73 ± 0.03 ^{a,2}	33.96 ± 0.03 ^{c,1}	0.001
	G_3 (20 mg/kg)	32.60 ± 0.03 ^{a,3}	30.30 ± 0.02 ^{b,4}	33.79 ± 0.02 ^{a,2}	34.14 ± 0.02 ^{b,1}	0.001
	G_4 (30 mg/kg)	32.49 ± 0.03 ^{a,3}	30.36 ± 0.03 ^{b,4}	33.76 ± 0.03 ^{a,2}	34.90 ± 0.03 ^{a,1}	0.001
	p-value	0.97	0.005	0.37	0.0005	
NOSE	G_1 (Saline)	22.62 ± 0.11 ^{a,3}	19.91 ± 0.11 ^{a,4}	23.30 ± 0.11 ^{b,2}	26.82 ± 0.11 ^{d,1}	0.001
	G_2 (10 mg/kg)	22.70 ± 0.10 ^{a,3}	19.97 ± 0.10 ^{a,4}	23.62 ± 0.10 ^{a,b,2}	29.02 ± 0.10 ^{c,1}	0.001
	G_3 (20 mg/kg)	22.55 ± 0.07 ^{a,3}	19.94 ± 0.07 ^{a,4}	23.76 ± 0.07 ^{a,2}	29.67 ± 0.07 ^{b,1}	0.001
	G_4 (30 mg/kg)	22.53 ± 0.05 ^{a,3}	20.08 ± 0.05 ^{a,4}	23.96 ± 0.05 ^{a,2}	32.29 ± 0.05 ^{a,1}	0.001
	p-value	0.88	0.69	0.0001	0.0001	
LIMB	G_1 (Saline)	27.19 ± 0.11 ^{a,3}	24.57 ± 0.11 ^{a,4}	28.05 ± 0.11 ^{a,2}	30.42 ± 0.11 ^{c,1}	0.001
	G_2 (10 mg/kg)	27.25 ± 0.11 ^{a,3}	24.64 ± 0.11 ^{a,4}	28.30 ± 0.11 ^{a,2}	31.38 ± 0.11 ^{b,1}	0.001
	G_3 (20 mg/kg)	27.06 ± 0.11 ^{a,3}	24.83 ± 0.11 ^{a,4}	28.41 ± 0.11 ^{a,2}	31.70 ± 0.11 ^{b,1}	0.001
	G_4 (30 mg/kg)	27.14 ± 0.11 ^{a,3}	24.79 ± 0.11 ^{a,4}	28.62 ± 0.11 ^{a,2}	33.49 ± 0.11 ^{a,1}	0.001
	p-value	0.93	0.95	0.60	0.001	

^{a,b,c,d} Different letters indicate significant differences between treatments in the same evaluation time. ^{1,2,3,4} Different numbers indicate significant differences between evaluation times in a single treatment. n = 120 piglets per each group. G_1: placebo (saline); G_2: low dose (10 mg/kg); G_3: medium dose (20 mg/kg); G_4: high dose (30 mg/kg).

Regarding LIMB, Table 3 shows that G_4 piglets had the highest temperatures at 24 h (33.49 ± 0.11 °C), differing by 2.11 °C from G_2 and by 1.79 °C from G_3 (p = 0.001). Between evaluation times, all treatments showed a temperature decrease from basal to 1 h of up to 2.62 °C. In contrast, from 1 h to 24 h, a progressive increase of 3.48–8.7 °C was registered in G_1, G_2, G_3, and G_4 (p = 0.001), where G_4 showed the greatest increase, from 24.79 ± 0.11 °C to 33.49 ± 0.11 °C (p = 0.001).

RT values are shown in Table 4, where at 24 h, G_4 was 0.58 °C higher in comparison to G_3, 0.66 °C higher than G_2, and 0.7 °C higher than G_1 (p = 0.004). Regarding the differences between evaluation times, G_4 presented the greatest increase in temperature at 1 h, 6 h, and 24 h, registering statistically significant increases of 0.13 °C, 0.14 °C, and 0.8 °C, respectively (p = 0.001).

Table 4. RT (°C) (mean ± SE) of newborn piglets with low birthweight and low vitality receiving different oral doses of caffeine.

Treatment	Basal	1 h	6 h	24 h	*p*-Value
G_1 (Saline)	36.18 ± 0.02 ^{a,2}	36.18 ± 0.01 ^{a,2}	36.27 ± 0.02 ^{a,1}	36.28 ± 0.03 ^{c,1}	0.001
G_2 (10 mg/kg)	36.18 ± 0.01 ^{a,2}	36.18 ± 0.02 ^{a,2}	36.27 ± 0.01 ^{a,1}	36.28 ± 0.02 ^{b,1}	0.001
G_3 (20 mg/kg)	36.12 ± 0.02 ^{a,1}	36.18 ± 0.01 ^{a,2}	36.27 ± 0.02 ^{a,2}	36.42 ± 0.01 ^{b,3}	0.001
G_4 (30 mg/kg)	36.14 ± 0.03 ^{a,1}	36.27 ± 0.02 ^{b,2}	36.28 ± 0.01 ^{a,2}	36.94 ± 0.01 ^{a,3}	0.001
p-value	0.85	0.001	0.98	0.004	

^{a,b,c} Different letters indicate significant differences between treatments in the same evaluation time. ^{1,2,3} Different numbers indicate significant differences between evaluation times in a single treatment. n = 120 piglets per each group. G_1: placebo (saline); G_2: low dose (10 mg/kg); G_3: medium dose (20 mg/kg); G_4: high dose (30 mg/kg).

Table 5 summarizes the correlations between the different thermal windows and RT. A weak positive correlation was observed for OCU (r = 0.26, p = 0.0001), EAR (r = 0.29,

p = 0.0001), and LIMB (r = 0.33, p = 0.0001), while NOSE had an intermediate positive correlation (r = 0.41, p = 0.0001).

Table 5. Correlation matrix.

	OCU	NOSE	EAR	LIMB	RT
OCU	1.00	0.68 p = 0.0001	0.92 p = 0.0001	0.76 p = 0.0001	0.26 p = 0.0001
NOSE	0.68 p = 0.0001	1.00	0.79 p = 0.0001	0.84 p = 0.0001	0.41 p = 0.0001
EAR	0.92 p = 0.0001	0.79 p = 0.0001	1.00	0.79 p = 0.0001	0.29 p = 0.0001
LIMB	0.76 p = 0.0001	0.84 p = 0.0001	0.79 p = 0.0001	1.00	0.33 p = 0.0001
RT	0.26 p = 0.0001	0.41 p = 0.0001	0.29 p = 0.0001	0.33 p = 0.0001	1.00

OCU = periocular area; EAR = auricular region; NOSE = nasal; LIMB = left hindlimb region; RT = rectal temperature.

4. Discussion

The results show that the administration of caffeine at a dose of 30 mg/kg (G_4) caused the most significant effects in low-birthweight and low-vitality newborn piglets, suggesting this dose as the more recommended one to improve the physiological, metabolic, thermal, and acid–base balance profile, in contrast to placebo (G_1), low dose (10 mg/kg) (G_2), and medium dose (20 mg/kg) (G_3). However, the benefits observed in all groups receiving caffeine show that, regardless of the dose, caffeine is a neuroprotective drug against the consequences of neonatal hypoxia-ischemia events.

The possible explanation for the differences in the therapeutic effect between the three doses is related to the pharmacokinetic characteristics of caffeine. Menozzi et al. [29] evaluated the pharmacokinetics of caffeine in sows receiving 25 mg/kg, finding that the maximum concentration (C_{Max}) was 20.02 ± 1.51 mg/mL at 9.51 h post-administration. Superchi et al. [30] reported that PO administration of caffeine at 27 mg/kg to parturient sows did not affect physiological parameters or behavior, although it reduced stillborn percentages. A similar result has been reported in sheep receiving 10 or 20 mg/kg of caffeine in Robertson et al. [31]'s study, where no significant effect on newborn survival or physiological parameters, such as RT, was reported. The authors attribute this variation to the differences in caffeine metabolism and distribution when comparing therapeutic and toxic ranges between humans and neonatal animals [38].

These results suggest that a lower dose of caffeine (10 or 20 mg/kg) would take longer to reach peak plasma concentrations and, consequently, have a possible delayed therapeutic effect in animals. Moreover, studies in neonate humans have shown that plasma half-life and metabolism of caffeine are longer compared with adults or animals [39]. In this sense, Murdock et al. [28] analyzed the optimal concentration and duration of caffeine to improve lamb viability, finding that 20 mg/kg administered at day 120 of gestation increased lambs' RT, as well as the attempts to suck colostrum and suction time. Accordingly, caffeine metabolism in the newborn can be considered less efficient due to a lower expression of the CYP1A2 protein enzymes, responsible for drug biotransformation, resulting in lower production of active metabolites such as theophylline and theobromine, the main caffeine-derived effectors. Therefore, although caffeine at 10 or 20 mg/kg might have a therapeutic effect in the newborn piglet (G_2 and G_3), the effect is limited and lasts for a short time, as observed in the results of this study.

In the present research, the use of caffeine at a dose of 30 mg/kg in G_4 piglets significantly improved the alterations observed in the physio-metabolic profile associated with hypoxia and hypothermia. Regardless of the treatment group, all animals presented blood gas imbalances such as respiratory acidosis, reduced pO_2, and hypercapnia at basal time before caffeine administration. Robertson et al. [40] mention that the main cause of acidemia in animals is hypercapnia, which is a recurrent event in newborn piglets.

Swinbourne et al. [2] describe that the decrease in maternal blood flow to the fetus significantly affects gas exchange, leading to suboptimal O_2 levels, elevated pCO_2 concentrations, and respiratory acidosis. The decrease in blood pH triggers compensatory mechanisms such as the formation of HCO_3^- when CO_2 and H_2O are metabolized by carbonic anhydrase [41,42]. It is known that caffeine stimulates the respiratory centers, improving gas exchange, increasing the tidal volume and HCO_3^- excretion at the renal level, helping to correct the acid–base imbalances in the newborn [32,43,44]. Thus, as observed in piglets from G_4, caffeine could have a protective effect against ischemia caused by hypoxia and acidemia [22,45,46].

Additionally, the administration of caffeine reduced pCO_2 and increased pO_2 levels, which has beneficial effects by improving vitality scores and survival rates in piglets. The antagonism of caffeine on A2a and A2b adenosine receptors present in pulmonary and cerebral tissue [47,48] stimulates the medullary respiratory centers, increasing sensitivity to CO_2 concentrations and improving respiratory function [26,49]. In the same way, stimulation of the central *vagus* nerve and the *medulla oblongata* help increase tidal volume, which would explain the increase in O_2 levels and the reduction in CO_2 [24,50]. According to this respiratory and physiological response, it can be suggested that high doses of caffeine (30 mg/kg) could be the best to improve the ventilatory activity of newborn piglets. The higher improvement observed with this dose is related to the pharmacokinetic properties of caffeine and that at high doses, plasmatic concentrations of the drug can be found even at 9 h after its administration [31].

Regarding the respiratory effect of caffeine and its impact on pCO_2 and pO_2 levels, the prepartum administration of caffeine in sows and its effect on piglets [51] and newborn lambs [52] has shown that caffeine lowers the incidence of hypoxia due to a short-term physio-metabolic and behavioral adaptation during birth, avoiding cerebral ischemic damage and improving the performance of the piglets during feeding and lactation. It is worth mentioning that since the pH levels depend on pCO_2 and pO_2 levels, lactate is a biomarker that is also associated with neonatal acidosis. Van Dijk et al. [53] explains that the decrease in oxygen supply, apart from reducing blood pH, also leads to states of metabolic acidosis due to the increase in the production of lactic acid, as could be observed in the basal events of all the treatments. In piglets suffering from fetal hypoxia, the production of lactic acid is an alternative anaerobic energy pathway that leads to an inefficient production of ATP [21,54]. As found in G_2, G_3, and G_4, caffeine reduced lactate concentrations at 1 h, 6 h, and 24 h, suggesting that it can restore aerobic pathways for energy production.

The observed results also showed that caffeine significantly increased glucose, obtaining the most significant increase at 24 h in G_4 ($p = 0.0001$). Perhaps, this is one of the most important effects of caffeine related to vitality. The increase in glucose observed in all groups receiving caffeine is due to the sympathetic stimulation, which allows the stimulation of the chromaffin cells of the adrenal medulla that consequently release adrenaline, leading to an increase in blood glucose due to hepatic glycogenolysis and gluconeogenesis [55]. The effect that caffeine has on glucose levels has been controversial among authors because it has been seen in previous reports that caffeine in newborn piglets has not shown a positive increasing effect in glucose. However, the same authors describe that a limitation could be the administration of caffeine prior to colostrum intake [10,27]. In humans, it has been seen that fasting before caffeine administration reduces the body's sensitivity to insulin, thus increasing blood glucose. The above may be due to the effect of adrenaline and glucagon, which promote the release of glucose stored in the liver, thus increasing blood glucose levels [56].

It would be worth adding that, according to the findings of other authors, caffeine treatment cannot replace colostrum intake, as colostrum is a valuable energy resource that is essential to thermoregulate, increase vitality, and obtain the therapeutic effect of caffeine [57]. Moreover, apart from colostrum and caffeine dose, birthweight highly influences the effect that this drug has on newborn piglets [27]. Thus, when considering caffeine as a neonatal drug for piglets, both the dose and the weight of the animal need to be considered to start a therapeutic protocol, as shown in the present study where a dose of 30 mg/kg could be considered an appropriate dose for piglets below 1 kg [10].

Regarding Ca^{++} values, caffeine administration at three doses significantly reduced Ca^{++} concentration at 24 h, especially in G_4. It has been reported that hypoxia alters calcium metabolism in animals [58] and that caffeine is known to reduce Ca++ concentrations by reducing renal and intestinal reabsorption [59]. However, in the present study, the decrease in Ca^{++} might suggest an adaptative mechanism to hypoxia in response to caffeine and its stimulant cardiovascular effects. Caffeine facilitates intracellular Ca^{++} mobilization to improve muscular performance and cardiac activity [60]. In vitro studies have shown that caffeine has benefits on cardiac contractility [61], possibly improving oxygen transportation to other cellular tissues in hypoxic piglets. This was reported in Merino newborn lambs, in which 20 mg/kg of caffeine administration reduced neonatal mortality in the first week post-farrowing due to hypoxia reduction. Moreover, the increased availability of intracellular Ca^{++} might also be used by piglets to increase their body temperature through muscle shivering or muscle-based non-shivering thermogenesis through sarcoendoplasmic reticulum calcium [62].

Regarding the thermal response in different thermal windows and on the RT of piglets, the progressive increases in OCU, EAR, NOSE, LIMB, and RT are due to caffeine boosting basal metabolism by interacting with adenosine receptors in the sympathetic nervous system [63]. Jarrat et al. [10] mentions that raising metabolic activity in newborn piglets that have limited energy resources at birth can help them increase body temperature [13]. When comparing the administration of caffeine alone or in combination with glucose, the combination prolonged energy intake and improved the thermoregulatory capacity of newborn piglets [10].

OCU reflects the sympathetic influence on the surface temperature because it is a region where blood supply from the infra-orbital artery is mediated by sympathetic fibers from the facial nerve. Therefore, the change in the thermal response could be an indirect way to assess sympathetic activation [20,57], as shown in bovines [64] and canids [65]. On the other hand, NOSE temperature is related to the respiratory activity of piglets, keeping a positive correlation with respiratory rate and heat loss—or the amount of heat radiated through this thermal window [66,67]. Considering that caffeine stimulates cardiorespiratory function, the administration of caffeine in G4 induced the greatest increase in NOSE values, possibly reflecting a positive correlation between respiratory rate and thermal exchange.

Finally, the improved metabolic activity in G_2, G_3, and G_4—see glucose discussion—is related to RT and the different thermal windows [63]. In addition, although it was not the main objective of the present study, it was possible to observe a difference between the thermal response of peripheral regions (LIMB) and central thermal windows (OCU, EAR, and NOSE). This could indicate that in hypothermic newborn piglets, blood microcirculation is shifted to vital body regions, as observed in other species [64,65]. Additionally, although the present study did not directly assess the effect of caffeine doses on the mortality rate of piglets (which could be considered a limitation of the study), further research needs to consider the effect that caffeine has on mortality rate and its association with blood gas parameters, physio-metabolic profile, and thermal response of piglets. To this extent, the findings of the present research suggest that IRT, along with blood biomarkers, could be a tool for monitoring the health status of newborn piglets [20]. Nonetheless, future research is needed.

5. Conclusions

According to the results, the administration of caffeine at 30 mg/kg reduced the alterations observed in piglets with low birthweight and low vitality scores. This was observed as a decrease in pCO_2 levels and increasing pO_2, which allows the recovery of the acid–base balance by increasing blood pH. Likewise, the administration of this drug enables a reduction in metabolic imbalances thanks to a reduction in lactic acid values and an increase in blood glucose. Finally, this drug increases infrared superficial thermal response capacity in the nasal, ocular, auricular, and nasal thermal windows in newborn piglets with low birthweight and low vitality scores. Further studies could consider the parameters evaluated in the present study and compare animals with low birthweight and low vitality scores days after farrowing with healthy piglets. Therefore, according to the attenuation of the metabolic, acid–base, and thermal changes observed in newborn piglets, it could be suggested that the use of high doses of caffeine (30 mg/kg) can improve the physiological response of piglets.

These results suggest that the administration of caffeine is a useful strategy to recover low vitality birth piglets, which is important for improving animal welfare and performance in the pig production system.

Author Contributions: Conceptualization, D.V.-G., A.D.-O. and M.M.-R.; Formal analysis, D.V.-G., I.H.-Á., A.D.-O. and M.M.-R.; Writing—original draft, M.G., P.M.-M., I.H.-Á., A.O.-H., A.C.-A., K.L.-G. and D.R.-G.; Writing—review & editing, D.V.-G., M.G., P.M.-M., I.H.-Á., A.O.-H., A.C.-A., K.L.-G., A.D.-O., D.R.-G. and M.M.-R.; Investigation: I.H.-Á. All authors have read and agreed to the published version of the manuscript.

Funding: This research received no external funding.

Institutional Review Board Statement: The experimental procedure of the current investigation received approval from the Laboratory of Animal Ethology and Neurophysiology (ETO-078-21) and the Postgraduate Veterinary Commission of the Universidad Autónoma Metropolitana in Mexico City, Mexico. The management of the animals was in accordance with FRAME's guidelines for animal research and the ethical considerations for applied ethological studies [32]. Acknowledging that the procedures of the present study might cause mild pain, suffering, and/or distress, the use of cell phones, televisions, and other sources of noise was restricted to minimize stressors. All animals were treated with humane care. The design and specifications of the farrowing facilities were in accordance with the Mexican Official Standard NOM-062-ZOO-1999 [29].

Informed Consent Statement: Informed consent was obtained from the owner of the animals and the Managing Director responsible for overseeing and authorizing scientific projects at the commercial pig farm.

Data Availability Statement: The data presented in this study are available on request from the corresponding author. The data are not publicly available due to the original data belonging to the swine breeding herd.

Acknowledgments: The authors want to express their gratitude to Daniel Mota-Rojas, an expert in animal behavior and welfare, for his role in conceiving and overseeing the project that gave rise to this manuscript. We are grateful for his consistent support and guidance.

Conflicts of Interest: The authors declare no conflict of interest.

References

1. Sadeghi, E.; Kappers, C.; Chiumento, A.; Derks, M.; Havinga, P. Improving Piglets Health and Well-Being: A Review of Piglets Health Indicators and Related Sensing Technologies. *Smart Agric. Technol.* **2023**, *5*, 100246. [CrossRef]
2. Swinbourne, A.M.; Kind, K.L.; Flinn, T.; Kleemann, D.O.; van Wettere, W.H.E.J. Caffeine: A potential strategy to improve survival of neonatal pigs and sheep. *Anim. Reprod. Sci.* **2021**, *226*, 106700. [CrossRef] [PubMed]
3. Farmer, C.; Edwards, S.A. Review: Improving the performance of neonatal piglets. *Animal* **2022**, *16*, 100350. [CrossRef]
4. Jovic, S.; Cupic, V.; Ristic, G.; Vakanjac, S.; Dimitrijevic, B.; Miladinovic, D.; Zivkovic, L. The influence of the induction of farrowing on live birth, body mass, appearance of dystocia, mortality and surviving of neonatal pigs in litter during the first ten days. *Vet. Glas.* **2016**, *70*, 13–29. [CrossRef]
5. Nam, N.H.; Sukon, P. Incidence of dystocia at piglet level in cloprostenol-induced farrowings and associated risk factors. *Arch. Anim. Breed.* **2022**, *65*, 97–103. [CrossRef]
6. Mota-Rojas, D.; López, A.; Martínez-Burnes, J.; Muns, R.; Villanueva-García, D.; Mora-Medina, P.; González-Lozano, M.; Olmos-Hernández, A.; Ramírez-Necoechea, R. Is vitality assessment important in neonatal animals? *CAB Rev. Perspect. Agric. Vet. Sci. Nutr. Nat. Resour.* **2018**, *13*, 1–13. [CrossRef]
7. Ward, S.A.; Kirkwood, R.N.; Plush, K.L. Effects of Oxytocin and Carbetocin on Farrowing Performance. *Anim. Reprod. Sci.* **2019**, *205*, 88–93. [CrossRef]

8. Mota-Rojas, D.; Villanueva-García, D.; Mota-Reyes, A.; Orihuela, A.; Hernández-Ávalos, I.; Domínguez-Oliva, A.; Casas-Alvarado, A.; Flores-Padilla, K.; Jacome-Romero, J.; Martínez-Burnes, J. Meconium aspiration syndrome in animal models: Inflammatory process, apoptosis, and surfactant inactivation. *Animals* **2022**, *12*, 3310. [CrossRef]
9. Santiago, P.R.; Martínez-Burnes, J.; Mayagoitia, A.L.; Ramírez-Necoechea, R.; Mota-Rojas, D. Relationship of vitality and weight with the temperature of newborn piglets born to sows of different parity. *Livest. Sci.* **2019**, *220*, 26–31. [CrossRef]
10. Jarratt, L.; James, S.E.; Kirkwood, R.N.; Nowland, T.L. Effects of caffeine and glucose supplementation at birth on piglet pre-weaning growth, thermoregulation, and survival. *Animals* **2023**, *13*, 435. [CrossRef] [PubMed]
11. Oliviero, C.; Junnikkala, S.; Peltoniemi, O. The challenge of large litters on the immune system of the sow and the piglets. *Reprod. Domest. Anim.* **2019**, *54*, 12–21. [CrossRef]
12. Le Dividich, J.; Noblet, J. Thermoregulation and energy metabolism in the neonatal pig. *Ann. Rech. Vet.* **1983**, *14*, 375–381.
13. Villanueva-García, D.; Mota-Rojas, D.; Martínez-Burnes, J.; Olmos-Hernández, A.; Mora-Medina, P.; Salmerón, C.; Gómez, J.; Boscato, L.; Gutiérrez-Pérez, O.; Cruz, V.; et al. Hypothermia in newly born piglets: Mechanisms of thermoregulation and pathophysiology of death. *J. Anim. Behav. Biometeorol.* **2021**, *9*, 2101. [CrossRef]
14. Alonso-Spilsbury, M.; Ramirez-Necoechea, R.; Gonzalez-Lozano, M.; Mota-Rojas, D.; Trujillo, M. Piglet survival in early lactation: A review. *J. Anim. Vet. Adv.* **2007**, *6*, 76–86.
15. Jiarpinitnun, P.; Loyawatananan, S.; Sangratkanjanasin, P.; Kompong, K.; Nuntapaitoon, M.; Muns, R.; De Rensis, F.; Tummaruk, P. Administration of Carbetocin after the First Piglet Was Born Reduced Farrowing Duration but Compromised Colostrum Intake in Newborn Piglets. *Theriogenology* **2019**, *128*, 23–30. [CrossRef]
16. Nuntapaitoon, M.; Tummaruk, P. Neonatal piglet survival associated with blood glucose concentration. *Thai J. Vet. Med.* **2014**, *44*, S159–S160.
17. Muns, R.; Nuntapaitoon, M.; Tummaruk, P. Non-infectious causes of pre-weaning mortality in piglets. *Livest. Sci.* **2016**, *184*, 46–57. [CrossRef]
18. Mota-Rojas, D.; Titto, C.G.; Orihuela, A.; Martínez-Burnes, J.; Gómez-Prado, J.; Torres-Bernal, F.; Flores-Padilla, K.; Carvajal-de la Fuente, V.; Wang, D.; la Fuente, V.C.; et al. Physiological and behavioral mechanisms of thermoregulation in mammals. *Animals* **2021**, *11*, 1733. [CrossRef] [PubMed]
19. Mellor, D.J.; Stafford, K.J. Animal welfare implications of neonatal mortality and morbidity in farm animals. *Vet. J.* **2004**, *168*, 118–133. [CrossRef] [PubMed]
20. Gómez-Prado, J.; Pereira, A.M.F.; Wang, D.; Villanueva-García, D.; Domínguez-Oliva, A.; Mora-Medina, P.; Hernández-Avalos, I.; Martínez-Burnes, J.; Casas-Alvarado, A.; Olmos-Hernández, A.; et al. Thermoregulation mechanisms and perspectives for validating thermal windows in pigs with hypothermia and hyperthermia: An overview. *Front. Vet. Sci.* **2022**, *9*, 1023294. [CrossRef]
21. Bienboire-Frosini, C.; Muns, R.; Marcet-Rius, M.; Gazzano, A.; Villanueva-García, D.; Martínez-Burnes, J.; Domínguez-Oliva, A.; Lezama-García, K.; Casas-Alvarado, A.; Mota-Rojas, D. Vitality in Newborn Farm Animals: Adverse Factors, Physiological Responses, Pharmacological Therapies, and Physical Methods to Increase Neonate Vigor. *Animals* **2023**, *13*, 1542. [CrossRef] [PubMed]
22. Villanueva-García, D.; Mota-Rojas, D.; Miranda-Cortés, A.E.; Mora-Medina, P.; Hernández-Avalos, I.; Casas-Alvarado, A.; Olmos-Hernández, A.; Martínez-Burnes, J. Neurobehavioral and neuroprotector effects of caffeine in animal models. *J. Anim. Behav. Biometeorol.* **2020**, *8*, 298–307. [CrossRef]
23. Lodha, A.; Seshia, M.; McMillan, D.D.; Barrington, K.; Yang, J.; Lee, S.K.; Shah, P.S. Association of early caffeine administration and neonatal outcomes in very preterm neonates. *JAMA Pediatr.* **2015**, *169*, 33. [CrossRef]
24. Villanueva-García, D.; Mota-Rojas, D.; Miranda-Cortés, A.; Ibarra-Ríos, D.; Casas-Alvarado, A.; Mora-Medina, P.; Martínez-Burnes, J.; Olmos-Hernández, A.; Hernández-Avalos, I. Caffeine: Cardiorespiratory effects and tissue protection in animal models. *Exp. Anim.* **2021**, *70*, 20–0185. [CrossRef]
25. Hsieh, E.M.; Hornik, C.P.; Clark, R.H.; Laughon, M.M.; Benjamin, D.K.; Smith, P.B. Medication use in the neonatal intensive care unit. *Am. J. Perinatol.* **2014**, *31*, 811–821. [CrossRef]
26. Kumar, V.H.S.; Lipshultz, S.E. Caffeine and cinical outcomes in premature neonates. *Children* **2019**, *6*, 118. [CrossRef]
27. Nowland, T.L.; Kind, K.; Hebart, M.L.; van Wettere, W.H.E.J. Caffeine supplementation at birth, but not 8 to 12 h post-birth, increased 24 h pre-weaning mortality in piglets. *Animal* **2020**, *14*, 1529–1535. [CrossRef]
28. Murdock, N.J.; Weaver, A.C.; Kelly, J.M.; Kleemann, D.O.; van Wettere, W.H.E.J.; Swinbourne, A.M. Supplementing pregnant Merino ewes with caffeine to improve neonatal lamb thermoregulation and viability. *Anim. Reprod. Sci.* **2021**, *226*, 106715. [CrossRef] [PubMed]
29. Menozzi, A.; Mazzoni, C.; Serventi, P.; Zanardelli, P.; Bertini, S. Pharmacokinetics of oral caffeine in sows: A pilot study. *Large Anim. Rev.* **2015**, *21*, 207–210.
30. Superchi, P.; Saleri, R.; Farina, E.; Cavalli, V.; Riccardi, E.; Sabbioni, A. Effects of oral administration of caffeine on some physiological parameters and maternal behaviour of sows at farrowing. *Res. Vet. Sci.* **2016**, *105*, 121–123. [CrossRef] [PubMed]
31. Robertson, S.M.; Edwards, S.H.; Doran, G.S.; Friend, M.A. Maternal caffeine administration to ewes does not affect perinatal lamb survival. *Anim. Reprod. Sci.* **2021**, *231*, 106799. [CrossRef]

32. Orozco, G.; Mota-Rojas, D.; Jaime, H.; Trujillo, M.E.; Becerril-Herrera, M.; Hernández-González, R.; Villanueva-García, D. Effects of Administration of Caffeine on Metabolic Variables in Neonatal Pigs with Peripartum Asphyxia. *Am. J. Vet. Res.* **2010**, *71*, 1214–1219. [CrossRef] [PubMed]
33. Gregorio, O.; Bonilla, H.; Mota-Rojas, D.; Trujillo-Ortega, M.E.; Roldan, P.; Martínez-Rodríguez, R.; Borderas-Tordesillas, F.; Flores-Peinado, S.; Mora-Medina, P.; Ramírez-Necoechea, R. Effects of Subcutaneous Administration of Caffeine on the Physiometabolic Profile of Low-Birthweight Neonate Piglets. *Anim. Prod. Sci.* **2012**, *52*, 981. [CrossRef]
34. *NOM-062-ZOO-1999*; Especificaciones Técnicas para la Producción, Cuidado y Uso de los Animales de Laboratorio. The Secretariat of Agriculture and Rural Development: Mexico City, Mexico, 1999. Available online: http://publico.senasica.gob.mx/?doc=743 (accessed on 8 August 2023).
35. Zaleski, H.M.; Hacker, R.R. Comparison of viability scoring and blood gas analysis as measures of piglet viability. *Can. J. Anim. Sci.* **1993**, *73*, 649–653. [CrossRef]
36. Mota-Rojas, D.; Martínez-Burnes, J.; Trujillo, M.E.; López, A.; Rosales, A.M.; Ramírez, R.; Orozco, H.; Merino, A.; Alonso-Spilsbury, M. Uterine and fetal asphyxia monitoring in parturient sows treated with oxytocin. *Anim. Reprod. Sci.* **2005**, *86*, 131–141. [CrossRef]
37. Sherwin, C.M.; Christiansen, S.B.; Duncan, I.J.; Erhard, H.W.; Lay, D.C.; Mench, J.A.; O'Connor, C.E.; Petherick, J.C. Guidelines for the ethical use of animals in applied ethology studies. *Appl. Anim. Behav. Sci.* **2003**, *81*, 291–305. [CrossRef]
38. Atik, A.; Cheong, J.; Harding, R.; Rees, S.; De Matteo, R.; Tolcos, M. Impact of daily high-dose caffeine exposure on developing white matter of the immature ovine brain. *Pediatr. Res.* **2014**, *76*, 54–63. [CrossRef]
39. Abdel-Hady, H. Caffeine therapy in preterm infants. *World J. Clin. Pediatr.* **2015**, *4*, 81. [CrossRef] [PubMed]
40. Robertson, S.A. Simple acid-base disorders. *Vet. Clin. North Am. Small Anim. Pract.* **1989**, *19*, 289–306. [CrossRef]
41. Day, T.K. Blood gas analysis. *Vet. Clin. N. Am. Small Anim. Pract.* **2002**, *32*, 1031–1048. [CrossRef] [PubMed]
42. Muir, W.W. Acid-Base Physiology. In *Veterinary Anesthesia and Analgesia*; John Wiley & Sons, Ltd.: Chichester, UK, 2017; pp. 355–371.
43. Monnig, A.A. Practical cid-Babse in veterinary patients. *Vet. Clin. N. Am. Small Anim. Pract.* **2013**, *43*, 1273–1286. [CrossRef] [PubMed]
44. Dearlove, B.A.; Kind, K.L.; Gatford, K.L.; van Wettere, W.H.E.J. Oral caffeine administered during late gestation increases gestation length and piglet temperature in naturally farrowing sows. *Anim. Reprod. Sci.* **2018**, *198*, 160–166. [CrossRef] [PubMed]
45. Rossor, T.; Bhat, R.; Ali, K.; Peacock, J.; Rafferty, G.F.; Greenough, A. The effect of caffeine on the ventilatory response to hypercarbia in preterm infants. *Pediatr. Res.* **2018**, *83*, 1152–1157. [CrossRef]
46. Carter, A.J.; O'Connor, W.T.; Carter, M.J.; Ungerstedt, U. Caffeine enhances acetylcholine release in the hippocampus in vivo by a selective interaction with adenosine A1 receptors. *J. Pharmacol. Exp. Ther.* **1995**, *273*, 637–642.
47. Fredholm, B.B.; IJzerman, A.P.; Jacobson, K.A.; Klotz, K.N.; Linden, J. International Union of Pharmacology. XXV. Nomenclature and classification of adenosine receptors. *Pharmacol. Rev.* **2001**, *53*, 527–552.
48. Jacobson, K.A.; Gao, Z.; Matricon, P.; Eddy, M.T.; Carlsson, J. Adenosine A 2A Receptor Antagonists: From Caffeine to Selective Non-xanthines. *Br. J. Pharmacol.* **2022**, *179*, 3496–3511. [CrossRef] [PubMed]
49. Synnes, A.; Grunau, R.E. Neurodevelopmental Outcomes after Neonatal Caffeine Therapy. *Semin. Fetal Neonatal Med.* **2020**, *25*, 101160. [CrossRef]
50. Kourtidou-Papadeli, C.; Papadelis, C.; Louizos, A.-L.; Guiba-Tziampiri, O. Maximum cognitive performance and physiological time trend measurements after caffeine intake. *Cogn. Brain Res.* **2002**, *13*, 407–415. [CrossRef]
51. Ortega, M.E.; Mota-Rojas, D.; Juárez, O.; Villanueva-García, D.; Santiago, P.; Becerril-Herrera, M.; Hernández-González, R.; Mora-Medina, P.; Alonso-Spilsbury, M.; Rosales, A.M.; et al. Porcine neonates failing vitality score: Physio-metabolic profile and latency to the first teat contact. *Czech J. Anim. Sci.* **2011**, *56*, 499–508. [CrossRef]
52. Robertson, S.M.; Friend, M.A.; Doran, G.S.; Edwards, S. Caffeine supplementation of ewes during lambing may increase lamb survival. *Animal* **2018**, *12*, 376–382. [CrossRef]
53. van Dijk, A.J.; van Loon, J.P.A.M.; Taverne, M.A.M.; Jonker, F.H. Umbilical cord clamping in term piglets: A useful model to study perinatal asphyxia? *Theriogenology* **2008**, *70*, 662–674. [CrossRef]
54. Lezama-García, K.; Mota-Rojas, D.; Martínez-Burnes, J.; Villanueva-García, D.; Domínguez-Oliva, A.; Gómez-Prado, J.; Mora-Medina, P.; Casas-Alvarado, A.; Olmos-Hernández, A.; Soto, P.; et al. Strategies for hypothermia compensation in altricial and precocial newborn mammals and their monitoring by infrared thermography. *Vet. Sci.* **2022**, *9*, 246. [CrossRef] [PubMed]
55. Alagbonsi, A.; Salman, T.M.; Salahdeen, H.M.; Alada, A.A. Effects of adenosine and caffeine on blood glucose levels in rats. *Niger. J. Exp. Clin. Biosci.* **2016**, *4*, 35–41. [CrossRef]
56. Mayo Clinic. Diabetes de Tipo 2. Available online: https://www.mayoclinic.org/es/diseases-conditions/type-2-diabetes/symptoms-causes/syc-20351193 (accessed on 9 August 2023).
57. Quesnel, H.; Farmer, C.; Devillers, N. Colostrum intake: Influence on piglet performance and factors of variation. *Livest. Sci.* **2012**, *146*, 105–114. [CrossRef]
58. Kanatous, S.B.; Mammen, P.P.A.; Rosenberg, P.B.; Martin, C.M.; White, M.D.; DiMaio, J.M.; Huang, G.; Muallem, S.; Garry, D.J. Hypoxia reprograms calcium signaling and regulates myoglobin expression. *Am. J. Physiol. Physiol.* **2009**, *296*, C393–C402. [CrossRef]

59. Al-Othman, A.; Al-Musharaf, S.; Al-Daghri, N.M.; Yakout, S.; Alkharfy, K.M.; Al-Saleh, Y.; Al-Attas, O.S.; Alokail, M.S.; Moharram, O.; Sabico, S.; et al. Tea and coffee consumption in relation to vitamin D and calcium levels in Saudi adolescents. *Nutr. J.* **2012**, *11*, 56. [CrossRef] [PubMed]
60. Mota-Rojas, D.; Villanueva-García, D.; Hernández-Ávalos, I.; Casas-Alvarado, A.; Domínguez-Oliva, A.; Lezama-García, K.; Miranda-Cortés, A.; Martínez-Burnes, J. Cardiorespiratory and neuroprotective effects of caffeine in neonate animal models. *Animals* **2023**, *13*, 1769. [CrossRef]
61. Rasmussen, C.A.F.; Sutko, J.L.; Barry, W.H. Effects of ryanodine and caffeine on contractility, membrane voltage, and calcium exchange in cultured heart cells. *Circ. Res.* **1987**, *60*, 495–504. [CrossRef]
62. Bal, N.C.; Periasamy, M. Uncoupling of sarcoendoplasmic reticulum calcium ATPase pump activity by sarcolipin as the basis for muscle non-shivering thermogenesis. *Philos. Trans. R. Soc. B Biol. Sci.* **2020**, *375*, 20190135. [CrossRef] [PubMed]
63. da Silva, L.A.; Wouk, J.; Weber, V.M.R.; da Luz Eltchechem, C.; de Almeida, P.; Martins, J.C.L.; Malfatti, C.R.M.; Osiecki, R. Mechanisms and biological effects of Caffeine on substrate metabolism homeostasis: A systematic review. *J. Appl. Pharm. Sci.* **2017**, *7*, 215–221. [CrossRef]
64. Mota-Rojas, D.; Martínez-Burnes, J.; Casas-Alvarado, A.; Gómez-Prado, J.; Hernández-Ávalos, I.; Domínguez-Oliva, A.; Lezama-García, K.; Jacome-Romero, J.; Rodríguez-González, D.; Pereira, A.M.F. Clinical usefulness of infrared thermography to detect sick animals: Frequent and current cases. *CABI Rev.* **2022**, *17*, 1–27. [CrossRef]
65. Casas-Alvarado, A.; Mota-Rojas, D.; Hernández-Avalos, I.; Martínez- Burnes, J.; Rosas, M.; Miranda-Cortés, A.E.; Domínguez-Oliva, A.; Mora- Medina, P. Assessment of thermal response, cardiorespiratory parameters and postoperative analgesia in dogs undergoing ovariohysterectomy with different combinations of epidural anesthesia and isoflurane. *J. Anim. Behav. Biometeorol.* **2023**, *11*, e2023009. [CrossRef]
66. Lowe, G.; Sutherland, M.; Waas, J.; Schaefer, A.; Cox, N.; Stewart, M. Infrared thermography—A non-invasive method of measuring respiration rate in calves. *Animals* **2019**, *9*, 535. [CrossRef] [PubMed]
67. Stewart, M.; Wilson, M.T.; Schaefer, A.L.; Huddart, F.; Sutherland, M.A. The use of infrared thermography and accelerometers for remote monitoring of dairy cow health and welfare. *J. Dairy Sci.* **2017**, *100*, 3893–3901. [CrossRef]

animals

MDPI

Review

Thermal Imaging as a Method to Indirectly Assess Peripheral Vascular Integrity and Tissue Viability in Veterinary Medicine: Animal Models and Clinical Applications

Daniel Mota-Rojas [1,*], Asahi Ogi [2], Dina Villanueva-García [3], Ismael Hernández-Ávalos [4], Alejandro Casas-Alvarado [1], Adriana Domínguez-Oliva [1], Pamela Lendez [5] and Marcelo Ghezzi [5,6]

1 Neurophysiology of Pain, Behavior and Assessment of Welfare in Domestic Animals, DPAA, Universidad Autónoma Metropolitana (UAM), Mexico City 14389, Mexico
2 Department of Neurobiology and Molecular Medicine, IRCCS Fondazione Stella Maris, 56128 Pisa, Italy
3 Division of Neonatology, Hospital Infantil de México Federico Gómez, Mexico City 06720, Mexico
4 Clinical Pharmacology and Veterinary Anesthesia, Biological Sciences Department, FESC, Universidad Nacional Autónoma de México, Cuautitlán 54714, Mexico
5 Anatomy Area, Faculty of Veterinary Sciences (FCV), Universidad Nacional del Centro de la Provincia de Buenos Aires (UNCPBA), University Campus, Tandil 7000, Argentina
6 Animal Welfare Area, Faculty of Veterinary Sciences (FCV), Universidad Nacional del Centro de la Provincia de Buenos Aires (UNCPBA), University Campus, Tandil 7000, Argentina
* Correspondence: dmota100@yahoo.com.mx

Simple Summary: Infrared thermography can indirectly assess peripheral vascular diseases because skin surface temperature depends on blood flow and heat dissipation through skin microvasculature. The present review aims to summarize and analyze the application of infrared thermography in veterinary medicine as a method to indirectly assess peripheral vascular integrity and its relation to the amount of radiated heat and as a diagnostic technique for tissue viability, degree of damage, and wound care. Although additional research is needed to establish the utility of thermal imaging to monitor vascular integrity in veterinary medicine, studies performed in animals diagnosed with thromboembolisms or thermal impairment due to neuropathies or are in need of burn wound management suggest a promising application of infrared thermography to evaluate peripheral vascular alterations and tissue viability.

Abstract: Infrared thermography (IRT) is a technique that indirectly assesses peripheral blood circulation and its resulting amount of radiated heat. Due to these properties, thermal imaging is currently applied in human medicine to noninvasively evaluate peripheral vascular disorders such as thrombosis, thromboembolisms, and other ischemic processes. Moreover, tissular damage (e.g., burn injuries) also causes microvasculature compromise. Therefore, thermography can be applied to determine the degree of damage according to the viability of tissues and blood vessels, and it can also be used as a technique to monitor skin transplant procedures such as grafting and free flaps. The present review aims to summarize and analyze the application of IRT in veterinary medicine as a method to indirectly assess peripheral vascular integrity and its relation to the amount of radiated heat and as a diagnostic technique for tissue viability, degree of damage, and wound care.

Keywords: thrombosis; ischemia; burn wounds; skin grafting; free flaps; wound healing

Citation: Mota-Rojas, D.; Ogi, A.; Villanueva-García, D.; Hernández-Ávalos, I.; Casas-Alvarado, A.; Domínguez-Oliva, A.; Lendez, P.; Ghezzi, M. Thermal Imaging as a Method to Indirectly Assess Peripheral Vascular Integrity and Tissue Viability in Veterinary Medicine: Animal Models and Clinical Applications. *Animals* **2024**, *14*, 142. https://doi.org/10.3390/ani14010142

Academic Editor: Michael Lindinger

Received: 2 December 2023
Revised: 24 December 2023
Accepted: 30 December 2023
Published: 31 December 2023

1. Introduction

Infrared thermography (IRT) is widely used in human and animal medicine as a noninvasive and complementary diagnostic tool for the early detection of inflammatory processes [1–4] and peripheral vascular disorders where blood flow is compromised [5,6]. This is due to its potential to indirectly evaluate the volume of blood flow to the skin [7]. Therefore, any disturbance to normal blood circulation can potentially alter the amount of radiated heat from the skin of mammals [5]. Examples of this in humans include thrombosis, thromboembolisms, vasculitis, vascular injuries, lupus, diabetic foot, and ischemia processes where peripheral blood supply is abnormal [6,8–10].

Contrary to human medicine, the application of IRT to assess vascular compromise in veterinary medicine is still developing, and limited studies have been published. However, it has been suggested that identifying an altered local temperature response in animals might help to recognize blood perfusion disorders such as vasculitis [11] or even discern between benign and malign neoplasia [12,13]. For example, aortic thromboembolism cases in cats show a hypothermic response in pelvic limbs, a method that can serve to differentiate between healthy animals and those affected by the pathology [14]. Another example is burn injuries, for which determining the degree of damage is essential to adopt treatment protocols and estimate recovery time [15–17].

Canine, swine, and murine models show that decreases in temperature in the burn wound site are associated with a higher burn degree (e.g., full-thickness) [18] and that this is also related to a poor prognosis when skin-grafting or free-flap surgeries are performed, due to vascular alterations [19–21]. Likewise, indocyanine green imaging and IRT were used to evaluate tissue blood flow in a porcine model of intestinal resection, and the authors found that IRT can differentiate well-blooded intestinal segments [22]. Moreover, IRT has been applied to determine skin transplant viability and wound healing progress [17,23,24].

Therefore, due to the wide field of research in which IRT can be used to identify medical processes where blood flow is altered, the present review aims to summarize and analyze the application of IRT to evaluate peripheral vascular disorders and blood flow integrity in skin transplant procedures in veterinary medicine, in both a clinical and an experimental setting.

2. Search Methodology

The literature search was performed using the Web of Science and PubMed databases. Major keywords used for the search were the combination of "infrared thermography" or "thermal imaging" with "peripheral vascular disorder animal models", "animal peripheral ischemic processes", "skin grafting in animals", and "animal models of burn injuries". The selected papers included studies where IRT—alone or together with other techniques—was used to evaluate vascular integrity or thermal changes in clinical cases or experimental models of peripheral ischemia. The search also considered human studies to compare with and use as a base, since current research regarding vascular disorders and veterinary medicine is limited. Due to this, the search included a range of different animal species.

3. Heat Radiation and Its Relationship with Tissular Perfusion

All objects with a temperature above absolute zero are known to emit thermal radiation [6,25]. IRT detects this infrared radiation, and the average emissivity of the skin (0.98) makes it a material appropriate for thermal imaging [26–30]. Therefore, distinctive thermal changes and local changes in heat production can be detected due to an affected blood flow pattern or certain pathologies [31].

To understand how IRT is related to heat production and tissue perfusion, it is necessary to address how circulation can affect surface heat radiation. According to Morrison and Nakamura [32], peripheral thermoreceptors in the skin detect thermal changes. Transient receptor potential (TRP) ion channels are the main thermoreceptors able to detect a wide range of temperatures from heat/cold to noxious heat/cold [33]. In particular, TRP of the vanilloid family (TRPV) can respond to both thermal (noxious/painful heat) and nonthermal inputs such as the release of chemical mediators [e.g., histamine, serotonin, interleukin (IL)-6, IL-10, and tumor necrosis factor-alpha (TNF-α)], elements present during the inflammatory process and tissular injury [34,35], triggering nocifensive behaviors (skin vasomotor responses or behavioral changes) to restore thermostability and increase or decrease heat loss [36–38]. After stimuli detection, hypothalamic centers such as the preoptic area, the anterior hypothalamus, and the dorsomedial hypothalamus are activated to process thermal information received from the lateral parabrachial nucleus of the pons [39,40]. For example, Figure 1 shows the facial thermal response of a febrile canine patient due to an infectious process. When comparing the temperature of the febrile animal with a healthy patient, a difference of up to 1.1 °C can be observed in the periocular region. Diego et al. [41] reported a similar pattern in febrile ovine experimentally infected with bluetongue virus serotypes 1 and 8. The authors found that ocular infrared temperature was moderately positively correlated with the rectal temperature (r = 0.504) and that IRT could discriminate between febrile and nonfebrile animals with a sensitivity of 85% and a specificity of 97%.

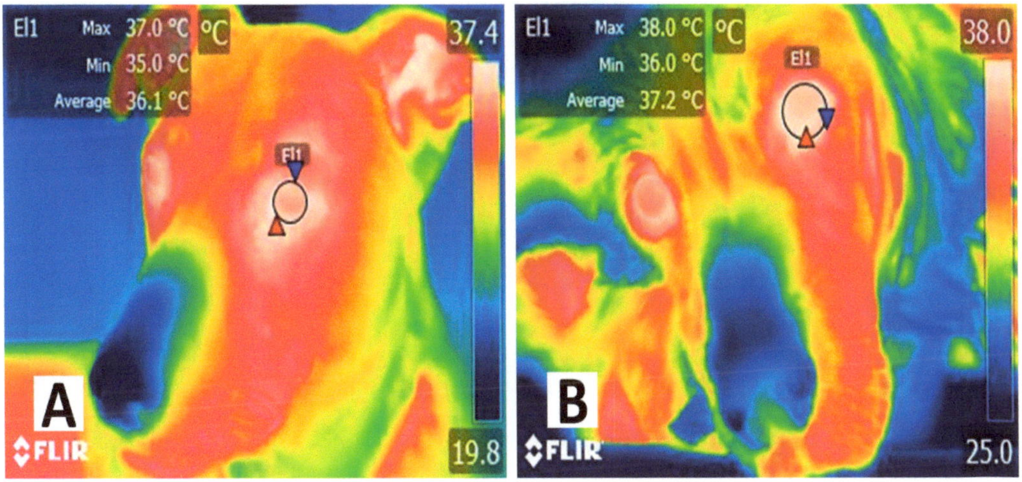

Figure 1. The surface temperature of a dog undergoing an infectious disease. (**A**) Thermal response in a healthy 3-year-old mixed female dog. The maximum (red triangle) surface temperature of the periocular region (El1) shows a value of 37 °C, while the minimum (blue triangle) and average temperatures were 36 °C and 35 °C, respectively. (**B**) Thermal response in a 7-year-old female English Setter dog diagnosed with pyometra. It is observed that the surface temperature of the periocular region (El1) is 1 °C higher than the maximum (red triangle) and average temperature of a healthy dog. Regarding the minimum temperature (blue triangle), a difference of +1.1 °C was found when compared to a healthy animal. The presence of cytokines, such as interleukin (IL)-1, IL-6, and prostaglandin F2 alpha, induces fever. The increase in body core temperature is reflected in the vasodilation of superficial blood vessels to increase heat dissipation. Therefore, IRT can help recognize febrile states in animals. Radiometric images were obtained using a T1020 FLIR thermal camera. Image resolution 1024 × 768; up to 3.1 MP with UltraMax. FLIR Systems, Inc. Wilsonville, OR, USA.

Apart from general changes in temperature, circulatory disorders in skin blood vessels can change heat distribution and, therefore, variations in heat emissions through the skin [21]. Finstad [31] reported this in 49 client-owned dogs undergoing laparotomy to correct foreign body obstructions, a pathology known to compress intestinal blood vessels and cause tissular necrosis if not treated correctly. In the study, IRT was assessed using an infrared camera held 1 m from the obstructed small intestine portion. The average surface temperature of the small intestine directly over the foreign body was 2.4 °C lower than the rest of the intestine (healthy tissue), particularly when it was a hard material (3.2 to 5.2 °C). Although these findings did not show a correlation with lactate or blood pressure, IRT was able to detect vascular compromise due to the foreign body.

Similarly, during other inflammatory processes, as shown in Figure 2 in the knee joint, differences in temperature in the affected limb can be observed when compared to the healthy side. In these cases, and during burn injuries, the inflammatory process is the result of the release of pro-inflammatory mediators (e.g., prostaglandin F2 alpha, histamine, serotonin, IL-6, IL-10, and TNFα), substances that produce vasodilation of the superficial blood capillaries and increase heat radiation through the skin [42].

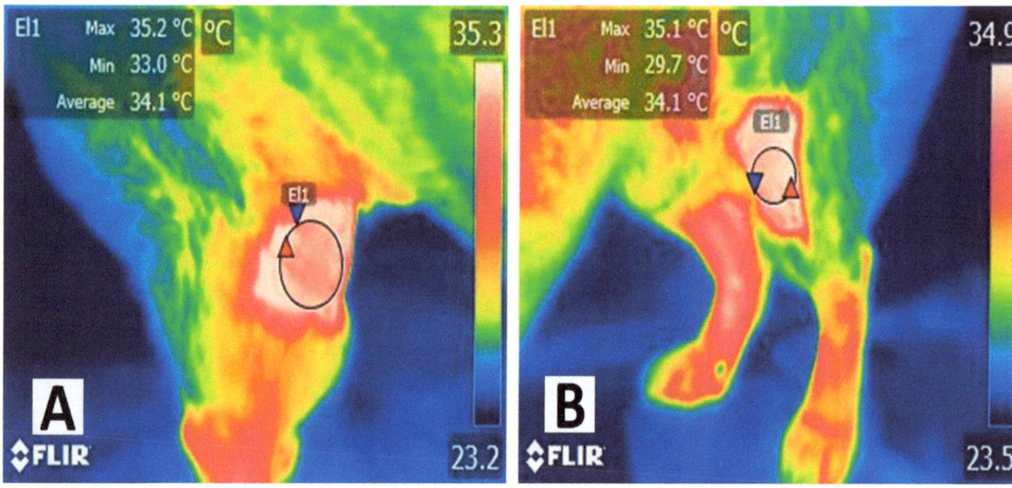

Figure 2. Thermal response associated with an inflammatory process in a dog with a knee fracture. (**A**) Thermal response in a 6-year-old female Shiba Inu dog with inflammation. Temperature differences can be observed in the right pelvic limb with a fracture in the femoral–tibial–patellar joint. The maximum surface temperature (red triangle) of the region (El1) was 35.2 °C, with an average temperature of 34.1 °C and a minimum of 33 °C (blue triangle). (**B**) Thermal response in a healthy knee. In the same animal, the thermal response of the femoral–tibial–patellar joint (El1) of the left hindlimb had a difference of up to 3.3 °C when comparing the average surface temperature of both knees. The release of pro-inflammatory mediators (e.g., IL-1, IL-6, IL-10, prostaglandins, serotonin, and histamine) produces vasodilation of capillaries in the dermal tissue, increasing heat dissipation and radiation.

Therefore, peripheral hemodynamic changes due to a general condition (e.g., fever) or to a local injury (e.g., inflammatory process or burn wound) will undoubtedly modify skin surface temperature, a parameter that is influenced by critical factors such as local blood circulation, heat conduction by deeper tissues, and heat loss [26]. Figure 3 schematizes the vascular changes observed during the ischemic and inflammatory processes.

In large animals such as ruminants, early detection of inflammatory processes such as laminitis could help to prevent complications in the hooves. In this sense, IRT could be applied as a complementary tool to traditional diagnostic methods, as reported by Lin et al. [43], who found an optimal threshold based on actual foot temperature of 23.3 °C to detect healthy limbs from those affected by laminitis, with a sensibility of 78.5% and specificity of 39.2%. Due to these properties, thermographic imaging could be a reliable and real-time clinical tool to detect inflammatory-related events or abnormal blood flow processes.

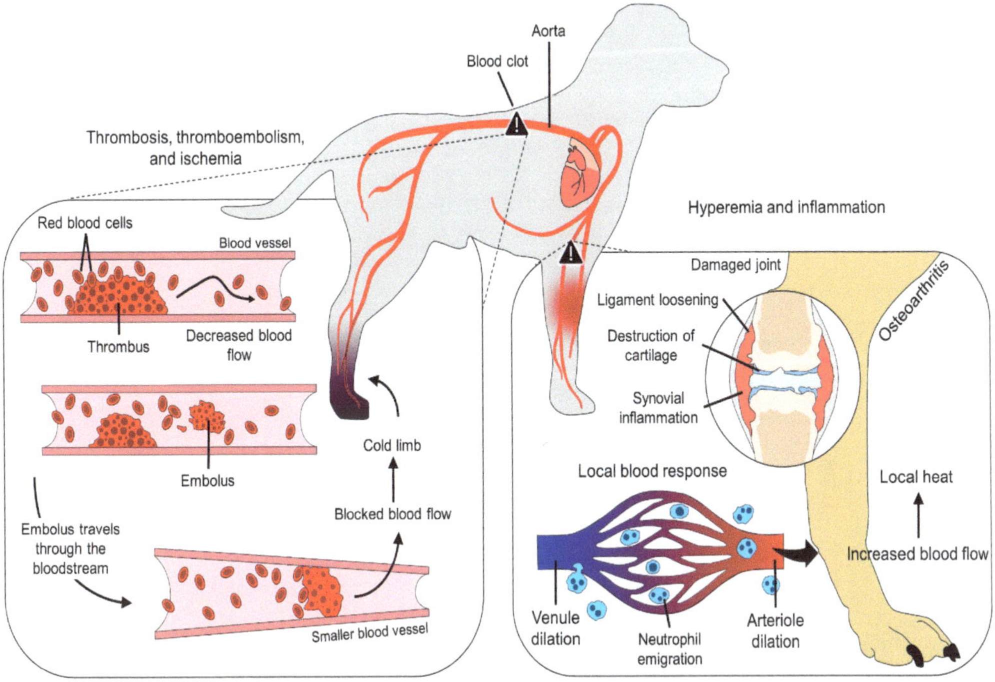

Figure 3. Peripheral vascular changes due to ischemic and inflammatory processes. Whether a decreased or increased blood flow is present, changes in the local temperature arise due to alterations in the normal blood flow. For example, during thromboembolisms, initially, the thrombus decreases blood flow by obstructing part of the blood vessel. However, the vessel can be completely blocked when an embolus travels through the bloodstream. Locally, this will stop or significantly reduce blood flow, decreasing the local surface temperature. On the contrary, during an inflammatory process such as osteoarthritis, vasodilation of venules and arterioles increases blood flow to promote neutrophil emigration. The release of inflammatory mediators also causes vasodilation, increasing local heat radiation.

4. Peripheral Vascular Disorders and Thermography

The association between the control of surface temperature and the vasomotor response suggests that any vascular involvement disorders can alter the thermal response of tissues [44]. An example of this is a diabetic foot, a pathology where vascular compromise is present, as mentioned by Ilo et al. [24], who compared the thermal response of human patients with suspected diabetic feet (n = 118) and healthy patients (n = 93). The surface temperature of the dorsal and plantar regions of the foot was significantly higher (33% higher) in patients with diabetic foot by 2 °C. Moreover, thermal findings coincided with decreases in pressure in the toes to 50 mmHg. A similar influence of blood pressure in limb surface temperature was reported in piglets, in whom a negative correlation was found between temperature and blood pressure [45]. These results suggest that the neuropathies occurring in diabetes patients generate microcirculatory alterations that have consequences on the thermoregulatory mechanisms of peripheral regions such as the foot. Furthermore, diabetes is also linked to arteriovenous shunting [46]. These arteriovenous anastomoses are innervated by sympathetic vasoconstrictor nerves that allow high flow rates in low-resistance vessels. States of local capillary hypoperfusion can lead to impairments in wound healing [47,48]. Thus, local and systemic disorders can alter peripheral blood flow and, in the case of hypoperfusion, these changes can be observed with IRT as a decrease in heat elimination.

Another pathology where IRT has been suggested to diagnose and monitor vascular alterations is arterial thromboembolism. A study carried out by Pouzot-Nevoret et al. [14] evaluated the application of IRT to differentiate between nonischemic and aortic thromboembolism in 16 cats (healthy cats = 10 and sick cats = 6). By assessing the temperature of the pelvic limbs, the authors found that in animals with aortic thromboembolism, limb temperature was significantly lower (by 3.3 °C) than the values recorded in healthy cats. Furthermore, a decrease of 2.4 °C was set as a reference to differentiate healthy from affected animals, with a sensitivity of 80% and a specificity of 100%. Since a thrombus obstructs normal blood circulation, the decrease in surface temperature is an expected response that can be detected through IRT [44]. Figure 4 schematizes a feline patient diagnosed with aortic thromboembolism and the thermal response evaluated through IRT.

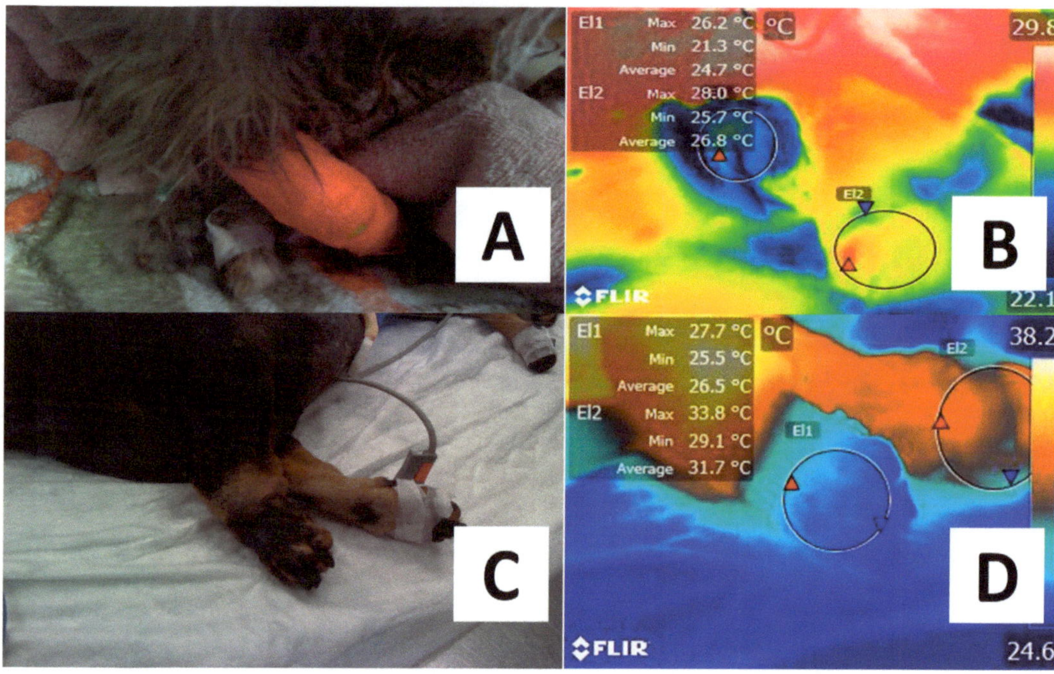

Figure 4. Difference in the thermal response of patients diagnosed with peripheral vascular alterations. (**A**) Persian male cat diagnosed with aortic thromboembolism after evaluating clinical signs such as pain on palpation in the right pelvic limb, absence of pulse, cold limb, and mobility difficulty. (**B**) The phalangeal region of the affected right hindlimb (El1) showed lower temperatures of up to 3.3 °C when compared to the healthy left hindlimb (El2). (**C**) A four-year-old male Dachshund dog diagnosed with thrombosis due to a secondary liver infection was presented with a reduced perfusion in the left forelimb. Necrosis can be observed. (**D**) With thermal imaging, it can be observed that the average surface temperature of the phalangeal region of the right forelimb (El2) is 5.2 °C higher than the same region in the affected limb (El1). The explanation for this thermal response is that the presence of a thrombus obstructs blood flow due to occlusion at the arterial level. Thus, the decrease in blood flow has an impact on local heat response. Maximal temperature is indicated with a red triangle and the minimal with a blue triangle. Radiometric images were obtained using a T1020 FLIR thermal camera. Image resolution 1024 × 768; up to 3.1 MP with UltraMax. FLIR Systems, Inc. Wilsonville, OR, USA.

Postmortem evaluations using IRT to detect thermal patterns during coronary heart disease have been studied in pigs. Thermal imaging coronarography enabled the evaluation of coronary vessels in ex situ swine hearts [49]. Although this was a postmortem evaluation, further analysis is required to apply this technique in veterinary medicine. Experimental hypoperfusion studies in rats also used IRT to visualize vasodilation of choked vessels in a model of delayed pedicled perforator flap rat model [50]. In 43 rats divided according to the experimental treatment, after the use of a flap and after strangulation of the thoracodorsal artery, a red zone was associated with a 1.4% increase in local temperature and a 0.3 mm increase in diameter of the same artery compared to the iliac artery [50]. According to the authors, this result is consistent with the changes due to the congestion process that occurs after the obstruction. The same occurs when a thrombus is present, leading to hypoperfusion and a subsequent decrease in local temperature. IRT was used in the same species to detect tissue perfusion disorders in rats [51]. In animals with unilateral or bilateral femoral vessel ligation, statistically significant differences in thigh, shank, and foot temperature were observed. A slow return in the temperature of the shank and foot was observed in the ligated limbs when compared with healthy limbs [51].

Similarly, a study aimed to evaluate variations in surface temperature after rapid hemodynamic changes in eight healthy piglets under general anesthesia. The animals were exposed to different blood pressure levels and monitored through IRT. The surface temperature of the left forelimb had a negative correlation with the temperature gradient and blood pressure ($r = -0.042$), which led the authors to conclude that IRT is a tool for detecting early changes in peripheral perfusion [45]. In this sense, Caramalac et al. [52] evaluated a 9-year-old dog with clinical signs of a cold limb without a pulse, proprioception, and deep pain. Due to the clinical signs, thromboembolism was a presumptive diagnosis. IRT was able to determine a difference of 3.7 °C between the affected limb and the healthy one.

Therefore, the scientific evidence suggests that IRT is a promising healthcare tool for the evaluation of vascular compromise in peripheral tissues. In addition, it could help to detect early changes before a chronic or severe stage can be developed. Table 1 summarizes the animal studies where vascular disorders were reported.

Table 1. Technical and experimental characteristics of animal studies using IRT to evaluate vascular disorders.

Species	Camera	Distance (cm)	Emissivity	Ambient Temperature (°C)	Aim	Outcome
Cats [14]	Flir C2	75	0.95	20.0	Identify aortic thrombosis.	2.4 °C to differentiate sick animals.
Dogs [52]	FlirOne Pro	30	NR	25.0	Diagnose thromboembolism.	Affected limb had lower temperature (31.3 °C vs. 35.0 °C).
Pigs [49]	FlirOne	NR	NR	NR	Establish coronary circulation.	Visible occlusion of coronaries appearing in bright yellow.
Pigs [45]	Flir-E6	50	0.98	NR	Assess changes in skin temperature in response to blood pressure.	Negative correlation between blood pressure and temperature gradient.
Rats [50]	Flir T650SC	45	NR	NR	Monitor pedicled island perforator flaps.	1.4% increase in red zones or hot spots.
Rats [51]	Flir 335	20	NR	NR	Detect tissue perfusion disorders in femoral vessels.	Return of warmth in the limb was slower.

NR: not reported.

5. Thermal Imaging Applied to Assess Tissular Damage Degree and Microvascular Repair

Because IRT can remotely evaluate the vascular integrity of tissues, thermal imaging has also been applied to monitor the degree of tissular damage and blood perfusion restoration [15–17]. Different studies in humans showed that blood flow assessment can predict the time of healing [15,53,54]. This is of clinical relevance in burn patients in whom IRT can help to assess burn depth (e.g., superficial, superficial partial-thickness, deep partial-, or full-thickness) and estimate the recovery time [16]. These studies show that IRT can help to categorize the deepness of the injury as a primary and real-time technology for burn wound management, including surgical repair using skin grafting or free flaps to monitor tissue viability and improve the healing process [55]. Although the common procedure to evaluate reconstructive surgery is by looking at granulation tissue and proper healing, the ability of IRT to detect vascular imbalances could help monitor the success of skin transplantation. Since skin transplants need an integrated microvascular blood flow to survive, vascular alterations are the main reason for failure. This is relevant because the failure rate of skin grafts and free flaps is around 2.5–23.7% and 0.5–23.8%, respectively [20].

In the case of burn patients, there are different types of burn injuries: first-degree burn or superficial wounds; second-degree or superficial partial-thickness burn; third-degree or deep partial-thickness wound; and fourth-degree or full-thickness injury [56]. According to the degree of damage, IRT can detect the zones where the integrity of vascular structures is maintained, differentiate between abnormal heat transfer of burn zones, and serve as a guide to performing grafts or flaps [19,53]. Identification of the degree and heat pattern changes is due to the different zones that are present in burns: (1) coagulation zone of irreversible necrosis; (2) stasis zone with damage of the vasculature; and (3) hyperemia zone of edematous nature and with tissue that can recover [53]. Prindeze et al. [57] mention that the hyperemia zone, using ADT in humans, can be observed as a zone with higher temperatures when compared to the coagulation and stasis zones. Moreover, after a burn injury, the permeability of capillaries and venules increases, causing edema and hypovolemia [58]. Additionally, the release of pro-inflammatory mediators such as histamine, bradykinin, cytokines, and platelet-activating factors alters vascular permeability [58].

The vascular damage and the altered heat transfer are physiological responses that can be monitored through IRT. Infrared evaluation applied to burn depth assessment in humans showed a sensitivity and specificity of 84% and 76%, respectively [16], while visual evaluation had an accuracy of 60% [59]. Moreover, the application of IRT could guide surgeons in deciding burn wound treatment, whether conservative or surgical [60].

In contrast to human medicine, burn wound management (including diagnosis and treatment) is challenging and uncommon in veterinary medicine [61]. In small animal medicine, most thermal burn cases are associated with accidental burns caused by supplemental heat (e.g., thermal mattresses) [62]. Some studies made in animal models have shown the application of IRT to burns, such as Lawson [18], who was the first researcher using thermal imaging to determine burn depth in dogs and measure cutaneous blood flow [63]. Porcine models of burn injury in Yorkshire pigs showed that a progressive increase and warmer surface temperatures (from 29.99 ± 0.63 °C to 30.93 ± 0.47 °C) during the first 7 days after burn injury was related to lower scar depth at 28 days after injury (1.29 ± 0.68 mm). Contrarily, pigs with surface hypothermia (from 31.00 ± 0.71 °C to 27.31 ± 0.37 °C) resulted in larger scar depth (4.68 ± 0.60 mm) [59]. These results suggest that the surface temperature of the injury during the first days can predict burn depth, and this is relevant because it is known that, in humans, first-degree and second-degree burns usually halt within three weeks, while third-degree ones require more than three weeks to heal and might develop hypertrophic scars [64]. Similarly, evaluation of the thermal time constant using active dynamic thermography (ADT) in pigs showed a high correlation between surface temperature and predicted healing time with an accuracy of 83.0%, close to the accuracy of histopathologic assessment (84.0%) [60]. The authors found that a shorter thermal time (49.1 ± 23 s) was associated with wound self-healing within three weeks, while a longer time (86.3 ± 32.3 s) resulted in an extended healing period [60].

In another swine model, high-resolution IRT was used to determine its usefulness in predicting burn severity during the first four days after injury [65]. In Yorkshire pigs, the surface temperature decreased with increasing burn severity, meaning that superficial burn wounds had the warmest temperatures (approximately 33.7 °C to 35 °C), while full-thickness wounds had progressive lower temperatures (approximately 32.4 °C to 33 °C). Additionally, IRT was able to predict histological changes such as necrosis, apoptosis, and vascular occlusion with correlation values between 0.62 and 0.13 [65]. The same pattern has been observed in human patients, in whom Hardwicke et al. [66] reported that when compared to nonburn patients, the skin temperatures of full-thickness and deep partial burns were significantly cooler (−2.3 °C and −1.2 °C, respectively), while superficial partial-thickness burns had no significant differences of 0.1 °C.

Between grafts and flaps, there are vascular differences that need to be considered when using IRT. On one side, grafts are a type of skin transplant (dermis, epidermis, or both) completely detached from the blood supply. Grafts need a well-vascularized bed wound to survive [67]. This means that the free segment of skin needs to reestablish a vascular supply to consider it a successful grafting procedure [68]. Grafts are put on a well-vascularized bed. In contrast, a flap has an integral blood supply [67].

Klama-Baryla et al.'s [21] study shows the application of IRT to evaluate the survivability of autologous split-thickness engraftment in patients suffering thermal injuries. Consecutive monitoring at days 0, 5, 14, and 21 showed that a progressive increase in surface temperature between 1.21 °C and 3.1 °C immediately after the graft indicated a normal healing process with angiogenesis (confirmed with angiography). Contrarily, graft loss occurred when low temperatures around 32.9 °C were observed. IRT can also help to predict treatment by either reepithelization, skin graft, or amputation [69,70]. This was determined in burn patients with partial- or full-thickness burns in extremities [69]. The difference of temperatures between measuring points (ΔT) showed an average ΔT of 1.77 ± 0.92 °C for superficial and deep partial-thickness wounds, while full-thickness burns had a ΔT of 5.45 ± 2.86 °C. Regarding the predictive value of IRT, ΔT was 1.75 ± 0.89 °C for the patients that healed by reepithelization. In contrast, patients who required grafts and those who underwent amputation had ΔT values of 3.28 ± 0.68 °C and 7.71 ± 1.89 °C, respectively. According to these findings, IRT had a prediction accuracy of 85.35% [69].

In the case of veterinary medicine, currently, skin grafting is performed in small animals, particularly when a large defect on a distal limb is present [68]. In murine animal models, thermography was used to evaluate the usefulness of therapeutic ultrasound on skin grafts. In the treated group, the authors found higher temperatures between 35 and 36 °C, while the control group had lower values between 33.3 and 34.3 °C. This could be interpreted as an improved tissue repair using ultrasound [71]. In another study in mongrel dogs, Lekas et al. [72] monitored coronary artery bypass grafting and found that thermal imaging can accurately identify ischemic processes due to stenosis or occlusion of the coronary artery. In this sense, ligation of the artery caused a rapid decrease in the epicardial surface temperature from 31 °C to 26–28 °C. Since the main cause of grafting failure, particularly in small animals [73], is the lack of adequate blood flow and the consequent ischemia and autophagy of the surrounding tissue, early detection of these processes using noninvasive equipment such as IRT is of utmost importance [20].

Contrary to skin grafting, flaps are skin transplants with an integral blood supply [67]. These types of surgeries have been monitored with thermal imaging to identify the so-called "hot spots" where correct vascularization is present, reaching a prediction accuracy of 100% of flap survival and ischemia identification [74]. Tools such as dynamic infrared thermography (DIRT) were used to monitor real-time skin perfusion of a pedicled thoracodorsal artery perforator flap in 21 patients in a Sjøberg et al. [75] study. Preoperative mapping of perforator vessels and evaluation of postsurgical perfusion showed that hot spots detected with DIRT corresponded to arterial Doppler sounds, while the disappearance of hot spots was associated with less audible or absent arterial sounds [75].

Another researcher who used DIRT was Miland et al. [76] to visualize skin perfusion in female patients undergoing abdominal skin flaps. DIRT correlated with angiography techniques to identify perfused areas of the flap. Likewise, in free flaps performed in breast reconstruction surgeries, surgeons have been using IRT, together with Doppler ultrasound and computed tomography angiography techniques, to monitor the viability of the flap but also as a novel tool to preoperatively locate perforator vessels in flaps [23]. In contrast, some studies reported that IRT was unable to detect a case of poorly perfused flap until the skin transplant was dead; nonetheless, the accuracy of the technique was estimated at 96.43% [77].

In the case of nonhuman animals, Miland et al. [78] performed skin flaps in Wistar rats and used DIRT and angiography to predict flap survival. The percentage of flap survival using both techniques was similar (63–74%), with an accuracy of 0.86–0.90%. The study also showed higher temperatures in the axial center of the flaps (around 33–34 °C) and a significant decrease in the distal ends (the lowest temperature was 30 °C), showing that the healing process starts at the edges of the flap, where vascular supply can be easily compromised. Tenorio et al. [79] used IRT in rats to recognize perfusion failure in free flaps. The authors performed ligation of one artery or vein to observe the thermal changes. The results showed that ligation of the blood vessel made the hot spot disappear, meaning a 100% chance of identifying permeable and occluded vessels. In the same way, in rats undergoing free flaps, localization of perforators in the abdominal skin was performed with IRT by identifying hot spots. Then, 24 h after flap placement, the correct blood flow through perforators was assessed with IRT and macroscopical (necropsy) observations. Necrosis occurred in places where IRT did not detect perforators in 12 of 31 cases [80].

Apart from determining the degree of damage and correct blood flow, IRT has also been used to assess the effect of several treatments on skin transplant success. This was reported in a study performed on rats receiving five different treatments (saline, capsaicin, methylprednisolone, mitomycin, and pentadecapeptide) on dorsal skin flaps [81]. Through IRT, it was found that methylprednisolone and pentadecapeptide reduced ischemic necrosis. These results were comparable to the Shejbal et al. [82] study, in which a full-thickness dorsal murine skin flap was treated with saline, capsaicin, and methylprednisolone to assess their effect on tissue vitality and survival area. Methylprednisolone resulted in a uniform temperature distribution of the skin associated with reduced flap necrosis by 56%.

Assessing the degree of damage of a tissular injury implicates not only cell death but also damage to the surrounding blood vessels [83]. Knowing that vascular trauma affects tissular viability and also alters local heat exchange is the basis of understanding why IRT is used to provide an early evaluation of burn wounds. IRT can help surgeons to identify the extension of the damage and the most suitable treatment and to monitor surgical procedures such as skin grafting and free flaps, as summarized in Table 2. Moreover, during the postoperative period, the healing process can also be monitored with IRT to improve recovery time and propose complementary treatments and evaluation devices. Nonetheless, this has not been extensively researched in veterinary medicine.

Table 2. Technical and experimental characteristics of animal studies using IRT to assess burn depth and skin transplant survival.

Species	Camera	Distance (cm)	Emissivity	Ambient Temperature (°C)	Aim	Outcome
Pigs [59]	Flir T300	30	NR	NR	Assess burn depth.	Surface hypothermia (27.31 ± 0.37 °C) predicts larger scar depth.
Pigs [60]	ADT	NR	NR	NR	Assess burn depth.	Shorter thermal time (49.1 ± 23 s) related to fast healing.
Pigs [65]	Flir A655sc	40	NR	NR	Assess burn severity.	Severe burns have lower temperatures (32.4–33 °C).

Table 2. *Cont.*

Species	Camera	Distance (cm)	Emissivity	Ambient Temperature (°C)	Aim	Outcome
Rats [76]	DIRT Flir ThermaCAM S65	NR	NR	NR	Predict flap survival.	74% flap survival.
Rats [79]	BioScanIR	NR	NR	NR	Detect flap failure.	Hot spot disappeared, followed by macroscopic congestion.
Red wolf [84]	eVs DTIS 500	NR	NR	NR	Monitor free skin graft.	Neovascularization in the graft becomes thermally equivalent to healthy tissue.

ADT: active dynamic thermography; DIRT: dynamic infrared thermography; NR: not reported.

6. Infrared Thermography Used to Monitor Healing and Surgical Wound Care

While repair refers to a process where substantial tissular damage is present and the normal architecture of the cells cannot be completely restored, healing refers to the proliferation process when cells can regenerate. The healing process after surgery is another field where IRT has been used. Healing implicates several phases such as the inflammatory phase (2–5 days), proliferative phase (5 days to 3 weeks), and remodeling phase (3 weeks to years) [54]. In the case of burn injuries and skin transplants, one of the primary things to ensure transplant success is to promote correct blood perfusion [54]. Authors such as Ramírez-GarciaLuna et al. [15] mention that regarding wound care, comparisons between the surface temperature of the wound bed and the proximate healthy tissue are useful because alterations in blood flow due to inflammation or obstructed blood flow might cause an increase or decrease, respectively, in skin surface temperature.

As reported by Li et al. [54], IRT can predict healing status with a sensitivity of 91.67% and a specificity of 85.71% by monitoring local blood perfusion in thoracic surgical incisions. According to the results, healed patients maintained a higher incisional temperature between 32.3 ± 1.6 °C and 33.8 ± 0.9 °C, while unhealed patients recorded hypothermic areas between 30.0 ± 1.2 °C and 32.4 ± 0.8 °C. Moreover, IRT detected hypothermic areas in the incision surface of a patient due to local infection, which could be used as a prognosis marker [54]. Similarly, higher periwound temperatures (more than 35 °C) in humans suffering from pressure ulcers were associated with a better healing process, while poor prognosis wounds had a temperature of less than 34 °C. Differences between the wound bed and periwound surface were also greater in more severe pressure wounds [85]. The hypothermia observed in these patients is known to delay recovery time [86]. Therefore, promptly identifying the surface decrease in temperature might be helpful to intervene and reduce postsurgical complications.

In the case of veterinary medicine, in a red wolf sustaining an injury in the left thoracic limb, Hurley-Sanders et al. [84] performed a free skin graft and evaluated through IRT the correct vascularization of the graft. The authors evaluated 2 days before surgery, immediately after surgery, and every 2–3 days within 14 postsurgical days. The results showed that graft adherence and healing were obtained when neovascularization caused a similar temperature between the graft site and surrounding healthy tissue [84]. Similarly, in Nigerian indigenous dogs, IRT was used in animals undergoing castration, otectomy, and gastrotomy [87]. The authors evaluated wound surface temperature and found that the initial temperature of castrated dogs was 33.34 ± 0.86 °C, a temperature that increased to 35.64 ± 0.57 °C at 6 h postsurgery, 35.14 ± 0.38 at 24 h, 35.82 ± 0.61 °C at 48 h, 35.24 ± 0.64 at 1 week, and progressively decreased in the second week postsurgery (32.84 ± 1.27 °C). In contrast, dogs undergoing otectomy had an initial temperature of 33.78 ± 0.86 °C that increased to 35.34 ± 1.13 °C, 34.58 ± 0.49 °C, 36.94 ± 0.31 °C, 36.58 ± 0.29, and 34.66 ± 0.74 °C at 6 h, 24 h, 48 h, 1 week, and the second week, respectively. A similar thermal response was observed in gastrotomy-subjected animals, in whom the initial value of 31.18 ± 0.86 °C progressively increased to 32.70 ± 1.02 °C, 34.10 ± 0.55 °C, 35.36 ± 0.33 °C, and 36.00 ± 1.37 °C in the same period. The higher temperatures recorded in these animals might be attributed to a higher pain and inflammatory response, as this process alters body temperature and is also a stressor that will modify peripheral blood flow [42].

Healing monitoring was also performed in piglets after CO_2 surgical laser for castration. Viscardi et al. [88] used IRT in these Yorkshire x Landrace piglets to evaluate if surgical laser helps to reduce pain and inflammation and improve the healing process. Laser-castrated piglets had lower temperatures at the incision site than scalpel-castrated animals ($p = 0.0005$). Due to the laser, thermal lesion was observed in the piglets, along with pain-related behaviors, which would make this technique not a refinement technique for the conventional techniques. Agents to promote healing (e.g., germicide, aluminum powder, liquid bandage) have also been evaluated with IRT, as shown by Marti et al. [89] in Angus bulls undergoing castration. The scrotal surface temperature was not significantly different among groups during the 49 days of evaluation (average values between 27.5 °C and 28.5 °C), and the time for incision healing did not differ. These results might be associated with a similar inflammatory and painful response in all bulls, showing that IRT can be an adjuvant to monitor the healing period and manage pain, an important sign that also must be considered [90]. In longer-lasting pain procedures such as rubber ring castration, usually performed in domestic animals, Nogues et al. [91] evaluated wound healing and inflammation in dairy calves after surgical or rubber ring castration. Rubber-ring-castrated animals did not fully heal within the eight evaluation weeks and, according to IRT, the thermal surface response around the lesion was 1.7 ± 0.35 °C higher than the surgical group. In contrast, surgically castrated calves fully healed at week four.

These studies, apart from showing that IRT can help to determine a sustained inflammatory response associated with healing delay, show that it is a tool that can help to refine current procedures not only because they are painful but because tissue damage might be greater with a greater inflammatory response. For example, analgesic options such as lidocaine in surgical incisions were evaluated through IRT to assess if the local anesthetic has a negative effect on wound healing, as studied in cats and dogs by Herlofson et al. [87]. Surface temperatures between lidocaine alone or NaCl did not differ ($p = 0.554$) with healing scores of 3 and 2.5, respectively, showing that lidocaine addition does not influence wound healing. Therefore, the effect that surgical stress, inflammation, and pain have on local tissue perfusion and heat production is relevant and is a field that can benefit from the application of IRT.

7. Technical Factors That Can Influence Thermal Imaging in Animals

It is important to consider some technical and physiological aspects that can influence the surface thermal response in animals. An example of this is the image resolution used in different studies, which might affect the quality of the data. In horse and cattle, resolutions of 80 × 90 pixels to 320 × 240 pixels were proposed [92–94]. In human medicine, Schuster et al. [95] reported that high-resolution thermography (320 × 340 pixels) helps to assess clinical examination in tungiasis-associated morbidity. This is due to the minimal changes in surface temperature that can be detected with high-resolution cameras. Thus, the technical properties of thermal cameras must be addressed when suggesting IRT to assess vascular disorders.

On the other hand, body emissivity is another element that is highly relevant when using IRT, since this parameter describes the ability of a body surface to emit radiation [25,96]. Several studies reported differences in skin emissivity according to the species. For example, in humans, skin emissivity is 0.98, a value that is close to pig skin (between 0.92 and 0.93) [97]. In horses and rabbits, values of 0.95 and 0.97, respectively, were mentioned [98,99]. These differences are related to the skin characteristics (e.g., presence of hair, feathers, glabrous skin) even within the same species, as discussed by Soerensen et al. [97] in pigs, in whom different emissivity values were reported in hairy skin areas (0.93), in dorsal areas (0.95), or in dead animals (0.8). In horses, Jørgensen et al. [100]

mention that hair acts as a thermal insulator, which might influence heat exchange with the environment, recording lower surface temperatures.

The presence or absence of hair and its characteristics (e.g., color, length, type, and distribution) need to be considered when using thermal imaging to assess peripheral blood flow disorders in domestic and nondomestic animals [101]. For example, Kwon and Brundage [102] compared the surface temperature of dogs with different types of hair coats: short, curly, long, and double-coat types. The authors found that animals with short hair had the highest surface temperatures (31.77 ± 0.19 °C) and that these values were approximately 2 °C above the temperatures of the other animals. This could be related to short hair not providing a thicker insulative layer, in contrast to long or double-coat animals, which would permit and make more easily recognizable an increase in IRT values. Nonetheless, another study performed on pregnant and nonpregnant mares with different hair-coat features did not find significant differences regarding hair length [103]. Furthermore, as mentioned by Favilla and Costa [104], thermoregulatory strategies also differ among species. In the case of marine mammals, heat dissipation is performed through the peripheral vascularization of the flippers and flukes. In contrast, birds exposed to heat stress tend to dissipate heat through the bill (beak) [105,106]. Moreover, Gauchte et al. [105] mention that it is not only the species or the presence of hair/feathers that influences the thermoregulatory response of animals but also external factors such as wind speed, ambient temperature, sun exposure, and shooting distance that can affect infrared readings.

To address the influence that hair presence and type have on IRT values, the use of certain thermal windows has been proposed. These regions are characterized by a rich network of superficial capillaries and arteriovenous anastomosis that facilitate heat exchange and also tend to be zones without hair [25]. The ocular surface temperature is one of the most-used thermal windows to evaluate the health state, stress process, and pain perception in several species [107–109]. Nonetheless, the evaluation of a peripheral ischemic process and its influence and other thermal windows that are not close to the ischemic zone has not been reported in animals.

The consideration of environmental elements that can influence thermal imagining is another issue that needs to be addressed. For example, in dairy cattle, it was reported that a wind speed of 7 km/h or 12 km/h can decrease the ocular surface temperature by up to 0.43 °C and 0.78 °C, respectively. Likewise, direct solar radiation can increase the surface temperature of animals by 0.56 °C [110,111]. To reduce the effect of these elements, signal-processing techniques are proposed to improve the reliability of IRT. The use of cameras that control the automatic brightness could provide clearer pictures when thermal imaging is performed under field conditions (e.g., in farm animals where environmental factors cannot always be controlled) [112]. Other processing techniques such as wavelet transforms and signal reconstruction, among others, are currently applied in industrial thermal imaging to improve IRT reliability [113]. Nonetheless, the application of these techniques in veterinary medicine is still needed to improve its use in animals.

8. Perspectives on the Use of IRT in Veterinary Medicine

The discussion presented shows that IRT is sensitive to detecting changes in both global and local perfusion. While it has been reported as a tool to recognize hypoperfusion states during pathological thromboembolism or diabetic foot, it is necessary to consider whether it is possible to apply this tool as a reliable and sensitive diagnostic method focused on any of these pathologies [14]. Nonetheless, since its application in veterinary medicine to monitor vascular diseases is limited, further research could aim to evaluate patients with lupus or vasculitis. Moreover, necrotic tissue monitoring observed in patients attacked by venomous animals such as snakes or spiders could be another field for investigation.

Because these peripheral circulatory alterations can affect extremities with vascular compromise, IRT could be implemented as a method to decide whether a limb needs amputation or not [44]. In veterinary medicine, this field of research is limited, and additional studies would be needed to validate the use of IRT for wound care and treatment decisions in animals. Moreover, a field where IRT has not been applied yet in veterinary medicine is the viability of organ transplant surgeries, where monitoring adequate blood perfusion could be related to the success of the surgical procedure.

Although IRT has been used in human medicine to evaluate orthopedic conditions and postoperative outcomes related to joint reconstructions or implant placements [70], in veterinary medicine, only a couple of studies have been published. Examples of these are Zhou et al.'s [114] preclinical study on dogs undergoing bone autografts, in whom an infrared camera was used to find the speed of drilling that resulted in less thermal-mechanical damage to the bone cell viability and proliferation. The authors found that local increases in the cortical surface temperature of 2–6 °C with 500 rpm drilling were associated with enhanced cell performance and osteogenic gene expression, showing that thermal evaluation of these procedures might help to improve the quality of bone autografts. Similarly, the cortical bone surface temperature of sheep receiving implants placed in the iliac crest was evaluated to determine the effect of different osteotomy drilling speed protocols. It was found that although 50 rpm drilling had the highest bone temperature and the highest cell stability, all drilling protocols showed implant stability regardless of the temperature changes [115]. Furthermore, since some studies report that bone implants can increase the local surface temperature within the first week after the surgical procedure as an indicator of the repairing phase, IRT could be applied in animal patients to monitor orthopedic interventions [116].

Likewise, it is necessary to consider the limitations of the technique. The administration of drugs such as local anesthetics can alter the thermal response and could be misinterpreted as a pathological effect instead of a drug-related response. Furthermore, IRT indirectly assesses the amount of radiated heat, and this depends on external elements such as environmental temperature, and wetness of the surface, among others [117]. Therefore, IRT is recommended as a complementary tool that should be used with other techniques (e.g., angiography, ultrasound, or histopathologic analysis), since some studies have mentioned that IRT does not have a higher sensitivity than traditional imaging techniques such as radiography, ultrasonography, or magnetic resonance in neurological patients [118]. For example, Han et al. [119] concluded that indocyanine green angiography performed in a rat perforator flap model predicts tissue necrosis with an accuracy of 92%, in contrast to thermal imaging (74%) and near-infrared spectroscopy (71%). Thus, a comprehensive evaluation of peripheral vascular alterations and tissue viability in animal medicine requires additional technologies.

9. Conclusions

IRT is a technique that can noninvasively and indirectly detect peripheral vascular disorders and the consequent alterations to the amount of radiated heat from the skin. Although the application of IRT to study vascular integrity and tissue viability is more frequent in human medicine, the current research performed in animals has shown that IRT can differentiate healthy from sick patients in cases of thrombosis, thromboembolisms, and ischemia processes.

Additionally, the microvascular alterations that can be found as a result of burn wounds is a field where IRT can help in determining the degree of damage and predict healing success in cases where conservative treatment or skin transplant is optional. In this sense, higher differences between wound bed surface temperature and healthy tissue have been associated with a higher degree of tissular damage. Likewise, a hypothermic local response is related to delayed healing and possible postsurgical complications such as incisional infection.

Since the lack of an integrated network of blood perfusion affects tissue viability and the wound healing process, IRT can be suggested as a complementary tool to other techniques such as angiography or Doppler ultrasound to assess vascular integrity or to promptly adopt treatment options.

Author Contributions: All authors contributed to the conceptualization, writing, and revision. All authors have read and agreed to the published version of the manuscript.

Funding: This research received no external funding.

Institutional Review Board Statement: Not applicable.

Informed Consent Statement: Not applicable.

Data Availability Statement: Not applicable.

Conflicts of Interest: The authors declare no conflicts of interest.

References

1. Schaefer, A.L.; Cook, N.; Tessaro, S.V.; Deregt, D.; Desroches, G.; Dubeski, P.L.; Tong, A.K.W.; Godson, D.L. Early Detection and Prediction of Infection Using Infrared Thermography. *Can. J. Anim. Sci.* **2004**, *84*, 73–80. [CrossRef]
2. Cook, N.J.N.; Chabot, B.; Lui, T.; Bench, C.C.J.; Schaefer, A.L. AL Infrared Thermography Detects Febrile and Behavioural Responses to Vaccination of Weaned Piglets. *Animal* **2015**, *9*, 339–346. [CrossRef] [PubMed]
3. Mota-Rojas, D.; Olmos-Hernández, A.; Verduzco-Mendoza, A.; Lecona-Butrón, H.; Martínez-Burnes, J.; Mora-Medina, P.; Gómez-Prado, J.; Orihuela, A. Infrared Thermal Imaging Associated with Pain in Laboratory Animals. *Exp. Anim.* **2021**, *70*, 1–12. [CrossRef] [PubMed]
4. Rekant, S.I.; Lyons, M.A.; Pacheco, J.M.; Arzt, J.; Rodriguez, L.L. Veterinary Applications of Infrared Thermography. *Am. J. Vet. Res.* **2016**, *77*, 98–107. [CrossRef] [PubMed]
5. Higashino, T.; Nakagami, G.; Kadono, T.; Ogawa, Y.; Iizaka, S.; Koyanagi, H.; Sasaki, S.; Haga, N.; Sanada, H. Combination of Thermographic and Ultrasonographic Assessments for Early Detection of Deep Tissue Injury. *Int. Wound J.* **2014**, *11*, 509–516. [CrossRef] [PubMed]
6. Lahiri, B.B.; Bagavathiappan, S.; Jayakumar, T.; Philip, J. Medical Applications of Infrared Thermography: A Review. *Infrared Phys. Technol.* **2012**, *55*, 221–235. [CrossRef] [PubMed]
7. Deng, F.; Tang, Q.; Zeng, G.; Wu, H.; Zhang, N.; Zhong, N. Effectiveness of Digital Infrared Thermal Imaging in Detecting Lower Extremity Deep Venous Thrombosis. *Med. Phys.* **2015**, *42*, 2242–2248. [CrossRef]
8. Lin, P.H.; Echeverria, A.; Poi, M.J. Infrared Thermography in the Diagnosis and Management of Vasculitis. *J. Vasc. Surg. Cases Innov. Tech.* **2017**, *3*, 112–114. [CrossRef]
9. Branco, J.H.L.; Branco, R.L.L.; Siqueira, T.C.; de Souza, L.C.; Dalago, K.M.S.; Andrade, A. Clinical Applicability of Infrared Thermography in Rheumatic Diseases: A Systematic Review. *J. Therm. Biol.* **2022**, *104*, 103172. [CrossRef]
10. Saygin, D.; Highland, K.B.; Tonelli, A.R. Microvascular Involvement in Systemic Sclerosis and Systemic Lupus Erythematosus. *Microcirculation* **2019**, *26*, e12440. [CrossRef]
11. Huang, C.-L.; Wu, Y.-W.; Hwang, C.-L.; Jong, Y.-S.; Chao, C.-L.; Chen, W.-J.; Wu, Y.-T.; Yang, W.-S. The Application of Infrared Thermography in Evaluation of Patients at High Risk for Lower Extremity Peripheral Arterial Disease. *J. Vasc. Surg.* **2011**, *54*, 1074–1080. [CrossRef] [PubMed]
12. Mocanu, J.; Militaru, M.; Ciobotaru, E.; Ionascu, I. Thermographic Aspects of Feline Fibrosarcoma Complex. *Vet. Dermatol.* **2004**, *15*, 62. [CrossRef]
13. Nitrini, A.G.C.; Cogliati, B.; Matera, J.M. Thermographic Assessment of Skin and Soft Tissue Tumors in Cats. *J. Feline Med. Surg.* **2021**, *23*, 513–518. [CrossRef] [PubMed]
14. Pouzot-Nevoret, C.; Barthélemy, A.; Goy-Thollot, I.; Boselli, E.; Cambournac, M.; Guillaumin, J.; Bonnet-Garin, J.-M.; Allaouchiche, B. Infrared Thermography: A Rapid and Accurate Technique to Detect Feline Aortic Thromboembolism. *J. Feline Med. Surg.* **2018**, *20*, 780–785. [CrossRef] [PubMed]
15. Ramirez-GarciaLuna, J.L.; Bartlett, R.; Arriaga-Caballero, J.E.; Fraser, R.D.J.; Saiko, G. Infrared Thermography in Wound Care, Surgery, and Sports Medicine: A Review. *Front. Physiol.* **2022**, *13*, 838528. [CrossRef]
16. Asif, A.; Poyiatzis, C.; Raheman, F.J.; Rojoa, D.M. The Use of Infrared Thermography (IRT) in Burns Depth Assessment: A Diagnostic Accuracy Meta-Analysis. *Eur. Burn J.* **2022**, *3*, 432–446. [CrossRef]
17. John, H.E.; Niumsawatt, V.; Rozen, W.M.; Whitaker, I.S. Clinical Applications of Dynamic Infrared Thermography in Plastic Surgery: A Systematic Review. *Gland Surg.* **2016**, *5*, 122–132. [CrossRef]
18. Lawson, R.N.; Wlodek, G.D.; Webster, D.R. Thermographic Assessment of Burns and Frostbite. *Can. Med. Assoc. J.* **1961**, *84*, 1129–1131.
19. Simmons, J.D.; Kahn, S.A.; Vickers, A.L.; Crockett, E.S.; Whitehead, J.D.; Krecker, A.K.; Lee, Y.-L.; Miller, A.N.; Patterson, S.B.; Richards, W.O.; et al. Early Assessment of Burn Depth with Far Infrared Time-Lapse Thermography. *J. Am. Coll. Surg.* **2018**, *226*, 687–693. [CrossRef]
20. Özsoylu, D.; Janus, K.A.; Achtsnicht, S.; Wagner, T.; Keusgen, M.; Schöning, M.J. (Bio-)Sensors for Skin Grafts and Skin Flaps Monitoring. *Sens. Actuators Rep.* **2023**, *6*, 100163. [CrossRef]
21. Klama-Baryła, A.; Kitala, D.; Łabuś, W.; Kraut, M.; Szapski, M.; Smętek, W. Infrared Thermal Imaging as a Method of Improving Skin Graft Qualification Procedure and Skin Graft Survivability. *Transplant. Proc.* **2020**, *52*, 2223–2230. [CrossRef] [PubMed]

22. Pokorná, J.; Staffa, E.; Čan, V.; Bernard, V.; Mornstein, V.; Farkašová, M.; Zetelolová, A.; Kala, Z. Intestinal Resection of a Porcine Model under Thermographic Monitoring. *Physiol. Meas.* **2019**, *40*, 014003. [CrossRef] [PubMed]
23. Hennessy, O.; Potter, S.M. Use of Infrared Thermography for the Assessment of Free Flap Perforators in Autologous Breast Reconstruction: A Systematic Review. *JPRAS Open* **2020**, *23*, 60–70. [CrossRef] [PubMed]
24. Ilo, A.; Romsi, P.; Mäkelä, J. Infrared Thermography and Vascular Disorders in Diabetic Feet. *J. Diabetes Sci. Technol.* **2020**, *14*, 28–36. [CrossRef] [PubMed]
25. Tattersall, G.J. Infrared Thermography: A Non-Invasive Window into Thermal Physiology. *Comp. Biochem. Physiol. Part A Mol. Integr. Physiol.* **2016**, *202*, 78–98. [CrossRef] [PubMed]
26. Jorge, J.; Harford, M.; Villarroel, M.; Chaichulee, S.; Davidson, S.; Finnegan, E.; Clark, S.H.; Young, J.D.; Watkinson, P.J.; Tarassenko, L. Non-Contact Assessment of Peripheral Artery Haemodynamics Using Infrared Video Thermography. *IEEE Trans. Biomed. Eng.* **2021**, *68*, 276–288. [CrossRef]
27. Mota-Rojas, D.; Pereira, A.M.F.; Martínez-Burnes, J.; Domínguez-Oliva, A.; Mora-Medina, P.; Casas-Alvarado, A.; Rios-Sandoval, J.; de Mira Geraldo, A.; Wang, D. Thermal Imaging to Assess the Health Status in Wildlife Animals under Human Care: Limitations and Perspectives. *Animals* **2022**, *12*, 3558. [CrossRef]
28. Gómez-Prado, J.; Pereira, A.M.F.; Wang, D.; Villanueva-García, D.; Domínguez-Oliva, A.; Mora-Medina, P.; Hernández-Avalos, I.; Martínez-Burnes, J.; Casas-Alvarado, A.; Olmos-Hernández, A.; et al. Thermoregulation Mechanisms and Perspectives for Validating Thermal Windows in Pigs with Hypothermia and Hyperthermia: An Overview. *Front. Vet. Sci.* **2022**, *9*, 1023294. [CrossRef]
29. Casas-Alvarado, A.; Martínez-Burnes, J.; Mora-Medina, P.; Hernández-Avalos, I.; Domínguez-Oliva, A.; Lezama-García, K.; Gómez-Prado, J.; Mota-Rojas, D. Thermal and Circulatory Changes in Diverse Body Regions in Dogs and Cats Evaluated by Infrared Thermography. *Animals* **2022**, *12*, 789. [CrossRef]
30. Verduzco-Mendoza, A.; Bueno-Nava, A.; Wang, D.; Martínez-Burnes, J.; Olmos-Hernández, A.; Casas, A.; Domínguez, A.; Mota-Rojas, D. Experimental Applications and Factors Involved in Validating Thermal Windows Using Infrared Thermography to Assess the Health and Thermostability of Laboratory Animals. *Animals* **2021**, *11*, 3448. [CrossRef]
31. Finstad, J.B.; Cooper, E.; Ten Cate, S.C.; Yaxley, P.; Her, J.; Guillaumin, J. Infrared Thermography Is a Novel Tool to Assess Small Intestinal Surface Temperature in Dogs Undergoing Laparotomy for Foreign Body Obstruction. *Am. J. Vet. Res.* **2023**, *84*, ajvr.23.04.0082. [CrossRef] [PubMed]
32. Morrison, S.F.; Nakamura, K. Central Mechanisms for Thermoregulation. *Annu. Rev. Physiol.* **2019**, *81*, 285–308. [CrossRef] [PubMed]
33. Kashio, M.; Tominaga, M. TRP Channels in Thermosensation. *Curr. Opin. Neurobiol.* **2022**, *75*, 102591. [CrossRef] [PubMed]
34. Montell, C.; Caterina, M.J. Thermoregulation: Channels That Are Cool to the Core. *Curr. Biol.* **2007**, *17*, R885–R887. [CrossRef] [PubMed]
35. Cenac, N.; Altier, C.; Motta, J.-P.; D'Aldebert, E.; Galeano, S.; Zamponi, G.W.; Vergnolle, N. Potentiation of TRPV4 Signalling by Histamine and Serotonin: An Important Mechanism for Visceral Hypersensitivity. *Gut* **2010**, *59*, 481–488. [CrossRef] [PubMed]
36. Lezama-García, K.; Mota-Rojas, D.; Pereira, A.M.F.; Martínez-Burnes, J.; Ghezzi, M.; Domínguez, A.; Gómez, J.; de Mira Geraldo, A.; Lendez, P.; Hernández-Ávalos, I.; et al. Transient Receptor Potential (TRP) and Thermoregulation in Animals: Structural Biology and Neurophysiological Aspects. *Animals* **2022**, *12*, 106. [CrossRef] [PubMed]
37. Mota-Rojas, D.; Titto, C.G.; Orihuela, A.; Martínez-Burnes, J.; Gómez-Prado, J.; Torres-Bernal, F.; Flores-Padilla, K.; Carvajal-de la Fuente, V.; Wang, D. Physiological and Behavioral Mechanisms of Thermoregulation in Mammals. *Animals* **2021**, *11*, 1733. [CrossRef]
38. Romanovsky, A.A. Skin Temperature: Its Role in Thermoregulation. *Acta Physiol.* **2014**, *210*, 498–507. [CrossRef]
39. Sanin, L.Y.; Cabrera, A.M.Z.; Morales, A.M.T. Adaptive Responses to Thermal Stress in Mammals. *Rev. Med. Vet.* **2016**, *31*, 121–135.
40. Sokolova, I. Temperature Regulation. In *Encyclopedia of Ecology*; Elsevier: Amsterdam, The Netherlands, 2019; pp. 633–639.
41. Pérez de Diego, A.C.; Sánchez-Cordón, P.J.; Pedrera, M.; Martínez-López, B.; Gómez-Villamandos, J.C.; Sánchez-Vizcaíno, J.M. The Use of Infrared Thermography as a Non-Invasive Method for Fever Detection in Sheep Infected with Bluetongue Virus. *Vet. J.* **2013**, *198*, 182–186. [CrossRef]
42. Whittaker, A.L.; Muns, R.; Wang, D.; Martínez-Burnes, J.; Hernández-Ávalos, I.; Casas-Alvarado, A.; Domínguez-Oliva, A.; Mota-Rojas, D. Assessment of Pain and Inflammation in Domestic Animals Using Infrared Thermography: A Narrative Review. *Animals* **2023**, *13*, 2065. [CrossRef] [PubMed]
43. Lin, Y.; Mullan, S.; Main, D.C.J. Optimising Lameness Detection in Dairy Cattle by Using Handheld Infrared Thermometers. *Vet. Med. Sci.* **2018**, *4*, 218–226. [CrossRef] [PubMed]
44. Casas-Alvarado, A.; Mota-Rojas, D.; Hernández-Ávalos, I.; Mora-Medina, P.; Olmos-Hernández, A.; Verduzco-Mendoza, A.; Reyes-Sotelo, B.; Martínez-Burnes, J. Advances in Infrared Thermography: Surgical Aspects, Vascular Changes, and Pain Monitoring in Veterinary Medicine. *J. Therm. Biol.* **2020**, *92*, 102664. [CrossRef] [PubMed]
45. Magnin, M.; Junot, S.; Cardinali, M.; Ayoub, J.Y.; Paquet, C.; Louzier, V.; Garin, J.M.B.; Allaouchiche, B. Use of Infrared Thermography to Detect Early Alterations of Peripheral Perfusion: Evaluation in a Porcine Model. *Biomed. Opt. Express* **2020**, *11*, 2431–2446. [CrossRef]

46. Abbott, C.A.; Carrington, A.L.; Ashe, H.; Bath, S.; Every, L.C.; Griffiths, J.; Hann, A.W.; Hussein, A.; Jackson, N.; Johnson, K.E.; et al. The North-West Diabetes Foot Care Study: Incidence of, and Risk Factors for, New Diabetic Foot Ulceration in a Community-based Patient Cohort. *Diabet. Med.* **2002**, *19*, 377–384. [CrossRef] [PubMed]
47. Nabuurs-Franssen, M.; Houben, A.; Tooke, J.; Schaper, N. The Effect of Polyneuropathy on Foot Microcirculation in Type II Diabetes. *Diabetologia* **2002**, *45*, 1164–1171. [CrossRef]
48. Abularrage, C.J.; Sidawy, A.N.; Aidinian, G.; Singh, N.; Weiswasser, J.M.; Arora, S. Evaluation of the Microcirculation in Vascular Disease. *J. Vasc. Surg.* **2005**, *42*, 574–581. [CrossRef] [PubMed]
49. Fais, P.; Mazzotti, M.C.; Montisci, M.; Palazzo, C.; Leone, O.; Cecchetto, G.; Viel, G.; Pelotti, S. Post-Mortem Thermal Angiography: A Pilot Study on Swine Coronary Circulation. *Int. J. Leg. Med.* **2019**, *133*, 571–581. [CrossRef]
50. Li, X.; Chen, M.; Maharjan, S.; Cui, J.; Lu, L.; Gong, X. Evaluating Surgical Delay Using Infrared Thermography in an Island Pedicled Perforator Flap Rat Model. *J. Reconstr. Microsurg.* **2017**, *33*, 518–525. [CrossRef]
51. Łokaj, M.; Czapla, N.; Falkowski, A.; Prowans, P. The Use of Thermography in Early Detection of Tissue Perfusion Disorders in Rats. *Videosurgery Other Miniinvasive Tech.* **2014**, *3*, 329–336. [CrossRef]
52. Caramalac, S.M.; De Bortoli, B.L.; Caramalac, S.M.; Souza, T.J.; Castilho, M.O.; Babo-terra, V.J.; Isa, M.; Palumbo, P. Thromboembolism in a Bitch—Diagnosis by Infrared Thermography. *Acta Sci. Vet.* **2023**, *51*, 871. [CrossRef]
53. Prindeze, N.J.; Fathi, P.; Mino, M.J.; Mauskar, N.A.; Travis, T.E.; Paul, D.W.; Moffatt, L.T.; Shupp, J.W. Examination of the Early Diagnostic Applicability of Active Dynamic Thermography for Burn Wound Depth Assessment and Concept Analysis. *J. Burn Care Res.* **2015**, *36*, 626–635. [CrossRef] [PubMed]
54. Li, F.; Wang, M.; Wang, T.; Wang, X.; Ma, X.; He, H.; Ma, G.; Zhao, D.; Yue, Q.; Wang, P.; et al. Smartphone-based Infrared Thermography to Assess Progress in Thoracic Surgical Incision Healing: A Preliminary Study. *Int. Wound J.* **2023**, *20*, 2000–2009. [CrossRef] [PubMed]
55. Dang, J.; Tan, C.; Pham, C.; Huang, S.; Yenikomshian, H.; Gillenwater, T.J. Use of Infrared Thermography for Flap Monitoring: A Systematic Review. *Plast. Reconstr. Surg.-Glob. Open* **2021**, *9*, 164. [CrossRef]
56. Abazari, M.; Ghaffari, A.; Rashidzadeh, H.; Badeleh, S.M.; Maleki, Y. A Systematic Review on Classification, Identification, and Healing Process of Burn Wound Healing. *Int. J. Low. Extrem. Wounds* **2022**, *21*, 18–30. [CrossRef] [PubMed]
57. Prindeze, N.J.; Hoffman, H.A.; Ardanuy, J.G.; Zhang, J.; Carney, B.C.; Moffatt, L.T.; Shupp, J.W. Active Dynamic Thermography Is a Sensitive Method for Distinguishing Burn Wound Conversion. *J. Burn Care Res.* **2016**, *37*, e559–e568. [CrossRef] [PubMed]
58. Chi, Y.; Liu, X.; Chai, J. A Narrative Review of Changes in Microvascular Permeability after Burn. *Ann. Transl. Med.* **2021**, *9*, 719. [CrossRef] [PubMed]
59. Miccio, J.; Parikh, S.; Marinaro, X.; Prasad, A.; McClain, S.; Singer, A.J.; Clark, R.A.F. Forward-Looking Infrared Imaging Predicts Ultimate Burn Depth in a Porcine Vertical Injury Progression Model. *Burns* **2016**, *42*, 397–404. [CrossRef]
60. Renkielska, A.; Kaczmarek, M.; Nowakowski, A.; Grudziński, J.; Czapiewski, P.; Krajewski, A.; Grobelny, I. Active Dynamic Infrared Thermal Imaging in Burn Depth Evaluation. *J. Burn Care Res.* **2014**, *35*, e294–e303. [CrossRef]
61. Albernaz, V.G.P.; Castro, J.L.C.; Santalucia, S.; Huppes, R.R.; De Nardi, A.B.; Pazzini, J.M. Reconstructive Surgical Repair of a Forth Degree Iatrogenic Burn in a Dog. *Acta Sci. Vet.* **2016**, *44*, 5. [CrossRef]
62. de Lima, J.G.; Bahr, A.M.V. Medical and Surgical Management of a Large Thermal Burn in a Dog. *Acta Sci. Vet.* **2015**, *43*, 103.
63. Wearn, C.; Lee, K.C.; Hardwicke, J.; Allouni, A.; Bamford, A.; Nightingale, P.; Moiemen, N. Prospective Comparative Evaluation Study of Laser Doppler Imaging and Thermal Imaging in the Assessment of Burn Depth. *Burns* **2018**, *44*, 124–133. [CrossRef] [PubMed]
64. Ye, H.; De, S. Thermal Injury of Skin and Subcutaneous Tissues: A Review of Experimental Approaches and Numerical Models. *Burns* **2017**, *43*, 909–932. [CrossRef] [PubMed]
65. Burmeister, D.M.; Cerna, C.; Becerra, S.C.; Sloan, M.; Wilmink, G.; Christy, R.J. Noninvasive Techniques for the Determination of Burn Severity in Real Time. *J. Burn Care Res.* **2017**, *38*, e180–e191. [CrossRef] [PubMed]
66. Hardwicke, J.; Thomson, R.; Bamford, A.; Moiemen, N. A Pilot Evaluation Study of High Resolution Digital Thermal Imaging in the Assessment of Burn Depth. *Burns* **2013**, *39*, 76–81. [CrossRef] [PubMed]
67. Dunn, R. Grafts and Flaps. *Surgery* **2006**, *24*, 27–32. [CrossRef]
68. Aisa, J.; Vernon, J. The Use of Skin Grafts in Small Animal Reconstructive Surgery. *Companion Anim.* **2016**, *21*, 642–648. [CrossRef]
69. Martínez-Jiménez, M.A.; Ramirez-GarciaLuna, J.L.; Kolosovas-Machuca, E.S.; Drager, J.; González, F.J. Development and Validation of an Algorithm to Predict the Treatment Modality of Burn Wounds Using Thermographic Scans: Prospective Cohort Study. *PLoS ONE* **2018**, *13*, e0206477. [CrossRef]
70. Kumar, P.; Gaurav, A.; Rajnish, R.K.; Sharma, S.; Kumar, V.; Aggarwal, S.; Patel, S. Applications of Thermal Imaging with Infrared Thermography in Orthopaedics. *J. Clin. Orthop. Trauma* **2022**, *24*, 101722. [CrossRef]
71. da Silveira, Y.G.; Kajiura, C.; Bernardes, F.J.L.; Maria, S.P.; Fernandes, M.P.; Frasson, M.T.; Cassino, P.; Moreira, S.H.; Gómez, J.L.Á.; Pazzini, J.M.; et al. Use of Thermography in Skin Grafts after the Application of Therapeutic Ultrasound in Wistar Rats. *Acta Cirúrgica Bras.* **2020**, *35*, e202000703. [CrossRef]
72. Lekas, R.; Jakuška, P.; Kriščiukaitis, A.; Veikutis, V.; Dzemyda, G.; Mickevičius, T.; Morkūnaitė, K.; Vilkė, A.; Treigys, P.; Civinskienė, G.; et al. Monitoring changes in heart tissue temperature and evaluation of graft function after coronary artery bypass grafting surgery. *Medicina* **2009**, *45*, 221. [CrossRef] [PubMed]
73. Probst, C.W. Grafting Techniques and Failures in Small Animal Surgery. *Probl. Vet. Med.* **1990**, *2*, 413–423. [PubMed]

74. Hallock, G.G. The Use of Smartphone Thermography to More Safely Unmask and Preserve Circulation to Keystone Advancement Flaps in the Lower Extremity. *Injury* **2020**, *51*, S121–S125. [CrossRef] [PubMed]
75. Sjøberg, T.; Mercer, J.B.; Weum, S.; de Weerd, L. The Value of Dynamic Infrared Thermography in Pedicled Thoracodorsal Artery Perforator Flap Surgery. *Plast. Reconstr. Surg. Glob. Open* **2020**, *8*, e2799. [CrossRef] [PubMed]
76. Miland, Å.O.; de Weerd, L.; Weum, S.; Mercer, J.B. Visualising Skin Perfusion in Isolated Human Abdominal Skin Flaps Using Dynamic Infrared Thermography and Indocyanine Green Fluorescence Video Angiography. *Eur. J. Plast. Surg.* **2008**, *31*, 235–242. [CrossRef]
77. Rabbani, M.J.; Bhatti, A.Z.; Shahzad, A. Flap Monitoring Using Thermal Imaging Camera: A Contactless Method. *J. Coll. Physicians Surg. Pak.* **2021**, *31*, 703–706. [CrossRef]
78. Miland, Å.O.; de Weerd, L.; Mercer, J.B. Intraoperative Use of Dynamic Infrared Thermography and Indocyanine Green Fluorescence Video Angiography to Predict Partial Skin Flap Loss. *Eur. J. Plast. Surg.* **2008**, *30*, 269–276. [CrossRef]
79. Tenorio, X.; Mahajan, A.L.; Wettstein, R.; Harder, Y.; Pawlovski, M.; Pittet, B. Early Detection of Flap Failure Using a New Thermographic Device. *J. Surg. Res.* **2009**, *151*, 15–21. [CrossRef]
80. Czapla, N.; Łokaj, M.; Falkowski, A.; Prowans, P. The Use of Thermography to Design Tissue Flaps–Experimental Studies on Animals. *Videosurgery Other Miniinvasive Tech.* **2014**, *3*, 319–328. [CrossRef]
81. Drvis, P.; Shejbal, D.; Svaic, S.; Boras, I.; Sundov, I.; Susa, M.; Bedekovic, V.; Kalogjera, L.; Petrovic, I.; Sikiric, P. Evaluation of Cutaneous Flap Survival by IR Thermography. In Proceedings of the 2004 International Conference on Quantitative InfraRed Thermography, Zagreb, Croatia, 5–8 January 2004; pp. 1–10.
82. Shejbal, D.; Drvis, P.; Bedekovic, V. Thermography-measured Effect of Capsaicin, Methylprednisolone and Mitomycin on the Survival of Random Skin Flaps in Rats. *Ski. Res. Technol.* **2012**, *18*, 157–161. [CrossRef]
83. Clark, R.A.F.; Fenner, J.; Sasson, A.; McClain, S.A.; Singer, A.J.; Tonnesen, M.G. Blood Vessel Occlusion with Erythrocyte Aggregates Causes Burn Injury Progression—Microvasculature Dilation as a Possible Therapy. *Exp. Dermatol.* **2018**, *27*, 625–629. [CrossRef] [PubMed]
84. Hurley-Sanders, J.L.; Sladky, K.K.; Nolan, E.C.; Loomis, M.R. Use of Cortical Bone Fenestration, Autogenous Free Skin Graft, and Thermography for Wound Treatment and Monitoring in a Red Wolf (Canis Rufus Gregoryi). *J. Zoo Wildl. Med.* **2015**, *46*, 617–620. [CrossRef] [PubMed]
85. Lin, Y.-H.; Chen, Y.-C.; Cheng, K.-S.; Yu, P.-J.; Wang, J.-L.; Ko, N.-Y. Higher Periwound Temperature Associated with Wound Healing of Pressure Ulcers Detected by Infrared Thermography. *J. Clin. Med.* **2021**, *10*, 2883. [CrossRef] [PubMed]
86. Driver, J.; Fielding, A.; Mullhi, R.; Chipp, E.; Torlinski, T. Temperature Management of Adult Burn Patients in Intensive Care: Findings from a Retrospective Cohort Study in a Tertiary Centre in the United Kingdom. *Anaesthesiol. Intensive Ther.* **2022**, *54*, 226–233. [CrossRef] [PubMed]
87. Saidu, A.M.; Olorunfemi, J.O.; Laku, D. Infrared Thermography Following Castration, Otectomy and Gastrotomy in Nigerian Indigenous Dogs. *Sahel J. Vet. Sci.* **2023**, *20*, 50–56. [CrossRef]
88. Viscardi, A.V.; Cull, C.A.; Kleinhenz, M.D.; Montgomery, S.; Curtis, A.; Lechtenberg, K.; Coetzee, J.F. Evaluating the Utility of a CO2 Surgical Laser for Piglet Castration to Reduce Pain and Improve Wound Healing: A Pilot Study. *J. Anim. Sci.* **2020**, *98*, skaa320. [CrossRef]
89. Marti, S.; Schwartzkopf-Genswein, K.S.; Janzen, E.D.; Meléndez, D.M.; Gellatly, D.; Pajor, E.A. Use of Topical Healing Agents on Scrotal Wounds after Surgical Castration in Weaned Beef Calves. *Can. Vet. J. La Rev. Vet. Can.* **2017**, *58*, 1081–1085.
90. Casas-Alvarado, A.; Mota-Rojas, D.; Hernández-Ávalos, I.; Martínez-Burnes, J.; Rosas, M.E.; Miranda-Cortés, A.; Domínguez-Oliva, A.; Mora-Medina, P. Assessment of Thermal Response, Cardiorespiratory Parameters and Post-Operative Analgesia in Dogs Undergoing Ovariohysterectomy with Different Combinations of Epidural Analgesia and Isoflurane. *J. Anim. Behav. Biometeorol.* **2023**, *11*, e2023009. [CrossRef]
91. Nogues, E.; von Keyserlingk, M.A.G.; Weary, D.M. Pain in the Weeks Following Surgical and Rubber Ring Castration in Dairy Calves. *J. Dairy Sci.* **2021**, *104*, 12881–12886. [CrossRef]
92. Loughin, C.A.; Marino, D.J. Evaluation of Thermographic Imaging of the Limbs of Healthy Dogs. *Am. J. Vet. Res.* **2007**, *68*, 1064–1069. [CrossRef]
93. Malafaia, O.; Brioschi, M.L.; Aoki, S.M.S.; Dias, F.G.; Gugelmin, B.S.; Aoki, M.S.; Aoki, Y.S. Infrared Imaging Contribution for Intestinal Ischemia Detection in Wound Healing. *Acta Cir. Bras.* **2008**, *23*, 511–519. [CrossRef] [PubMed]
94. Ringer, S.K.; Lischer, C.J.; Ueltschi, G. Assessment of Scintigraphic and Thermographic Changes after Focused Extracorporeal Shock Wave Therapy on the Origin of the Suspensory Ligament and the Fourth Metatarsal Bone in Horses without Lameness. *Am. J. Vet. Res.* **2005**, *66*, 1836–1842. [CrossRef] [PubMed]
95. Schuster, A.; Thielecke, M.; Raharimanga, V.; Ramarokoto, C.E.; Rogier, C.; Krantz, I.; Feldmeier, H. High-Resolution Infrared Thermography: A New Tool to Assess Tungiasis-Associated Inflammation of the Skin. *Trop. Med. Health* **2017**, *45*, 23. [CrossRef] [PubMed]
96. Galvin, E.M.; Niehof, S.; Medina, H.J.; Zijlstra, F.J.; van Bommel, J.; Klein, J.; Verbrugge, S.J.C. Thermographic Temperature Measurement Compared with Pinprick and Cold Sensation in Predicting the Effectiveness of Regional Blocks. *Anesth. Analg.* **2006**, *102*, 598–604. [CrossRef]
97. Soerensen, D.D.; Clausen, S.; Mercer, J.B.; Pedersen, L.J. Determining the Emissivity of Pig Skin for Accurate Infrared Thermography. *Comput. Electron. Agric.* **2014**, *109*, 52–58. [CrossRef]

98. Autio, E.; Neste, R.; Airaksinen, S.; Heiskanen, M.-L. Measuring the Heat Loss in Horses in Different Seasons by Infrared Thermography. *J. Appl. Anim. Welf. Sci.* **2006**, *9*, 211–221. [CrossRef]
99. Ludwig, N.; Gargano, M.; Luzi, F.; Carenzi, C.; Verga, M. Technical Note: Applicability of Infrared Thermography as a Non Invasive Measurements of Stress in Rabbit. *World Rabbit Sci.* **2010**, *15*, 588. [CrossRef]
100. Jørgensen, G.H.; Mejdell, C.M.; Bøe, K.E. Effects of Hair Coat Characteristics on Radiant Surface Temperature in Horses. *J. Therm. Biol.* **2020**, *87*, 102474. [CrossRef]
101. Mota-Rojas, D.; Titto, C.G.; de Mira Geraldo, A.; Martínez-Burnes, J.; Gómez, J.; Hernández-Ávalos, I.; Casas, A.; Domínguez, A.; José, N.; Bertoni, A.; et al. Efficacy and Function of Feathers, Hair, and Glabrous Skin in the Thermoregulation Strategies of Domestic Animals. *Animals* **2021**, *11*, 3472. [CrossRef]
102. Kwon, C.J.; Brundage, C.M. Quantifying Body Surface Temperature Differences in Canine Coat Types Using Infrared Thermography. *J. Therm. Biol.* **2019**, *82*, 18–22. [CrossRef]
103. Maśko, M.; Witkowska-Piłaszewicz, O.; Jasiński, T.; Domino, M. Thermal Features, Ambient Temperature and Hair Coat Lengths: Limitations of Infrared Imaging in Pregnant Primitive Breed Mares within a Year. *Reprod. Domest. Anim.* **2021**, *56*, 1315–1328. [CrossRef] [PubMed]
104. Favilla, A.B.; Costa, D.P. Thermoregulatory Strategies of Diving Air-Breathing Marine Vertebrates: A Review. *Front. Ecol. Evol.* **2020**, *8*, 555509. [CrossRef]
105. Gauchet, L.; Jaeger, A.; Grémillet, D. Using Facial Infrared Thermography to Infer Avian Body Temperatures in the Wild. *Mar. Biol.* **2022**, *169*, 57. [CrossRef]
106. McQueen, A.; Barnaby, R.; Symonds, M.R.E.; Tattersall, G.J. Birds Are Better at Regulating Heat Loss through Their Legs than Their Bills: Implications for Body Shape Evolution in Response to Climate. *Biol. Lett.* **2023**, *19*, 20230373. [CrossRef] [PubMed]
107. Lowe, G.; McCane, B.; Sutherland, M.; Waas, J.; Schaefer, A.; Cox, N.; Stewart, M. Automated Collection and Analysis of Infrared Thermograms for Measuring Eye and Cheek Temperatures in Calves. *Animals* **2020**, *10*, 292. [CrossRef] [PubMed]
108. Domínguez-Oliva, A.; Hernández-Ávalos, I.; Olmos-Hernández, A.; Villegas-Juache, J.; Verduzco-Mendoza, A.; Mota-Rojas, D. Thermal Response of Laboratory Rats (*Rattus norvegicus*) during the Application of Six Methods of Euthanasia Assessed by Infrared Thermography. *Animals* **2023**, *13*, 2820. [CrossRef] [PubMed]
109. Arfuso, F.; Acri, G.; Piccione, G.; Sansotta, C.; Fazio, F.; Giudice, E.; Giannetto, C. Eye Surface Infrared Thermography Usefulness as a Noninvasive Method of Measuring Stress Response in Sheep during Shearing: Correlations with Serum Cortisol and Rectal Temperature Values. *Physiol. Behav.* **2022**, *250*, 113781. [CrossRef]
110. Church, J.S.; Hegadoren, P.R.; Paetkau, M.J.; Miller, C.C.; Regev-Shoshani, G.; Schaefer, A.L.; Schwartzkopf-Genswein, K.S. Influence of Environmental Factors on Infrared Eye Temperature Measurements in Cattle. *Res. Vet. Sci.* **2014**, *96*, 220–226. [CrossRef]
111. de Lima, V.; Piles, M.; Rafel, O.; López-Béjar, M.; Ramón, J.; Velarde, A.; Dalmau, A. Use of Infrared Thermography to Assess the Influence of High Environmental Temperature on Rabbits. *Res. Vet. Sci.* **2013**, *95*, 802–810. [CrossRef]
112. Diakides, N.A.; Bronzino, J.D. *Medical Infrared Imaging*; CRC Press: Boca Raton, FL, USA, 2007.
113. Chung, Y.; Lee, S.; Kim, W. Latest Advances in Common Signal Processing of Pulsed Thermography for Enhanced Detectability: A Review. *Appl. Sci.* **2021**, *11*, 12168. [CrossRef]
114. Zhou, T.; Gan, Z.; Zhang, H.; Liu, Z.; Pu, Y.; Rong, M. A Novel Technique to Harvest Bone Autografts with Mild Local Hyperthermia and Enhanced Osteogenic Bone Quality: A Preclinical Study in Dogs. *BMC Oral Health* **2023**, *23*, 838. [CrossRef] [PubMed]
115. Ozcan, M.; Salimov, F.; Temmerman, A.; Turer, O.U.; Alkaya, B.; Haytac, M.C. The Evaluation of Different Osteotomy Drilling Speed Protocols on Cortical Bone Temperature, Implant Stability and Bone Healing: An Experimental Study in an Animal Model. *J. Oral Implantol.* **2022**, *48*, 3–8. [CrossRef] [PubMed]
116. Krasnikov, A.V.; Krasnikova, E.S. Use of Infrared Thermography to Control Osteoreparative and Integrative Processes during Implantation in Animals. *J. Phys. Conf. Ser.* **2020**, *1515*, 052011. [CrossRef]
117. Ilo, A.; Romsi, P.; Mäkelä, J. Infrared Thermography as a Diagnostic Tool for Peripheral Artery Disease. *Adv. Ski. Wound Care* **2020**, *33*, 482–488. [CrossRef]
118. Grossbard, B.P.; Loughin, C.A.; Marino, D.J.; Marino, L.J.; Sackman, J.; Umbaugh, S.E.; Solt, P.S.; Afruz, J.; Leando, P.; Lesser, M.L.; et al. Medical Infrared Imaging (Thermography) of Type I Thoracolumbar Disk Disease in Chondrodystrophic Dogs. *Vet. Surg.* **2014**, *43*, 869–876. [CrossRef]
119. Han, T.; Khavanin, N.; Wu, J.; Zang, M.; Zhu, S.; Chen, B.; Li, S.; Liu, Y.; Sacks, J.M. Indocyanine Green Angiography Predicts Tissue Necrosis More Accurately Than Thermal Imaging and Near-Infrared Spectroscopy in a Rat Perforator Flap Model. *Plast. Reconstr. Surg.* **2020**, *146*, 1044–1054. [CrossRef]

animals MDPI

Review

Utilization of Infrared Thermography in Assessing Thermal Responses of Farm Animals under Heat Stress

Marcelo Daniel Ghezzi [1], Fabio Napolitano [2,†], Alejandro Casas-Alvarado [3], Ismael Hernández-Ávalos [4], Adriana Domínguez-Oliva [3,*], Adriana Olmos-Hernández [5] and Alfredo M. F. Pereira [6,*]

1 Faculty of Veterinary Sciences, Veterinary Research Center (CIVETAN), Universidad Nacional del Centro de la Provincia de Buenos Aires (UNCPBA), CONICET-CICPBA, Tandil 7000, Argentina
2 Scuola di Scienze Agrarie, Forestali, Alimentari ed Ambientali, Università degli Studi della Basilicata, 85100 Potenza, Italy
3 Neurophysiology, Behavior and Animal Welfare Assessment, DPAA, Universidad Autónoma Metropolitana (UAM), Unidad Xochimilco, Mexico City 04960, Mexico
4 Facultad de Estudios Superiores Cuautitlán, Universidad Nacional Autónoma de México (UNAM), Cuautitlan Izcalli 54714, Mexico
5 Division of Biotechnology—Bioterio and Experimental Surgery, Instituto Nacional de Rehabilitación-Luis Guillermo Ibarra (INR-LGII), Mexico City 14389, Mexico
6 Mediterranean Institute for Agriculture, Environment and Development (MED), Institute for Advanced Studies and Research, Universidade de Évora, Pólo da Mitra, Ap. 94, 7006-554 Évora, Portugal
* Correspondence: mvz.freena@gmail.com (A.D.-O.); apereira@uevora.pt (A.M.F.P.)
† In honor of his memory.

Simple Summary: Heat stress is an event that causes health alterations and decreases the productive performance of farm animals. Therefore, it is important to establish methods that can help to evaluate the thermal state of animals noninvasively. According to the physiological response of vasodilation, when the brain detects an increase in body temperature, infrared thermography can be used to detect these peripheral changes and predict heat stress. The present review aims to analyze the neurobiological response associated with heat stress and how thermal imaging in different thermal windows can be used to recognize heat stress in farmed ungulates.

Abstract: Heat stress is a condition that can affect the health, performance, and welfare of farm animals. The perception of thermal stress leads to the activation of the autonomic nervous system to start a series of physiological and behavioral mechanisms to restore thermostability. One of these mechanisms is vasodilation of peripheral blood vessels to increase heat loss through the skin. Due to this aspect, infrared thermography has been suggested as a method to assess the thermal state of animals and predict rectal temperature values noninvasively. However, it is important to consider that predicting rectal temperature is challenging, and its association with IRT is not always a direct linear relationship. The present review aims to analyze the neurobiological response associated with heat stress and how thermal imaging in different thermal windows can be used to recognize heat stress in farmed ungulates.

Keywords: heat stress; farm animals; thermoregulations; thermal imaging

Citation: Ghezzi, M.D.; Napolitano, F.; Casas-Alvarado, A.; Hernández-Ávalos, I.; Domínguez-Oliva, A.; Olmos-Hernández, A.; Pereira, A.M.F. Utilization of Infrared Thermography in Assessing Thermal Responses of Farm Animals under Heat Stress. *Animals* **2024**, *14*, 616. https://doi.org/10.3390/ani14040616

Academic Editor: Michael Lindinger

Received: 18 December 2023
Revised: 11 February 2024
Accepted: 12 February 2024
Published: 14 February 2024

1. Introduction

Maintaining the thermal comfort of animals is one of the top priorities inside food production facilities [1]. High ambient temperatures cause heat stress (HS), triggering thermoregulatory changes that include physiological and behavioral responses, such as covering themselves in mud or using shade to reduce body temperature [2–4]. Continuous exposure to direct solar radiation can lead livestock to develop HS. In this context, although HS and febrile conditions cause an increase in body temperature, fever is caused by pyrogenic substances such as cytokines that act on hypothalamic centers, while HS does not necessarily involve pyrogens. Among the physiological changes that promote heat elimination, sweating and vasodilation of dermal blood vessels increase heat exchange with the environment and decrease body core temperature [5,6]. HS also causes alterations in the health and productive performance of farm animals, which is a threat to their welfare and the demand for animal food products [7]. Nonetheless, although these modifications aim to recover thermostability (a term used to define a state where heat production and loss are under balance), chronic exposure to HS has negative effects on animals' health and productive performance.

These alterations are associated with the activation of stress-mediated pathways, such as the hypothalamic–pituitary–adrenal (HPA) axis and its connection to peripheral thermoreceptors that identify thermal variations [8,9]. The HPA axis facilitates the availability of energy resources and increases body heat production [10]. This effect leads animals to reduce feed intake and, consequently, daily weight gain (DWG) and milk yield [11,12]. These examples show the importance of recognizing and proposing strategies to monitor and maintain the thermal neutrality of farm animals.

One of the suggested tools to evaluate the thermal stability of animals is infrared thermography (IRT). Our previous studies have shown that IRT has benefits over the gold standard of rectal temperature (RT) since it is a noninvasive tool that evaluates peripheral vasomotor changes closely related to body core temperature when the animal is in a thermal neutral state and nondiseased [13]. Several studies have reported that increases in surface temperature can be related to increases in core temperature due to vasodilation of dermal blood vessels [14,15]. For example, zones with rich arteriovenous anastomoses, glabrous skin, or sparse hair are called "thermal windows" because these areas promote and facilitate heat exchange through vasomotor modulation [16], providing an approximation of body core temperature [17]. Particularly, ocular surface temperature is one of the most used regions to perform thermal imaging because it meets the previous characteristics and is also not affected by body size [18,19]. Eye surface temperature has shown positive and strong correlations with RT in healthy cats ($r = 0.93$) [20], and Bland–Altman testing has shown agreement between ocular temperature and RT in horses undergoing exercise and sheep during shearing [21,22]. However, no correlation between both temperatures was found in anesthetized mongrel dogs, probably due to the effect of anesthesia [19]. Therefore, thermal imaging and its interpretation highly depend on the characteristics of the evaluated body region, and currently, there is no consensus on the most practical thermal window to evaluate thermostability in domestic species. For this reason, this review aims to analyze the neurobiological response associated with thermal stress, the application of IRT, and the current thermal windows that have been used in farmed ungulates to recognize HS.

2. Neurobiological Response to Heat Stress

One of the current challenges for farms is maintaining the animals' thermostability without affecting the productive performance of the species. The main reason why HS is associated with decreased productive performance is the physiological consequences on animals. These physiological consequences also differ among species due to the species-specific characteristics [23]. An example is the water buffalo and *Bos taurus/indicus* bovines, both ruminants with notable differences. On one hand, the water buffalo's skin is thicker (6.5 mm) and has a smaller number of hair follicles (between 135 and 145 hair follicles/cm^2), while conventional cattle have an average skin thickness of 4.3 mm and 3000 hair follicles/cm^2 [24]. Furthermore, water buffaloes have a small number of functional sweat glands (394 vs. 2633 number/cm^2) and dark skin, making them susceptible to heat retention [24–26]. Therefore, the susceptibility of animals to HS also depends on the species and even the breed.

To understand the effect of HS on the neurobiological response of animals to maintain their body core temperature within the thermoneutrality zone, it is necessary to define the meaning of thermoneutrality. According to Taylor [27] and Aggarwal and Upadhyay [28], this term refers to the range of environmental temperatures where animals do not require additional energy use to maintain body core temperature and where the physiological cost is minimal while productivity is maintained. Moreover, thermal neutrality can change depending on the temperature conditioning an animal is exposed to [29]. In the case of sheep, thermoneutrality is achieved within 12–32 °C [30,31]. In buffaloes, it is 13–24 °C [32,33], similar to cattle at 25 °C [34], and for most farm animals, it is considered to be between 4 and 25 °C [35]. A concept derived from the understanding of thermoneutrality zone is upper critical temperature. This refers to a point where the maximum body temperature that can be tolerated by an organism is reached (around 30–45 °C in mammals) [2,36]. This level of tolerance can be lower/higher according to the anatomical and physiological traits of animals. When this upper limit is exceeded, thermoregulatory mechanisms become ineffective and animals are not able to dissipate the excess heat, culminating in HS [37,38].

The recognition of these thermal changes (both environmental temperature and rise in body core temperature as a response to HS) is performed through thermoreceptors located in the skin, as mentioned by our research team [39]. These receptors, called transient receptor potential (TRP) channels, transduce thermal information into an electrical signal that is transmitted to the dorsal horn of the spinal cord [40,41]. Each of these TRPs is activated according to the temperature degree. For example, TRPV1, TRPV2, and TRPV3 subunits respond to temperature above 43, 52, and 33 °C, respectively [42,43]. The activation of these receptors initiates the physiological response of animals to HS. The integration of thermal signaling also activates other cerebral regions, such as the lateral parabrachial nucleus of the pons [44], where neurons synapse in other areas, such as the preoptic area, the anterior hypothalamus, and the dorsomedial hypothalamus, to start the thermoregulatory pathways, as shown in Figure 1 [5,45,46].

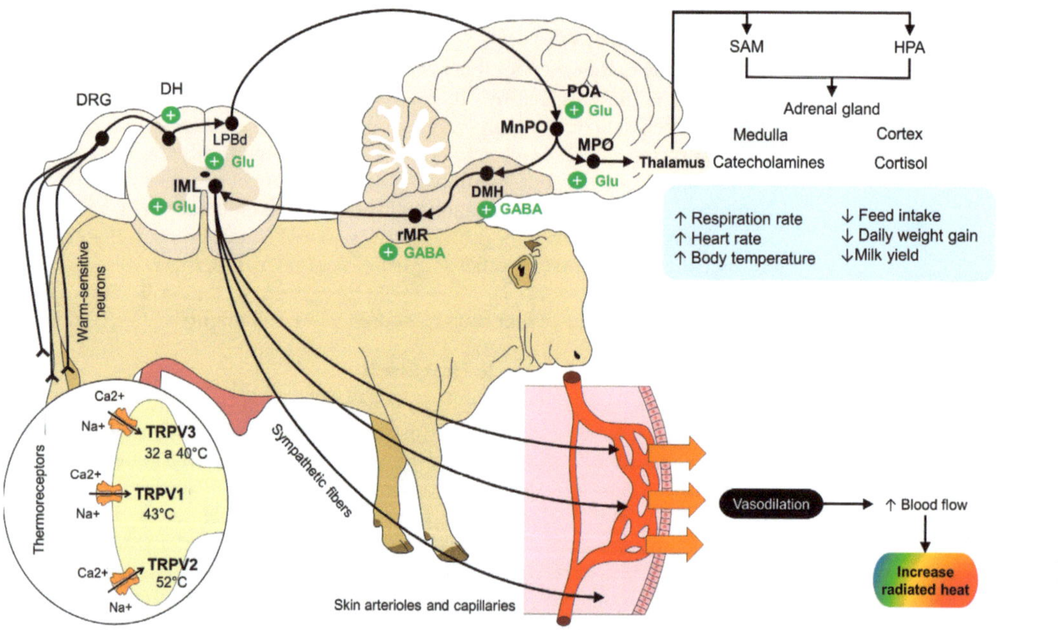

Figure 1. Neurophysiology of mammals under heat stress. The perception of noxious and non-noxious heat by peripheral thermoreceptors (TRP) triggers a series of electrical signals aiming to thermoregulate. Temperatures between 32 and 52 °C activate TRPV3, TRPV1, and TRPV2 ion channels. Warm-sensitive neurons transmit the thermal signaling into the DH, where the neurons synapse with the LPBd nucleus to start a physiological response against heat stress. After reaching the main thermoregulatory center (the POA, MnPO, and MPO), projections to the DMH, the rMR, and the IML activate the sympathetic fibers that innervate the skin arterioles and capillaries. The

compensatory vasodilation response increases blood flow to, consequently, increase the amount of radiated heat through the skin to decrease body core temperature. Simultaneously, the connection of the MPO with the thalamus and sympathetic-mediated axis (SAM and HPA) causes the release of catecholamines and cortisol, which highly influence the physiological and behavioral mechanisms of thermoregulation in animals. DH: dorsal horn of the spinal cord; DMH: dorsomedial hypothalamus; DRG: dorsal root ganglion; HPA: hypothalamic–pituitary–adrenal axis; IML: intermediolateral laminae; LPBd: dorsal part of the parabrachial nucleus; MPOa: medial preoptic area; MnPO: median preoptic area; POA: preoptic area; rMR: rostral medullary raphe; SAM: symptom-adrenomedullary axis; TRP: transient receptor potential.

Vasomotor response is one of the compensatory mechanisms used by most animals because modifying the diameter of the peripheral vessels closest to the skin dissipates heat by promoting a greater thermal exchange with the environment [47]. This was studied in buffaloes during transport by our team in Rodríguez-Gonzáles et al. [48], where IRT was used to evaluate the surface temperature in different body areas. We found an increase of up to 5 °C in the surface temperature of the pelvic limbs after transport. Likewise, the temperature of the parietal-frontal region increased by up to 10 °C after mobilization. These authors explain that the increase in surface temperature might be due to transport being a known stressor for farm animals. Moreover, this study showed that regions with a significant blood supply facilitate the elimination of heat to regulate body temperature. Something similar has been reported in camels, in whom increases in the surface temperature of the axilla, flanks, and hump were directly correlated with the rate of sweating due to high environmental temperature [49]. Other examples have been reported in pigs, in which sows exposed to noncooled pens (air temperature of 25.6 °C) had the highest temperature in the mammary gland (35.9 °C), a value that differed from that in sows kept under cooler pens (air temperature of 23.3 °C) [50]. These studies suggest that, for animals exposed to HS, vasomotor response is an important mechanism for heat elimination and that peripheral thermal response depends on the body region where it is evaluated [23,51]. Figure 2 shows the thermal response obtained by the authors of the present review from healthy farm species.

The vasomotor mechanism started from the perception of environmental temperatures is integrated in hypothalamic brain centers, such as the lateral parabrachial nucleus (LPB) [52]. From the LPB, depending on the stimulus, neurons in the dorsal subnucleus of the LPB are activated during HS [53]. The information on these neurons is projected to the median preoptic nucleus (MnPO) modulates' heat loss by exciting or inhibiting neurons in other structures, such as the rostral magnus raphe, which transmits the signal to the efferent sympathetic preganglionic fibers of the intermediolateral (IML) columns of the spinal cord that innervate and provide vasomotor control to cutaneous blood vessels and brown adipose tissue (BAT) [54,55].

Due to the sympathetic cutaneous innervation, two fundamental thermoregulatory responses can be observed: vasodilation (by cholinergic fibers) and vasoconstriction (by noradrenergic fibers) [5,56]. Therefore, when animals are exposed to high ambient temperatures, vasodilation of the capillaries increases blood flow at the superficial level to increase passive heat dissipation through the skin [55]. This effect is more prominent in those regions with rich subdermal capillaries and with a lower density of hair or glabrous skin, zones that facilitate thermal exchange (also called thermal windows) [23,57]. Direct exposure to solar radiation when animals remain in unshaded areas significantly increases the surface temperature of animals as an effort to decrease heat load. This was reported in Santa Ines sheep and their crosses with Dorper sheep maintained under direct solar radiation for 8 h [58]. Both rectal and ocular temperature increased by up to 39.23 and 39 °C, respectively, reaching an average of 1.54 °C above initial temperatures. This effect is also shown in Figure 3, where the effect of solar radiation on the ocular surface temperature in healthy water buffaloes was assessed by the authors of the present review.

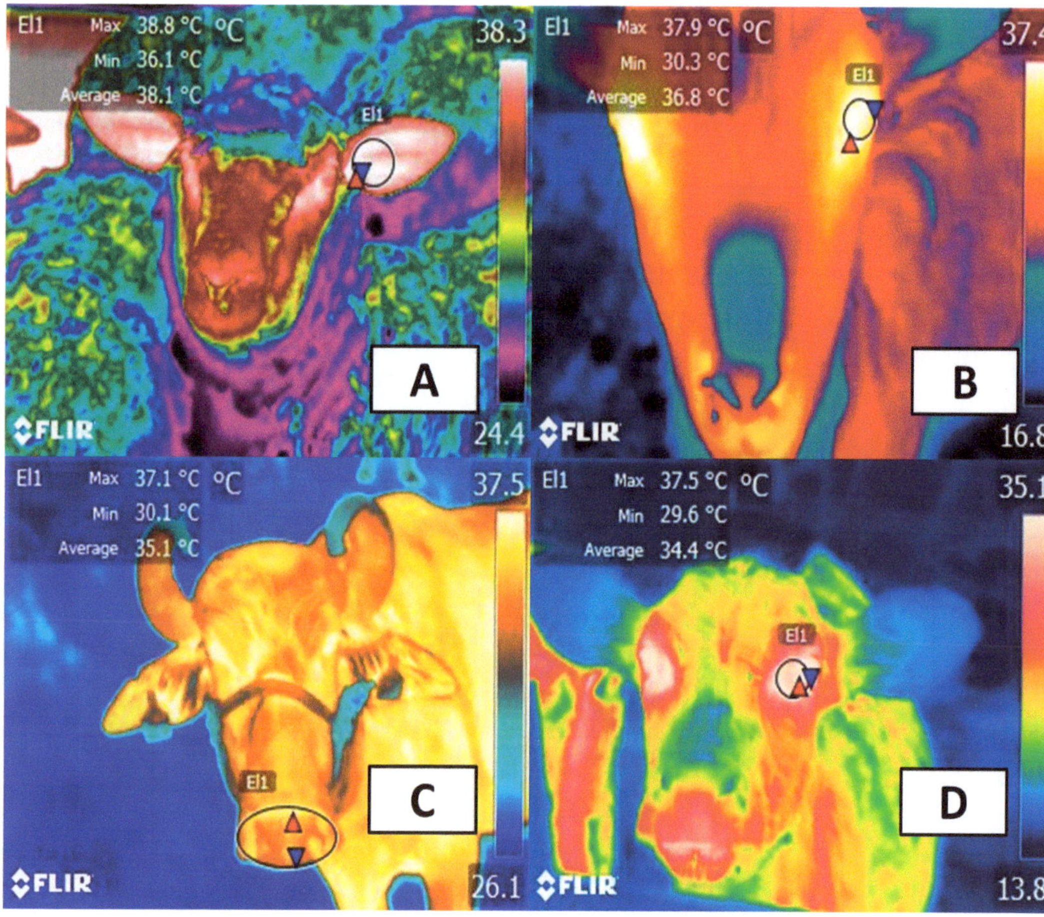

Figure 2. Suggested thermal windows in farm animals. (**A**) Auricular (El1). This window indirectly records the radiated temperature of the tympanic membrane, which is supplied by the auricularis medialis and the marginal arteries. (**B**) Ocular. This region considers the temperature coming from the eyeball and the eyelids (El1). Both are irrigated by the supraorbital vein and artery, the angular oculi vein, and branches of the ophthalmic artery. (**C**) Nasal. This window assesses the changes in the temperature of the exhaled air, also having irrigation by the superficial capillaries of the maxillary artery (El1). (**D**) Lacrimal caruncle. This region, like the ocular thermal window, reflects the circulation of the blood capillaries of the maxillary and infraorbital arteries, together with sympathetic innervation in the medial commissure of the eyelids (El1). Red triangles represent the maximum temperatures of the thermal windows. Blue triangles represent the minimum temperatures of the region of interest. FLIR thermal camera, model E60, lens FOL 18 mm, IR resolution 320 × 240.

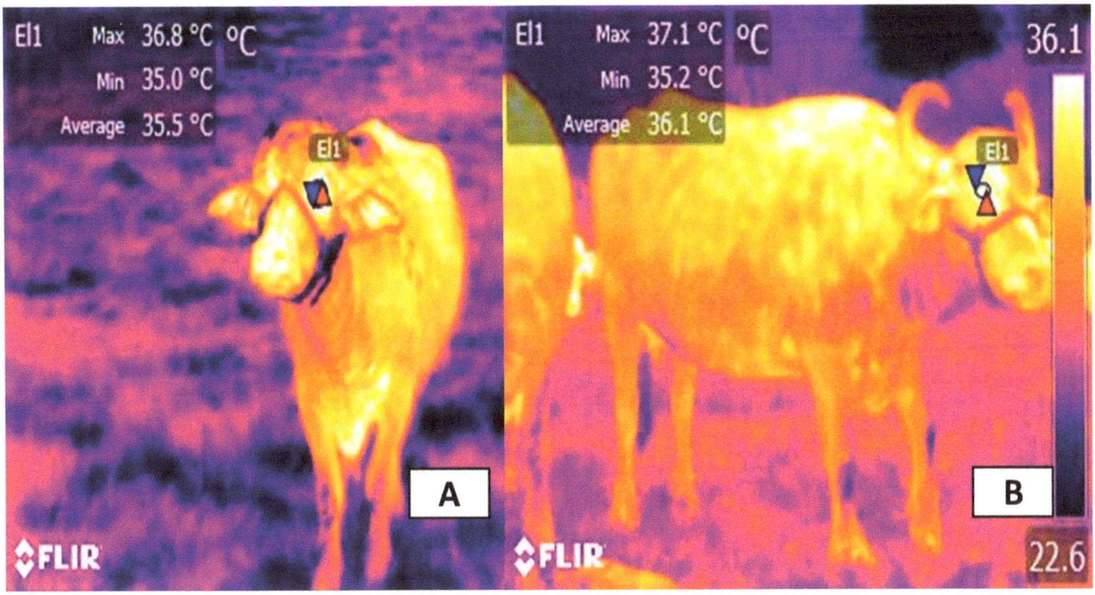

Figure 3. Effect of solar radiation on the surface temperature of the periocular region of water buffaloes. (**A**) Morning eye temperature. The surface temperature of the periocular region (El1) in the morning has a maximum value of 36.8 °C, while the average and minimum temperatures are 35.5 and 35 °C. (**B**) Eye temperature at midday. In comparison with the temperature shown in the morning, the maximum temperature increased by 0.3 °C, the average temperature increased by 0.2 °C, and the minimum temperature decreased by 0.6 °C (El1), which is possibly due to the solar radiation during the time of the day, where animals are most susceptible to HS. Red triangles represent the maximum temperatures of the thermal windows. Blue triangles represent the minimum temperatures of the region of interest. FLIR thermal camera, model E60, lens FOL 18 mm, IR resolution 320 × 240.

3. Physiological Responses to Mitigate Heat Stress

Apart from these vasomotor changes, Gupta and Mondal [59] reported that thermal imbalance due to HS causes a consequent HPA-induced release of cortisol and adrenocorticotropic hormone (ACTH), as well as catecholamines. These substances cause physiological and metabolic alterations. On one hand, catecholamines start the "flight-and-fight" response that is often accompanied with tachypnea, tachycardia, and increases in body core temperature, while cortisol promotes catecholamine release, mobilization of energy, and immunosuppression when released chronically [60,61]. For example, sheep chronically exposed to HS (THI 27.9) had higher hair cortisol levels when deprived of water for 3 h (2.78 pg/mg) than those animals without water restriction (2.63 pg/mg) due to the endocrine response to the stressor [62].

Catecholamines not only act on the cardiorespiratory system, but also have effects on other sympathetic-mediated changes, such as pupillary diameter. Recently, pupillometry has been used to assess the response of animals exposed to different stressors, including thermal stress. In this context, Lopes Neto et al. [63] evaluated pupillary dilation response as an indicator of thermal stress in six Boer crossbred goats. The findings showed that an increase of 26.96 mm2 in the pupillary area was related to heart rate (HR) increasing by 16 beats per minute (bpm) and respiratory rate (RR) increasing by 20 breaths per minute when the air temperature was 33 °C. The authors explain that HS triggers sympathetic changes such as tachycardia and tachypnea to increase blood flow circulation and heat dissipation by evaporative heat loss. A similar study was performed by Marques et al. [64], who evaluated the effect of exposure to hot air at different temperatures (26, 29, and

33 °C) in crossbred goats of the Boer breed ($^3/_4$ Boer and $\frac{1}{4}$ no defined racial pattern). A significant increase in HR and RR of up to 50% was recorded. Similarly, there was a significant decrease of 0.65 kg in daily weight gain (DWG) when exposed to the hottest air temperature (33 °C). Furthermore, Yamin et al. [65] evaluated the association of THI with physiological and biochemical responses in Malabari crossbred and Attappady black goats. They found positive correlations between the increase in THI and HR (r = 0.29) and RR (0.23). Additionally, they observed that in the animals of the Malabari breed, blood pH decreased by 0.08 and CO_2 increased by 2 mmHg, in contrast with the animals of the Attappady black breed. For these authors, this shows that there is a difference between breeds in their capacity for habituation to HS.

Although these changes aim to maintain homeostasis by activating several compensating mechanisms (e.g., vasomotor changes and increases in evaporative heat loss), they also have consequences when animals are chronically exposed to HS. In this context, changes in respiratory pattern and at the vasomotor level can affect the animal's acid–base balance. Due to the vasodilation of blood capillaries, gas exchange becomes difficult in peripheral tissues by decreasing CO_2 elimination despite the increase in RR [66]. This is possibly due to the affinity of oxygen to hemoglobin, which can cause difficulty in the release of this molecule, decreasing blood pH and activating buffer systems to restore homeostasis after HS [67,68].

Farm animals also adopt several thermoregulatory behaviors to increase heat dissipation by radiation, convection, or evaporation to cool themselves [5,48]. Shade seeking, changes in body posture (standing or lying time), reduced feed intake, wallowing, water intake, reduced activity, and urinating frequently are among the most frequent behavioral responses that livestock use to respond to HS [69,70]. For example, shade seeking is one of the first behavioral responses of farm animals because it has been shown to reduce radiant heat load in cattle by about 45% [71]. Species like sheep have been reported to have a preference for shaded areas of approximately 56–43% [72], and the provision of shade in the same species effectively reduces the compensatory tachypnea that sheep use to maintain thermostability [73]. Nonetheless, another study performed by our research team [74] found that water buffaloes resort to wallowing in mud or water to decrease their body core temperature.

In the case of pigs, Bonneau et al. [75] reported an increase in the amount of time the animals were lying down on the side—to dissipate heat by contact with cool surfaces—by up to 5.9% when the temperature increased by 1 °C. The authors also determined that creole pigs (dark skin) spent more time lying down than large white pigs (lighter skin) (4%), probably due to the anatomical differences between breeds, which could help to identify heat-tolerant breeds for certain pig production units where ambient temperatures are high. In small-tail Han sheep raised indoors without climate control, Li et al. [76] observed that ewe and ram lying behavior increased according to the THI increase (up to 45.10 ± 2.42%). Nonetheless, ewes seem to have a higher tolerance to HS than ram based on fewer incidences of standing (23.94 ± 1.98% vs. 38.42 ± 2.25%) and lying (11.56 ± 3.28% vs. 44.38 ± 10.60%) during extremely severe HS. Likewise, goat and sheep resort to other behaviors such as increasing the frequency of water intake or shade seeking to mitigate the effects of HS [77]. Consequently, HS triggers a cascade of physiological and behavioral responses to decrease body core temperature and restore thermostability, and the degree of the said response is highly influenced by the species.

The aforementioned physiological responses and the activation of the HPA axis are addressed in the case of heat stress. Nonetheless, this stress response is not unique to thermal stress. It can be triggered by other stressors inherent in livestock farming, such as transport, social isolation, or having diseases [8,9,48]. This could be a potential limitation of using IRT to detect heat stress, since IRT may capture not only thermal stress but the generalized stress response. Thus, the combination of physiological and behavioral measurements in combination with IRT is suggested as the best alternative for thermoregulatory imbalance detection.

4. Productive and Physiological Consequences of Heat Stress

Thermoregulatory changes to mitigate HS aim to maintain body core temperature within acceptable ranges according to the species [78]. Since HS can be considered a stressor that challenges the homeostasis of animals, it has several consequences that might alter the health and fitness of animals.

Regarding the effects on the productive performance of animals, it has been reported that HS decreases DWG. For example, goats exposed to HS have lower plasma glucose and cholesterol levels [79]. This might be due to a reduction in digestibility, utilization efficiency, and feed consumption and the role of catabolic hormones that are present during HS [80,81]. Moreover, decreases in glucose levels might be related to a decrease in nutrient availability when animals reduce their feeding intake, while cholesterol levels can decrease as a response to low acetate concentration or increased body water [79]. Reduced absorption from nutrient digestion also causes a lower growth rate. According to a comprehensive review by Lima et al. [82], although goats are considered a resilient species to HS when chronically exposed, their energy metabolism is affected. The activation of the HPA axis facilitates the availability of energy resources due to the secretion of cortisol, which allows the increase in body temperature. Therefore, this increase in temperature limits feed consumption and growth rate [83]. This coincides with what was observed by Pragna et al. [84] in 36 goats from three native breeds, where the level of growth performance during HS in the summer was evaluated. The results showed a reduction of 10%, 8%, and 6% in growth for the Osmanabadi, Malabari, and Salem Black breeds, respectively, suggesting that thermal stress has a direct impact on the productive level of the animal [85].

In addition to DWG alterations, decreases in milk yield have also been reported [86–89]. This is related to the reduction in feed consumption, which can negatively affect energy balance, leading to a decrease in milk synthesis and ejection. According to Berman [90], maintaining dairy cattle under ambient temperatures above 35 °C activates the stress-related systems and reduces milk yield. Interestingly, the reduction in feed consumption increases the energy maintenance requirements of dairy cattle by 30% [91].

Likewise, a comparative study conducted by Gaafar et al. [92] in Holstein dairy cattle found that an increase in THI reduced total milk production by 39% in the summer, in contrast with a reduction of 31% in the spring and 29% in the winter. Joksimovic-Todorovic et al. [93] also found that the average total milk yield was 3 L higher in the spring compared with the production during the summer due to a lower food intake and alterations in the absorption of carbohydrates, lipids, and proteins [94]. Therefore, HS has a direct Impact on both the physiological and productive functions of animals, causing acid–base imbalances or a decrease in DWG, feed intake, or milk yield. Therefore, these consequences must be considered in management strategies to reduce the effects of HS on livestock.

5. Infrared Thermography as a Tool to Assess Thermostability in Farm Animals

IRT has been suggested as a tool to evaluate the thermal stability of animals under HS due to peripheral vasomotor changes [95]. Vasodilation increases blood flow to dissipate heat. This also increases the amount of radiated heat from the skin. Due to this, IRT can detect these vasomotor changes [96,97]. A study carried out by Marques et al. [14] evaluated the potential of IRT as a tool to assess thermal changes in six male Boer crossbred goats exposed to three different temperatures of 26, 30, and 34 °C. By assessing the surface temperature of the ocular region, head, shoulder, and hindlimbs, the authors observed that the surface temperature of all regions significantly increased by 2 °C after exposure to the highest temperature. Moreover, it was also observed that the ocular temperature was positively correlated with the rectal temperature of the animals ($r = 0.95$). Similarly, in 24 Blanca Serrana goats where surface eye temperature was evaluated, increases in IRT response (+1.1 °C) and in RR (+20 breaths per minute) were observed after exposure to hot air, which were positively correlated with each other ($r = 0.56$) [15]. These results coincide with what was found in domestic cattle by Peng et al. [98], who assessed the relation between rectal temperature and surface temperature under different THIs in 488 Holstein dairy cows. The authors observed that an increase in THI significantly increased the surface temperature of the eye, ear, cheek, forehead, flank, rump, fore udder, and rear udder. In particular, the forehead was considered the most suitable region to detect changes in body core temperature with a THI of 66.8–71.4. Furthermore, they found a significant correlation between rectal temperature and surface temperature (r = 0.55).

These studies show the association between surface thermal response and rectal temperature due to hyperthermia caused by stress.

Thermal Windows Used to Assess Thermostability of Animals

Scientific evidence indicates that IRT is a useful tool to establish the thermal status of animals and its relation to other physiological parameters associated with HS [13]. However, it is essential to establish which thermal window could be more suitable to show sympathetic activity after HS [99,100]. One of the most used thermal windows to find associations with rectal temperature is the ocular region [14,15,98]. This anatomical region, in particular, the lacrimal caruncle blood supply from the infraorbital artery, is a region highly sensible to vasomotor changes [9]. Moreover, this artery is related to sympathetic adrenergic fibers, so an increase in ocular temperature might be a predictor of a stress response.

Mincu et al. [101] found that surface temperature also changes when exposed to other psychological stressors. In dairy Holstein-Friesian cows, an increase of 2 °C in ocular surface temperature was reported after social isolation. Similarly, a study performed in water buffaloes showed that increases in eye temperature might be related to transport stress. This coincides with what was reported by Stewart et al. [102], who evaluated six Holstein cattle under six different random treatments (control, ACTH, corticotropin-releasing factor at 20 or 40 µg, epinephrine, and social isolation). The authors found that cortisol increases were accompanied by a significant increase of 0.3 °C in eye temperature during catheterization and with the administration of epinephrine (0.5 °C). In contrast, the administration of ACTH and CRH did not influence thermal imaging despite the increased level of cortisol. Due to these results, some other authors have concluded that ocular temperature can indirectly assess the response of animals subjected to stress and even pain [103–105].

Another thermal window that has been suggested is the nasal region because the heat eliminated in this region is related to RR. In this context, Lowe et al. [106] evaluated the thermal fluctuations of the nasal region in 5 Hereford calves of 27 days. It was observed that the nasal thermal window had a high correlation with RR evaluated through flank movements (r = 0.93). This coincides with what was reported in a study performed on 22 cows (Friesian and Friesian x Jersey breed) exposed to a startle. In these animals, RR assessed with the technique of the study above recorded an average difference of -0.01 ± 0.87. Moreover, the thermal response of the nasal region reliably predicted the heart rate variability due to changes in respiratory rate [107]. These results suggest that the nasal region is useful for evaluating RR, and in animals exposed to HS, RR tends to increase (+4.3 breaths per minute) as a mechanism to dissipate up to 30% of heat production [108,109].

Ear surface temperature has also been assessed to monitor body temperature in livestock. In cows and calves, rectal and vaginal temperatures were compared with the surface temperature of the back of the ear, eye, shoulder, and vulva [110]. The results showed that ocular and ear maximum temperature were suitable regions to assess body temperature. Other thermal windows, such as the back and flank regions, have also been suggested to evaluate animals' thermoregulation. An example of this was reported by Sevegnani et al. [111], who evaluated the thermoregulatory response of 15 daily Murrah buffaloes during the premilking and postmilking periods. The surface temperatures of the ear, neck, and back increased between 5 and 12 °C during postmilking. Nonetheless, the authors discussed that the thermal response of regions such as the flanks and neck is highly influenced by environmental conditions such as THI and air temperature. Therefore, the response on these windows might not be directly related to the stressor per se. An alternative is the assessment of pelvic or thoracic limbs, where vasodilation has been reported in animals undergoing HS [23]. Likewise, another possible thermal window is the frontal-parietal region, which is possibly associated with the circulation present in the horns and might serve as a region to dissipate heat [112]. Theusme et al. [113] reported that the anatomical region with the highest correlation with rectal temperature was the forehead (r = 0.58) in Holstein heifers during the summer.

Consequently, the ocular and nasal thermal windows are suggested to be the most acceptable regions to recognize and evaluate thermal stress in animals. Nonetheless, it would be suitable to consider other regions such as the horns or limbs that are also related to peripheral thermoregulatory mechanisms and might give an appreciation to HS.

6. Perspectives

The neurobiological response associated with thermoregulation indicates the importance of free nerve endings present in the skin. In this context, a possible area of further research could aim to compare if, between species, there are differences regarding TRP activation threshold or the presence/absence of certain thermoreceptors. Likewise, another area of research is to apply IRT in farms where the use of natural or artificial shadows is being implemented. This might reduce the influence that direct solar radiation has on IRT readings, as shown in agroforest systems [114]. Comparisons between farms or between animals from the same unit could help to establish the benefits of these additions to promote thermostability in animals.

Finally, one aspect that needs emphasis is that IRT is suggested as a complementary tool to other current technologies and physiological/behavioral evaluations due to its limitations. For example, Unruh et al. [115] did not find an association between increases in surface temperature (+2 °C) and panting score in crossbred beef heifers subjected to thermal stress. This lack of association between rectal temperature and other physiological parameters suggests that IRT might have some limitations that need to be considered when applying thermal imaging to assess the thermal state of animals.

The possible limitations of IRT are related to the factors that can affect the evaluation of the thermal response. For example, ocular surface temperature has shown to be highly affected by environmental temperature [116–118]. Moreover, apart from external factors, individual characteristics such as the presence/absence of hair or coat length/color/type also influence the amount of heat loss/gain [17]. In this regard, the length and hair color can cause a difference in temperature of up to 2 °C [119–121]. Another factor is wind speed. In the case of dairy cattle, it is reported that a wind speed of 7 km/h can reduce ocular surface temperature by 0.43 $\pm$ 0.13 °C, while a speed of 12 km/h can alter the temperature by 0.78 $\pm$ 0.33 °C [122]. Although this tool has been shown to have a direct relationship with physiological parameters such as rectal temperature, HR, and RR, the selection of the thermal window needs to be validated and the different factors that can alter IRT must be considered to precisely assess HS [123].

7. Conclusions

Heat stress in farm animals is a condition where high ambient temperatures exceed the thermostability wherein they can maintain a balance between produced and dissipated heat. When animals are exposed to HS, peripheral vasomotor changes increase heat loss, together with other physiological responses, such as tachypnea and tachycardia. Due to this vasomotor mechanism, IRT can be used as a method to identify changes in body surface temperature due to vasodilation. Currently, ocular surface temperature is considered the region that is highly correlated with body core temperature. However, adopting other thermal windows requires additional research to determine the differences between species and under different temperatures to establish a window that accurately evaluates the thermal state of animals.

Author Contributions: All authors contributed to the conceptualization, writing, and reading. All authors have read and agreed to the published version of the manuscript.

Funding: This research received no external funding.

Acknowledgments: The authors want to express their gratitude to Daniel Mota-Rojas, an expert in animal behavior and welfare, for his role in conceiving and overseeing the project that gave rise to this manuscript. We are grateful for his consistent support and guidance.

Conflicts of Interest: The authors declare no conflicts of interest.

References

1. Chala, M.E.; Cherukuri, V.R.; Lemma, F.; Rama, P.J.; Suhas, S.J. Global Warming: Genesis, Facts and Impacts on Livestock Farming and Mitigation Strategies. *Int. J. Agric. Innov. Res.* **2015**, *3*, 1494–1503.
2. Angilletta, M.J.; Youngblood, J.P.; Neel, L.K.; VandenBrooks, J.M. The Neuroscience of Adaptive Thermoregulation. *Neurosci. Lett.* **2019**, *692*, 127–136. [CrossRef]
3. Unger, S.D.; Rollins, M.A.; Thompson, C.M. Hot- or Cold-Blooded? A Laboratory Activity That Uses Accessible Technology to Investigate Thermoregulation in Animals. *Am. Biol. Teach.* **2020**, *82*, 227–233. [CrossRef]
4. Coria-Avila, G.A.; Herrera-Covarrubias, D.; García, L.I.; Toledo, R.; Hernández, M.E.; Paredes-Ramos, P.; Corona-Morales, A.A.; Manzo, J. Neurobiology of Maternal Behavior in Nonhuman Mammals: Acceptance, Recognition, Motivation, and Rejection. *Animals* **2022**, *12*, 3589. [CrossRef]
5. Mota-Rojas, D.; Titto, C.G.; Orihuela, A.; Martínez-Burnes, J.; Gómez-Prado, J.; Torres-Bernal, F.; Flores-Padilla, K.; Carvajal-de la Fuente, V.; Wang, D.; la Fuente, V.C.; et al. Physiological and Behavioral Mechanisms of Thermoregulation in Mammals. *Animals* **2021**, *11*, 1733. [CrossRef]
6. McCafferty, D.J.; Pandraud, G.; Gilles, J.; Fabra-Puchol, M.; Henry, P.-Y. Animal Thermoregulation: A Review of Insulation, Physiology and Behaviour Relevant to Temperature Control in Buildings. *Bioinspir. Biomim.* **2017**, *13*, 011001. [CrossRef]
7. Montgomery, H.; Haughey, S.A.; Elliott, C.T. Recent Food Safety and Fraud Issues within the Dairy Supply Chain (2015–2019). *Glob. Food Sec.* **2020**, *26*, 100447. [CrossRef]
8. Das, K.S.; Singh, J.K.; Singh, G.; Upadhyay, R.C.; Malik, R.; Oberoi, P.S. Heat Stress Alleviation in Lactating Buffaloes: Effect on Physiological Response, Metabolic Hormone, Milk Production and Composition. *Indian J. Anim. Sci.* **2014**, *84*, 275–280. [CrossRef]
9. Idris, M.; Uddin, J.; Sullivan, M.; McNeill, D.M.; Phillips, C.J.C. Non-Invasive Physiological Indicators of Heat Stress in Cattle. *Animals* **2021**, *11*, 71. [CrossRef]
10. Ko, J.C.; Krimins, R.A. Thermoregulation. In *Zoo Animal and Wildlife Immobilization and Anesthesia*; Wiley: Hoboken, NJ, USA, 2014; pp. 65–68.
11. Gourdine, J.L.; Rauw, W.M.; Gilbert, H.; Poullet, N. The Genetics of Thermoregulation in Pigs: A Review. *Front. Vet. Sci.* **2021**, *8*, 770480. [CrossRef]
12. Cymbaluk, N.F.; Christison, G.I. Environmental Effects on Thermoregulation and Nutrition of Horses. *Vet. Clin. N. Am. Equine Pract.* **1990**, *6*, 355–372. [CrossRef] [PubMed]
13. Mota-Rojas, D.; Martínez-Burnes, J.; Casas-Alvarado, A.; Gómez-Prado, J.; Hernández-Ávalos, I.; Domínguez-Oliva, A.; Lezama-García, K.; Jacome-Romero, J.; Rodríguez-González, D.; Pereira, A.M.F. Clinical Usefulness of Infrared Thermography to Detect Sick Animals: Frequent and Current Cases. *CABI Rev.* **2022**, *2022*, 202217040. [CrossRef]
14. Marques, J.I.; Leite, P.G.; Lopes Neto, J.P.; Furtado, D.A.; de M. Lopes, F.F. Estimation of rectal temperature of goats based on surface temperature. *Eng. Agríc.* **2021**, *41*, 591–598. [CrossRef]
15. Bartolomé, E.; Azcona, F.; Cañete-Aranda, M.; Perdomo-González, D.I.; Ribes-Pons, J.; Terán, E.M. Testing Eye Temperature Assessed with Infrared Thermography to Evaluate Stress in Meat Goats Raised in a Semi-Intensive Farming System: A Pilot Study. *Arch. Anim. Breed.* **2019**, *62*, 199–204. [CrossRef]
16. Weissenböck, N.M.; Weiss, C.M.; Schwammer, H.M.; Kratochvil, H. Thermal Windows on the Body Surface of African Elephants (Loxodonta Africana) Studied by Infrared Thermography. *J. Therm. Biol.* **2010**, *35*, 182–188. [CrossRef]
17. Mota-Rojas, D.; Titto, C.G.; de Mira Geraldo, A.; Martínez-Burnes, J.; Gómez, J.; Hernández-Ávalos, I.; Casas, A.; Domínguez, A.; José, N.; Bertoni, A.; et al. Efficacy and Function of Feathers, Hair, and Glabrous Skin in the Thermoregulation Strategies of Domestic Animals. *Animals* **2021**, *11*, 3472. [CrossRef]
18. Piccione, G.; Giannetto, C.; Fazio, F.; Giudice, E. Accuracy of Auricular Temperature Determination as Body Temperature Index and Its Daily Rhythmicity in Healthy Dog. *Biol. Rhythm Res.* **2011**, *42*, 437–443. [CrossRef]
19. Kapcak, A.B.; Dogan, E. Short Communication: Correlation of Thermographic Ocular and Auricular Temperatures with Rectal Temperature in Anesthetized Dogs. *Vet. Med. Int.* **2023**, *2023*, 9939580. [CrossRef]
20. Giannetto, C.; Di Pietro, S.; Falcone, A.; Pennisi, M.; Giudice, E.; Piccione, G.; Acri, G. Thermographic Ocular Temperature Correlated with Rectal Temperature in Cats. *J. Therm. Biol.* **2021**, *102*, 103104. [CrossRef]
21. Arfuso, F.; Acri, G.; Piccione, G.; Sansotta, C.; Fazio, F.; Giudice, E.; Giannetto, C. Eye Surface Infrared Thermography Usefulness as a Noninvasive Method of Measuring Stress Response in Sheep during Shearing: Correlations with Serum Cortisol and Rectal Temperature Values. *Physiol. Behav.* **2022**, *250*, 113781. [CrossRef]

22. Aragona, F.; Di Pietro, S.; Arfuso, F.; Fazio, F.; Piccione, G.; Giudice, E.; Giannetto, C. Correlation between Ocular and Rectal Temperature with Intra Ocular Pressure in Horse during Exercise. *Animals* **2022**, *12*, 1850. [CrossRef]
23. Mota-Rojas, D.; Pereira, A.M.F.; Wang, D.; Martínez-Burnes, J.; Ghezzi, M.; Hernández-Avalos, I.; Lendez, P.; Mora-Medina, P.; Casas, A.; Olmos-Hernández, A.; et al. Clinical Applications and Factors Involved in Validating Thermal Windows Used in Infrared Thermography in Cattle and River Buffalo to Assess Health and Productivity. *Animals* **2021**, *11*, 2247. [CrossRef] [PubMed]
24. Hafez, E.S.E.; Badreldin, A.L.; Shafei, M.M. Skin Structure of Egyptian Buffaloes and Cattle with Particular Reference to Sweat Glands. *J. Agric. Sci.* **1955**, *46*, 19–30. [CrossRef]
25. Abdul Raheem, H.M.; Elias, A.M.; Ahmed, N.S. The Correction Factor of Hair Density in the Skin of Buffalo. *Iraqi J. Vet. Sci.* **2006**, *20*, 101–106. [CrossRef]
26. Debbarma, D.; Uppal, V.; Bansal, N.; Gupta, A. Histomorphometrical Study on Regional Variation in Distribution of Sweat Glands in Buffalo Skin. *Dermatol. Res. Pract.* **2018**, *2018*, 5345390. [CrossRef]
27. Taylor, R.E. Adaptation to the Environment. In *Scientific Farm Animal Production*; Macmillan Publishing Company: New York, NY, USA, 1992; pp. 326–332.
28. Aggarwal, A.; Upadhyay, R. Thermoregulation. In *Heat Stress and Animal Productivity*; Springer: Delhi, India, 2013; pp. 1–25.
29. Schaefer, A.L.L.; Ominski, K.; Thompson, S.; Crow, G.; Bench, C.; Colyn, J.; Rodas-Gonzalez, A.; Maharjan, D.; Bollum, R.; Cook, N.J.J.; et al. Energy Utilization in Cattle with Steady State and Non-Steady State Methods: The Importance of Thermal Neutrality. *Heliyon* **2018**, *4*, e00843. [CrossRef]
30. Ramón, M.; Díaz, C.; Pérez-Guzman, M.D.; Carabaño, M.J. Effect of Exposure to Adverse Climatic Conditions on Production in Manchega Dairy Sheep. *J. Dairy Sci.* **2016**, *99*, 5764–5779. [CrossRef]
31. Blaxter, K.L.; Graham, N.M.; Wainman, F.W. Environmental Temperature, Energy Metabolism and Heat Regulation in Sheep. III. The Metabolism and Thermal Exchanges of Sheep with Fleeces. *J. Agric. Sci.* **1959**, *52*, 41–49. [CrossRef]
32. Napolitano, F.; De Rosa, G.; Chay-Canul, A.; Álvarez-Macías, A.; Pereira, A.M.F.; Bragaglio, A.; Mora-Medina, P.; Rodríguez-González, D.; García-Herrera, R.; Hernández-Ávalos, I.; et al. The Challenge of Global Warming in Water Buffalo Farming: Physiological and Behavioral Aspects and Strategies to Face Heat Stress. *Animals* **2023**, *13*, 3103. [CrossRef]
33. Mota-Rojas, D.; Habeeb, A.; Ravikanth, R.P.; Ghezzi, M.D.; Napolitano, F.; Lendez, P.; Cuibus, A.; Ceriani, C.; Sarubbi, J.; Braghieri, A.; et al. Chapter 32. Water Buffalo Thermoregulation: Neurobiological Mechanisms, Microcirculatory Changes, and Practical Applications of Infrared Thermography. In *El Búfalo de Agua en las Américas. Comportamiento y Productividad*, 4th ed.; Napolitano, F., Mota-Rojas, D., Orihuela, A., Braghieri, A., Hufana-Duran, D., Strappini, A., Pereira, A.M., Ghezzi, M., Guerrero-Legarreta, I., Martínez-Burnes, J., Eds.; B.M. Editores: Mexico City, Mexico, 2023; pp. 864–888.
34. Robinson, J.B.; Ames, D.R.; Milliken, G.A. Heat Production of Cattle Acclimated to Cold, Thermoneutrality and Heat When Exposed to Thermoneutrality and Heat Stress. *J. Anim. Sci.* **1986**, *62*, 1434–1440. [CrossRef] [PubMed]
35. Mota-Rojas, D.; Ghezzi, M.D.; Pereira, A.M.F.; Martínez-Burnes, J.; Napolitano, F.; Orihuela, A.; Domínguez, A.; Lezama, K.; Casas, A.; Rodríguez, D.; et al. El búfalo de agua recién nacido y sus mecanismos de compensación térmica. In *El Búfalo de Agua en las Américas. Comportamiento y Productividad*, 4th ed.; Napolitano, F., Mota-Rojas, D., Orihuela, A., Braghieri, A., Hufana-Duran, D., Strappini, A., Pereira, A.M., Ghezzi, M., Guerrero-Legarreta, I., Martínez-Burnes, J., Eds.; B.M. Editores: Mexico City, Mexico, 2023; pp. 552–593.
36. Hine, R. *A Dictionary of Biology*, 8th ed.; Oxford University Press: Oxford, UK, 2019.
37. Kovács, L.; Kézér, F.L.; Póti, P.; Boros, N.; Nagy, K. Short Communication: Upper Critical Temperature-Humidity Index for Dairy Calves Based on Physiological Stress Variables. *J. Dairy Sci.* **2020**, *103*, 2707–2710. [CrossRef]
38. Foroushani, S.; Amon, T. Thermodynamic Assessment of Heat Stress in Dairy Cattle: Lessons from Human Biometeorology. *Int. J. Biometeorol.* **2022**, *66*, 1811–1827. [CrossRef]
39. Lezama-García, K.; Mota-Rojas, D.; Pereira, A.M.F.; Martínez-Burnes, J.; Ghezzi, M.; Domínguez, A.; Gómez, J.; de Mira Geraldo, A.; Lendez, P.; Hernández-Ávalos, I.; et al. Transient Receptor Potential (TRP) and Thermoregulation in Animals: Structural Biology and Neurophysiological Aspects. *Animals* **2022**, *12*, 106. [CrossRef]
40. McKinley, M.J.; Martelli, D.; Pennington, G.L.; Trevaks, D.; McAllen, R.M. Integrating Competing Demands of Osmoregulatory and Thermoregulatory Homeostasis. *Physiology* **2018**, *33*, 170–181. [CrossRef]
41. Morrison, S.F.; Nakamura, K. Central Mechanisms for Thermoregulation. *Annu. Rev. Physiol.* **2019**, *81*, 285–308. [CrossRef] [PubMed]
42. Wetsel, W.C. Sensing Hot and Cold with TRP Channels. *Int. J. Hyperth.* **2011**, *27*, 388–398. [CrossRef] [PubMed]
43. Benham, C.D.; Gunthorpe, M.J.; Davis, J.B. TRPV Channels as Temperature Sensors. *Cell Calcium* **2003**, *33*, 479–487. [CrossRef]
44. Verduzco-Mendoza, A.; Bueno-Nava, A.; Wang, D.; Martínez-Burnes, J.; Olmos-Hernández, A.; Casas, A.; Domínguez, A.; Mota-Rojas, D. Experimental Applications and Factors Involved in Validating Thermal Windows Using Infrared Thermography to Assess the Health and Thermostability of Laboratory Animals. *Animals* **2021**, *11*, 3448. [CrossRef]
45. Sokolova, I. Temperature Regulation. In *Encyclopedia of Ecology*; Elsevier: Amsterdam, The Netherlands, 2019; pp. 633–639.
46. Sanín, L.Y.; Cabrera, A.M.Z.; Morales, A.M.T. Adaptive Responses to Thermal Stress in Mammals. *Rev. Med. Vet.* **2016**, *31*, 121–135.
47. Romanovsky, A.A. Skin Temperature: Its Role in Thermoregulation. *Acta Physiol.* **2014**, *210*, 498–507. [CrossRef] [PubMed]

48. Rodríguez-González, D.; Guerrero-Legarreta, I.; Cruz Monterrosa, R.G.; Napolitano, F.; Gonçalves-Titto, C.; El-Aziz, A.H.A.; Hernández- Ávalos, I.; Casas-Alvarado, A.; Oliva- Domínguez, A.; Mota-Rojas, D. Assessment of Thermal Changes in Water Buffalo Mobilized from the Paddock and Transported by Short Journeys. *Front. Vet. Sci.* **2023**, *12*, 1184577. [CrossRef]
49. Abdoun, K.A.; Samara, E.M.; Okab, A.B.; Al-haidary, A.A. Regional and Circadian Variations of Sweating Rate and Body Surface Temperature in Camels (*Camelus Dromedarius*). *Anim. Sci. J.* **2012**, *83*, 556–561. [CrossRef]
50. Dela Ricci, G.; Silva-Miranda, K.O.; Titto, C.G. Infrared Thermography as a Non-Invasive Method for the Evaluation of Heat Stress in Pigs Kept in Pens Free of Cages in the Maternity. *Comput. Electron. Agric.* **2019**, *157*, 403–409. [CrossRef]
51. Gómez-Prado, J.; Pereira, A.M.F.; Wang, D.; Villanueva-García, D.; Domínguez-Oliva, A.; Mora-Medina, P.; Hernández-Avalos, I.; Martínez-Burnes, J.; Casas-Alvarado, A.; Olmos-Hernández, A.; et al. Thermoregulation Mechanisms and Perspectives for Validating Thermal Windows in Pigs with Hypothermia and Hyperthermia: An Overview. *Front. Vet. Sci.* **2022**, *9*, 1023294. [CrossRef]
52. Yahiro, T.; Kataoka, N.; Nakamura, Y.; Nakamura, K. The Lateral Parabrachial Nucleus, but Not the Thalamus, Mediates Thermosensory Pathways for Behavioural Thermoregulation. *Sci. Rep.* **2017**, *7*, 5031. [CrossRef]
53. Scarpellini, C.D.S.; Cristina-Silva, C.; Biancardi, V.; Gargaglioni, L.H.; Almeida, M.C.; Bícego, K.C. Hypothalamic TRPV4 Channels Participate in the Medial Preoptic Activation of Warmth-Defence Responses in Wistar Male Rats. *Pflügers Arch. Eur. J. Physiol.* **2019**, *471*, 1191–1203. [CrossRef]
54. Abbott, S.B.G.; Saper, C.B. Role of the Median Preoptic Nucleus in the Autonomic Response to Heat-Exposure. *Temperature* **2018**, *5*, 4–6. [CrossRef]
55. McAllen, R.M.; McKinley, M.J. Efferent Thermoregulatory Pathways Regulating Cutaneous Blood Flow and Sweating. In *Handbook of Clinical Neurology*; Elsevier: Amsterdam, The Netherlands, 2018; pp. 305–316.
56. Alba, B.K.; Castellani, J.W.; Charkoudian, N. Cold-induced Cutaneous Vasoconstriction in Humans: Function, Dysfunction and the Distinctly Counterproductive. *Exp. Physiol.* **2019**, *104*, 1202–1214. [CrossRef]
57. Ootsuka, Y.; Tanaka, M. Control of Cutaneous Blood Flow by Central Nervous System. *Temperature* **2015**, *2*, 392–405. [CrossRef] [PubMed]
58. Pulido-Rodríguez, L.F.; Titto, C.G.; de Bruni, G.A.; Froge, G.A.; Fuloni, M.F.; Payan-Carrera, R.; Henrique, F.L.; Geraldo, A.C.A.P.D.M.; Pereira, A.M.F. Effect of Solar Radiation on Thermoregulatory Responses of Santa Inês Sheep and Their Crosses with Wool and Hair Dorper Sheep. *Small Rumin. Res.* **2021**, *202*, 106470. [CrossRef]
59. Gupta, M.; Mondal, T. Heat Stress and Thermoregulatory Responses of Goats: A Review. *Biol. Rhythm Res.* **2021**, *52*, 407–433. [CrossRef]
60. Thau, L.; Gandhi, J.; Sharma, S. *Physiology, Cortisol*; StatPearls Publishing: Boca Raton, FL, USA, 2022.
61. Mohankumar, S.M.J.; Balasubramanian, P.; Dharmaraj, M.; Mohankumar, P.S. Neuroendocrine Regulation of Adaptive Mechanisms in Livestock. In *Environmental Stress and Amelioration in Livestock Production*; Springer: Berlin/Heidelberg, Germany, 2012; Volume 9783642292, pp. 263–298.
62. Ghassemi Nejad, J.; Lohakare, J.D.; Son, J.K.; Kwon, E.G.; West, J.W.; Sung, K.I. Wool Cortisol Is a Better Indicator of Stress than Blood Cortisol in Ewes Exposed to Heat Stress and Water Restriction. *Animal* **2014**, *8*, 128–132. [CrossRef]
63. Lopes Neto, J.P.; Marques, J.I.; Furtado, D.A.; Lopes, F.F.M.; Borges, V.P.; Araújo, T.G.P. Pupillary Stress Index: A New Thermal Comfort Index for Crossbred Goats | Índice de Estresse Pupilar: Um Novo Índice de Conforto Térmico Para Caprinos Mestiços. *Rev. Bras. Eng. Agric. Ambient.* **2018**, *22*, 866–871. [CrossRef]
64. Marques, J.I.; Lopes Neto, J.P.; do Nascimento, J.W.B.; Talieri, I.C.; de Medeiros, G.R.; Furtado, D.A. Pupillary Dilation as a Thermal Stress Indicator in Boer Crossbred Goats Maintained in a Climate Chamber. *Small Rumin. Res.* **2018**, *158*, 26–29. [CrossRef]
65. Yamin, D.; Beena, V.; Ramnath, V.; Zarina, A.; Harikumar, S.; Venkatachalapathy, R.T.; Gleeja, V.L. Impact of Thermal Stress on Physiological, Behavioural and Biochemical Parameters in Native and Crossbred Goats. *Small Rumin. Res.* **2022**, *216*, 106794. [CrossRef]
66. Nagashima, K. Central Mechanisms for Thermoregulation in a Hot Environment. *Ind. Health* **2006**, *44*, 359–367. [CrossRef]
67. Monnig, A.A. Practical Acid-Base in Veterinary Patients. *Vet. Clin. N. Am. Small Anim. Pract.* **2013**, *43*, 1273–1286. [CrossRef] [PubMed]
68. Hopper, K. Respiratory Acid–Base Disorders in the Critical Care Unit. *Vet. Clin. N. Am. Small Anim. Pract.* **2017**, *47*, 351–357. [CrossRef]
69. Ratnakaran, A.P.; Sejian, V.; Jose, V.S.; Vaswani, S.; Bagath, M.; Krishnan, G.; Beena, V.; Devi, P.I.; Varma, G.; Bhatta, R. Behavioral Responses to Livestock Adaptation to Heat Stress Challenges. *Asian J. Anim. Sci.* **2016**, *11*, 1–13. [CrossRef]
70. Herbut, P.; Hoffmann, G.; Angrecka, S.; Godyń, D.; Vieira, F.M.C.; Adamczyk, K.; Kupczyński, R. The Effects of Heat Stress on the Behaviour of Dairy Cows—A Review. *Ann. Anim. Sci.* **2021**, *21*, 385–402. [CrossRef]
71. Kamal, R.; Dutt, T.; Patel, M.; Dey, A.; Bharti, P.K.; Chandran, P.C. Heat Stress and Effect of Shade Materials on Hormonal and Behavior Response of Dairy Cattle: A Review. *Trop. Anim. Health Prod.* **2018**, *50*, 701–706. [CrossRef]
72. Taylor, D.; Brown, W.; Price, I.; Trotter, M.; Lamb, D.; Hinch, G. Gps Tracking of Sheep To Investigate Shelter and Shade Use. In Proceedings of the 10th International Conference Precision Agriculture, Denver, CO, USA, 18–20 July 2010; pp. 247–248.
73. Knight, M.I.; Linden, N.P.; Butler, K.L.; Rice, M.; Ponnampalam, E.N.; Behrendt, R.; Jongman, E.C. The Effect of Shade on Sheep Grazing Pasture during Summer Conditions. *J. Vet. Behav.* **2023**, *64*, 16–24. [CrossRef]

74. Bertoni, A.; Mota-Rojas, D.; Álvarez-Macias, A.; Mora-Medina, P.; Guerrero-Legarreta, I.; Morales-Canela, A.; Gómez-Prado, J.; José-Pérez, N.; Martínez-Burnes, J. Scientific Findings Related to Changes in Vascular Microcirculation Using Infrared Thermography in the River Buffalo. *J. Anim. Behav. Biometeorol.* **2020**, *8*, 288–297. [CrossRef]
75. Bonneau, M.; Poullet, N.; Beramice, D.; Dantec, L.; Canario, L.; Gourdine, J.-L. Behavior Comparison During Chronic Heat Stress in Large White and Creole Pigs Using Image-Analysis. *Front. Anim. Sci.* **2021**, *2*, 784376. [CrossRef]
76. Li, F.K.; Yang, Y.; Jenna, K.; Xia, C.H.; Lv, S.J.; Wei, W.H. Effect of Heat Stress on the Behavioral and Physiological Patterns of Small-Tail Han Sheep Housed Indoors. *Trop. Anim. Health Prod.* **2018**, *50*, 1893–1901. [CrossRef]
77. Berihulay, H.; Abied, A.; He, X.; Jiang, L.; Ma, Y. Adaptation Mechanisms of Small Ruminants to Environmental Heat Stress. *Animals* **2019**, *9*, 75. [CrossRef] [PubMed]
78. Mota-Rojas, D.; Wang, D.; Titto, C.G.; Gómez-Prado, J.; Carvajal-de la Fuente, V.; Ghezzi, M.; Boscato-Funes, L.; Barrios-García, H.; Torres-Bernal, F.; Casas-Alvarado, A.; et al. Pathophysiology of Fever and Application of Infrared Thermography (IRT) in the Detection of Sick Domestic Animals: Recent Advances. *Animals* **2021**, *11*, 2316. [CrossRef]
79. Srangi, S. Adaptability of Goats to Heat Stress: A Review. *Pharma. Innov. J.* **2018**, *7*, 1114–1126.
80. Hirayama, T.; Katoh, K. Effects of Heat Exposure and Restricted Feeding on Behavior, Digestibility and Growth Hormone Secretion in Goats. *Asian-Australas. J. Anim. Sci.* **2004**, *17*, 655–658. [CrossRef]
81. Popoola, M.A.; Bolarinwa, M.O.; Yahaya, M.O.; Adebisi, G.L.; Saka, A.A. Thermal Comfort Effects on Physiological Adaptations and Growth Performance of West African Dwarf Goats Raised in Nigeria. *Eur. Sci. J.* **2014**, *3*, 275–382.
82. Lima, A.R.C.; Silveira, R.M.F.; Castro, M.S.M.; De Vecchi, L.B.; da Fernandes, M.H.M.R.; Resende, K.T. de Relationship between Thermal Environment, Thermoregulatory Responses and Energy Metabolism in Goats: A Comprehensive Review. *J. Therm. Biol.* **2022**, *109*, 103324. [CrossRef] [PubMed]
83. Al-Dawood, A. Towards Heat Stress Management in Small Ruminants—A Review. *Ann. Anim. Sci.* **2017**, *17*, 59–88. [CrossRef]
84. Pragna, P.; Sejian, V.; Bagath, M.; Krishnan, G.; Archana, P.R.; Soren, N.M.; Beena, V.; Bhatta, R. Comparative Assessment of Growth Performance of Three Different Indigenous Goat Breeds Exposed to Summer Heat Stress. *J. Anim. Physiol. Anim. Nutr.* **2018**, *102*, 825–836. [CrossRef]
85. Das, R.; Sailo, L.; Verma, N.; Bharti, P.; Saikia, J.; Imtiwati; Kumar, R. Impact of Heat Stress on Health and Performance of Dairy Animals: A Review. *Vet. World* **2016**, *9*, 260–268. [CrossRef]
86. Bouraoui, R.; Lahmar, M.; Majdoub, A.; Djemali, M.; Belyea, R. The Relationship of Temperature-Humidity Index with Milk Production of Dairy Cows in a Mediterranean Climate. *Anim. Res.* **2002**, *51*, 479–491. [CrossRef]
87. West, J.W. Effects of Heat-Stress on Production in Dairy Cattle. *J. Dairy Sci.* **2003**, *86*, 2131–2144. [CrossRef]
88. Spiers, D.E.; Spain, J.N.; Sampson, J.D.; Rhoads, R.P. Use of Physiological Parameters to Predict Milk Yield and Feed Intake in Heat-Stressed Dairy Cows. *J. Therm. Biol.* **2004**, *29*, 759–764. [CrossRef]
89. Wheelock, J.B.; Rhoads, R.P.; VanBaale, M.J.; Sanders, S.R.; Baumgard, L.H. Effects of Heat Stress on Energetic Metabolism in Lactating Holstein Cows. *J. Dairy Sci.* **2010**, *93*, 644–655. [CrossRef]
90. Berman, A. Estimates of Heat Stress Relief Needs for Holstein Dairy Cows1. *J. Anim. Sci.* **2005**, *83*, 1377–1384. [CrossRef] [PubMed]
91. NRC. *Nutrient Requirements of Small Ruminants, Sheep, Goats, Cervids, and New World Camelids*; National Academies Press: Washington, DC, USA, 2007.
92. Gaafar, H.M.A.; Gendy, M.E.; Bassiouni, M.I.; Shamiah, S.M.; Halawa, A.A.; Hamd, M.A. Effect of Heat Stress on Performance of Dairy Friesian Cow's Milk Production and Composition. *Researcher* **2011**, *3*, 85–93.
93. Joksimovic-Todorovic, M.; Davidovic, V.; Hristov, S.; Stankovic, B. Effect of Heat Stress on Milk Production in Dairy Cows. *Biotechnol. Anim. Husb.* **2011**, *27*, 1017–1023. [CrossRef]
94. Baumgard, L.H.; Rhoads, R.P. Effects of Heat Stress on Postabsorptive Metabolism and Energetics. *Annu. Rev. Anim. Biosci.* **2013**, *1*, 311–337. [CrossRef]
95. Mota-Rojas, D.; Ogi, A.; Villanueva-García, D.; Hernández-Ávalos, I.; Casas-Alvarado, A.; Domínguez-Oliva, A.; Lendez, P.; Ghezzi, M. Thermal Imaging as a Method to Indirectly Assess Peripheral Vascular Integrity and Tissue Viability in Veterinary Medicine: Animal Models and Clinical Applications. *Animals* **2023**, *14*, 142. [CrossRef]
96. Mota-Rojas, D.; Wang, D.D.-H.; Titto, C.G.; Martínez-Burnes, J.; Villanueva-García, D.; Lezama, K.; Domínguez, A.; Hernández-Avalos, I.; Mora-Medina, P.; Verduzco, A.; et al. Neonatal Infrared Thermography Images in the Hypothermic Ruminant Model: Anatomical-Morphological-Physiological Aspects and Mechanisms for Thermoregulation. *Front. Vet. Sci.* **2022**, *9*, 963205. [CrossRef]
97. Villanueva-García, D.; Mota-Rojas, D.; Martínez-Burnes, J.; Olmos-Hernández, A.; Mora-Medina, P.; Salmerón, C.; Gómez, J.; Boscato, L.; Gutiérrez-Pérez, O.; Cruz, V.; et al. Hypothermia in Newly Born Piglets: Mechanisms of Thermoregulation and Pathophysiology of Death. *J. Anim. Behav. Biometeorol.* **2021**, *9*, 1–10. [CrossRef]
98. Peng, D.; Chen, S.; Li, G.; Chen, J.; Wang, J.; Gu, X. Infrared Thermography Measured Body Surface Temperature and Its Relationship with Rectal Temperature in Dairy Cows under Different Temperature-Humidity Indexes. *Int. J. Biometeorol.* **2019**, *63*, 327–336. [CrossRef]
99. Casas-Alvarado, A.; Martínez-Burnes, J.; Mora-Medina, P.; Hernández-Avalos, I.; Domínguez-Oliva, A.; Lezama-García, K.; Gómez-Prado, J.; Mota-Rojas, D. Thermal and Circulatory Changes in Diverse Body Regions in Dogs and Cats Evaluated by Infrared Thermography. *Animals* **2022**, *12*, 789. [CrossRef]

100. Casas-Alvarado, A.; Mota-Rojas, D.; Hernández-Ávalos, I.; Mora-Medina, P.; Olmos-Hernández, A.; Verduzco-Mendoza, A.; Reyes-Sotelo, B.; Martínez-Burnes, J. Advances in Infrared Thermography: Surgical Aspects, Vascular Changes, and Pain Monitoring in Veterinary Medicine. *J. Therm. Biol.* **2020**, *92*, 102664. [CrossRef]
101. Mincu, M.; Nicolae, I.; Gavojdian, D. Infrared Thermography as a Non-Invasive Method for Evaluating Stress in Lactating Dairy Cows during Isolation Challenges. *Front. Vet. Sci.* **2023**, *10*, 1236668. [CrossRef]
102. Stewart, M.; Webster, J.R.; Verkerk, G.A.; Schaefer, A.L.; Colyn, J.J.; Stafford, K.J. Non-Invasive Measurement of Stress in Dairy Cows Using Infrared Thermography. *Physiol. Behav.* **2007**, *92*, 520–525. [CrossRef]
103. Stewart, M.; Webster, J.; Schaeder, A. Infrared Thermography and Heart Rate Variability for Non-Invasive Assessment of Animal Welfare. *ANZCCART Hum. Sci. News* **2008**, *1*, 1–6.
104. Sutherland, M.A.; Worth, G.M.; Dowling, S.K.; Lowe, G.L.; Cave, V.M.; Stewart, M. Evaluation of Infrared Thermography as a Non-Invasive Method of Measuring the Autonomic Nervous Response in Sheep. *PLoS ONE* **2020**, *15*, e0233558. [CrossRef]
105. Stewart, M.; Stafford, K.J.; Dowling, S.K.; Schaefer, A.; Webster, J. Eye Temperature and Heart Rate Variability of Calves Disbudded with or without Local Anaesthetic. *Physiol. Behav.* **2008**, *93*, 789–797. [CrossRef]
106. Lowe, G.; Sutherland, M.; Waas, J.; Schaefer, A.; Cox, N.; Stewart, M. Infrared Thermography—A Non-Invasive Method of Measuring Respiration Rate in Calves. *Animals* **2019**, *9*, 535. [CrossRef]
107. Stewart, M.; Wilson, M.T.; Schaefer, A.L.; Huddart, F.; Sutherland, M.A. The Use of Infrared Thermography and Accelerometers for Remote Monitoring of Dairy Cow Health and Welfare. *J. Dairy Sci.* **2017**, *100*, 3893–3901. [CrossRef]
108. Leroy Hahn, G. Environmental Management for Improved Livedstock Performance, Health and Well-Beging. *Jpn. J. Livest. Manag.* **1995**, *30*, 113–127.
109. Nienaber, J.A.; Hahn, G.L. Livestock Production System Management Responses to Thermal Challenges. *Int. J. Biometeorol.* **2007**, *52*, 149–157. [CrossRef]
110. Hoffmann, G.; Schmidt, M.; Ammon, C.; Rose-Meierhöfer, S.; Burfeind, O.; Heuwieser, W.; Berg, W. Monitoring the Body Temperature of Cows and Calves Using Video Recordings from an Infrared Thermography Camera. *Vet. Res. Commun.* **2013**, *37*, 91–99. [CrossRef]
111. Sevegnani, K.B.; Fernandes, D.P.B.; Silva, S.H.M.-G. da Evaluation of Thermorregulatory Capacity of Dairy Buffaloes Using Infrared Thermography. *Eng. Agrícola* **2016**, *36*, 1–12. [CrossRef]
112. Algra, M.; de Keijzer, L.; Arndt, S.S.; van Eerdenburg, F.J.C.M.; Goerlich, V.C. Evaluation of the Thermal Response of the Horns in Dairy Cattle. *Animals* **2023**, *13*, 500. [CrossRef]
113. Theusme, C.; Avendaño-Reyes, L.; Macías-Cruz, U.; Castañeda-Bustos, V.; García-Cueto, R.; Vicente-Pérez, R.; Mellado, M.; Meza-Herrera, C.; Vargas-Villamil, L. Prediction of Rectal Temperature in Holstein Heifers Using Infrared Thermography, Respiration Frequency, and Climatic Variables. *Int. J. Biometeorol.* **2022**, *66*, 2489–2500. [CrossRef]
114. Karvatte, N.; Miyagi, E.S.; Carvalho de Oliveira, C.; Mastelaro, A.P.; de Aguiar Coelho, F.; Bayma, G.; Bungenstab, D.J.; Alves, F.V. Spatiotemporal Variations on Infrared Temperature as a Thermal Comfort Indicator for Cattle under Agroforestry Systems. *J. Therm. Biol.* **2021**, *97*, 102871. [CrossRef]
115. Unruh, E.M.; Theurer, M.E.; White, B.J.; Larson, R.L.; Drouillard, J.S.; Schrag, N. Evaluation of Infrared Thermography as a Diagnostic Tool to Predict Heat Stress Events in Feedlot Cattle. *Am. J. Vet. Res.* **2017**, *78*, 771–777. [CrossRef]
116. Vieira, R.A.; Dias, E.A.; Stumpf, M.T.; Pereira, G.R.; Barcellos, J.O.J.; Kolling, G.J.; McManus, C. Use of Thermography and Physiological Rate to Assess Heat Tolerance in Cattle Breeds. *Trop. Anim. Health Prod.* **2023**, *55*, 223. [CrossRef]
117. Joy, A.; Taheri, S.; Dunshea, F.R.; Leury, B.J.; DiGiacomo, K.; Osei-Amponsah, R.; Brodie, G.; Chauhan, S.S. Non-Invasive Measure of Heat Stress in Sheep Using Machine Learning Techniques and Infrared Thermography. *Small Rumin. Res.* **2022**, *207*, 106592. [CrossRef]
118. Kim, N.Y.; Moon, S.H.; Kim, S.J.; Kim, E.K.; Oh, M.; Tang, Y.; Jang, S.Y. Summer Season Temperature-Humidity Index Threshold for Infrared Thermography in Hanwoo (*Bos Taurus Coreanae*) Heifers. *Asian-Australas. J. Anim. Sci.* **2020**, *33*, 1691–1698. [CrossRef]
119. Nomura, R.H.C.; de Freitas, I.B.; Guedes, R.L.; Araújo, F.F.; Mafra, A.C.D.N.; Ibañez, J.F.; Dornbusch, P.T. Thermographic Images from Healthy Knees between Dogs with Long and Short Hair. *Ciência Rural.* **2018**, *48*, 1–7. [CrossRef]
120. Maśko, M.; Witkowska-Piłaszewicz, O.; Jasiński, T.; Domino, M. Thermal Features, Ambient Temperature and Hair Coat Lengths: Limitations of Infrared Imaging in Pregnant Primitive Breed Mares within a Year. *Reprod. Domest. Anim.* **2021**, *56*, 1315–1328. [CrossRef]
121. Meisfjord Jørgensen, G.H.; Mejdell, C.M.; Bøe, K.E. Effects of Hair Coat Characteristics on Radiant Surface Temperature in Horses. *J. Therm. Biol.* **2020**, *87*, 102474. [CrossRef]
122. Church, J.S.; Hegadoren, P.R.; Paetkau, M.J.; Miller, C.C.; Regev-Shoshani, G.; Schaefer, A.L.; Schwartzkopf-Genswein, K.S. Influence of Environmental Factors on Infrared Eye Temperature Measurements in Cattle. *Res. Vet. Sci.* **2014**, *96*, 220–226. [CrossRef]
123. Sejian, V.; Shashank, C.G.; Silpa, M.V.; Madhusoodan, A.P.; Devaraj, C.; Koenig, S. Non-Invasive Methods of Quantifying Heat Stress Response in Farm Animals with Special Reference to Dairy Cattle. *Atmosphere* **2022**, *13*, 1642. [CrossRef]

animals

MDPI

Communication

Associations between Surface and Rectal Temperature Profiles of Low-Birth-Weight Piglets

Bryony S. Tucker [1,*], Kiro R. Petrovski [1,2], Jessica R. Craig [3], Rebecca S. Morrison [3], Robert J. Smits [4,†] and Roy N. Kirkwood [1]

1 School of Animal and Veterinary Sciences, The University of Adelaide, Roseworthy, SA 5371, Australia; kiro.petrovski@adelaide.edu.au (K.R.P.); roy.kirkwood@adelaide.edu.au (R.N.K.)
2 Davies Livestock Research Centre, The University of Adelaide, Roseworthy, SA 5371, Australia
3 Rivalea Australia Pty. Ltd., JBS Australia Pork Division, Corowa, NSW 2646, Australia; jcraig@rivalea.com.au (J.R.C.); rmorrison@rivalea.com.au (R.S.M.)
4 Research and Innovation, Australian Pork Limited, Barton, ACT 2600, Australia; robsmits@triadvice.au
* Correspondence: bryony.tucker@sa.gov.au
† Current address: TRI Advice Pty Ltd., Macgregor, ACT 2615, Australia.

Simple Summary: Piglets experience a significant temperature drop soon after birth and the time taken to recover can impact their survival and growth. Rectal temperature is the best method currently used to monitor temperature change, however, it is invasive and requires handling which can be stressful. Infrared camera technology has improved and surface temperature has been suggested to be a good substitute for rectal temperature. A study was undertaken to investigate the utility of infrared measurements of piglets from immediately after birth to 24 h old. Results showed that surface temperature had a similar temperature pattern to rectal temperature but was highly affected by the environment and cannot be recommended as a substitute for rectal temperature measurement.

Abstract: The use of infrared cameras to record surface temperature has shown some promise in older pigs, but neonatal piglets are metabolically less mature and experience rapid temperature changes during their first 24 h. The present experiment aimed to compare rectal temperature to surface temperature at the base of the ear, measured using an infrared camera, for piglets of different birth weights. During farrowing, 48 multiparous sows were monitored, and rectal and surface temperatures were recorded for their lower-birth-weight (≤1.2 kg) piglets within 3 min of birth and at 0.25, 0.50, 0.75, 1, 1.25, 1.50, 2, 3, 4, and 24 h. Piglet birth weights were assigned to one of three categories (BWC): BWC1 (≤0.80 kg), BWC2 (0.81 to 1.10 kg), or BWC3 (1.11 to 1.20 kg). Piglet rectal temperatures at 1.25 h after birth were assigned to one of three categories: RC1 (≤32.0 °C), RC2 (32.1 to 35.0 °C), or RC3 (≥35.1 °C). Surface temperatures showed a similar recovery pattern to rectal temperatures in the first 24 h across all piglet birth weights, although large and variable differences seen in the current study militate against surface temperature being an appropriate replacement for neonatal rectal temperature for use in production.

Keywords: piglet; birth weight; temperature; rectal; surface; infrared camera; thermoregulation

Citation: Tucker, B.S.; Petrovski, K.R.; Craig, J.R.; Morrison, R.S.; Smits, R.J.; Kirkwood, R.N. Associations between Surface and Rectal Temperature Profiles of Low-Birth-Weight Piglets. *Animals* **2023**, *13*, 3259. https://doi.org/10.3390/ani13203259

Academic Editor: Daniel Mota-Rojas

Received: 12 September 2023
Revised: 15 October 2023
Accepted: 16 October 2023
Published: 19 October 2023

1. Introduction

Piglets are born wet and into an environment that is significantly cooler than conditions in utero (38–40 °C) [1,2], leading piglets to experience a dramatic reduction in body temperature during the first 15–90 min after birth [3–5]. The cooler ambient temperature and physical environment, combined with the presence of amniotic fluid, contribute to this decrease in temperature [6,7]. The average temperature recovery time from birth exceeds 4 h, during which it is critical for piglets to actively seek teats and consume sufficient colostrum to obtain nutrients for thermoregulation and acquire passive immunity [8,9]. In addition, failure to obtain adequate colostrum may result in hypoglycaemia and chilling, which can predispose piglets to being overlain by the sow [10].

Smaller piglets have a greater surface area to volume ratio than heavier piglets, and thus have a proportionately greater potential for heat loss and reduction in the ability to thermoregulate [4]. Consequently, heavier piglets experience a less severe postnatal drop in temperature than smaller piglets [11]. As piglets are born with little adipose tissue and no brown fat, they have a limited energy supply to maintain efficient thermoregulation [12]. As such, they rely on Muscle Non-shivering Thermogenesis to produce heat, which rapidly depletes their energy supplies [13]. Farrowing house ambient temperatures are invariably below the lower critical temperature of piglets, and so piglets must rely on the use of supplemental heat sources or interventions to assist in regulating their temperature. Previous studies have tested a variety of interventions with varying effects on temperature change and subsequent survival, such as energy supplements, drying, and warm intraperitoneal saline injections [2,14–16].

Rectal temperature is the gold standard for measuring the internal temperature of pigs as it is relatively unaffected by external factors. However, repetitive rectal thermometer readings to monitor temperature changes and to identify critical periods for the piglets is an invasive procedure which can become stressful for the piglet [17]. This includes removal from the crate environment and restraint, resulting in exposure to different environmental temperatures, handling stress, and potential loss of suckling time. An alternative method published recently in the literature is the use of infrared cameras for the monitoring of piglet thermal status [18–20]. Automatic systems generating a whole image analysis would be useful to monitor piglet thermal changes, but handheld machines are more practical and easily employed in practice. There is evidence that rectal and surface temperatures show similar diurnal trends, and that surface temperatures interact with piglet age and growth rate [19]. It has been suggested that the ocular, auricular, and nasal regions are the most commonly monitored to determine changes in heat [6]; however, many of these also require manipulation of the piglet to obtain clear images. Interestingly, the use of an aural thermometer has shown a high correlation with rectal temperature, but also requires restraint, thus offering little benefit beyond that of rectal thermometers [20].

A variety of different infrared thermometers and cameras can indicate core temperature, but are affected by factors such as distance from the animal [20]. In that study, the best targets for the infrared camera were the inner thigh and abdomen, but these too required manipulation of the piglet. The aim of the present experiment was to compare rectal temperatures with surface temperatures measured using an infrared thermal camera at the base of the ear of piglets with different body weights. We hypothesised that rectal temperature would not be highly impacted by body weight, but that surface temperature would be more sensitive to this factor.

2. Methods

This experiment was conducted during September and October at a commercial piggery (Rivalea, Corowa, NSW, Australia) and was approved by the Rivalea Animal Care and Ethics Committee (Protocol number 20R027) in accordance with the Australian Code for the Care and Use of Animals for Scientific Purposes (National Health and Medical Research Council, 2013).

2.1. Housing and Management

Multiparous Large White × Landrace sows (N = 48; mean parity 3.4 ± 1.2) were moved into their farrowing accommodations at 110 d (±2 d) of gestation, being housed in individual farrowing crates equipped with sow- and piglet-level nipple drinkers and a heat lamp positioned centrally over one side (creep area). The sows were fed 2.5 kg/day of a standard lactation diet formulated to provide 15 MJ DE/kg, 16.7% protein, and 0.90% SID lysine from entry into the farrowing house until farrowing (116 ± 1.5 d of gestation). Farrowings were observed and measurements were recorded daily from 0600 h to 1900 h.

Once farrowed, sows were fed to appetite via hand feeding twice per day until weaning (21 ± 2 d of lactation). The farrowing rooms were semi-enclosed with natural ventilation and a dripper cooling system was set to automatically activate at 28 °C. Additional portable evaporative coolers were placed in the shed at sow entry and used until 2 weeks after farrowing to aid in temperature control.

2.2. Experimental Design

Piglets weighing ≤ 1.2 kg at birth and born during normal working hours over a 3-day farrowing period were included in this trial. Based on these constraints, 67 piglets (36 female and 31 male) born to 48 sows were included in the experiment. Piglet birth weights were assigned to one of three categories: BWC1 (≤0.80 kg), BWC2 (0.81 kg to 1.10 kg), or BWC3 (1.11 to 1.20 kg). Within 3 min of birth (0 h), piglet surface and rectal temperatures were recorded after they were moved into a semi-enclosed container positioned over a scale and were individually ear tagged, weighed, and their sex recorded. Rectal temperatures were measured using a standard digital thermometer with a lowest possible reading of 32 °C (0.3 °C sensitivity). Surface temperature was measured at the base of the right ear (Figure 1) using a FLIR E8-XT thermal camera (Teledyne FLIR, Wilsonville, OR, USA; <0.06 °C thermal sensitivity), with an emissivity value of 0.98 pre-set in the camera, held 30 cm from the piglet (using a generic ruler), and all images were saved for further confirmation [21]. Surface and rectal temperatures were recorded at 0, 0.25, 0.50, 0.75, 1, 1.25, 1.5, 2, 3, 4, and 24 h. All attempts were made to minimise further manipulation of the piglet when obtaining temperatures, though occasional placement of hand next to pig in the container was required to stabilise the animal prior to taking the photo when the piglet was wet.

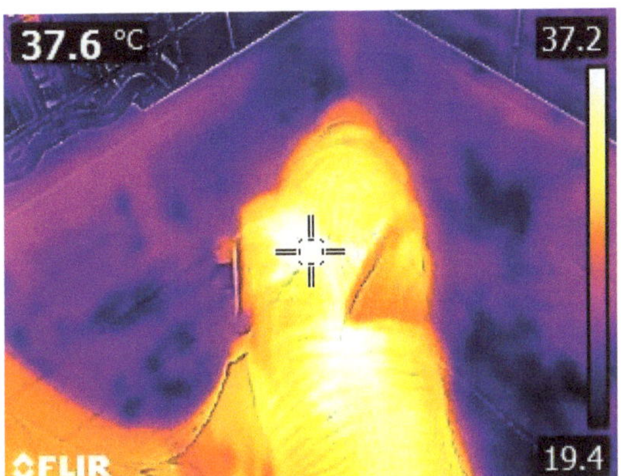

Figure 1. Example of a newborn piglet in the semi-enclosed container while the base-of-ear surface temperature measurement was obtained with minimal handling.

2.3. Statistical Analysis

All statistical analyses were performed using SAS version 9.4 (Statistical Analysis Software, Cary, NC, USA). There were 737 piglet-related data point observations available for analysis. After bootstrapping the data at a root of 24 possibilities per piglet ID, 11,196 piglet observations were available for analysis. For the data analysis:

Piglet birth weights were assigned to one of three categories: BWC1 (≤0.80 kg), BWC2 (0.81 kg to 1.10 kg), or BWC3 (1.11 to 1.20 kg).

Piglet rectal temperatures at 1.25 h after birth were assigned to one of three categories: RC1 (≤32.0 °C), RC2 (32.1 °C to 35.0 °C), or RC3 (>35.0 °C).

The correlation between the rectal and surface temperatures across body weight categories were estimated using PROC CORR, with the output being the Pearson correlation coefficient (r) and the respective 95% confidence intervals. Correlation was deemed very high if $r \geq 0.90$, high if r was between 0.70 and 0.89, moderate if r was between 0.50 and 0.69, low if r was between 0.30 and 0.49, and negligible if $r < 0.30$ [22]. The level of significance for analysis was set at $p < 0.05$.

The effects of the birth weight, time relative to birth, and the interaction of birth weight and time relative to birth on the rectal and surface temperatures of piglets were estimated using a mixed model in PROC MIXED, as presented in Equation (1)

$$temperature = \text{ } birth\ weight\ category_{fr,ls,p,\ s} + time\ related\ to\ birth_{fr,ls,p,\ s} + (birth\ weight\ category \times time\ related\ to\ birth)_{fr,ls,p,\ s} \quad (1)$$

where fr = farrowing room; ls = litter size; p = sow parity; and s = sow. The preliminary model also tested the effects of piglet sex, but it was found to be not significant and was removed from the final model. The outputs of the model were least-square means and their respective standard errors.

3. Results

Relationships between Rectal and Surface Temperatures

Overall, rectal and surface temperatures showed a moderate correlation of 68.7% (67.7–69.7; $p < 0.001$), although the initial surface temperatures recorded were at least 4 °C lower than the rectal temperatures (Figure 2).

Rectal and surface temperatures were higher in accordance with higher body weight across time, but the surface temperature profile appeared more variable than the rectal temperature profiles (Figure 2A,B). Heavier piglets had higher rectal temperatures from 0.25 h to 24 h ($p < 0.001$; Figure 2A).

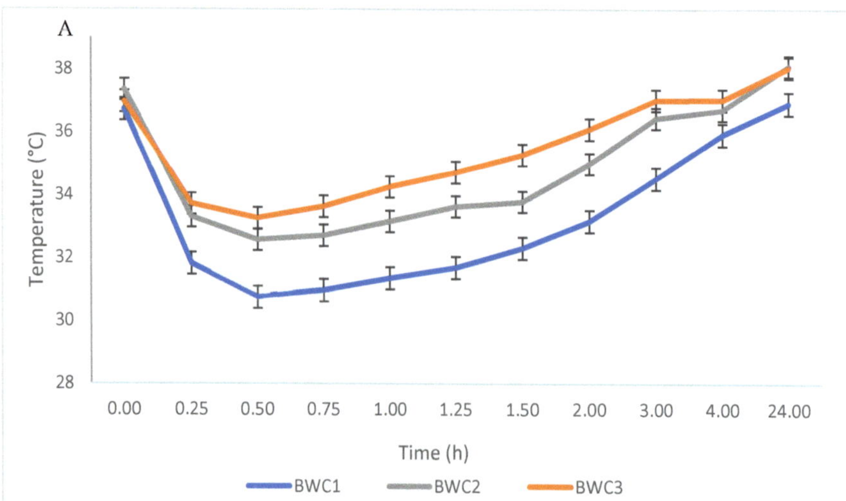

Figure 2. *Cont.*

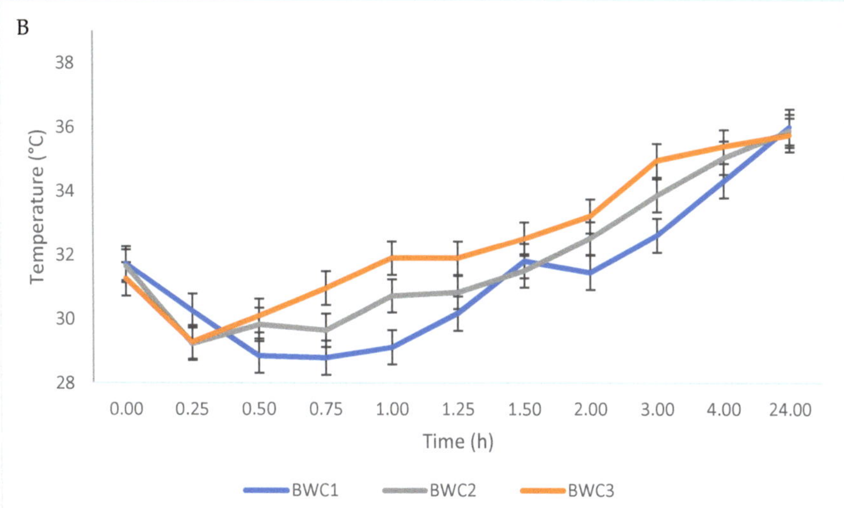

Figure 2. Least-square mean ± standard error for piglet rectal temperatures (**A**) and surface temperatures measured at the base of the right ear (**B**) across body weight category (BWC); adjusted for the effect of the sow, parity, litter size, and farrowing room, in 11,196 piglet observations. BWC1: ≤0.80 kg; BWC2: 0.81 kg–1.10 kg; BWC3: 1.11–1.20 kg.

4. Discussion

From our data, the skin surface temperature, at the base of the ear, and rectal temperature of neonatal piglets in the postpartum period appear, at most, only moderately correlated. This is at odds with previous research (reviewed by [19]) that suggested the base of the ear is one of the best thermal windows, with a high correlation between surface and rectal temperatures. A thermal window is a skin/surface area which is well perfused by blood and, as a result, is a good "window" into the core body temperature [23]. The ear is highly vascularised and is relatively easy to access with minimal interference; as such, changes in the heat exchange can be monitored easily [3,6]. However, previous studies using infrared thermal technology in pigs have largely focused on older pigs that have thicker skin, established thermoregulation, and greater fat deposits [23]. Small piglets have little natural insulation, a higher surface area to volume ratio, and limited energy reserves to assist in thermoregulation, compared to their heavier-born counterparts [1,12]. Furthermore, piglets are born wet, moving from a regulated temperature in utero of approximately 39 °C to a significantly cooler external environment [24]. Therefore, previously validated methods of measuring surface temperature using infrared thermal technology for older pigs are not as applicable to neonatal piglets.

Birth weight significantly influenced the rectal and surface temperature profiles of piglets during their first 24 h of life. None of the smallest piglets (BWC1) had rectal temperatures above 35 °C at 1.25 h. However, if they did have a relatively higher temperature, <35 °C, their rate of survival increased. A relationship between temperature (and hypothermia) and birth weight has been established for piglet preweaning survival [1]. That the observed effects might have been different if the temperatures were measured at a different time cannot be discounted, but we rationalised that by 1.25 h, any prenatal influence of the sow should have worn off as the piglet would have dried and sucked. Further, by 1.25 h, thermoregulation should be improved, and piglets would be on the upwards curve of temperature recovery, as previously documented [4,5].

Similar to previous research, piglet temperatures had decreased by 0.25 h post-birth, although the start of recovery was evident from 0.5 h rather than 1 h as previously documented [4]. This was followed by a gradual increase until 24 h. The difference in recovery patterns between our study and that of Caldara et al. [4] presumably reflects differences in the farrowing environments, including, but not limited to, ambient temperatures, heat management, and piglet drying [4]. In our study, the difference between the rectal and surface temperature recovery curves highlights the influence that external factors have on surface temperature. It is known that smaller piglets have a greater surface area to volume ratio, and therefore presumably lose heat more rapidly and would be affected more quickly by environmental changes.

These findings, along with the physiology of the piglet, highlight the importance of good management from early life to ensure optimal survival. Providing an optimal environment from birth is critical to support this; however, arguably the most important strategy to improve survival is to ensure fast and adequate access to colostrum, particularly for low-birth-weight piglets. Colostrum intake is positively related to rectal temperature and provides the critical energy required to thermoregulate and build an immune response, thus setting the piglets up for survival [25,26]. Being able to monitor this temperature change using technology would be a valuable tool for piglet management and welfare.

Other studies which have focused on the use of infrared thermal technology in piglets have concluded that this technology has the potential to be an effective, less invasive tool for temperature measurement, albeit not identical to rectal temperature [18]. There are many factors affecting the surface temperature using infrared technology, such as the angle of the camera, background lighting, ambient temperature, humidity, and stress to the piglet [27]. One major challenge of recording surface temperature is the manipulation of the piglet required to obtain an accurate reading. The most effective method is to pick up the piglet and manoeuvre it so that exact and repeatable measures can be recorded. However, in doing so, the piglet is exposed to different environmental conditions within the crate which may trigger a stress response, which has been shown to have a substantial impact on thermal window temperature readings, in addition to using energy from their limited supply [27]. These challenges, and our results, suggest that the use of a handheld infrared camera is impractical for direct application by producers. However, with increasing interest in machine learning and automation in pig production, there is potential for further research and application in this area [28].

5. Conclusions

Surface temperature measured using infrared thermal camera technology at the base of the ear produces large variations compared to rectal temperature, although the recovery pattern was similar. Therefore, these devices cannot be recommended for use in commercial settings at this stage. Birth weight and temperature are directly related and should be considered together when monitoring piglets.

Author Contributions: Conceptualisation, B.S.T. and R.N.K.; methodology, B.S.T., R.N.K., R.S.M. and J.R.C.; validation, B.S.T., J.R.C., R.S.M., R.J.S. and R.N.K.; formal analysis, B.S.T., J.R.C. and K.R.P.; investigation, B.S.T.; resources, R.S.M. and R.J.S.; writing—original draft preparation, B.S.T.; writing—review and editing, all authors; visualisation, B.S.T.; supervision, J.R.C., R.S.M., R.J.S. and R.N.K. All authors have read and agreed to the published version of the manuscript.

Funding: This work was supported in part by Australian Pork Limited (APL Project 2018/0068).

Institutional Review Board Statement: This experiment was conducted during September and October at a commercial piggery (Rivalea Corowa, NSW, Australia) and was approved by the Rivalea Animal Care and Ethics Committee (Protocol number 20R027) in accordance with the Australian Code for the Care and Use of Animals for Scientific Purposes (National Health and Medical Research Council, 2013).

Data Availability Statement: The data presented in this study are available on reasonable request from the corresponding author.

Acknowledgments: The first author acknowledges partial sponsorship by the Australian Pork Limited Study award (2018/0086) and the University of Adelaide. The authors thank the Rivalea animal technical team and farming staff for their assistance.

Conflicts of Interest: The authors declare no conflict of interest.

References

1. Kammersgaard, T.S.; Pedersen, L.J.; Jørgensen, E. Hypothermia in neonatal piglets: Interactions and causes of individual differences. *J. Anim. Sci.* **2011**, *89*, 2073–2085. [CrossRef] [PubMed]
2. Andersen, H.M.; Pedersen, L.J. Effect of radiant heat at the birth site in farrowing crates on hypothermia and behaviour in neonatal piglets. *Animal* **2016**, *10*, 128–134. [CrossRef] [PubMed]
3. Villanueva-Garcíaa, D.; Mota-Rojasb, D.; Martínez-Burnesc, J.; Mora-Medinae, P.; Salmerónf, C.; Gómezb, J.; Boscatob, L.; Gutiérrez-Pérezf, O.; Cruzf, V.; Reyesb, B.; et al. Hypothermia in newly born piglets: Mechanisms of thermoregulation and pathophysiology of death. *J. Anim. Behav. Biometerology* **2021**, *9*, 2101. [CrossRef]
4. Caldara, F.R.; Dos Santos, L.S.; Machado, S.T.; Moi, M.; de Alencar Naas, I.; Foppa, L.; Garcia, R.G.; de Kassia Silva Dos Santos, R. Piglets' surface temperature change at different weights at birth. *Asian-Australas. J. Anim. Sci.* **2014**, *27*, 431–438. [CrossRef]
5. Vande Pol, K.D.; Cooper, N.; Tolosa, A.; Ellis, M.; Shull, C.M.; Brown, K.; Alencar, S. Effect of method of drying piglets at birth on rectal temperature over the first 24 hours after birth. *Transl. Anim. Sci.* **2020**, *97*, 4–5. [CrossRef]
6. Gómez-Prado, J.; Pereira, A.M.F.; Wang, D.; Villanueva-García, D.; Domínguez-Oliva, A.; Mora-Medina, P.; Hernández-Avalos, I.; Martínez-Burnes, J.; Casas-Alvarado, A.; Olmos-Hernández, A.; et al. Thermoregulation mechanisms and perspectives for validating thermal windows in pigs with hypothermia and hyperthermia: An overview. *Front. Vet. Sci.* **2022**, *9*, 1023294. [CrossRef]
7. Bienboire-Frosini, C.; Muns, R.; Marcet-Rius, M.; Gazzano, A.; Villanueva-García, D.; Martínez-Burnes, J.; Domínguez-Oliva, A.; Lezama-García, K.; Casas-Alvarado, A.; Mota-Rojas, D. Vitality in Newborn Farm Animals: Adverse Factors, Physiological Responses, Pharmacological Therapies, and Physical Methods to Increase Neonate Vigor. *Animals* **2023**, *13*, 1542. [CrossRef]
8. Rooke, J.A.; Bland, I.M. The acquisition of passive immunity in the new-born piglet. *Livest. Prod. Sci.* **2002**, *78*, 13–23. [CrossRef]
9. Theil, P.K.; Lauridsen, C.; Quesnel, H. Neonatal piglet survival: Impact of sow nutrition around parturition on fetal glycogen deposition and production and composition of colostrum and transient milk. *Animal* **2014**, *8*, 1021–1030. [CrossRef]
10. Pandolfi, F.; Edwards, S.A.; Robert, F.; Kyriazakis, I. Risk factors associated with the different categories of piglet perinatal mortality in French farms. *Prev. Vet. Med.* **2017**, *137*, 1–12. [CrossRef]
11. Cooper, N.; Vande Pol, K.D.; Ellis, M.; Xiong, Y.; Gates, R. Effect of piglet birth weight and drying on post-natal changes in rectal temperature. *J. Anim. Sci.* **2019**, *97*, 4. [CrossRef]
12. Herpin, P.; Damon, M.; Le Dividich, J. Development of thermoregulation and neonatal survival in pigs. *Livest. Prod. Sci.* **2002**, *78*, 25–45. [CrossRef]
13. Bienboire-Frosini, C.; Wang, D.; Marcet-Rius, M.; Villanueva-García, D.; Gazzano, A.; Domínguez-Oliva, A.; Olmos-Hernández, A.; Hernández-Ávalos, I.; Lezama-García, K.; Verduzco-Mendoza, A.; et al. The Role of Brown Adipose Tissue and Energy Metabolism in Mammalian Thermoregulation during the Perinatal Period. *Animals* **2023**, *13*, 2173. [CrossRef]
14. Declerck, I.; Dewulf, J.; Decaluwé, R.; Maes, D. Effects of energy supplementation to neonatal (very) low birth weight piglets on mortality, weaning weight, daily weight gain and colostrum intake. *Livest. Sci.* **2016**, *183*, 48–53. [CrossRef]
15. Vande Pol, K.; Tolosa, A.; Shull, C.; Brown, C.; Alencar, S.; Ellis, M. Effect of drying and warming piglets at birth on pre-weaning mortality. *Transl. Anim. Sci.* **2021**, *5*, txab016. [CrossRef] [PubMed]
16. Tucker, B.S.; Petrovski, K.R.; Kirkwood, R.N. Neonatal piglet temperature changes: Effect of intraperitoneal warm saline injection. *Animals* **2022**, *12*, 1312. [CrossRef] [PubMed]
17. Gimsa, U.; Brückmann, R.; Tuchscherer, A.; Tuchscherer, M.; Kanitz, E. Early-life maternal deprivation affects the mother-offspring relationship in domestic pigs, as well as the neuroendocrine development and coping behavior of piglets. *Front. Behav. Neurosci.* **2022**, *16*, 980350. [CrossRef]
18. Kammersgaard, T.S.; Malmkvist, J.; Pedersen, L.J. Infrared thermography--a non-invasive tool to evaluate thermal status of neonatal pigs based on surface temperature. *Animal* **2013**, *7*, 2026–2034. [CrossRef]
19. Zakari, F.O.; Akefe, I.O.; Uchendu, C. Comparison of diurnal rectal and body surface temperatures in large white piglets during the hot-dry season in a tropical Guinea savannah. *J. Therm. Biol.* **2021**, *99*, 102953. [CrossRef]
20. Schmid, S.M.; Büscher, W.; Steinhoff-Wagner, J. Suitability of different thermometers for measuring body core and skin temperatures in suckling piglets. *Animals* **2021**, *11*, 1004. [CrossRef]
21. Soerensen, D.; Clausen, S.; Mercer, J.; Pedersen, L. Determining the emissivity of pig skin for accurate infrared thermography. *Comput. Electron. Agric.* **2014**, *109*, 52–58. [CrossRef]
22. Hue, D.T.; Williams, J.L.; Petrovski, K.; Bottema, C.D.K. Predicting colostrum and calf blood components based on refractometry. *J. Dairy Res.* **2021**, *88*, 194–200. [CrossRef] [PubMed]
23. Soerensen, D.D.; Pedersen, L.J. Infrared skin temperature measurements for monitoring health in pigs: A review. *Acta Vet. Scand.* **2015**, *57*, 5. [CrossRef] [PubMed]
24. Muns, R.; Nuntapaitoon, M.; Tummaruk, P. Non-infectious causes of pre-weaning mortality in piglets. *Livest. Sci.* **2016**, *184*, 46–57. [CrossRef]
25. Decaluwe, R.; Maes, D.; Wuyts, B.; Cools, A.; Piepers, S.; Janssens, G.P.J. Piglets' colostrum intake associates with daily weight gain and survival until weaning. *Livest. Sci.* **2014**, *162*, 185–192. [CrossRef]

26. Devillers, N.; Le Dividich, J.; Prunier, A. Influence of colostrum intake on piglet survival and immunity. *Animal* **2011**, *5*, 1605–1612. [CrossRef]
27. Magnani, D.; Gatto, M.; Cafazzo, S.; Stelletta, C.; Morgante, M.; Nanni Costa, L. Difference of surface body temperature in piglets due to the backtest and environmental condition. In Proceedings of the Animal Hygiene and Sustainable Livestock Production; the XVth International Congress of the International Scociety for Animal Hygiene, Vienna, Austria, 3–7 July 2011; pp. 1029–1032.
28. Lu, M.; He, J.; Chen, C.; Okinda, C.; Shen, M.; Liu, L.; Yao, W.; Norton, T.; Berckmans, D. An automatic ear base temperature extraction method for top view piglet thermal image. *Comput. Electron. Agric.* **2018**, *155*, 339–347. [CrossRef]

animals

MDPI

Review

Hypothalamic Neuromodulation and Control of the Dermal Surface Temperature of Livestock during Hyperthermia

Juliana Sarubbi [1,*], Julio Martínez-Burnes [2], Marcelo Daniel Ghezzi [3], Adriana Olmos-Hernandez [4], Pamela Anahí Lendez [5], María Carolina Ceriani [5] and Ismael Hernández-Avalos [6,*]

1 Department of Animal Science, Federal University of Santa Maria, Av. Independência, Palmeira das Missões 3751, RS, Brazil
2 Facultad de Medicina Veterinaria y Zootecnia, Universidad Autónoma de Tamaulipas, Victoria City 87000, Mexico
3 Animal Welfare Area, Faculty of Veterinary Sciences (FCV), Universidad Nacional del Centro de la Provincia de Buenos Aires (UNCPBA), University Campus, Tandil 7000, Argentina
4 Division of Biotechnology—Bioterio and Experimental Surgery, Instituto Nacional de Rehabilitación-Luis Guillermo Ibarra Ibarra (INR-LGII), Mexico City 14389, Mexico
5 Faculty of Veterinary Sciences (FCV), Universidad Nacional del Centro de la Provincia de Buenos Aires, CIVETAN, UNCPBA-CICPBA-CONICET (UNCPBA), University Campus, Tandil 7000, Argentina
6 Facultad de Estudios Superiores Cuautitlán (FESC), Universidad Nacional Autónoma de Mexico (UNAM), Cuautitlán 54714, Mexico
* Correspondence: jusarubbi.ufsm@hotmail.com (J.S.); mvziha@hotmail.com (I.H.-A.)

Simple Summary: This review aims to discuss the hypothalamic control of hyperthermia in livestock, including the main physiological and behavioral changes that animals adopt to maintain their thermal stability. This review analyzes the main changes that can be observed in livestock when exposed to intense heat. Vasodilation, sweating, and behavioral modifications (e.g., standing posture, shade seeking, feed intake, among others) are the results of neuronal projections from peripheral thermoreceptors to the preoptic area of the hypothalamus and effector neurons. Understanding these mechanisms can help to visualize their effect to restore thermal neutrality in livestock.

Abstract: Hyperthermia elicits several physiological and behavioral responses in livestock to restore thermal neutrality. Among these responses, vasodilation and sweating help to reduce core body temperature by increasing heat dissipation by radiation and evaporation. Thermoregulatory behaviors such as increasing standing time, reducing feed intake, shade-seeking, and limiting locomotor activity also increase heat loss. These mechanisms are elicited by the connection between peripheral thermoreceptors and cerebral centers, such as the preoptic area of the hypothalamus. Considering the importance of this thermoregulatory pathway, this review aims to discuss the hypothalamic control of hyperthermia in livestock, including the main physiological and behavioral changes that animals adopt to maintain their thermal stability.

Keywords: heat stress; livestock; heat defense behavior

Citation: Sarubbi, J.; Martínez-Burnes, J.; Ghezzi, M.D.; Olmos-Hernandez, A.; Lendez, P.A.; Ceriani, M.C.; Hernández-Avalos, I. Hypothalamic Neuromodulation and Control of the Dermal Surface Temperature of Livestock during Hyperthermia. *Animals* **2024**, *14*, 1745. https://doi.org/10.3390/ani14121745

Academic Editor: Mariangela Caroprese

Received: 17 April 2024
Revised: 31 May 2024
Accepted: 6 June 2024
Published: 9 June 2024

1. Introduction

Exposure to high ambient temperatures can be implicit inside production systems due to the design of the units or the geographical location of farms, making livestock susceptible to heat stress and hyperthermia [1,2]. Moreover, climate change and global warming are current challenges for farm animals [3,4]. Preventing increases in core body temperature is essential because several physiological and behavioral mechanisms are activated when animals are under thermal stress [5]. When not modulated, these mechanisms can negatively affect the health and productive performance of livestock. Thus, understanding these responses is essential to promote thermal stability in animals.

Maintaining a regular body temperature depends on the interaction between internal and external factors, which is also influenced by the size, weight, morphology, physiological state (e.g., gestation, estrous cycle), and the metabolic rate of animals [6–8]. Generally, when mammals experience an increase in core body temperature, two main mechanisms are activated: physiological and behavioral. In the first, physiological responses include involuntary autonomic adaptations such as sweating, respiratory evaporation, and changes in blood perfusion to facilitate thermal exchange (e.g., vasodilation) [9–11]. On the other hand, behavioral modifications include compensatory changes in body posture (e.g., increasing their standing time or lying on the side) or shade-seeking and increasing water intake to promote heat loss and limit heat production [10,12,13].

Regardless of the physiological and behavioral adaptation, in mammals, the main thermoregulatory center is the hypothalamus, particularly the preoptic area (POA) of the hypothalamus, where peripheral thermal afferents and thermoregulatory efferents are processed [5,14,15]. This review aims to discuss the hypothalamic control of hyperthermia in livestock, including the main physiological and behavioral changes that animals adopt to maintain their thermal stability.

2. Search Methodology

A literature review was conducted using Web of Science, Scopus, Science Direct, and PubMed. The keywords used for the search were "hyperthermia", "heat stress", "transport stress", and "infrared thermography". The selected articles were studies or review papers published from 2000 to 2024 on animals with heat stress where IRT was applied as a complementary monitoring tool.

3. Hypothalamic Control of Hyperthermia: Peripheral Perception of Thermal Stimuli and Supraspinal Processing

3.1. Peripheral Thermal Receptors

Specialized receptors that transduce cold and heat stimuli into electrical signals (called thermoreceptors) detect changes in the external and internal temperature of the organism [16,17]. When animals are outside their thermo-neutral zone (a temperature range where mammals can maintain a balance between heat production and dissipation) and exposed to temperatures that exceed their maximum temperature, several thermoregulatory mechanisms are triggered to increase the amount of heat dissipated from the body or to decrease the amount of heat produced by the animals [18].

Both physiological and behavioral changes are directly related to the activation of thermoreceptors located mainly in the skin, where free nerve endings detect thermal variations [19,20]. Lezama-García et al. [21] describe that thermosensory neurons respond to both hot and cold stimuli, activating ion channels so the neurons can transmit the information to spinal and supraspinal structures [22]. In particular, the main thermoreceptors comprise the transient receptor potential (TRP). TRPs are cation channels formed by transmembrane proteins that act as transducers through changes in membrane potential due to the intracellular Ca^{2+} concentration. Most react to fluctuations in temperature, pH, and osmolarity; however, injury, pain, pheromones, flavors, taste, ionic imbalances (Ca^{2+}), volatile compounds, mechanosensations, and cytokines can also trigger them. Since their discovery, about 30 trp genes and over 100 TRP channels have been identified and categorized into their respective families and subfamilies [21].

Six subfamilies are identified as thermosensors (TRPV1-4, TRPM2-5,8, TRPA1, and TRPC5). Neuronal afferents that perceive heat and cold can be differentiated by presenting specific TRP channels for each stimulus because their activation depends on specific thermal properties [23]. Most neurons that allow the identification of differences in the range of −10 to 60 °C are thermosensitive receptors located in nerve endings that are found in the trigeminal ganglion that innervates the face and head, as well as in the ganglia of the dorsal horn of the spinal cord [24].

Based on the properties of TRP thermosensors, four basic thermal sensations are recognized: (1) cold, −10 to 15 °C; (2) cool, between 16 and 30 °C; (3) warm, 31 to 42 °C; and (4) hot, 43–60 °C. Depending on the stimulus, PRTs are classified into hot- and cold-sensitive receptors [21,25]. Regarding the heat-sensitive receptors, TRP vanilloid V1 and V1 perceive harmful heat above 43 °C and 52 °C, respectively [23]. Moreover, TRPV3 can also perceive harmless heat with warm temperatures above 34 °C (ranging between 23 and 39 °C) [26], while TRPV4 is activated with moderately warm temperatures of 27 to 34 °C [23]. Molecular studies have shown that TRPM8 participates in cold-defensive responses, while TRPM2, TRPM3, TRPM4, and TRPM5 are activated by heat. Particularly, TRPM3 responds to harmful heat. One difference between TRPM4 and TRPM5 and the others is that they are not permeable to Ca2+, though they participate in membrane depolarization [27,28].

Current research regarding warm-sensitive thermoreceptors is still ongoing. However, it has been suggested that TRPM2 may be involved in the activation of thermoregulatory behavioral responses in warm environments [29]. Nozadze et al. [30] suggest that the co-expression of TRPV1 and TRPA1 have a synergistic relationship to process thermal stimuli. This was evaluated in adult male Wistar rats, in whom the effect of mustard oil and capsaicin increased heat hyperalgesia at 30 °C. In another study performed by Cheng et al. [31], it was identified that heteromeric TRPV1 and TRPV3 channels actively participate in the detection of these stimuli. The authors observed that the activation threshold for TRPV1 channels is at 37.7 ± 0.7 °C, while TRPV3 channels activate at 31.3 ± 0.6 °C. This suggests that the co-expression of these channels contributes to heat perception, as also described by Lei et al. [32]: when identified in cell cultures of 4–5 week-old mice, the expression of a transmembrane protein called TMEM79 acted as a negative regulator of TRPV3. They also observed that in cells lacking TMEM79, the flow of Ca^{2+} ions was enhanced and was a process mediated by TRPV3. They also observed that mice with null expression of TMEM79 preferred higher temperatures than mice that presented this protein due to elevated TRPV3 function. This possibly supports the role of this molecule in the transduction of hot thermal signals and even possibly suggests that TRPV1/TRPV3 are unique heat sensors that contribute to the adjustment of sensitivity to the sensation of heat.

These channels transduce thermal signals from the skin to subsequent central regions such as the dorsal root ganglion to thermoregulate. In the spinal cord, it has been described that TRPV1, TRPV3, and TRPA1 channels are also expressed [33,34]. The activation of these ion channels is the first step to trigger the compensatory responses to dissipate heat by its connections to the main thermoregulatory center, the POA.

3.2. *Hypothalamic Modulation of Hyperthermia*

Physiological and behavioral changes depend on the activation of peripheral thermoreceptors, hypothalamic integration, and subsequent connections to the mesencephalon and spinal cord [6]. Thermoreceptors transduce the information and synapse with spinal dorsal horn neurons, where specialized warm-sensitive and cold-sensitive neurons located in the lateral parabrachial nucleus (LPB) respond to thermal stimuli [28,35]. The neurons located in the dorsal part of the LPB (LPBd) are activated by warm inputs (~36 °C) to elicit autonomic heat-defense responses [5,35].

LPBd neurons are glutamatergic, and their inactivation causes impaired autonomous thermoregulatory responses to avoid heat or cold stimuli [35,36]. Through optogenetics, it has been observed that LPB glutamatergic neurons express dynorphins and cholecystokinin to defend the organism from hyperthermia [37]. In rats, these neurons inhibit thermogenic mechanisms such as the activation of the brown adipose tissue while promoting tail vasodilation [37,38].

To modulate these responses, the LPBd projects to the POA of the hypothalamus. This area is the primary integrative center of thermoregulation, where 30% of the thermosensitive neurons are located [14,15]. These neurons are GABAergic and have glutamatergic and prostaglandin (PG) receptors [5], which are stimulated by changes in hypothalamic and central temperature and the temperature recorded in other anatomical regions such as the spinal cord, blood, and viscera [39–41]. In this sense, the role of other substances such as PG, nitric oxide, histamine, and tumor necrosis factor needs to be highlighted since, for example, PGE2 reduces the activity of GABAergic neurons in the hypothalamus and can elicit vasoconstriction and thermogenesis in rats [42]. In the same species, Nakamura et al. [43] found PG EP3 receptors in the POA. When stimulating PG EP3-expressing neurons, the body temperature decreases, while hyperthermia can be triggered by inhibition of the same receptors, showing that GABAergic signaling in the POA is fundamental to maintaining thermal homeostasis. Furthermore, the release of nitric oxide from endothelial cells is a potent vasodilator that can contribute to heat loss in animals suffering from hyperthermia [44].

Thermal information from these structures is transmitted by the vagus nerve (X cranial nerve) to the POA through the nucleus of the solitary tract in the caudal portion of the medulla oblongata for projection to the hypothalamus [36,45]. The spinal-LPB-POA circuit is considered the main thermosensory pathway to modulate the adaptative mechanisms to environmental thermal challenges [35,46,47]. In mouse models, opioidergic connections are present between the LPB to the POA, and studies have revealed that warm stimuli activate neurons expressing dynorphin and enkephalin, which participate in heat defense behaviors [36,48,49]. Nonetheless, other regions also participate in thermal perception, such as the afferent pathways from the LPB to the dorsomedial hypothalamus (DMH) and the rostral medullary raphe region (rRM) that includes the rostral raphe pallidus nucleus (rRPa), the adjacent raphe magnus nucleus (RMg), and the lateral pyramidal nucleus [14,50].

The LPB projects to the median preoptic nucleus (MnPO), mediating ascending heat-related responses [50]. The GABAergic projection neurons of the POA connect two nuclei: the MnPO and the medial preoptic area (MPO). It has been observed that neurons in these two regions are sensitive to heat, although MPO activity is greater at thermoneutral temperatures [51]. Furthermore, connections from the LPB to the lateral hypothalamus (LH) are required to modulate thermal stimuli [46]. There are also connections from the LPB to the periventricular nucleus (PVN) through the DMH [36]. The DMH projects glutamatergic neurons to the rRM and its different nuclei, which contain sympathetic premotor neurons that project to the sympathetic preganglionic neurons of the intermediate-lateral column of lamina VII of the spinal cord (IML). The activity of IML neurons depends on the secretion of glutamate and serotonin (5-HT) in the rRM. However, 5-HT modulates the excitatory activity of glutamate [14,52].

The projections of the POA connect with the neurons of the DMH and to the rRM without depending on the activity of the DMH, so it is thought that these direct or indirect connections of the POA-rRM trigger different responses. For example, the direct synapse of POA neurons to those of the rRM results in cutaneous vasoconstriction as a febrile response against cold, while the POA-DMH-rRM connection produces tachycardia and thermogenesis [4]. Both adrenergic and cholinergic fibers modulate blood flow, producing vasoconstriction and vasodilation, respectively [44]. Therefore, sympathetic cholinergic nerves produce cutaneous active vasodilation to dissipate heat [53]. Some reports have shown that transdermal application of acetylcholine and local heat (41 °C) cause hyperemia of the microvasculature of the skin [54]. Moreover, these nuclei are involved in the cutaneous vasomotor response and sweating response, controlling skin blood flow and BAT metabolism. These aspects will be addressed in the following sections (Figure 1) [55–57].

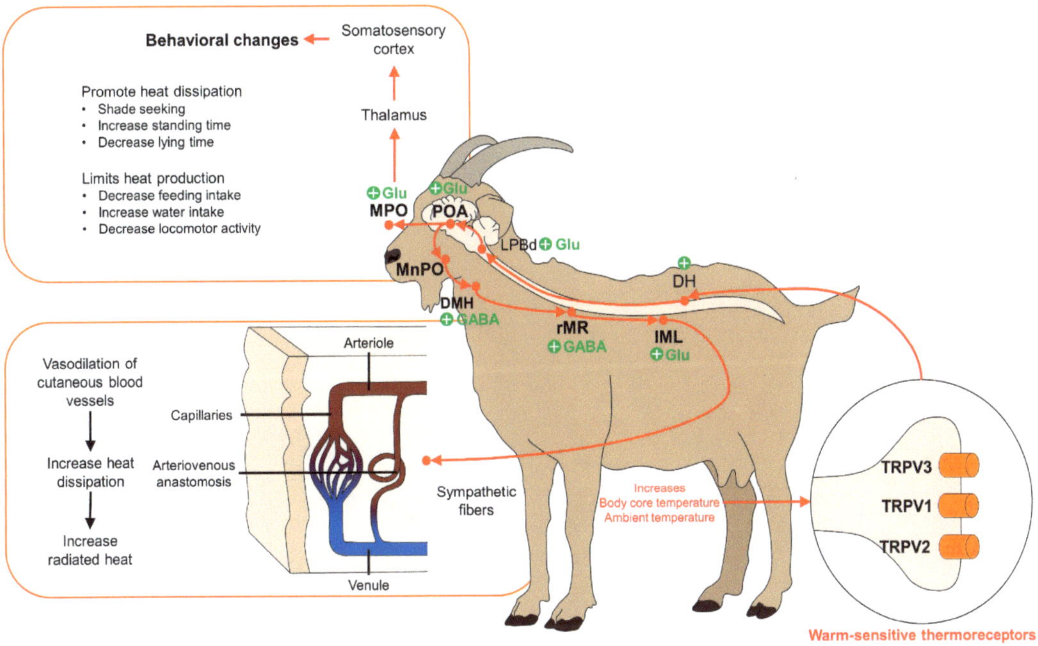

Figure 1. Hypothalamic modulation of temperature during hyperthermia. Warm-sensitive thermoreceptors detect increases in core body temperature or ambient temperature. These stimuli are transmitted to the DH of the spinal cord to subsequently project to the LPBd, the region that is particularly activated with heat-related signaling. From the LPBd, neurons project to the POA, the MnPO, the DMH, rMR, and IML to trigger vasodilation in cutaneous blood vessels. On the other hand, connections between the POA, MPO, thalamus, and somatosensory cortex are responsible for producing behavioral changes to promote heat dissipation and limit heat production. DH: dorsal horn of the spinal cord; DMH: dorsomedial hypothalamus; DRG: dorsal root ganglion; HPA: hypothalamic–pituitary–adrenal axis; IML: intermediolateral laminae; LPBd: dorsal part of the parabrachial nucleus; MPOa: medial preoptic area; MnPO: median preoptic area; POA: preoptic area; rMR: rostral medullary raphe; SAM: symptom–adrenomedullary axis; TRP: transient receptor potential.

4. Response to Hyperthermia: Systemic and Local Changes

In various circumstances, it has been documented that an increase in the body temperature of animals is associated with warm environmental conditions with high humidity and solar radiation [58]. Livestock shows a greater susceptibility to thermal stress when exposed to these climatic conditions and when maintained in intensive production systems where growth rates are sought to be maximized. Particularly in livestock, the consequences of hyperthermia include health problems, productivity, and deaths during extreme events, as well as serious economic losses [59].

Several POA neurons are known to modulate heat defense responses, including vasodilation and behavioral modifications such as reduction of physical activity and energy expenditure, body posture changes, and suppression of thermogenesis both by brown adipose tissue or shivering [37]. Cattle exposed to heat stress can also alter their endocrine and immune responses [60]. These mechanisms prevent further increases in core body temperature to restore homeostasis.

4.1. Vasomotor Response to Hyperthermia: Vasodilation

The immediate physiological response of animals to thermal stress involves altering the diameter of superficial blood vessels, facilitating heat exchange in skin tissue with the environment through radiation [49,61]. This effect is facilitated in regions with glabrous skin and dense anastomosis (regions called thermal windows that are schematized in Figure 2) [5].

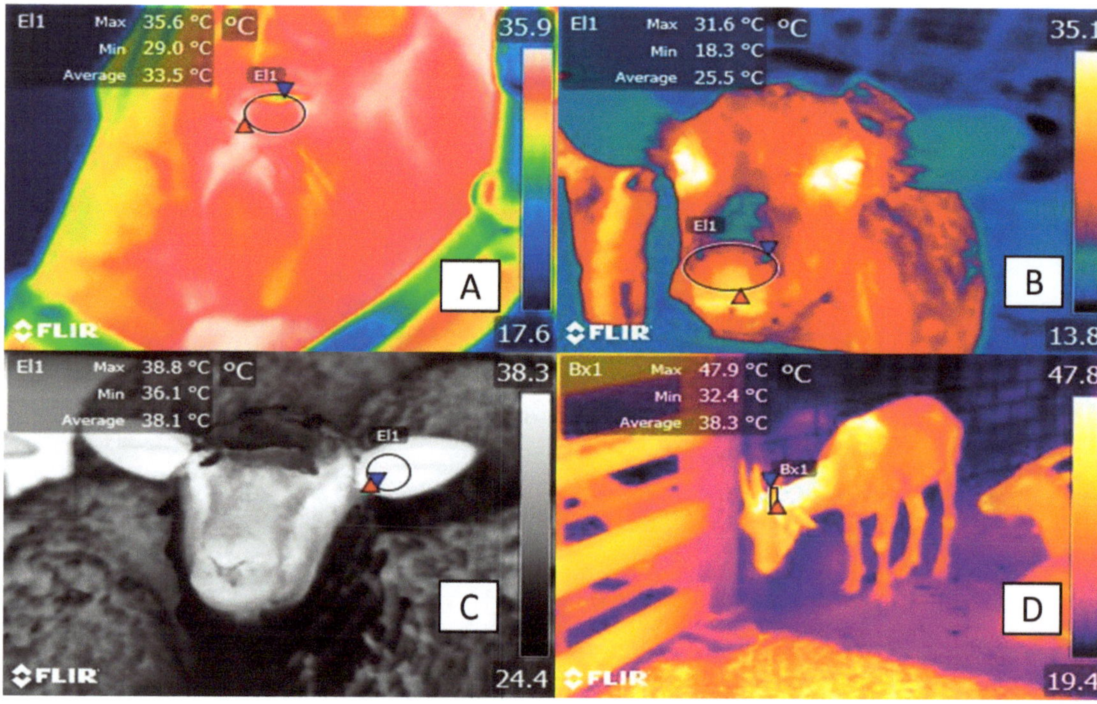

Figure 2. Recommend thermal windows in domestic animals for the recognition of thermal stress. (**A**) Ocular. This thermal window is recommended due to its irrigation of the infra and supraorbital artery, which helps detect changes in perfusion during states of hyperthermia. Likewise, this window has been shown to have a positive correlation with core body temperature. (**B**) Nasal. This region has been used to recognize changes in the respiratory pattern and frequency that are regularly modified during the perception of thermal stress. (**C**) Auricular. This window is irrigated by branches of the internal auricular artery (the marginal auricular and lateral auricular arteries). (**D**) Horn. This window has been recently proposed to evaluate the thermal state of animals due to the perfusion from the temporal artery that might represent a region for thermal exchange in livestock.

An example of this can be observed in Rodríguez-González et al.'s [62] study, where the effect of 2 h transport in the surface temperature of buffaloes was evaluated. In general, increases in the surface temperature above 4 °C were recorded for the lacrimal caruncle, periocular, lower eyelid, auditory canal, nostrils, and body regions such as the abdominal, lumbar, and parietal region after transport. A similar study made by the same researchers compared the effect of short (50.33 min.) and long journeys (13.31 h) on the thermal response of 1516 water buffaloes. They observed that the surface temperature of the periocular, lacrimal caruncle, nasal, auricular, frontal–parietal regions of the head, pelvic limb, abdominal and thoracic regions, and lumbar region increased by 3 °C during short journeys, a response that was not observed during long ones [62].

The physiological explanation for the increases in surface temperature is due to the peripheral vasodilation that is present when animals are exposed to intense heat. Peripheral vasodilation enhances heat dissipation and, therefore, helps to reduce core temperature [63,64]. This vasomotor response also responds to acetylcholine release and other co-transmitters from sympathetic cholinergic nerves, increasing cutaneous blood flow up to 7700 mL/min [19,63]. Moreover, other mediators such as vasoactive intestinal peptide, substance P, histamine, prostaglandins, and TRPV1 activation participate in vasodilation [65,66].

Understanding this mechanism allows us to explain the differences found in body regions for heat elimination. In a study where short and long journeys were compared to evaluate the thermal response of water buffaloes, the authors also found that the surface temperature of central regions was 1 °C higher compared to regions such as abdominal or lumbar [62]. The thermal capacity of certain regions could be related to the difference in the arrangement of blood capillaries, results that have been reported when comparing the surface temperature between the eye region and peripheral regions, finding that central regions are at least 1.5 °C above the temperature at the limbs of dogs and large ruminants [67,68] or the eye region when compared to the tail temperature in rodents [69]. Figure 3 shows the comparison between central and peripheral thermal windows in water buffalo.

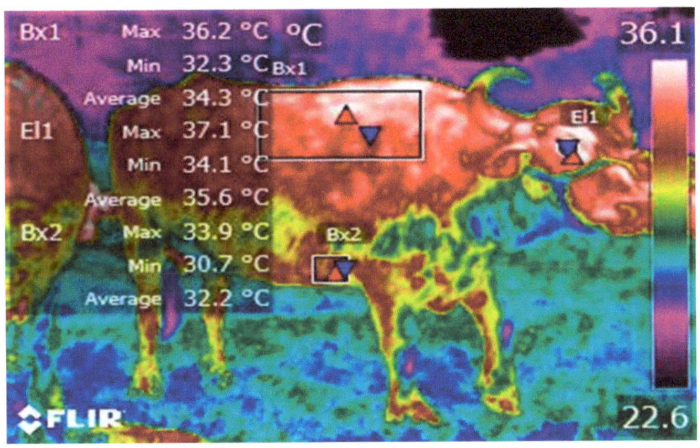

Figure 3. Thermal difference between superficial and central regions. The surface temperature of the dorsal (B × 1), axillary (B × 2), and ocular region (El1) is shown. In the dorsal region, a maximum temperature (red triangle) of 36.2 °C is observed, as well as an average of 34.3 °C and a minimum (blue triangle) of 32.3 °C. When comparing the average temperature of the dorsal region with the axillary region (B × 2), a difference of −2.1 °C was observed. In the orbital region (El1), the highest temperature can be observed, where the maximum, average, and minimum temperatures are above dorsal values by 0.9 °C, 1.3 °C, and 1.8 °C, respectively. This thermal difference between the regions may be due to the arrangement of the number of sweat glands and even the amount of blood capillaries present at a superficial level that affects the heat elimination capacity, which would facilitate the animals' ability to recover thermoneutrality.

Another possible example of this is the parietal region in different species of ruminants, where it has been suggested that horns might participate in thermoregulation [70]. A study conducted by Algra et al. [71] evaluated the thermoregulatory function of the horns in 18 cows from 3 farms in a temperate climate area. They observed that the temperature in the horns increased by 0.18 °C for each unit of the heat load index. They also reported that dehorned animals had a significantly higher eye temperature than animals with horns (by 1 °C). These results support the idea of the participation of horns in thermoregulation in this species. However, they also show the importance of the arrangement of the blood vessels to thermoregulate. In this sense, the parietal region is irrigated by the superficial temporal artery (*temporalis superficialis*), *transversa faciei*, *auricularis rostralis*, *palpebralis inferior lateralis*, and *palpebralis superior lateralis*, blood vessels that provide circulation to the horns [72].

One of the tools used to evaluate dermal surface temperature as an indicator of well-being in livestock is infrared thermography (IRT). Thus, various studies have exposed the possible uses and applications of IRT—for example, in the phenomena of hyperthermia associated with stress in lactating dairy cows during isolation challenges [73], diagnostic support of subclinical mastitis not only in cows but also camels (*Camelus dromedarius*) [74,75], and febrile processes due to diseases and in the evaluation of negative emotions such as fear and anxiety [9,76]. All these cases were evaluated indirectly by the vasomotor response associated with the increase in surface temperature.

Regarding other ruminants, in sheep, this tool has been used to determine and evaluate hoof injuries, where a difference of 8.5 °C could be observed between healthy and diseased hooves, showing a sensitivity of 92% and a specificity of 91% in the diagnosis of this disease [77]. In this same species, IRT of the ocular surface has been useful as a non-invasive method to measure the level of thermal stress developed during shearing, resulting in a tool that was positively correlated with rectal temperature and serum cortisol levels [78].

Likewise, similar findings have been reported in other species, where the possibility of correlating the environment's high temperatures with rabbits' health can be highlighted by correlating physiological responses with fecal cortisol levels [79]. Another example of application is in horses, where IRT seems to be a good indicator of physiological stress, which is why it can be useful to evaluate the effectiveness of training in endurance horses by measuring the temperature of the ocular surface, crown, gluteal muscles, and *longissimus dorsi* since the eye temperature could be correlated with the heart rate (r = 0.42 left eye; r = 0.48 right eye), while the frontal crown temperature had a positive correlation with cortisol levels ($p < 0.01$) [80]. Thus, IRT can become a useful tool for the early identification of horses that are not suitable to compete or continue in competition.

On the other hand, during routine procedures such as teeth clipping, dehorning, or tail docking, local vasodilation can also be observed as a result of possible acute pain perception, as observed in 1-day-old crossbred piglets (White Large × Landrace) undergoing teeth clipping [81]. A progressive increase in the upper lip temperature (by 0.7 °C) was recorded, possibly associated with vasomotor changes. Contrarily, although no differences were found in the soft tissue of piglets subjected to teeth grinding, the tooth temperature during the procedure reached up to 90 °C [82]. In dehorned animals, regardless of the use of oral and local analgesics, the temperature on the dehorning wound showed a progressive increase from day one (38.83 ± 0.42 °C) to day seven (40.30 ± 0.40 °C), a response that might be associated with inflammatory substances (e.g., PG) and local vasodilation [83]. Regarding tail docking, an increase in the surface temperature of the ocular region of crossbred piglets (by 1.1 °C) docked with side-cutter pliers was observed in these animals when comparing the before and after periods [81]. A similar response was observed in daily cattle subjected to tail docking. The surface temperature of the tail of docked animals was 1.43 °C warmer than the values registered in non-docked cattle, possibly as a result of local vasodilation and hypersensitivity [84].

In summary, the general vasomotor response in domestic animals can be considered a primary and passive method that increases blood flow to promote heat loss. The capacity of the different body regions to dissipate heat seems to be closely related to the arrangement of blood vessels, which is why it is suggested that in some species, the use of horns may be a key element in thermoregulation during hyperthermia.

4.2. Sweating

Another thermoregulation mechanism present in livestock during hyperthermia is sweating to lose heat through evaporation [19]. Sweat is released by eccrine glands distributed on all body surfaces. Their activation is mediated by sympathetic cholinergic fibers [85]. Although it is a mechanism particularly used to increase heat loss, it is not present in all species due to the difference in the arrangement and development of the sweat glands. An example is the Egyptian water buffalo, in whom the structure and dimension of sweat glands were studied and compared to *Bos* cattle by Hafez et al. [86]. The number of glands per cm^2 of skin was lower in buffalo (394 gland/cm^2) than in cattle (2633 gland/cm^2). This trait might be related to the thermal tolerance in the animals, as addressed by a study performed in 10 female Murrah buffaloes exposed to hot and humid environments. A negative correlation between respiratory frequency and skin thickness was observed ($r = -0.73$, $p = 0.015$). The authors conclude that water buffaloes resort to polypnea as a heat-related tolerance response that also needs to be enhanced by providing shade [7].

This study shows the importance of sweating as an essential mechanism for thermoregulation in animals during hyperthermia. However, the differences in distribution between species limit its effectiveness, and it is suggested that equidae, bovidae, and primates are the only species that have sweat glands [87]. In particular, for horses, sweating dissipates 70% of the heat loss during exercise due to evaporation. In this species, it has even been found that the sweat rate ($L/h/m^2$) is three times greater in the horse compared to the human with a similar exercise intensity [17,88,89].

In this regard, there are not only distribution differences between species but also differences in their location site. A histological analysis of water buffalo's sweat glands was performed by Debbarma et al. [90]. It was reported that the number of sweat glands/cm^2 in the head, neck, and tail regions was 20–30% higher compared to the thorax and abdomen region. This difference might impact not only the effectiveness of thermoregulation strategies in animals but also make it evident that animals use specific anatomical regions to thermoregulate. Thus, Barros et al. [91] evaluated the surface temperature in ten Murrah buffalo bulls in four anatomical parts (ocular, right flank, left flank, and scrotum). Although they observed a strong positive correlation between temperature and the temperature humidity index (THI) in all regions (ocular = 0.72, right flank = 0.77, left flank = 0.75, and scrotum = 0.41), the highest temperature was in the ocular region. This could be explained by the difference in the distribution of the number of sweat glands between regions and even in the same distribution of superficial blood capillaries, which would facilitate the thermoregulation capacity in the animal.

Therefore, sweating is an essential method used to recover thermoneutrality in animals in species that have an abundant arrangement of sweat glands in the skin. For this same reason, the presence of these same glands helps us understand why some species are more susceptible to thermal stress or have less-effective thermoregulation systems compared to other species.

4.3. Endocrine and Immunological Response Associated with Hyperthermia

Endocrinological changes play an important role in the metabolic response to heat stress through the neurosecretion and modulation of hormones, especially glucocorticoids, antidiuretic hormone, growth hormone, thyroxine, prolactin, and aldosterone, since they represent an adaptive mechanism of the organism regarding this condition [92,93]. For example, in hyperthermia due to acute heat stress, the concentrations of antidiuretic hormone, prolactin, glucocorticoids, and catecholamines increase, while those of aldosterone decrease. However, during chronic heat stress, levels of cortisol, growth hormone, and thyroxine may remain unchanged or may even decrease [60,94]. Changes in hormone concentrations during acute heat stress events are associated with decreased electrolyte concentrations and fluid balance, as water and electrolyte losses occur during panting and sweating due to glucocorticoid secretion. In this regard, it has been reported that elevated prolactin concentrations play an important role in sodium and potassium balance at elevated ambient temperatures, given its relationship with aldosterone secretion [95,96]. Derived from the action of glucocorticoids, catecholamine concentrations have been reported to be elevated during chronic heat stress events, possibly due to increased activity of sweat glands regulated by sympathetic innervation [97].

Hyperthermia due to stress and climate change means that dairy cows may be more susceptible to diseases that are usually positively associated with the survival of microorganisms and/or their vectors. These diseases can increase heat stress-related immunosuppression, negatively affecting the immune system through cellular and humoral immune responses mediated by the hypothalamic–pituitary–adrenal axis (HPA) and the sympathetic–adrenal medulla axis (SAM) [98]. The main responses are associated with an increase in blood cortisol levels, which is associated with the greater production of cytokines such as interleukin-4 (IL-4), IL-5, IL-6, and IL-12 [99]. Similarly, stress hyperthermia has been shown to inhibit interferon-gamma (IFNγ) and tumor necrosis factor α (TNF-α). The effects of heat stress on the immune response of livestock can be reduced by improving current animal-selection methods, as well as the further development of breeds resilient to climate change, which can support the sustainability of future livestock systems of animal production [100].

5. Behavioral Changes Observed in Domestic Animals Experiencing Hyperthermia

Apart from the physiological modifications, animals adopt certain behavioral changes or body postures to face hyperthermia and increase heat dissipation via radiation, convection, or evaporation [101–106]. These behaviors can be classified into those that promote heat dissipation (e.g., shade-seeking, changes in body posture) or those that limit heat production (e.g., reducing feed intake and locomotor activity) [35,107–110].

Regarding shade-seeking, several studies have reported the benefits that shaded areas have on animals' thermoregulation, reducing the heat load in cattle by about 45% [111,112]. Gaughan et al. [113] reported that providing shade reduced the rectal temperature of Angus heifers (between 38.8 ± 0.8 and 39.4 ± 0.8 °C) in comparison with heifers in unshaded pens (between 39.1 ± 0.8 and 39.7 ± 1.8 °C). Similarly, crossbred calves maintained under agro-net shading spaces had lower rectal temperatures (minimum of 38.99 °C) and respiratory rates (minimum of 20.83 rpm) than animals housed in places with roofs of other materials [114]. The same results were observed in sheep, animals who prefer shaded areas (56–43%), and this reduces the respiratory rate, a sign of thermal stability [115,116].

Changes in standing and lying position are also observed in animals facing intense heat to facilitate heat dissipation. Standing increases the exposure of skin to airflow, maximizing evaporation [104,110], and animals can increase this posture by up to 10% when exposed to a heat load above 15% [117,118]. So, cattle exposed to a THI > 72 (considered intense heat) and with core body temperatures above 38.8 °C increased the likelihood of cows standing by 50% [119]. Similarly, Herbut and Angrecka [120] reported in Holstein-Friesian dairy cows that a THI > 73 had a significant and inversely proportional correlation with lying posture, where cows decreased their lying time from 11.3 to 9.4 h/day. The same observation was discussed by Hut et al. [121] in dairy cows; however, the authors reported that cows start to show adaptative behaviors (increased time standing and reduced rumination time) from an average ambient temperature of 12 °C and a THI of 56, showing that animals do not need to reach critical temperatures to start thermoregulatory behaviors when their rectal temperature increases.

In Suffolk and Dorset sheep, the same reaction was observed in Reis et al.'s [122] study when comparing the behavioral responses of animals at ambient temperatures of 20 and 35 °C. The authors found that higher temperatures (35 °C) increased the total time of standing (355 vs. 486 min) and decreased eating (932 vs. 727 min). Likewise, the lying behavior of ewes and rams increased by up to 45.10 ± 2.42% during severe heat stress [123]. In the case of pigs, the animals increased their lying time on the side by up to 5.9% to promote a larger contact area of the skin surface with the floor to dissipate heat through conduction [124]. In these latter cases, the behavior may be contradictory; however, heat transfer by conduction and evaporation helps explain why the rest time was longer in these animals, as discussed by Gómez-Prado et al. [125], who describes that pigs have a very low density of sweat glands (30/cm^2) compared to species such as cattle (800–2000/cm^2), which causes the pig to lose little heat when sweating. Another factor to consider is subcutaneous fat; for example, pregnant or lactating sows and commercial-weight pigs require greater energy consumption during their productive stage, so this effect decreases the heat dissipation capacity.

The importance of coat/akin color has been previously reported by Arp et al. [126], who highlighted that the surface temperature of black-haired cattle can be 21 °C above the values registered in light-haired animals due to higher absorptivity. Another example is dark-skinned pigs (Creole breed), animals that spend 4% more time lying down than light-skinned pigs with lighter skin. In multiparous Holstein cows, Anzures-Olvera et al. [127] found that when comparing black- and white-colored coats, those with a black coat color had a higher RT (0.1 °C) and also had increased mean corpuscular volumes (dark cows: 54.7 fL vs. white cows: 53.8). Moreover, cows with a white coat produced 394 kg more fat-corrected milk.

Among the strategies to reduce heat production, reducing energy consumption and locomotor activity stand out. Another strategy to reduce heat production is limiting feed intake [128]. Ruminants reduce their grazing time and feed intake when they are outside their comfort zone (12–24 °C) [110]. This behavior has been noted in 90% of water buffaloes, which opt to graze at 7 am—seen as the cooler part of the day—and allocate around 76% of their time budget for resting during the hottest hours of the day [129]. In Holstein dairy cows under an ambient temperature of 28.9–30.5 °C, Spiers et al. [130] found significant correlations between increases in rectal temperature, milk yield, and feed intake. When dairy cows had an average rectal temperature of 39 °C, reduced dry matter intakes (DMI) by 14.6 kg/day and milk yields by 11.8 kg/day were recorded. Similarly, in mid-lactation cows, Chen et al. [131] determined that for each unit increase in THI, dry matter intake (kg/day) decreased from 21.3% in thermoneutral cows to 17.3% in heat-stressed animals, which also reduced milk yield (32.5 kg/day vs. 27.0 kg/day). Decreasing feeding intake is a strategy of mammals to reduce their metabolic rate and heat load, limiting heat production and preventing a further increase in core body temperature [132].

Contrarily to feed intake, animals significantly increase their water intake [133], as observed in Angus crossbred yearling steers, in whom water intake increased by 6.8 L/animal [134]. Moreover, another activity that involves water is wallowing. Wallowing is another behavior that is particularly adopted by water buffalo when exposed to heat stress [103]. Wallowing helps to decrease body temperature by increasing heat dissipation via evaporation, as shown in a study assessing the thoracoabdominal surface temperature of water buffaloes [135]. The authors observed that ambient temperatures of 38 °C increase the average surface temperature of buffaloes up to 36.4 °C, a value that can decrease up to 4 °C after wallowing. Water buffalo spend an average of 2.35–2.91 h a day performing this behavior [105,136], and some studies have found increases to 4.06 h when buffalo heifers are exposed to intense heat (>35 °C) in silvopastoral systems [105]. These examples help to understand how wallowing is a thermoregulatory behavior that enhances heat loss via evaporation and vasodilation and prevents further increases in body temperature [136]. Table 1 shows the main physiological and behavioral changes in animals experiencing hyperthermia.

Table 1. Physiological and behavioral changes in animals experiencing hyperthermia.

Species Studied	Physiological Response and Behavioral Modifications	References
Vasodilation, sweating		
Water Buffaloes	The number of glands per cm^2 of skin is lower in buffalo (394 gland/cm^2) than in cattle (2633 gland/cm^2). This difference makes animals resort to polypnea as a heat-related tolerance response that also needs to be enhanced by providing shade.	[7]
Water buffalo	After 2 h of transport, increases in the surface temperature above 4 °C were recorded with IRT for thermal windows: the head regions, such as the lacrimal caruncle, the periocular subregion, the lower eyelid, the ear canal, the nasal passages, and the parietal subregion, as well as body regions, such as the abdominal and lumbar regions.	[62]
Review study	Peripheral vasodilation enhances heat dissipation and, therefore, helps to reduce core temperature.	[63,64]

Table 1. *Cont.*

Species Studied	Physiological Response and Behavioral Modifications	References
Review study	Sympathetic cholinergic nerves, vasoactive intestinal peptide, substance P, histamine, prostaglandins, and TRPV1 activation participate in vasodilation.	[65,66]
Cow	The horns have a thermoregulatory function. In a temperate climate zone, the temperature in the horns increased by 0.18 °C for each unit of the heat load index.	[71]
Cow	Arrangement of the blood vessels to thermoregulate is very important. For example, the parietal region is irrigated by the superficial temporal artery (*temporalis superficialis*), *transversa faciei*, *auricularis rostralis*, *palpebralis inferior lateralis*, and *palpebralis superior lateralis*, blood vessels that provide circulation to the horns, which can favor thermoregulation.	[72]
Cattle	Stress-associated hyperthermia in lactating dairy cows during isolation challenges	[73]
Sheep	hoof thermoregulation. A difference of 8.5 °C was determined between healthy and diseased hooves (diagnostic sensitivity of 92% and specificity of 91%).	[77]
Sheep	The recording of thermal stress during shearing. Ocular surface IRT was positively correlated with rectal temperature and cortisol levels.	[78]
Horses	Sweating dissipates 70% of the heat loss during exercise due to evaporation. Sweat rate ($L/h/m^2$) in horses is three times greater in horses compared to humans with a similar exercise intensity.	[74,75]
Water buffalo	The number of sweat glands/cm^2 in the head, neck, and tail regions was 20–30% higher compared to the thorax and abdomen region. This difference might impact not only the effectiveness of thermoregulation strategies in animals but also make it evident that animals use specific anatomical regions to thermoregulate.	[90]
Endocrine and immunological response		
Cattle	In hyperthermia due to acute heat stress, the concentrations of antidiuretic hormone, prolactin, glucocorticoids, and catecholamines increase, while those of aldosterone decrease. During chronic heat stress, levels of cortisol, growth hormone, and thyroxine may remain unchanged or may even decrease.	[60,94]
Miniature pigs	Blood cortisol levels, which are associated with greater production of cytokines such as interleukin-4 (IL-4), IL-5, IL-6, and IL-12.	[99]
Thermoregulatory behaviors		
Cows, goats, elephants	Behaviors that promote heat dissipation (e.g., shade-seeking, changes in body posture) or those that limit heat production (e.g., reducing feed intake and locomotor activity).	[84–87].
Cows	Adaptative behaviors (increase time standing and reduce rumination time) from an average ambient temperature of 12 °C and a temperature–humidity index of 56.	[121]
Ewes and Rams	Higher temperatures environments (35 °C) increased the total time of standing (355 vs. 486 min) and decreased eating (932 vs. 727 min). Likewise, the lying behavior of ewes and rams increased by up to 45.10 ± 2.42% during severe heat stress.	[123]
Cows	An average rectal temperature of 39 °C, reduced dry matter intake (DMI) by 14.6 kg/day, and reduced and milk yields by 11.8 kg/day were recorded.	[106]
Water Buffaloes	Wallowing helps to decrease body temperature by increasing heat dissipation via evaporation, as shown in a study assessing the thoracoabdominal surface temperature.	[135]

Management of Heat Stress in Livestock Production Systems

Stress-induced hyperthermia events in livestock are characteristic of a process of maladaptation to environmental factors or deficiencies in thermoregulation; therefore, the veterinarian's attention should focus on prevention and risk reduction. Among the main suggested actions for the management of heat stress are the cessation of livestock movement, provision of supplemental water sources in pens, cooling of the pen surface, improvement and management of nutrition and feeding plans, removal of manure from pens, and the provision of shade, which have been demonstrated to be beneficial strategies for livestock exposed to hot climates or thermal stress events [134,137–140].

6. Perspectives

It has been discussed that animals undergoing hyperthermia trigger hypothalamic thermoregulatory mechanisms to coordinate vasomotor and sweating activity, as well as behavioral modifications. However, there are still remaining areas of research, such as the multi-modal participation of thermoreceptors TRPV1, TRPA1, and TRPV3 [31,141]; studying the activation threshold of said receptors to detect thermal stimuli and how this is different from when these receptors act as nociceptors could help to understand their role during both physiological conditions.

An interesting aspect to address could be the implementation of biomarkers such as amino acids and neuropeptides that can modulate the tolerance to heat stress in birds [142]. Regarding this, in ruminants, so-called heat shock proteins (HSF1, HSP70, and HSP90) have been proposed as a method to differentiate between hyper and fever, together with other elements such as glutathione peroxidases, superoxide dismutase, and malondialdehyde [143].

On the contrary, animal temperature-evaluation methods currently rely heavily on invasive techniques like rectal thermometers, which can induce acute stress during handling. Incorporating infrared thermography could aid in non-invasively detecting temperature rises [144–146]. Thus, it is crucial to validate this technique to recognize and diagnose states of hyperthermia or thermal stress, possibly allowing the development of an automatic tool with the considerable advantage of being non-invasive. In this sense, with this same tool, it is possible to identify a more appropriate anatomical region or thermal window for each species to adequately monitor their thermal state.

7. Conclusions

The main thermoregulatory responses in animals during hyperthermia are physiological and behavioral changes modulated by the thermal circuitry between peripheral thermoreceptors, the POA, and afferent fibers to effector organs. The connection between the POA and efferent pathways is essential for animals to thermoregulate. Among the physiological responses, vasodilation is a mechanism that increases heat dissipation to reduce heat load and decrease the core body temperature. Sweating also promotes heat loss by evaporation together with polypnea. Regarding behavioral changes, livestock increase their standing time to increase their skin surface's exposure to airflow. Moreover, they engage in other behaviors to limit metabolic heat production, such as decreasing feed intake and locomotor activity. These responses aim to restore thermoneutrality and prevent further increases in the core body temperature. Understanding that livestock has a thermoneutral zone where these responses are not triggered is essential to preventing hyperthermia in farm animals.

Author Contributions: All authors contributed to the conceptualization, writing, and reading. All authors have read and agreed to the published version of the manuscript.

Funding: This research received no external funding.

Institutional Review Board Statement: Not applicable.

Informed Consent Statement: Not applicable.

Data Availability Statement: Data sharing is not applicable.

Acknowledgments: The authors wish to extend their heartfelt gratitude to Daniel Mota-Rojas, an esteemed authority in animal behavior and welfare, for his instrumental role in conceptualizing and supervising the project from which this manuscript emerged. We are sincerely appreciative of his unwavering support and invaluable guidance throughout the process.

Conflicts of Interest: The authors declare no conflicts of interest.

References

1. Thornton, P.; Nelson, G.; Mayberry, D.; Herrero, M. Impacts of Heat Stress on Global Cattle Production during the 21st Century: A Modelling Study. *Lancet Planet. Health* **2022**, *6*, e192–e201. [CrossRef]
2. Fuquay, J.W. Heat Stress as It Affects Animal Production. *J. Anim. Sci.* **1981**, *52*, 164–174. [CrossRef]
3. North, M.A.; Franke, J.A.; Ouweneel, B.; Trisos, C.H. Global Risk of Heat Stress to Cattle from Climate Change. *Environ. Res. Lett.* **2023**, *18*, 094027. [CrossRef]
4. Napolitano, F.; De Rosa, G.; Chay-Canul, A.; Álvarez-Macías, A.; Pereira, A.M.F.; Bragaglio, A.; Mora-Medina, P.; Rodríguez-González, D.; García-Herrera, R.; Hernández-Ávalos, I.; et al. The Challenge of Global Warming in Water Buffalo Farming: Physiological and Behavioral Aspects and Strategies to Face Heat Stress. *Animals* **2023**, *13*, 3103. [CrossRef]
5. Mota-Rojas, D.; Titto, C.G.; Orihuela, A.; Martínez-Burnes, J.; Gómez-Prado, J.; Torres-Bernal, F.; Flores-Padilla, K.; Carvajal-de la Fuente, V.; Wang, D. Physiological and Behavioral Mechanisms of Thermoregulation in Mammals. *Animals* **2021**, *11*, 1733. [CrossRef]
6. Hankenson, F.C.; Marx, J.O.; Gordon, C.J.; David, J.M. Effects of Rodent Thermoregulation on Animal Models in the Research Environment. *Comp. Med.* **2018**, *68*, 425–438. [CrossRef]
7. Vilela, R.A.; Lourenço Junior, J.d.B.; Jacintho, M.A.C.; Barbosa, A.V.C.; Pantoja, M.H.d.A.; Oliveira, C.M.C.; Garcia, A.R. Dynamics of Thermolysis and Skin Microstructure in Water Buffaloes Reared in Humid Tropical Climate—A Microscopic and Thermographic Study. *Front. Vet. Sci.* **2022**, *9*, 871206. [CrossRef]
8. Domínguez-Oliva, A.; Ghezzi, M.D.; Mora-Medina, P.; Hernández-Ávalos, I.; Jacome, J.; Castellón, A.; Falcón, I.; Reséndiz, F.; Romero, N.; Ponce, R.; et al. Anatomical, Physiological, and Behavioral Mechanisms of Thermoregulation in Elephants. *J. Anim. Behav. Biometeorol.* **2022**, *10*, 2233. [CrossRef]
9. Mota-rojas, D.; Wang, D.; Titto, C.G.; Gómez-prado, J.; Fuente, V.C.; Ghezzi, M.; Boscato-funes, L.; Barrios-garcía, H.; Torres-bernal, F.; Casas-alvarado, A.; et al. Pathophysiology of Fever and Application of Infrared Thermography (IRT) in the Detection of Sick Domestic Animals: Recent Advances. *Animals* **2021**, *11*, 2316. [CrossRef]
10. Shelton, D.S.; Alberts, J.R. Development of Behavioral Responses to Thermal Challenges. *Dev. Psychobiol.* **2018**, *60*, 5–14. [CrossRef]
11. Athaíde, L.G.; Joset, W.C.L.; de Almeida, J.C.F.; Pantoja, M.H.d.A.; Noronha, R.d.P.P.; Bezerra, A.S.; Barbosa, A.V.C.; Martorano, L.G.; da Silva, J.A.R.; Lourenço Júnior, J.d.B. Thermoregulatory and Behavioral Responses of Buffaloes With and Without Direct Sun Exposure During Abnormal Environmental Condition in Marajó Island, Pará, Brazil. *Front. Vet. Sci.* **2020**, *7*, 522551. [CrossRef] [PubMed]
12. Nagashima, K. Central Mechanisms for Thermoregulation in a Hot Environment. *Ind. Health* **2006**, *44*, 359–367. [CrossRef] [PubMed]
13. Kadzere, C.T.; Murphy, M.R.; Silanikove, N.; Maltz, E. Heat Stress in Lactating Dairy Cows: A Review. *Livest. Prod. Sci.* **2002**, *77*, 59–91. [CrossRef]
14. Nakamura, K.; Morrison, S.F. A Thermosensory Pathway Mediating Heat-Defense Responses. *Proc. Natl. Acad. Sci. USA* **2010**, *107*, 8848–8853. [CrossRef] [PubMed]
15. Dimicco, J.A.; Zaretsky, D.V. The Dorsomedial Hypothalamus: A New Player in Thermoregulation. *Am. J. Physiol. Regul. Integr. Comp. Physiol.* **2007**, *292*, R47–R63. [CrossRef] [PubMed]
16. Autio, E.; Heiskanen, M.-L.; Mononen, J. Thermographic Evaluation of the Lower Critical Temperature in Weanling Horses. *J. Appl. Anim. Welf. Sci.* **2007**, *10*, 207–216. [CrossRef]
17. Kang, H.; Zsoldos, R.R.; Sole-Guitart, A.; Narayan, E.; Cawdell-Smith, A.J.; Gaughan, J.B. Heat Stress in Horses: A Literature Review. *Int. J. Biometeorol.* **2023**, *67*, 957–973. [CrossRef] [PubMed]
18. Herbut, P.; Angrecka, S.; Godyń, D.; Hoffmann, G. The Physiological and Productivity Effects of Heat Stress in Cattle—A Review. *Ann. Anim. Sci.* **2019**, *19*, 579–593. [CrossRef]
19. Tansey, E.A.; Johnson, C.D. Recent Advances in Thermoregulation. *Adv. Physiol. Educ.* **2015**, *39*, 139–148. [CrossRef]
20. Rosania, K. Blocking the Body's Response to Cold. *Lab Anim.* **2012**, *41*, 89. [CrossRef]
21. Lezama-García, K.; Mota-Rojas, D.; Pereira, A.M.F.; Martínez-Burnes, J.; Ghezzi, M.; Domínguez, A.; Gómez, J.; de Mira Geraldo, A.; Lendez, P.; Hernández-Ávalos, I.; et al. Transient Receptor Potential (TRP) and Thermoregulation in Animals: Structural Biology and Neurophysiological Aspects. *Animals* **2022**, *12*, 106. [CrossRef] [PubMed]
22. Reid, G.; Flonta, M.-L. Cold Current in Thermoreceptive Neurons. *Nature* **2001**, *413*, 480. [CrossRef]
23. Lamas, J.A.; Rueda-Ruzafa, L.; Herrera-Pérez, S. Ion Channels and Thermosensitivity: TRP, TREK, or Both? *Int. J. Mol. Sci.* **2019**, *20*, 2371. [CrossRef] [PubMed]
24. Wang, H.; Siemens, J. TRP Ion Channels in Thermosensation, Thermoregulation and Metabolism. *Temperature* **2015**, *2*, 178–187. [CrossRef] [PubMed]

25. Rosenbaum, T.; Simon, S.A. TRPV1 Receptors and Signal Transduction. In *TRP Ion Channel Function in Sensory Transduction and Cellular Signaling Cascades*; Liedtke, W.B., Heller, S., Eds.; CRC Press/Taylor & Francis: Boca Raton, FL, USA, 2007; Chapter 5, ISBN 0849340489. Available online: https://www.ncbi.nlm.nih.gov/books/NBK5260/ (accessed on 16 April 2024).
26. Desai, B.N.; Clapham, D.E. TRP Channels and Mice Deficient in TRP Channels. *Pflügers Arch.—Eur. J. Physiol.* **2005**, *451*, 11–18. [CrossRef]
27. Vriens, J.; Nilius, B.; Voets, T. Peripheral Thermosensation in Mammals. *Nat. Rev. Neurosci.* **2014**, *15*, 573–589. [CrossRef]
28. Morrison, S.F.; Nakamura, K. Central Mechanisms for Thermoregulation. *Annu. Rev. Physiol.* **2019**, *81*, 285–308. [CrossRef]
29. Togashi, K.; Hara, Y.; Tominaga, T.; Higashi, T.; Konishi, Y.; Mori, Y.; Tominaga, M. TRPM2 Activation by Cyclic ADP-Ribose at Body Temperature Is Involved in Insulin Secretion. *EMBO J.* **2006**, *25*, 1804–1815. [CrossRef]
30. Nozadze, I.; Tsiklauri, N.; Gurtskaia, G.; Tsagareli, M.G. Role of Thermo TRPA1 and TRPV1 Channels in Heat, Cold, and Mechanical Nociception of Rats. *Behav. Pharmacol.* **2016**, *27*, 29–36. [CrossRef]
31. Cheng, W.; Yang, F.; Liu, S.; Colton, C.K.; Wang, C.; Cui, Y.; Cao, X.; Zhu, M.X.; Sun, C.; Wang, K.; et al. Heteromeric Heat-Sensitive Transient Receptor Potential Channels Exhibit Distinct Temperature and Chemical Response. *J. Biol. Chem.* **2012**, *287*, 7279–7288. [CrossRef]
32. Lei, J.; Yoshimoto, R.U.; Matsui, T.; Amagai, M.; Kido, M.A.; Tominaga, M. Involvement of Skin TRPV3 in Temperature Detection Regulated by TMEM79 in Mice. *Nat. Commun.* **2023**, *14*, 4104. [CrossRef]
33. Vandewauw, I.; De Clercq, K.; Mulier, M.; Held, K.; Pinto, S.; Van Ranst, N.; Segal, A.; Voet, T.; Vennekens, R.; Zimmermann, K.; et al. A TRP Channel Trio Mediates Acute Noxious Heat Sensing. *Nature* **2018**, *555*, 662–666. [CrossRef]
34. Paricio-Montesinos, R.; Schwaller, F.; Udhayachandran, A.; Rau, F.; Walcher, J.; Evangelista, R.; Vriens, J.; Voets, T.; Poulet, J.F.A.; Lewin, G.R. The Sensory Coding of Warm Perception. *Neuron* **2020**, *106*, 830–841.e3. [CrossRef]
35. Yahiro, T.; Kataoka, N.; Nakamura, Y.; Nakamura, K. The Lateral Parabrachial Nucleus, but Not the Thalamus, Mediates Thermosensory Pathways for Behavioural Thermoregulation. *Sci. Rep.* **2017**, *7*, 5031. [CrossRef]
36. Childs, C. Body temperature and clinical thermometry. *Handb. Clin. Neurol.* **2018**, *157*, 467–482. [CrossRef] [PubMed]
37. Yang, W.Z.; Du, X.; Zhang, W.; Gao, C.; Xie, H.; Xiao, Y.; Jia, X.; Liu, J.; Xu, J.; Fu, X.; et al. Parabrachial Neuron Types Categorically Encode Thermoregulation Variables during Heat Defense. *Sci. Adv.* **2020**, *6*, eabb9414. [CrossRef] [PubMed]
38. Bienboire-Frosini, C.; Wang, D.; Marcet-Rius, M.; Villanueva-García, D.; Gazzano, A.; Domínguez-Oliva, A.; Olmos-Hernández, A.; Hernández-Ávalos, I.; Lezama-García, K.; Verduzco-Mendoza, A.; et al. The Role of Brown Adipose Tissue and Energy Metabolism in Mammalian Thermoregulation during the Perinatal Period. *Animals* **2023**, *13*, 2173. [CrossRef] [PubMed]
39. Tan, C.L.; Cooke, E.K.; Leib, D.E.; Lin, Y.-C.; Daly, G.E.; Zimmerman, C.A.; Knight, Z.A. Warm-Sensitive Neurons That Control Body Temperature. *Cell* **2016**, *167*, 47–59.e15. [CrossRef] [PubMed]
40. Zhao, Z.-D.; Yang, W.Z.; Gao, C.; Fu, X.; Zhang, W.; Zhou, Q.; Chen, W.; Ni, X.; Lin, J.-K.; Yang, J.; et al. A Hypothalamic Circuit That Controls Body Temperature. *Proc. Natl. Acad. Sci. USA* **2017**, *114*, 2042–2047. [CrossRef]
41. Kaiyala, K.J.; Ogimoto, K.; Nelson, J.T.; Muta, K.; Morton, G.J. Physiological Role for Leptin in the Control of Thermal Conductance. *Mol. Metab.* **2016**, *5*, 892–902. [CrossRef]
42. Keringer, P.; Furedi, N.; Gaszner, B.; Miko, A.; Pakai, E.; Fekete, K.; Olah, E.; Kelava, L.; Romanovsky, A.A.; Rumbus, Z.; et al. The Hyperthermic Effect of Central Cholecystokinin Is Mediated by the Cyclooxygenase-2 Pathway. *Am. J. Physiol.-Endocrinol. Metab.* **2022**, *322*, E10–E23. [CrossRef] [PubMed]
43. Nakamura, Y.; Yahiro, T.; Fukushima, A.; Kataoka, N.; Hioki, H.; Nakamura, K. Prostaglandin EP3 Receptor–Expressing Preoptic Neurons Bidirectionally Control Body Temperature via Tonic GABAergic Signaling. *Sci. Adv.* **2022**, *8*, eadd5463. [CrossRef] [PubMed]
44. Burke, S.; Hanani, M. The Actions of Hyperthermia on the Autonomic Nervous System: Central and Peripheral Mechanisms and Clinical Implications. *Auton. Neurosci.* **2012**, *168*, 4–13. [CrossRef] [PubMed]
45. Romanovsky, A.A. Thermoregulation: Some Concepts Have Changed. Functional Architecture of the Thermoregulatory System. *Am. J. Physiol.-Regul. Integr. Comp. Physiol.* **2007**, *292*, R37–R46. [CrossRef] [PubMed]
46. Jung, S.; Lee, M.; Kim, D.-Y.; Son, C.; Ahn, B.H.; Heo, G.; Park, J.; Kim, M.; Park, H.-E.; Koo, D.-J.; et al. A Forebrain Neural Substrate for Behavioral Thermoregulation. *Neuron* **2022**, *110*, 266–279.e9. [CrossRef] [PubMed]
47. Wang, T.A.; Teo, C.F.; Åkerblom, M.; Chen, C.; Tynan-La Fontaine, M.; Greiner, V.J.; Diaz, A.; McManus, M.T.; Jan, Y.N.; Jan, L.Y. Thermoregulation via Temperature-Dependent PGD2 Production in Mouse Preoptic Area. *Neuron* **2019**, *103*, 309–322.e7. [CrossRef] [PubMed]
48. Norris, A.J.; Shaker, J.R.; Cone, A.L.; Ndiokho, I.B.; Bruchas, M.R. Parabrachial Opioidergic Projections to Preoptic Hypothalamus Mediate Behavioral and Physiological Thermal Defenses. *eLife* **2021**, *10*, e60779. [CrossRef] [PubMed]
49. Nakamura, K. Central Circuitries for Body Temperature Regulation and Fever. *Am. J. Physiol.-Regul. Integr. Comp. Physiol.* **2011**, *301*, R1207–R1228. [CrossRef] [PubMed]
50. Yahiro, T.; Kataoka, N.; Nakamura, K. Two Ascending Thermosensory Pathways from the Lateral Parabrachial Nucleus That Mediate Behavioral and Autonomous Thermoregulation. *J. Neurosci.* **2023**, *43*, 5221–5240. [CrossRef]
51. Morrison, S.F. Central Neural Pathways for Thermoregulation. *Front. Biosci.* **2011**, *16*, 74. [CrossRef]
52. Morrison, S.F.; Nakamura, K.; Madden, C.J. Central Control of Thermogenesis in Mammals. *Exp. Physiol.* **2008**, *93*, 773–797. [CrossRef] [PubMed]

53. Francisco, M.A.; Gibson, B.M.; Simmons, G.H.; Halliwill, J.R.; Minson, C.T. Cholinergic Nerve Contribution to Cutaneous Active Vasodilation during Exercise Is Similar to Whole Body Passive Heating. *J. Appl. Physiol.* **2023**, *134*, 933–940. [CrossRef] [PubMed]
54. Christen, S.; Declachaux, A.; Dischl, B.; Golay, S.; Liaudet, L.; Feihl, F.; Waeber, N. Dose-Dependent Vasodilatory Effects of Acetylcholine and Local Warming on Skin Microcirculation. *J. Cardiovasc. Pharmacol.* **2004**, *44*, 659–664. [CrossRef] [PubMed]
55. Nazari, S.; Kourosh-Arami, M.; Komaki, A.; Hajizadeh, S. Relative Contribution of Central and Peripheral Factors in Superficial Blood Flow Regulation Following Cold Exposure. *Physiol. Pharmacol.* **2020**, *24*, 89–100. [CrossRef]
56. Francisco, M.A.; Minson, C.T. Cutaneous active vasodilation as a heat loss thermoeffector. *Handb. Clin. Neurol.* **2018**, *156*, 193–209 [CrossRef] [PubMed]
57. Mota-Rojas, D.; Ogi, A.; Villanueva-García, D.; Hernández-Ávalos, I.; Casas-Alvarado, A.; Domínguez-Oliva, A.; Lendez, P.; Ghezzi, M. Thermal Imaging as a Method to Indirectly Assess Peripheral Vascular Integrity and Tissue Viability in Veterinary Medicine: Animal Models and Clinical Applications. *Animals* **2023**, *14*, 142. [CrossRef]
58. da Silva, W.C.; da Silva, J.A.R.; Martorano, L.G.; da Silva, É.B.R.; Sousa, C.E.L.; Neves, K.A.L.; de Araújo, C.V.; Joaquim, L.A.; Rodrigues, T.C.G.d.C.; Belo, T.S.; et al. Thermographic Profiles in Livestock Systems under Full Sun and Shaded Pastures during an Extreme Climate Event in the Eastern Amazon, Brazil: El Niño of 2023. *Animals* **2024**, *14*, 855. [CrossRef] [PubMed]
59. Lees, A.M.; Sejian, V.; Wallage, A.L.; Steel, C.C.; Mader, T.L.; Lees, J.C.; Gaughan, J.B. The Impact of Heat Load on Cattle. *Animals* **2019**, *9*, 322. [CrossRef] [PubMed]
60. Idris, M.; Uddin, J.; Sullivan, M.; McNeill, D.M.; Phillips, C.J.C. Non-Invasive Physiological Indicators of Heat Stress in Cattle. *Animals* **2021**, *11*, 71. [CrossRef] [PubMed]
61. Robinson, E.B. Thermoregulation. In *Textbook of Veterinary Physiology*; Klein, B.G., Ed.; Elsevier: Amsterdam, The Netherlands, 2014; pp. 559–567.
62. Rodríguez-González, D.; Guerrero-Legarreta, I.; Cruz Monterrosa, R.G.; Napolitano, F.; Gonçalves-Titto, C.; El-Aziz, A.H.A.; Hernández-Ávalos, I.; Casas-Alvarado, A.; Oliva-Domínguez, A.; Mota-Rojas, D. Assessment of Thermal Changes in Water Buffalo Mobilized from the Paddock and Transported by Short Journeys. *Front. Vet. Sci.* **2023**, *10*, 1184577. [CrossRef]
63. Johnson, J.M.; Minson, C.T.; Kellogg, D.L. Cutaneous Vasodilator and Vasoconstrictor Mechanisms in Temperature Regulation. In *Comprehensive Physiology*; Wiley: Hoboken, NJ, USA, 2014; pp. 33–89.
64. Hodges, G.J.; Johnson, J.M. Adrenergic Control of the Human Cutaneous Circulation. *Appl. Physiol. Nutr. Metab.* **2009**, *34*, 829–839. [CrossRef]
65. Holowatz, L.A. Aging and the Control of Human Skin Blood Flow. *Front. Biosci.* **2010**, *15*, 718. [CrossRef] [PubMed]
66. Holowatz, L.A.; Kenney, W.L. Peripheral Mechanisms of Thermoregulatory Control of Skin Blood Flow in Aged Humans. *J. Appl. Physiol.* **2010**, *109*, 1538–1544. [CrossRef] [PubMed]
67. Bertoni, A.; Mota-Rojas, D.; Álvarez-Macias, A.; Mora-Medina, P.; Guerrero-Legarreta, I.; Morales-Canela, A.; Gómez-Prado, J.; José-Pérez, N.; Martínez-Burnes, J. Scientific Findings Related to Changes in Vascular Microcirculation Using Infrared Thermography in the River Buffalo. *J. Anim. Behav. Biometeorol.* **2020**, *8*, 288–297. [CrossRef]
68. Lezama-García, K.; Martínez-Burnes, J.; Pérez-Jiménez, J.C.; Domínguez-Oliva, A.; Mora-Medina, P.; Olmos-Hernández, A.; Hernández-Ávalos, I.; Mota-Rojas, D. Relation between the Dam's Weight on Superficial Temperature of Her Puppies at Different Stages of the Post-Partum. *Vet. Sci.* **2022**, *9*, 673. [CrossRef] [PubMed]
69. Verduzco-Mendoza, A.; Bueno-Nava, A.; Wang, D.; Martínez-Burnes, J.; Olmos-Hernández, A.; Casas, A.; Domínguez, A.; Mota-Rojas, D. Experimental Applications and Factors Involved in Validating Thermal Windows Using Infrared Thermography to Assess the Health and Thermostability of Laboratory Animals. *Animals* **2021**, *11*, 3448. [CrossRef] [PubMed]
70. Taylor, C.R. The Vascularity and Possible Thermoregulatory Function of the Horns in Goats. *Physiol. Zool.* **1966**, *39*, 127–139. [CrossRef]
71. Algra, M.; de Keijzer, L.; Arndt, S.S.; van Eerdenburg, F.J.C.M.; Goerlich, V.C. Evaluation of the Thermal Response of the Horns in Dairy Cattle. *Animals* **2023**, *13*, 500. [CrossRef]
72. ICVGAN. *Nómina Anatómica Veterinaria*, 6th ed.; Editorial Committee: Hanover, Germany; Ghent, Belgium; Columbia, MO, USA; Rio de Janeiro, Brazil, 2017.
73. Mincu, M.; Nicolae, I.; Gavojdian, D. Infrared Thermography as a Non-Invasive Method for Evaluating Stress in Lactating Dairy Cows during Isolation Challenges. *Front. Vet. Sci.* **2023**, *10*, 1236668. [CrossRef] [PubMed]
74. Velasco-Bolaños, J.; Ceballes-Serrano, C.C.; Velásquez-Mejía, D.; Riaño-Rojas, J.C.; Giraldo, C.E.; Carmona, J.U.; Ceballos-Márquez, A. Application of Udder Surface Temperature by Infrared Thermography for Diagnosis of Subclinical Mastitis in Holstein Cows Located in Tropical Highlands. *J. Dairy Sci.* **2021**, *104*, 10310–10323. [CrossRef]
75. Samara, E.M.; Ayadi, M.; Aljumaah, R.S. Feasibility of Utilising an Infrared-Thermographic Technique for Early Detection of Subclinical Mastitis in Dairy Camels (*Camelus dromedarius*). *J. Dairy Res.* **2014**, *81*, 38–45. [CrossRef]
76. Uddin, J.; McNeill, D.M.; Phillips, C.J.C. Infrared Thermography as a Tool for the Measurement of Negative Emotions in Dairy Cows. *Int. J. Biometeorol.* **2023**, *67*, 219–231. [CrossRef] [PubMed]
77. Byrne, D.T.; Berry, D.P.; Esmonde, H.; McGovern, F.; Creighton, P.; McHugh, N. Infrared Thermography as a Tool to Detect Hoof Lesions in Sheep. *Transl. Anim. Sci.* **2019**, *3*, 577–588. [CrossRef] [PubMed]
78. Arfuso, F.; Acri, G.; Piccione, G.; Sansotta, C.; Fazio, F.; Giudice, E.; Giannetto, C. Eye Surface Infrared Thermography Usefulness as a Noninvasive Method of Measuring Stress Response in Sheep during Shearing: Correlations with Serum Cortisol and Rectal Temperature Values. *Physiol. Behav.* **2022**, *250*, 113781. [CrossRef] [PubMed]

79. de Lima, V.; Piles, M.; Rafel, O.; López-Béjar, M.; Ramón, J.; Velarde, A.; Dalmau, A. Use of Infrared Thermography to Assess the Influence of High Environmental Temperature on Rabbits. *Res. Vet. Sci.* **2013**, *95*, 802–810. [CrossRef] [PubMed]
80. Redaelli, V.; Luzi, F.; Mazzola, S.; Bariffi, G.; Zappaterra, M.; Nanni Costa, L.; Padalino, B. The Use of Infrared Thermography (IRT) as Stress Indicator in Horses Trained for Endurance: A Pilot Study. *Animals* **2019**, *9*, 84. [CrossRef] [PubMed]
81. Whittaker, A.L.; Muns, R.; Wang, D.; Martínez-Burnes, J.; Hernández-Ávalos, I.; Casas-Alvarado, A.; Domínguez-Oliva, A.; Mota-Rojas, D. Assessment of Pain and Inflammation in Domestic Animals Using Infrared Thermography: A Narrative Review. *Animals* **2023**, *13*, 2065. [CrossRef] [PubMed]
82. Redaelli, V.; Luzi, F.; Farish, M.; Nanni Costa, L. The Use of Thermography to Assess the Teeth Temperature during Resection by Grinding in Piglets. In Proceedings of the 2012 International Conference on Quantitative InfraRed Thermography, Naples, Italy, 11–14 June 2012; QIRT Council: Naples, Italy, 2012; pp. 1–2.
83. Van der Saag, D.; Lomax, S.; Windsor, P.A.; Taylor, C.; White, P.J. Evaluating Treatments with Topical Anaesthetic and Buccal Meloxicam for Pain and Inflammation Caused by Amputation Dehorning of Calves. *PLoS ONE* **2018**, *13*, e0198808. [CrossRef] [PubMed]
84. Eicher, S.D.; Cheng, H.W.; Sorrells, A.D.; Schutz, M.M. Short Communication: Behavioral and Physiological Indicators of Sensitivity or Chronic Pain Following Tail Docking. *J. Dairy Sci.* **2006**, *89*, 3047–3051. [CrossRef] [PubMed]
85. Shibasaki, M.; Wilson, T.E.; Crandall, C.G. Neural Control and Mechanisms of Eccrine Sweating during Heat Stress and Exercise. *J. Appl. Physiol.* **2006**, *100*, 1692–1701. [CrossRef]
86. Hafez, E.S.E.; Badreldin, A.L.; Shafei, M.M. Skin Structure of Egyptian Buffaloes and Cattle with Particular Reference to Sweat Glands. *J. Agric. Sci.* **1955**, *46*, 19–30. [CrossRef]
87. McCutcheon, L.J.; Geor, R.J. Thermoregulation and exercise-associated heat illnesses. In *Equine Sports Medicine and Surgery*; Elsevier: Amsterdam, The Netherlands, 2014; pp. 901–918.
88. Kingston, J.K.; Geor, R.J.; McCutcheon, L.J. Rate and Composition of Sweat Fluid Losses Are Unaltered by Hypohydration during Prolonged Exercise in Horses. *J. Appl. Physiol.* **1997**, *83*, 1133–1143. [CrossRef] [PubMed]
89. Guthrie, A.J.; Lund, R.J. Thermoregulation: Base Mechanisms and Hyperthermia. *Vet. Clin. N. Am. Equine Pract.* **1998**, *14*, 45–59. [CrossRef] [PubMed]
90. Debbarma, D.; Uppal, V.; Bansal, N.; Gupta, A. Histomorphometrical Study on Regional Variation in Distribution of Sweat Glands in Buffalo Skin. *Dermatol. Res. Pract.* **2018**, *2018*, 1–7. [CrossRef] [PubMed]
91. Barros, D.V.; Silva, L.K.X.; Kahwage, P.R.; Lourenço Júnior, J.B.; Sousa, J.S.; Silva, A.G.M.; Franco, I.M.; Martorano, L.G.; Garcia, A.R. Assessment of Surface Temperatures of Buffalo Bulls (*Bubalus bubalis*) Raised under Tropical Conditions Using Infrared Thermography. *Arq. Bras. De Med. Vet. E Zootec.* **2016**, *68*, 422–430. [CrossRef]
92. Bauman, D.E.; Bruce Currie, W. Partitioning of Nutrients During Pregnancy and Lactation: A Review of Mechanisms Involving Homeostasis and Homeorhesis. *J. Dairy Sci.* **1980**, *63*, 1514–1529. [CrossRef]
93. Koenneker, K.; Schulze, M.; Pieper, L.; Jung, M.; Schmicke, M.; Beyer, F. Comparative Assessment of the Stress Response of Cattle to Common Dairy Management Practices. *Animals* **2023**, *13*, 2115. [CrossRef] [PubMed]
94. Afsal, A.; Sejian, V.; Bagath, M.; Krishnan, G.; Devaraj, C.; Bhatta, R. Heat Stress and Livestock Adaptation: Neuroendocrine Regulation. *Int. J. Vet. Anim. Med.* **2018**, *1*, 108.
95. Ochoa-Amaya, J.E.; Malucelli, B.E.; Cruz-Casallas, P.E.; Nasello, A.G.; Felicio, L.F.; Carvalho-Freitas, M.I.R. Acute and Chronic Stress and the Inflammatory Response in Hyperprolactinemic Rats. *Neuroimmunomodulation* **2010**, *17*, 386–395. [CrossRef] [PubMed]
96. Levine, S.; Muneyyirci-Delale, O. Stress-Induced Hyperprolactinemia: Pathophysiology and Clinical Approach. *Obstet. Gynecol. Int.* **2018**, *2018*, 1–6. [CrossRef]
97. Chen, Y.; Arsenault, R.; Napper, S.; Griebel, P. Models and Methods to Investigate Acute Stress Responses in Cattle. *Animals* **2015**, *5*, 1268–1295. [CrossRef]
98. Sejian, V.; Bhatta, R.; Gaughan, J.B.; Dunshea, F.R.; Lacetera, N. Review: Adaptation of Animals to Heat Stress. *Animal* **2018**, *12*, S431–S444. [CrossRef] [PubMed]
99. Ju, X.-H.; Xu, H.-J.; Yong, Y.-H.; An, L.-L.; Jiao, P.-R.; Liao, M. Heat Stress Upregulation of Toll-like Receptors 2/4 and Acute Inflammatory Cytokines in Peripheral Blood Mononuclear Cell (PBMC) of Bama Miniature Pigs: An in Vivo and in Vitro Study. *Animal* **2014**, *8*, 1462–1468. [CrossRef] [PubMed]
100. Bagath, M.; Krishnan, G.; Devaraj, C.; Rashamol, V.P.; Pragna, P.; Lees, A.M.; Sejian, V. The Impact of Heat Stress on the Immune System in Dairy Cattle: A Review. *Res. Vet. Sci.* **2019**, *126*, 94–102. [CrossRef] [PubMed]
101. El Sabry, M.; Almasri, O. Space Allowance: A Tool for Improving Behavior, Milk and Meat Production, and Reproduction Performance of Buffalo in Different Housing Systems—A Review. *Trop. Anim. Health Prod.* **2022**, *54*, 7–10. [CrossRef]
102. Galloso-Hernández, M.A.; Rodríguez-Estévez, V.; Alvarez-Díaz, C.A.; Soca-Pérez, M.; Dublin, D.; Iglesias-Gómez, J.; Simon Guelmes, L. Effect of Silvopastoral Systems in the Thermoregulatory and Feeding Behaviors of Water Buffaloes Under Different Conditions of Heat Stress. *Front. Vet. Sci.* **2020**, *7*, 393. [CrossRef] [PubMed]
103. Mota-Rojas, D.; Habeeb, A.; Ghezzi, M.D.; Kanth, P.; Napolitano, F.; Lendez, P.A.; Cuibus, A.; Ceriani, M.C.; Sarubbi, J.; Braghieri, A.; et al. Termorregulación del búfalo de agua: Mecanismos neurobiológicos, cambios microcirculatorios y aplicaciones prácticas de la termografía infrarroja. In *El Búfalo de Agua en Latinoamérica, Hallazgos Recientes*; Napolitano, F., Mota-Rojas, D., Guerrero-Legarreta, I., Orihuela, A., Eds.; BM Editores: Ciudad de México, México, 2020; pp. 923–958.

104. Amit, M.D.; Dash, S.S.; Sahu, S.; Sarangi, A.; Singh, M. Effect of Microclimate on Feeding, Drinking and Physiological Parameters of Buffalo: A Review. *Pharma Innov. J.* **2021**, *10*, 2416–2419.
105. Galloso-Hernández, M.A.; Soca-Pérez, M.; Dublin, D.; Alvarez-Díaz, C.A.; Iglesias-Gómez, J.; Díaz-Gaona, C.; Rodríguez-Estévez, V. Thermoregulatory and Feeding Behavior under Different Management and Heat Stress Conditions in Heifer Water Buffalo (*Bubalus bubalis*) in the Tropics. *Animals* **2021**, *11*, 1162. [CrossRef]
106. Garcia, A.R.; Silva, L.K.X.; Barros, D.V.; Junior, J.d.B.L.; Martorano, L.G.; Lisboa, L.S.S.; da Silva, J.A.R.; de Sousa, J.S.; da Silva, A.O.A. Key Points for the Thermal Comfort of Water Buffaloes in Eastern Amazon. *Ciência Rural* **2023**, *53*, e20210544. [CrossRef]
107. Joksimovic-Todorovic, M.; Davidovic, V.; Hristov, S.; Stankovic, B. Effect of Heat Stress on Milk Production in Dairy Cows. *Biotechnol. Anim. Husb.* **2011**, *27*, 1017–1023. [CrossRef]
108. Wright, P. Why Do Elephants Flap Their Ears? *South Afr. J. Zool.* **1984**, *19*, 266–269. [CrossRef]
109. Phillips, P.; Heath, J. Heat Loss in Dumbo: A Theoretical Approach. *J. Therm. Biol.* **2001**, *26*, 117–120. [CrossRef] [PubMed]
110. Gupta, M.; Mondal, T. Heat Stress and Thermoregulatory Responses of Goats: A Review. *Biol. Rhythm Res.* **2021**, *52*, 407–433. [CrossRef]
111. Kamal, R.; Dutt, T.; Patel, M.; Dey, A.; Bharti, P.K.; Chandran, P.C. Heat Stress and Effect of Shade Materials on Hormonal and Behavior Response of Dairy Cattle: A Review. *Trop. Anim. Health Prod.* **2018**, *50*, 701–706. [CrossRef]
112. Berihulay, H.; Abied, A.; He, X.; Jiang, L.; Ma, Y. Adaptation Mechanisms of Small Ruminants to Environmental Heat Stress. *Animals* **2019**, *9*, 75. [CrossRef]
113. Gaughan, J.B.; Tait, L.A.; Eigenberg, R.; Bryden, W.L. Effect of Shade on Respiration Rate and Rectal Temperature of Angus Heifers. *Anim. Prod. Aust.* **2004**, *25*, 69–72.
114. Kamal, R.; Dutt, T.; Patel, B.H.M.; Singh, G.; Chandran, P.C.; Dey, A.; Barari, S.K. Effect of Shade Materials on Rectal Temperature, Respiration Rate and Body Surface Temperature of Crossbred Calves during Rainy Season. *Indian J. Anim. Sci.* **2016**, *86*, 75–81. [CrossRef]
115. Taylor, D.; Brown, W.; Price, I.; Trotter, M.; Lamb, D.; Hinch, G. Gps Tracking of Sheep to Investigate Shelter and Shade Use in Relation to Weather Conditions. In Proceedings of the 10th International Conference Precision Agriculture, Denver, CO, USA, 18–21 July 2010; pp. 247–248.
116. Knight, M.I.; Linden, N.P.; Butler, K.L.; Rice, M.; Ponnampalam, E.N.; Behrendt, R.; Jongman, E.C. The Effect of Shade on Sheep Grazing Pasture during Summer Conditions. *J. Vet. Behav.* **2023**, *64–65*, 16–24. [CrossRef]
117. Tucker, C.B.; Rogers, A.R.; Schütz, K.E. Effect of Solar Radiation on Dairy Cattle Behaviour, Use of Shade and Body Temperature in a Pasture-Based System. *Appl. Anim. Behav. Sci.* **2008**, *109*, 141–154. [CrossRef]
118. Herbut, P.; Hoffmann, G.; Angrecka, S.; Godyń, D.; Vieira, F.M.C.; Adamczyk, K.; Kupczyński, R. The Effects of Heat Stress on the Behaviour of Dairy Cows—A Review. *Ann. Anim. Sci.* **2021**, *21*, 385–402. [CrossRef]
119. Allen, J.D.; Hall, L.W.; Collier, R.J.; Smith, J.F. Effect of Core Body Temperature, Time of Day, and Climate Conditions on Behavioral Patterns of Lactating Dairy Cows Experiencing Mild to Moderate Heat Stress. *J. Dairy Sci.* **2015**, *98*, 118–127. [CrossRef]
120. Herbut, P.; Angrecka, S. The Effect of Heat Stress on Time Spent Lying by Cows in a Housing System. *Ann. Anim. Sci.* **2018**, *18*, 825–833. [CrossRef]
121. Hut, P.R.; Scheurwater, J.; Nielen, M.; van den Broek, J.; Hostens, M.M. Heat Stress in a Temperate Climate Leads to Adapted Sensor-Based Behavioral Patterns of Dairy Cows. *J. Dairy Sci.* **2022**, *105*, 6909–6922. [CrossRef] [PubMed]
122. Reis, B.R.d.; Nguyen, T.; Sujani, S.; White, R.R. Open-Source Wearable Sensors for Behavioral Analysis of Sheep Undergoing Heat Stress. *Appl. Sci.* **2023**, *13*, 9281. [CrossRef]
123. Li, F.K.; Yang, Y.; Jenna, K.; Xia, C.H.; Lv, S.J.; Wei, W.H. Effect of Heat Stress on the Behavioral and Physiological Patterns of Small-Tail Han Sheep Housed Indoors. *Trop. Anim. Health Prod.* **2018**, *50*, 1893–1901. [CrossRef] [PubMed]
124. Bonneau, M.; Poullet, N.; Beramice, D.; Dantec, L.; Canario, L.; Gourdine, J.-L. Behavior Comparison During Chronic Heat Stress in Large White and Creole Pigs Using Image-Analysis. *Front. Anim. Sci.* **2021**, *2*, 784376. [CrossRef]
125. Gómez-Prado, J.; Pereira, A.M.F.; Wang, D.; Villanueva-García, D.; Domínguez-Oliva, A.; Mora-Medina, P.; Hernández-Avalos, I.; Martínez-Burnes, J.; Casas-Alvarado, A.; Olmos-Hernández, A.; et al. Thermoregulation Mechanisms and Perspectives for Validating Thermal Windows in Pigs with Hypothermia and Hyperthermia: An Overview. *Front. Vet. Sci.* **2022**, *9*, 1023294. [CrossRef] [PubMed]
126. Arp, S.; Owens, F.; Armbruster, S.; Schmidt, D. Effect of Animal Density, Coat Color and Heat Stress on Performance of Feedlot Steers. *Anim. Sci. Res. Rep.* **1983**, *114*, 79–81.
127. Anzures-Olvera, F.; Véliz, F.G.; de Santiago, A.; García, J.E.; Mellado, J.; Macías-Cruz, U.; Avendaño-Reyes, L.; Mellado, M. The Impact of Hair Coat Color on Physiological Variables, Reproductive Performance and Milk Yield of Holstein Cows in a Hot Environment. *J. Therm. Biol.* **2019**, *81*, 82–88. [CrossRef] [PubMed]
128. Terrien, J.; Perret, M.; Aujard, F. Behavioral Thermoregulation in Mammals: A Review. *Front. Biosci.* **2011**, *16*, 1428. [CrossRef]
129. Fundora, O.; Sánchez, R.R.R. Datos Preliminares de La Conducta Alimentaria de Búfalos de Río En Pastoreo. *Rev. Cuba. De Cienc. Agrícol.* **2001**, *35*, 15–17.
130. Spiers, D.E.; Spain, J.N.; Sampson, J.D.; Rhoads, R.P. Use of Physiological Parameters to Predict Milk Yield and Feed Intake in Heat-Stressed Dairy Cows. *J. Therm. Biol.* **2004**, *29*, 759–764. [CrossRef]
131. Chen, L.; Thorup, V.M.; Kudahl, A.B.; Østergaard, S. Impact of Heat Stress on Feed Intake, Milk Yield, Milk Composition, and Feed Efficiency in Dairy Cows: A Meta-Analysis. *J. Dairy Sci.* **2023**, *in press*. [CrossRef] [PubMed]

132. Becker, C.A.; Stone, A.E. Graduate Student Literature Review: Heat Abatement Strategies Used to Reduce Negative Effects of Heat Stress in Dairy Cows. *J. Dairy Sci.* **2020**, *103*, 9667–9675. [CrossRef] [PubMed]
133. Galloso-Hernández, M.A. Potencial de Los Sistemas Silvopastoriles Para La Producción Bufalina En Ambientes Tropicales; Universidad de Córdoba, UCO Press, Spain. 2021.
134. Mader, T.L.; Davis, M.S. Effect of Management Strategies on Reducing Heat Stress of Feedlot Cattle: Feed and Water Intake1. *J. Anim. Sci.* **2004**, *82*, 3077–3087. [CrossRef] [PubMed]
135. Mota-Rojas, D.; Napolitano, F.; Braghieri, A.; Guerrero-Legarreta, I.; Bertoni, A.; Martínez-Burnes, J.; Cruz-Monterrosa, R.; Gómez, J.; Ramírez-Bribiesca, E.; Barrios-García, H.; et al. Thermal Biology in River Buffalo in the Humid Tropics: Neurophysiological and Behavioral Responses Assessed by Infrared Thermography. *J. Anim. Behav. Biometeorol.* **2021**, *9*, 1–12. [CrossRef]
136. Napolitano, F.; Pacelli, C.; Grasso, F.; Braghieri, A.; De Rosa, G. The Behaviour and Welfare of Buffaloes (*Bubalus bubalis*) in Modern Dairy Enterprises. *Animal* **2013**, *7*, 1704–1713. [CrossRef]
137. Sullivan, K.F.; Mader, T.L. Managing Heat Stress Episodes in Confined Cattle. *Vet. Clin. N. Am. Food Anim. Pract.* **2018**, *34*, 325–339. [CrossRef] [PubMed]
138. Maia, A.S.C.; Moura, G.A.B.; Fonsêca, V.F.C.; Gebremedhin, K.G.; Milan, H.M.; Chiquitelli Neto, M.; Simão, B.R.; Campanelli, V.P.C.; Pacheco, R.D.L. Economically Sustainable Shade Design for Feedlot Cattle. *Front. Vet. Sci.* **2023**, *10*, 1110671. [CrossRef]
139. Dean, L.; Tarpoff, A.J.; Nickles, K.; Place, S.; Edwards-Callaway, L. Heat Stress Mitigation Strategies in Feedyards: Use, Perceptions, and Experiences of Industry Stakeholders. *Animals* **2023**, *13*, 3029. [CrossRef]
140. Mader, T.L.; Grffin, D. Management of Cattle Exposed to Adverse Environmental Conditions. *Vet. Clin. Food Anim. Pract.* **2015**, *31*, 247–258. [CrossRef]
141. St. Pierre, M.; Reeh, P.W.; Zimmermann, K. Differential Effects of TRPV Channel Block on Polymodal Activation of Rat Cutaneous Nociceptors in Vitro. *Exp. Brain Res.* **2009**, *196*, 31–44. [CrossRef] [PubMed]
142. Chowdhury, V.S. Heat Stress Biomarker Amino Acids and Neuropeptide Afford Thermotolerance in Chicks. *J. Poult. Sci.* **2019**, *56*, 1–11. [CrossRef] [PubMed]
143. Abdelnour, S.A.; Abd El-Hack, M.E.; Khafaga, A.F.; Arif, M.; Taha, A.E.; Noreldin, A.E. Stress Biomarkers and Proteomics Alteration to Thermal Stress in Ruminants: A Review. *J. Therm. Biol.* **2019**, *79*, 120–134. [CrossRef] [PubMed]
144. Marques, J.I.; Leite, P.G.; Lopes Neto, J.P.; Furtado, D.A.; Lopes, F.F.d.M. Estimation of Rectal Temperature of Goats Based on Surface Temperature. *Eng. Agrícola* **2021**, *41*, 591–598. [CrossRef]
145. Church, J.S.; Hegadoren, P.R.; Paetkau, M.J.; Miller, C.C.; Regev-Shoshani, G.; Schaefer, A.L.; Schwartzkopf-Genswein, K.S. Influence of Environmental Factors on Infrared Eye Temperature Measurements in Cattle. *Res. Vet. Sci.* **2014**, *96*, 220–226. [CrossRef]
146. Schaefer, A.L.; Cook, N.J.; Bench, C.; Chabot, J.B.; Colyn, J.; Liu, T.; Okine, E.K.; Stewart, M.; Webster, J.R. The Non-Invasive and Automated Detection of Bovine Respiratory Disease Onset in Receiver Calves Using Infrared Thermography. *Res. Vet. Sci.* **2012**, *93*, 928–935. [CrossRef]

animals

MDPI

Review

Use of Infrared Thermography and Heart Rate Variability to Evaluate Autonomic Activity in Domestic Animals

Marcelo Daniel Ghezzi [1], María Carolina Ceriani [2], Adriana Domínguez-Oliva [3,*], Pamela Anahí Lendez [1,2], Adriana Olmos-Hernández [4], Alejandro Casas-Alvarado [3] and Ismael Hernández-Avalos [5,*]

1 Anatomy Area, Faculty of Veterinary Sciences (FCV), Universidad Nacional del Centro de la Provincia de Buenos Aires (UNCPBA), University Campus, Tandil 7000, Argentina
2 Centro de Investigación Veterinaria de Tandil CIVETAN, UNCPBA-CICPBA-CONICET (UNCPBA), University Campus, Tandil 7000, Argentina
3 Neurophysiology, Behavior and Animal Welfare Assessment, DPAA, Universidad Autónoma Metropolitana, Xochimilco Campus, Mexico City 04960, Mexico
4 Division of Biotechnology—Bioterio and Experimental Surgery, Instituto Nacional de Rehabilitación-Luis Guillermo Ibarra Ibarra (INR-LGII), Tlalpan, Mexico City 14389, Mexico
5 Clinical Pharmacology and Veterinary Anesthesia, Facultad de Estudios Superiores Cuautitlán, Universidad Nacional Autónoma de México (UNAM), Cuautitlán Izcalli 54714, Mexico
* Correspondence: mvz.freena@gmail.com (A.D.-O.); mvziha@hotmail.com (I.H.-A.)

Simple Summary: The following review is based on the use of infrared thermography as a method to assess stress-related autonomic activity and its association with cardiovascular and heart rate variability in domestic animals. Stress-induced hyperthermia is one of the main physiological responses of animals to a stressor. The increase in body temperature activates compensatory peripheral mechanisms to increase heat dissipation and prevent further temperature increases. Changes in skin blood flow cause alterations in cutaneous temperature, alterations that can be assessed through infrared thermography and that will be discussed in the present review.

Citation: Ghezzi, M.D.; Ceriani, M.C.; Domínguez-Oliva, A.; Lendez, P.A.; Olmos-Hernández, A.; Casas-Alvarado, A.; Hernández-Avalos, I. Use of Infrared Thermography and Heart Rate Variability to Evaluate Autonomic Activity in Domestic Animals. *Animals* **2024**, *14*, 1366. https://doi.org/10.3390/ani14091366

Academic Editor: Kenneth McKeever

Received: 3 April 2024
Revised: 24 April 2024
Accepted: 27 April 2024
Published: 1 May 2024

Abstract: Most of the responses present in animals when exposed to stressors are mediated by the autonomic nervous system. The sympathetic nervous system, known as the one responsible for the "fight or flight" reaction, triggers cardiovascular changes such as tachycardia or vasomotor alterations to restore homeostasis. Increase in body temperature in stressed animals also activates peripheral compensatory mechanisms such as cutaneous vasodilation to increase heat exchange. Since changes in skin blood flow influence the amount of heat dissipation, infrared thermography is suggested as a tool that can detect said changes. The present review aims to analyze the application of infrared thermography as a method to assess stress-related autonomic activity, and their association with the cardiovascular and heart rate variability in domestic animals.

Keywords: autonomic nervous system; infrared thermography; skin blood flow; surface temperature; stress-induced hyperthermia

1. Introduction

The autonomic stress response to events that challenge the homeostasis of animals—including distress, diseases, or pain—is modulated by both branches of the autonomic nervous system (ANS): the sympathetic (SNS) and parasympathetic nervous system (PNS) [1,2]. The coordination of the central and peripheral responses to a stressor involves the hypothalamic–pituitary–adrenal (HPA) and the sympathetic–adrenomedular (SAM) axes [3]. These systems activate physiological, endocrine, and metabolic adaptive mechanisms to ensure the animal's survival [4]. Among these changes, modifications in the skin's sympathetic blood flow can occur due to physical, mental, and thermal stress [5–7], which can be non-invasively evaluated through infrared thermography (IRT) [8,9].

The activation of the SNS during the perception of a stressor leads to cardiovascular changes such as increasing heart rate (HR), vascular resistance, and body temperature mainly by the action of catecholamines (e.g., norepinephrine (NE)) released from the adrenal medulla as part of the "fight or flight" reaction [2,10]. Stress-induced hyperthermia (SIH) is a phenomenon where body temperature increases due to cardiovascular changes (tachycardia) inducing thermogenesis [11]. Increases in core temperature also activate vasodilator mechanisms to increase heat exchange in skin blood vessels [12]. Since IRT is a tool that can detect changes in the surface temperature of animals, it can indirectly assess sympathetic-mediated vasomotor changes [13–16]. This article aims to analyze the application of IRT as a method to assess stress-related autonomic activity, and their association with cardiovascular and heart rate variability in domestic animals.

2. Importance of Evaluating the Activity of the Sympathetic/Parasympathetic Pathways

The stress response is mediated by the ANS and the balance between the SNS and PNS [1]. Immediately after the perception of a stressor or a stimulus that threatens the animal's homeostasis [4,17,18], several physiological, endocrine, and metabolic changes are triggered to restore homeostasis. The SNS modulates the activity of the cardiovascular, gastrointestinal, respiratory, renal, and endocrine systems, among others [19]. Its activation increases the HR, cardiac output, and blood flow to key organs (Figure 1) [1,20]. Increased body temperature (BT) can also be observed as well as changes in the pupillary diameter as the integrated fight–flight response [21–23].

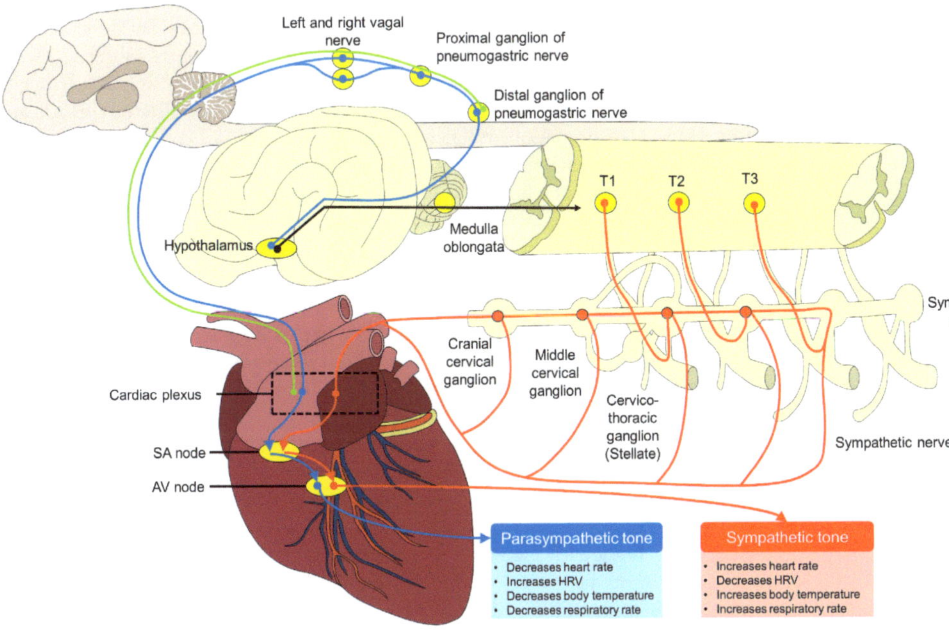

Figure 1. Schematic representation of the autonomic control of the cardiovascular function. Parasympathetic (blue lines) and sympathetic nerves (red lines) control the responses of animals to stress. Sympathetic-mediated responses involving the cardiorespiratory system (e.g., tachycardia, tachypnea, or hyperthermia) are triggered by the action of pre- and post ganglionic nerves (black lines) and the projection of the stimulus through the sympathetic chain. In contrast, parasympathetic-related responses, mediated by cardiac ganglia neurons, include decreases in heart rate, body temperature, and respiratory rate, and increases in heart rate variability (HRV) depend on the regulation of the vagal nerve on the sinoatrial (SA) node and the atrioventricular node (AV node). Green lines represent the participation of sensory neurons.

The sympathetic response is mediated by the locus coeruleus (LC) and the noradrenergic system. LC increases the release of NE to increase arousal and alertness [19,24]. Additionally, SAM releases the catecholamines epinephrine (E) and NE [1,20]. For example, when rats are exposed to immobilization as the stressor, Pacak et al. [25] reported an increase in the synthesis, release, and metabolism of NE in the paraventricular nucleus, the central nucleus of the amygdala, the bed nucleus of the stria terminalis (BNST), and the hypothalamus. Similarly, mice experiencing five days of restraint stress increased their NE release in the BNST by 3.3 to 5-fold [26]. In the case of pigs, Bozzo et al. [27] studied the activation patterns of the SAM and HPA axes during transport stress to determine the effect that the journey duration and loading position have on the endocrine profile of animals (plasma E, NE, and cortisol (CORT)). The authors found that pigs undergoing long journeys (11 h long) had significantly higher concentrations of stress hormones (E: 19,414 ng/L; NE: 17,469 ng/L; CORT: 139 nmol/L) than those transported for three hours (E: 9232 ng/L; NE: 6227 ng/L; CORT: 94 nmol/L), which is associated with a higher level of stress. In cattle, Rodríguez et al. [28] reported that Holstein–Friesian cows with lameness (mobility score of 3) had higher NE concentrations (901.28 ± 33.66 pg/mL) than those with mobility scores of 0 (655.83 ± 24.09 pg/mL) as a result of pain (an event that is also considered a stressor).

One of the main noradrenergic responses to psychogenic stressors is tachycardia due to increased plasma NE concentrations. Both E and NE can bind to adrenoreceptors ($\alpha1$, $\alpha2$, $\beta1$, $\beta2$, and $\beta3$) present in the endothelial and cardiac tissue, modifying the cardiovascular function [1,29,30]. In particular, $\beta1$ is mainly found in the heart and it regulates HR, contractility, and plasma renin release, being directly involved in HR increases; however, during stress reactions, $\beta2$ also contributes to tachycardia [31,32]. This has been observed in dogs during veterinary examination, an event that is frequently associated with other negative responses such as tachypnea, hypertension, mydriasis, hyperthermia, and increased muscular tension [33]. This was studied in 30 dogs of different breeds and ages during physical examination. When comparing average basal values in the waiting room of 97.7 ± 19.6 beats per minute (bpm) against 123.21 ± 21.2 bpm in the consulting room, a significant increase in the heart rate (HR) was observed, particularly when patted by the examiner (138.0 ± 25.2 bpm) and during a simulated vaccination (133.8 ± 28.7 bpm) [34].

Another event that can increase HR in animals is thermal stress since increases in HR and respiratory rate (RR) occur to promote heat dissipation, as reported in Boer goats exposed to high environmental temperatures (26, 29, and 33 °C) [35]. In this study, when animals were under air temperatures of 33 °C, HR and RR increased by 50% and pupillary diameter significantly increased by 37%. Moreover, a strong positive correlation between the increase in physiological parameters and pupillary diameter was found (r = 0.80) [35]. In cattle, Holstein bull calves maintained in non-shaded areas with an average temperature of 44.3 °C had higher HR, particularly in the afternoon (22.3 ± 1.4 bpm). Furthermore, RR also increased in non-shaded bulls [36].

Tachypnea increases the oxygen supply required by the high oxygen demand of muscle tissue during stressful events [1]. In the case of sporting animals such as dogs and horses, a higher RR increases the availability of arterial O_2, which is facilitated by the increase of cardiac activity and, consequently, the volume of blood ejected [37,38]. Martins-Pinge [39] mentions that the sympathetic branch of the ANS plays a key role in modulating the cardiorespiratory system during aerobic exercise, adjusting its function to the animal's requirements. Therefore, the ANS is vital to achieving stability in adverse conditions but also under physiological conditions of high demand such as exercise.

Another autonomous response that can be assessed in animals under stress is the changes in the pupil diameter of the species. Pupils can contract or dilate according to the level of activity of the ANS [40]. Noradrenergic and cholinergic circuits control the pupil diameter and size by direct innervation to the sphincter and the dilator muscle [41–43]. In this sense, pupillary constriction is modulated by postganglionary parasympathetic fibers through short ciliary nerves that innervate the pupillary sphincter diameter [44]. In contrast, pupil dilation is associated with sympathetic predominance and innervation to the iris dilator muscle and the release of NE after the activation of the ANS [45–47]. Due to the connection between the NE and pupil control, pupil diameter could serve as an indirect method to evaluate the activation of the ANS during stressful or painful events in animals [46,47].

Machado et al. [48] evaluated the pupil area, BT, and CORT concentrations in 360 piglets subjected to long-distance transport (75 km). The pupil area increased by 19.3%, along with a significant increase in CORT by 2.37 ng/mL and BT by 1.24 °C. Additionally, high positive correlations between pupil area, CORT, and BT were reported (r = 0.92, 0.90, and 0.82, respectively). In cats, the reaction of the animals to restraint during a physical examination was analyzed using the pupil dilation ratio, RR, and behavioral responses. In comparison with animals receiving passive restraint, cats subjected to full body restraint resulted in a larger pupil dilation ratio (approximately 0.7 vs. 0.45) and presented higher RR (29 vs. 18 bpm) and lip licking (2.3 vs. 1.5 licks/min). This possibly reaffirms the idea that activation of the ANS helps animals to compensate for changes during stressors and in this way could recover their stability [49].

As shown above, CORT release is also associated with the activation of the ANS because the response to stressors is mediated by the simultaneous activation of the SNS, SAM, and HPA axes [19,50,51]. Several studies have reported increases in CORT levels when animals are exposed to stressors such as heat stress. In this sense, Kim et al. [52] found that Korean native beef calves exposed to high THI levels (87.72) had serum CORT increases (17.1 ng/mL) as well as higher HR (73 bpm) and RT (39.9 °C) than calves under a THI of 70.01 (4.8 ng/mL, 60.3 bpm, and 38.9 °C, respectively). Another example is during equestrian events where horses can be exposed to potential stressors. Olvera-Maneau et al. [53] studied this response in pure-breed Menorca stallions and the effect that these festivals have on the salivary CORT concentration of animals. The authors found that CORT significantly increased to up to 3.8 nmol/L during the event, suggesting that it represents an acute stressor for the animals. Therefore, assessment of autonomic activity can be an additional parameter to monitor health and the current state of the animals.

3. Heart Rate Variability and Its Association with Autonomic Activity

During the perception of a stressor, the activation of the ANS modifies the cardiovascular activity to mobilize energy resources [54]. As described by von Borell et al. [55], the cardiovascular system is modulated by the ANS and its two branches: sympathetic and parasympathetic, where both the SNS and PNS can increase or decrease HR, respectively [1,56]. A predominant vagal (or parasympathetic) tone regulates cardiac activity and maintains the beat-to-beat interval. A healthy heartbeat is characterized by irregular time intervals between consecutive beats. This variability is consequent to the rhythmic oscillation of the regulatory components of cardiac activity to maintain cardiovascular homeostasis. When fluctuations in vagal tone are present, these can be related to variations in the vagal nuclei activity, which are mediated by the baroreceptors. Thus, HR variability (HRV) is affected by the responsiveness to O_2 and CO_2 levels in baroreceptors during the respiratory cycle [57,58].

This could be related to Ille et al.'s [59] findings in 16 sport horses divided into two groups according to their level of exercise experience. Salivary CORT, HR, HRV, and standard deviation of the successive R-R interval were assessed. The authors observed that the HR in both groups increased by 50 bpm after the exercise; however, HRV and standard deviation of the R-R interval decreased by 20 ms. The decrease in both parameters is indicative of increased sympathetic tone that shortens the interval between R-R complexes and results in a greater number of heartbeats.

The presence of these oscillations can reflect the level of ANS activity—whether SNS or PNS—by assessing the oscillation changes of HRV [60]. During potentially stressful events such as pain or nociception, the shortening between R-R complexes occurs due to tachycardia. Additionally, changes in the respiratory pattern stimulate baroreceptors and shift to a predominant sympathetic tone [61–64]. Currently, a numerical index to assess the level of parasympathetic tone through HRV has been developed, known as the Parasympathetic Tone Activity (PTA) index [64–66]. The PTA index is calculated using the following formula: PTA = (100 + [α + aucMIN + β]/12.8) + 100/161, where the value for α is 5.1 and β is 1.2, predetermined values to maintain the coherence of the respiratory influence on the series of RR intervals and the quantitative value of PTA. Thus, 100/161, 100/171, and 100/163 are predetermined coefficients for dogs, cats, and horses, respectively [67,68]. The PTA index uses a 0–100 scale, where values close to 100 indicate a relative parasympathetic tone, while those close to 0 represent diminished parasympathetic activity or a sympathetic predominance [69]. Figure 2 shows a PTA index greater than 60% in a feline patient due to the effect of the administration of anxiolytics.

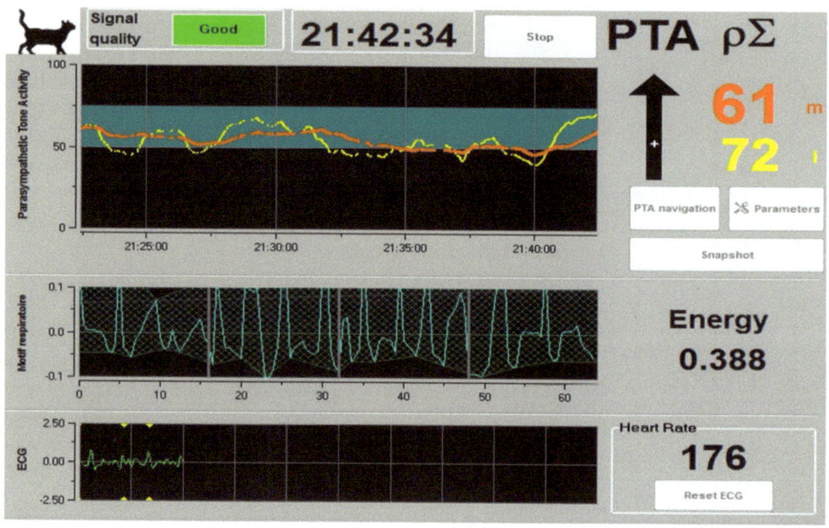

Figure 2. Screenshot of the parasympathetic tone index (PTA) monitor evaluating a 3-year-old shorthair domestic cat. The feline patient was premedicated orally with 5 mg/kg of pregabalin before undergoing an ovary hysterectomy. It can be observed that the PTA index is greater than 60%. This indicates a predominant parasympathetic tone due to the anxiolytic effect of pregabalin, which reduces stress during transport and routine consultation. Orange lines and number represent real time values; yellow lines and numbers represent 2-min average values. Picture taken by the authors.

Some studies in both dogs and horses have suggested that PTA can help to recognize the level of pain or nociception in animals during surgical procedures and assess the analgesic efficacy of drugs [67,70]. For example, Leitão et al. [71] applied this index in healthy and anesthetized Large White pigs receiving a noxious stimulus and two types of analgesics (ketorolac and ketorolac/tramadol). Pigs receiving both analgesic treatments had a PTA mean value that was 30 units lower than pigs to whom drugs were not administered. Moreover, the HR of animals without analgesia was 45 bpm higher than medicated subjects, which reflects the ANS activation due to the painful stimulus. Similarly, Hernández-Ávalos et al. [72] used the PTA index as a method to assess the level of postoperative analgesia in 30 healthy bitches undergoing ovariohysterectomy. Although no variation was observed in the initial and postoperative value of the PTA index, a clinical relationship was observed between an index value of 40–49 with a score below 10 on the University of Melbourne Pain Scale, showing a sensitivity and specificity of 40% and 98.46%, respectively. Thus, through the evaluation of HRV due to its regulation by the ANS, it is possible to determine the predominant tone during adverse states such as pain or stress. It is therefore possible that the evaluation of the same is likely to determine the degree of analgesia during surgical procedures.

In other studies, the PTA index has been used as a method to predict changes in arterial pressure following the idea that changes in the sympathetic/parasympathetic balance alter cardiovascular function. In anesthetized horses, Mansour et al. [68] determined that values in the PTA decreased five minutes before hypotension, recognizing this event with a sensitivity and specificity of 62.5% and 94.6%, respectively. Similarly, intraoperative nociception and cardiovascular changes have been evaluated in Beagle dogs [73]. In these animals, it was reported that the PTA index was able to detect nociception even in the absence of cardiovascular changes, reaching PTA values as low as 25 ± 15. In the same species, Mansour et al. [74] compared the effect of three anesthetic protocols (morphine, morphine + medetomidine, and morphine + acepromazine) through the PTA index to identify intraoperative hemodynamic reactions of animals. During cutaneous incision, PTA values significantly decreased (range −21 to −23%) while increases in median arterial pressure (MAP) (between +27–28%) and HR (between +3.5–10.8%) were registered, except on animals receiving morphine + medetomidine. In contrast, Ruíz-López et al. [75] reported that PTA values below 51 were associated with hemodynamic response in bitches undergoing ovariectomy (increases in MAP but not in HR) but the sensitivity and specificity of PTA were considered poor (69 and 52%) to predict these changes. Therefore, further studies are required to establish the usefulness of PTA in assessing the cardiovascular activity of the ANS when no significant changes are observed in the HR, as described by Lima et al. [76] in cats undergoing ovariectomy. In this study, PTA was effective in detecting nociception when increases in HR above 20% reflect the activation of the SNS.

4. Stress-Mediated Thermal Response and Its Assessment through Infrared Thermography

A phenomenon known as stress-induced hyperthermia occurs when the BT of animals increases after the perception of a stressor [56,77]. When psychological stress is perceived by the amygdala, projections to the dorsomedial hypothalamus and the preoptic area of the hypothalamus (POA) modulate the mechanisms to increase core temperature [78]. Through the activation of the hypothalamic–medullary–sympathetic axis, catecholamines are released, increasing cardiovascular activity [79,80]. Neurons located in the rostral raphe pallidus nucleus and the intermediolateral cell column promote thermogenesis in brown adipose tissue and cutaneous vasoconstriction [78,81,82]. Increases in NE can be accompanied by increases in adrenocorticotropic hormone, CORT, and glucose, which are known as stress hormones that increase the activity of the cardiovascular system and promote heat production [83–85].

Increases in BT also involve peripheral thermoregulation mechanisms mediated by the POA because mammals regulate their BT by balancing the amount of metabolically produced heat and the one that is exchanged with the environment [86]. During the perception of a stressor—and as a response to SIH—the diameter of cutaneous blood vessels changes, promoting vasodilation to increase blood flow and heat dissipation [12]. Estimation of these vasomotor changes can be assessed through IRT in different body regions called thermal windows [13,14]. Thermal windows are characterized by regions with poor insulation or glabrous skin, rich vascular beds, and arteriovenous anastomoses that facilitate heat transfer [86] and might be a tool that indirectly assesses the activity of the SNS [15,16].

Ocular surface temperature has been correlated with autonomic modulation and glucocorticoid concentrations, suggesting it as a biomarker for indirect assessment of ANS activity [56]. In particular, the lacrimal caruncle is considered a region that reflects the autonomic activity [87–89]. In this sense, Shu et al. [90] determined the correlation of the ocular temperature of dairy cows under heat stress assessed in five points: medial canthus, lateral canthus, eyeball, whole eye, and lacrimal sac. The highest correlation between RT and RR was found in the lacrimal sac (up to 0.60), concluding that this region should be considered as an adequate predictor of heat stress in animals. According to Mota Rojas et al. [8], the blood flow of the lacrimal caruncle is provided by the orbital artery and its branches (supraorbital and infraorbital artery) [91]. The suborbital artery receives sympathetic innervation from the facial nerve [92]. Therefore, changes in the surface temperature of this region can reflect sympathetic-mediated responses on skin blood flow, as shown in Figure 3. A pilot study made by the present authors evaluated the response of six dogs during an exercise test, where the average temperature of the lacrimal caruncle and the ocular surface was compared before and after exercise (jogging and 1000 m hiking). In this study, increases in the temperature of both thermal windows were recorded (of up to 1 °C) after 1000 m jogging. Additionally, the HR and RR of dogs also increased from control measurements to the end of the exercise, reaching an increase of up to 111.4 bpm and 50.6 rpm, respectively. This response is associated with a predominant sympathetic tone that increases BT and triggers cutaneous vasodilation, as well as tachycardia and tachypnea to compensate for the increased cardiac and muscular activity.

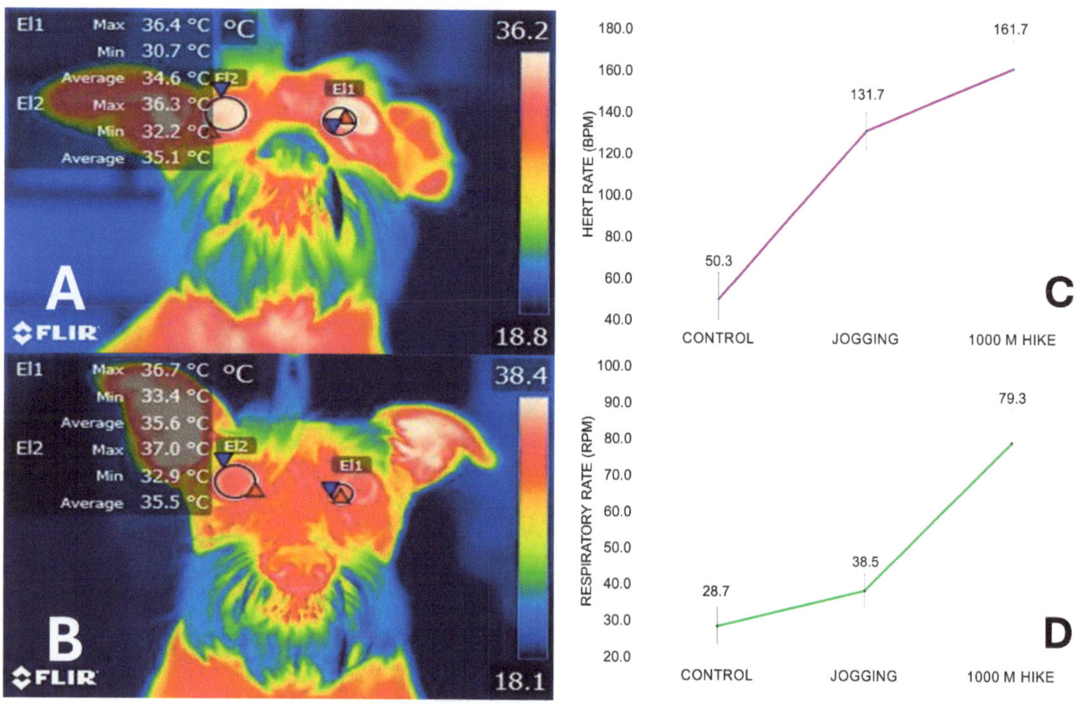

Figure 3. Surface thermal response of dogs during exercise. (**A**) Control measurements of a Schnauzer dog. The maximum, minimum, and average temperature of the lacrimal caruncle (El1) and the ocular surface (El2) can be observed, being 36.4, 30.7, and 34.6 °C. (**B**) Surface temperature after jogging. When comparing the average values of the lacrimal caruncle and ocular temperature, a difference of 1 °C and 0.4 °C, respectively, was reported, showing the increase in these thermal windows due to peripheral vasodilation. Graphs (**C**,**D**) show the increase in the heart rate and respiratory rate of the dogs included in the pilot study. Red triangle: maximum temperatures; blue triangle: minimum temperatures. Images taken with a FLIR E60 thermal camera, lens FOL 18 mm, IR resolution of 240 × 180, emissivity of 0.95, distance from object 30 cm.

Arfuso et al. [93] used IRT to evaluate the response of sheep to acute stress during a shearing procedure. By collecting blood samples, RT, and ocular surface temperature, the authors found that RT and the temperature of the medial canthus after shearing significantly increased by 0.5 °C. Simultaneously, increases in blood concentration of CORT were registered, values that were positively correlated with surface and rectal temperature (r = 0.70). Similarly, ocular surface temperature was assessed in thirty bull calves undergoing surgical castration with local anesthesia and sham handling control surgical castration.

In the control animals, increases in HR (by 15 bpm) and eye temperature (by 0.47 °C followed by a decrease of 1 °C) were found after the castration procedure. This coincided with the significant increase of 7.3 ng/mL of CORT after castration, changes that are after autonomic activity and the surface thermal response [2]. Figure 4 shows the preliminary results obtained by the present authors when evaluating ocular thermal response of cats exposed to a stressor (dog) and to the effect of a cat appeasing pheromone, stimulus known to activate the sympathetic and parasympathetic system, respectively.

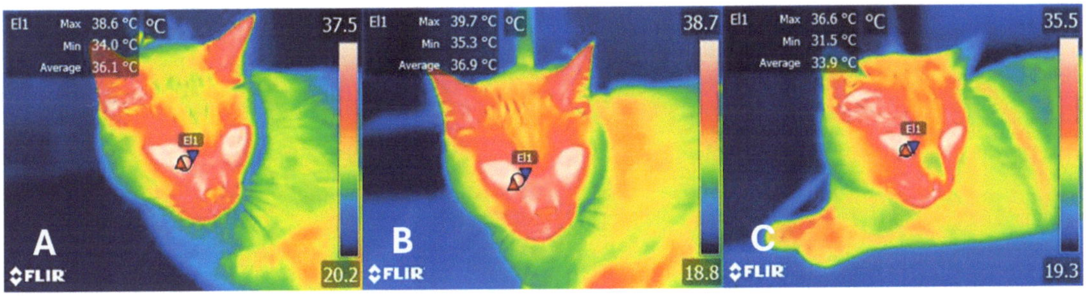

Figure 4. Thermal imaging to assess the autonomic response of cats exposed to different stressors. (**A**) Basal measurements in a male domestic shorthair cat. When the cat was not exposed to any stressor, a maximum temperature of the lacrimal caruncle (El1) of 38.6 °C, a minimum of 34.0 °C, and an average value of 36.1 °C was recorded. (**B**) After the exposure to an unfamiliar dog, an increase in the three temperatures was observed, registering increases of 1.1 °C, 1.3 °C, and 0.8 °C, respectively. (**C**) After the use of a cat appeasing pheromone, it can be observed that the surface temperature decreased when compared to B values (average of 3 °C). Furthermore, the cat pheromones caused lower temperatures than basal values (as low as −2.5 °C) and this might be related to the calming effect of the pheromones due to a predominant parasympathetic tone. Images taken with a FLIR E60 thermal camera, lens FOL 18 mm, IR resolution of 240 × 180, emissivity of 0.95, distance from object 30 cm. Red triangles: maximum temperatures; blue triangles: minimum temperatures.

In calves subjected to disbudding, the maximum eye temperature measurements (located at the lacrimal caruncle) detected changes between animals receiving local anesthetics and those that did not receive analgesics. A sudden drop in temperature (−0.27 °C) followed by an increase in eye temperature of up to 0.60 °C was recorded in calves without local anesthetics, an alteration related to an acute sympathetic response to pain [94]. During heat stress Chikkagoudara et al. [95] reported that the eye temperature of young buffalo bulls increased significantly when the temperature humidity index increased (up to 37.41 °C when the index was 48.34), showing that surface temperature can be used as an indicator of stress.

Although these studies show that IRT can indirectly assess ANS activity, Sutherland et al. [96] mentions that HRV is a more sensitive method than IRT, as observed in their study evaluating the effect of E infusion on the temperature of the lacrimal gland and HRV in 20 Romney ewes. The authors found that the infusion of epinephrine increased the ocular temperature (by 0.3 °C). However, increases in HR and changes in HRV were considered more sensitive to measure ANS activity. Thus, the cardiovascular changes that occur during stressful events can be detected through IRT. Using thermal imaging as a complementary tool to evaluate the response of animals to stress might help to recognize the predominant tone of the ANS.

5. Perspectives about Thermal Imagining

Due to the low sensitivity of IRT in comparison to other techniques, it has been suggested that further studies should focus on the difference and comparison of the different thermal windows or anatomical regions [7,89]. Likewise, since pain perception is highly modulated by the SNS, IRT could be a useful tool to assess the analgesic effect of certain drugs, as observed in dogs under controlled procedures [97]. Moreover, Bergamasco et al. [98] reported that eye surface temperature is mediated by the ANS but the response can differ according to the age of the animal in calves undergoing unmitigated castration Therefore, considering the age of the animals might be needed in future studies.

Currently, laser Dopler flowmetry is one of the main methods to non-invasively measure skin blood flow and recognize changes in microvascular perfusion [99]. In humans, several studies have applied the combined use of IRT with Doppler flowmetry to assess the angiospatic dysfunction of peripheral blood flow [100]. Authors such as Merla et al. [101] established that combining laser Doppler, thermal imagining, and bio-heat transfer modeling helps to discriminate healthy patients from those with impaired blood flow, but that thermal imaging is considered a faster technique for real-time detection of cutaneous perfusion alterations.

Other studies propose a thermal-associated pain index using skin conductance, HRV, and peripheral perfusion to optimize pain assessment in patients. Although this index has only been studied in humans, it has shown a sensitivity above 70% to detect pain, suggesting that it could be adapted to veterinary medicine. Furthermore, estimation of the HRV using facial thermal imaging is proposed as a contactless method to identify cardiovascular changes affecting HRV and HR, showing strong positive correlations ($r = 0.7$ and 0.67, respectively). Using IRT in combination with other techniques is suggested as a reliable protocol to recognize the activation of the ANS.

We used parallel or series measurement to increase the sensitivity and specificity of IRT measurements, and even compared the predictive value of thermal imaging to assess activation of the ANS [102]. For example, in humans, IRT as a diagnostic tool has a positive predictive value of 72.2, a negative predictive value of 93.3, and an overall sensitivity, specificity, and accuracy of 87.1, 87.2, and 89.7, respectively [103]. However, most studies have been performed in humans and a large variation in sensitivity and specificity due to operating characteristics has been reported. Therefore, its application in veterinary medicine, together with other tools, needs to be evaluated. Moreover, automatic systems adopting infrared thermography could be another strategy that could help recognize shifts in the activity of the ANS that might be related to stress or pain [104,105]. This could be useful in clinical or farm settings where a prompt recognition of autonomic alterations might prevent further alterations in animals.

6. Conclusions

Changes in the skin blood flow—vasoconstriction or vasodilation– due to stress are mediated by the ANS. Oscillations in skin blood flow modify the skin temperature, a parameter that can be assessed through IRT. During the perception of a stressor, the sympathetic predominance causes an increase in RT as well as in the surface temperature of animals. In particular, thermal imaging focused on the lacrimal caruncle is considered an adequate predictor to evaluate the activity of the ANS due to the predominant sympathetic innervation in this region. However, since IRT interpretation depends on the thermal window of animals, thermal imagining is recommended as a complementary method that should be used together with other sensitive and specific techniques (e.g., laser Doppler) to determine the autonomic control of skin blood flow.

Author Contributions: All authors contributed to the conceptualization, writing, and reading. All authors have read and agreed to the published version of the manuscript.

Funding: This research received no external funding.

Institutional Review Board Statement: Not applicable.

Informed Consent Statement: Not applicable.

Data Availability Statement: Data sharing not applicable.

Acknowledgments: The authors express profound gratitude to Daniel Mota-Rojas, whose expertise in cognition, emotions, animal behavior, and welfare greatly contributed to the inception and oversight of this manuscript. We sincerely appreciate his steadfast support and guidance in the fields of hypothalamic neuromodulation and infrared thermography, as well as his generosity in supplying the thermograms utilized in this article.

Conflicts of Interest: The authors declare no conflicts of interest.

References

1. Gaidica, M.; Dantzer, B. Quantifying the Autonomic Response to Stressors—One Way to Expand the Definition of "Stress" in Animals. *Integr. Comp. Biol.* **2020**, *60*, 113–125. [CrossRef]
2. Stewart, M.; Verkerk, G.A.A.; Stafford, K.J.J.; Schaefer, A.L.L.; Webster, J.R.R. Noninvasive Assessment of Autonomic Activity for Evaluation of Pain in Calves, Using Surgical Castration as a Model. *J. Dairy Sci.* **2010**, *93*, 3602–3609. [CrossRef] [PubMed]
3. Mueller, B.; Figueroa, A.; Robinson-Papp, J. Structural and Functional Connections between the Autonomic Nervous System, Hypothalamic–Pituitary–Adrenal Axis, and the Immune System: A Context and Time Dependent Stress Response Network. *Neurol. Sci.* **2022**, *43*, 951–960. [CrossRef] [PubMed]
4. Karaer, M.C.; Čebulj-Kadunc, N.; Snoj, T. Stress in Wildlife: Comparison of the Stress Response among Domestic, Captive, and Free-Ranging Animals. *Front. Vet. Sci.* **2023**, *10*, 1167016. [CrossRef] [PubMed]
5. Muller, M.D.; Sauder, C.L.; Ray, C.A. Mental Stress Elicits Sustained and Reproducible Increases in Skin Sympathetic Nerve Activity. *Physiol. Rep.* **2013**, *1*, e00002. [CrossRef] [PubMed]
6. Metzler-Wilson, K.; Toma, K.; Sammons, D.L.; Mann, S.; Jurovcik, A.J.; Demidova, O.; Wilson, T.E. Augmented Supraorbital Skin Sympathetic Nerve Activity Responses to Symptom Trigger Events in Rosacea Patients. *J. Neurophysiol.* **2015**, *114*, 1530–1537. [CrossRef] [PubMed]
7. Luzi, F.; Mitchell, M.; Nanni Costa, L.; Radaelli, V. *Thermography: Current Status and Advances in Livestock Animals and in Veterinary Medicine*; Fondazione Iniziative Zooprofilattiche e Zootecniche: Brescia, Italy, 2013.
8. Mota-Rojas, D.; Martínez-Burnes, J.; Casas-Alvarado, A.; Gómez-Prado, J.; Hernández-Ávalos, I.; Domínguez-Oliva, A.; Lezama-García, K.; Jacome-Romero, J.; Rodríguez-González, D.; Pereira, A.M.F. Clinical Usefulness of Infrared Thermography to Detect Sick Animals: Frequent and Current Cases. *CABI Rev.* **2022**, *2022*, 1–27. [CrossRef]
9. Mota-Rojas, D.; Ogi, A.; Villanueva-García, D.; Hernández-Ávalos, I.; Casas-Alvarado, A.; Domínguez-Oliva, A.; Lendez, P.; Ghezzi, M. Thermal Imaging as a Method to Indirectly Assess Peripheral Vascular Integrity and Tissue Viability in Veterinary Medicine: Animal Models and Clinical Applications. *Animals* **2023**, *14*, 142. [CrossRef] [PubMed]
10. Ziemssen, T.; Siepmann, T. The Investigation of the Cardiovascular and Sudomotor Autonomic Nervous System—A Review. *Front. Neurol.* **2019**, *10*, 53. [CrossRef]
11. Mota-Rojas, D.; Miranda-Córtes, A.; Casas-Alvarado, A.; Mora-Medina, P.; Boscato, L.; Hernández-Ávalos, I. Neurobiology and Modulation of Stress-Induced Hyperthermia and Fever in Animals. *Abanico Vet.* **2021**, *11*, 1–17.
12. Mota-Rojas, D.; Wang, D.; Titto, C.G.; Gómez-Prado, J.; Carvajal-de la Fuente, V.; Ghezzi, M.; Boscato-Funes, L.; Barrios-García, H.; Torres-Bernal, F.; Casas-Alvarado, A.; et al. Pathophysiology of Fever and Application of Infrared Thermography (IRT) in the Detection of Sick Domestic Animals: Recent Advances. *Animals* **2021**, *11*, 2316. [CrossRef]
13. Childs, C. Body Temperature and Clinical Thermometry. *Handb. Clin. Neurol.* **2018**, *157*, 467–482. [CrossRef]
14. Romanovsky, A.A. Skin Temperature: Its Role in Thermoregulation. *Acta Physiol.* **2014**, *210*, 498–507. [CrossRef]
15. Casas-Alvarado, A.; Mota-Rojas, D.; Hernández-Ávalos, I.; Mora-Medina, P.; Olmos-Hernández, A.; Verduzco-Mendoza, A.; Reyes-Sotelo, B.; Martínez-Burnes, J. Advances in Infrared Thermography: Surgical Aspects, Vascular Changes, and Pain Monitoring in Veterinary Medicine. *J. Therm. Biol.* **2020**, *92*, 102664. [CrossRef] [PubMed]
16. Casas-Alvarado, A.; Martínez-Burnes, J.; Mora-Medina, P.; Hernández-Avalos, I.; Domínguez-Oliva, A.; Lezama-García, K.; Gómez-Prado, J.; Mota-Rojas, D. Thermal and Circulatory Changes in Diverse Body Regions in Dogs and Cats Evaluated by Infrared Thermography. *Animals* **2022**, *12*, 789. [CrossRef] [PubMed]
17. Koolhaas, J.M.; Bartolomucci, A.; Buwalda, B.; de Boer, S.F.; Flügge, G.; Korte, S.M.; Meerlo, P.; Murison, R.; Olivier, B.; Palanza, P.; et al. Stress Revisited: A Critical Evaluation of the Stress Concept. *Neurosci. Biobehav. Rev.* **2011**, *35*, 1291–1301. [CrossRef] [PubMed]
18. Shimizu, H.; Okabe, M. Evolutionary Origin of Autonomic Regulation of Physiological Activities in Vertebrate Phyla. *J. Comp. Physiol. A* **2007**, *193*, 1013–1019. [CrossRef]
19. Murison, R. The Neurobiology of Stress. In *Neuroscience of Pain, Stress, and Emotion*; Al'Absi, M., Flaten, M.A., Eds.; Academic Press: London, UK, 2016; pp. 29–49.
20. Oyama, M.A. Mechanisms of Heart Failure. In *Small Animal Critical Care Medicine*; Silverstein, D.C., Hopper, K., Eds.; Elsevier: Maryland Heights, MO, USA, 2023; pp. 238–242.
21. Hernández-Avalos, I.; Mota-Rojas, D.; Mendoza-Flores, J.E.; Casas-Alvarado, A.; Flores-Padilla, K.; Miranda-Cortes, A.E.; Torres-Bernal, F.; Gómez-Prado, J.; Mora-Medina, P. Nociceptive Pain and Anxiety in Equines: Physiological and Behavioral Alterations. *Vet. World* **2021**, *14*, 2984–2995. [CrossRef] [PubMed]

22. José-Pérez, N.; Mota-Rojas, D.; Ghezzi, M.D.; Rosmini, M.R.; Mora-Medina, P.; Bertoni, A.; Rodríguez-González, D.; Domínguez-Oliva, A.; Guerrero-Legarreta, I. Effects of Transport on Water Buffaloes (*Bubalus bubalis*): Factors Associated with the Frequency of Skin Injuries and Meat Quality. *J. Anim. Behav. Biometeorol.* **2022**, *10*, e2216. [CrossRef]
23. Mota-Rojas, D.; Lezama-García, K.; Domínguez-Oliva, A.; Olmos-Hernández, A.; Verduzco-Mendoza, A.; Casas-Alvarado, A.; Torres-Bernal, F.; Martínez-Burnes, J. Neurobiology of Emotions in Animal Relationships: Facial Expressions and Their Biological Functions in Mammals. *J. Anim. Behav. Biometeorol.* **2023**, *11*, e2023ss01. [CrossRef]
24. Kenney, M.J. Animal Aging and Regulation of Sympathetic Nerve Discharge. *J. Appl. Physiol.* **2010**, *109*, 951–958. [CrossRef] [PubMed]
25. Pacak, K.; Palkovits, M.; Kopin, I.J.; Goldstein, D.S. Stress-Induced Norepinephrine Release in the Hypothalamic Paraventricular Nucleus and Pituitary-Adrenocortical and Sympathoadrenal Activity: In Vivo Microdialysis Studies. *Front. Neuroendocrinol.* **1995**, *16*, 89–150. [CrossRef] [PubMed]
26. Schmidt, K.T.; Makhijani, V.H.; Boyt, K.M.; Cogan, E.S.; Pati, D.; Pina, M.M.; Bravo, I.M.; Locke, J.L.; Jones, S.R.; Besheer, J.; et al. Stress-Induced Alterations of Norepinephrine Release in the Bed Nucleus of the Stria Terminalis of Mice. *ACS Chem. Neurosci.* **2019**, *10*, 1908–1914. [CrossRef] [PubMed]
27. Bozzo, G.; Padalino, B.; Bonerba, E.; Barrasso, R.; Tufarelli, V.; Zappaterra, M.; Ceci, E. Pilot Study of the Relationship between Deck Level and Journey Duration on Plasma Cortisol, Epinephrine and Norepinephrine Levels in Italian Heavy Pigs. *Animals* **2020**, *10*, 1578. [CrossRef]
28. Rodríguez, A.R.; Herzberg, D.E.; Werner, M.P.; Müller, H.Y.; Bustamante, H.A. Plasma Concentration of Norepinephrine, β-Endorphin, and Substance P in Lame Dairy Cows. *J. Vet. Res.* **2018**, *62*, 193–197. [CrossRef] [PubMed]
29. Fujii, A.M.; Vatneer, S.F. Autonomic Mechanisms Regulating Myocardial Contractility in Conscious Animals. *Pharmacol. Ther.* **1985**, *29*, 221–238. [CrossRef] [PubMed]
30. Jänig, W. Autonomic Nervous System. In *Human Physiology*; Springer: Berlin/Heidelberg, Germany, 1989; pp. 333–370.
31. Khan, M.G. Beta-Blockers. In *Encyclopedia of Heart Diseases*; Khan, G., Ed.; Academic Press: Burlington, VT, USA, 2006; pp. 159–167.
32. Motiejunaite, J.; Amar, L.; Vidal-Petiot, E. Adrenergic Receptors and Cardiovascular Effects of Catecholamines. *Ann. Endocrinol.* **2021**, *82*, 193–197. [CrossRef] [PubMed]
33. Lefman, S.H.; Prittie, J.E. Psychogenic Stress in Hospitalized Veterinary Patients: Causation, Implications, and Therapies. *J. Vet. Emerg. Crit. Care* **2019**, *29*, 107–120. [CrossRef] [PubMed]
34. Edwards, P.T.; Smith, B.P.; McArthur, M.L.; Hazel, S.J. At the Heart of a Dog's Veterinary Experience: Heart Rate Responses in Dogs Vary across a Standard Physical Examination. *J. Vet. Behav.* **2022**, *51*, 23–34. [CrossRef]
35. Marques, J.I.; Lopes Neto, J.P.; do Nascimento, J.W.B.; Talieri, I.C.; de Medeiros, G.R.; Furtado, D.A. Pupillary Dilation as a Thermal Stress Indicator in Boer Crossbred Goats Maintained in a Climate Chamber. *Small Rumin. Res.* **2018**, *158*, 26–29. [CrossRef]
36. Kovács, L.; Kézér, F.L.; Ruff, F.; Jurkovich, V.; Szenci, O. Heart Rate, Cardiac Vagal Tone, Respiratory Rate, and Rectal Temperature in Dairy Calves Exposed to Heat Stress in a Continental Region. *Int. J. Biometeorol.* **2018**, *62*, 1791–1797. [CrossRef] [PubMed]
37. Franklin, S.H.; Van Erck-Westergren, E.; Bayly, W.M. Respiratory Responses to Exercise in the Horse. *Equine Vet. J.* **2012**, *44*, 726–732. [CrossRef] [PubMed]
38. Rovira, S.; Munoz, A.; Benito, M. Effect of Exercise on Physiological, Blood and Endocrine Parameters in Search and Rescue-Trained Dogs. *Vet. Med.* **2008**, *53*, 333–346. [CrossRef]
39. Martins-Pinge, M.C. Cardiovascular and Autonomic Modulation by the Central Nervous System after Aerobic Exercise Training. *Braz. J. Med. Biol. Res.* **2011**, *44*, 848–854. [CrossRef]
40. Mathôt, S. Pupillometry: Psychology, Physiology, and Function. *J. Cogn.* **2018**, *1*, 16. [CrossRef] [PubMed]
41. Hall, C.A.; Chilcott, R.P. Eyeing up the Future of the Pupillary Light Reflex in Neurodiagnostics. *Diagnostics* **2018**, *8*, 19. [CrossRef] [PubMed]
42. Larsen, R.S.; Waters, J. Neuromodulatory Correlates of Pupil Dilation. *Front. Neural Circuits* **2018**, *12*, 21. [CrossRef] [PubMed]
43. Lee, S.-H.; Dan, Y. Neuromodulation of Brain States. *Neuron* **2012**, *76*, 209–222. [CrossRef] [PubMed]
44. Larson, M.D.; Behrends, M. Portable Infrared Pupillometry. *Anesth. Analg.* **2015**, *120*, 1242–1253. [CrossRef] [PubMed]
45. Zele, A.J.; Gamlin, P.D. Editorial: The Pupil: Behavior, Anatomy, Physiology and Clinical Biomarkers. *Front. Neurol.* **2020**, *11*, 211. [CrossRef] [PubMed]
46. Larson, M.D.; Tayefeh, F.; Sessler, D.I.; Daniel, M.; Noorani, M. Sympathetic Nervous System Does Not Mediate Reflex Pupillary Dilation during Desflurane Anesthesia. *Anesthesiology* **1996**, *85*, 748–754. [CrossRef] [PubMed]
47. Payen, J.-F.; Isnardon, S.; Lavolaine, J.; Bouzat, P.; Vinclair, M.; Francony, G. La Pupillométrie En Anesthésie-Réanimation. *Ann. Françaises d'Anesthésie Réanim.* **2012**, *31*, e155–e159. [CrossRef] [PubMed]
48. Machado, N.A.F.; Barbosa-Filho, J.A.D.; Souza-Junior, J.B.F.; Ramalho, G.L.B.; Parente, M.d.O.M. Use of Pupillometry in the Diagnosis of Stress in Piglets Transported in a Tropical Climate. *Eng. Agrícola* **2021**, *41*, 402–408. [CrossRef]
49. Moody, C.M.; Picketts, V.A.; Mason, G.J.; Dewey, C.E.; Niel, L. Can You Handle It? Validating Negative Responses to Restraint in Cats. *Appl. Anim. Behav. Sci.* **2018**, *204*, 94–100. [CrossRef]
50. Tsigos, C.; Kyrou, I.; Kassi, E.; Chrousos, G.P. *Stress, Endocrine Physiology and Pathophysiology*; MDText.com, Inc.: South Dartmouth, MA, USA, 2000.

51. Noushad, S.; Ahmed, S.; Ansari, B.; Mustafa, U.-H.; Saleem, Y.; Hazrat, H. Physiological Biomarkers of Chronic Stress: A Systematic Review. *Int. J. Health Sci.* **2021**, *15*, 46–59.
52. Kim, W.S.; Lee, J.-S.; Jeon, S.W.; Peng, D.Q.; Kim, Y.S.; Bae, M.H.; Jo, Y.H.; Lee, H.G. Correlation between Blood, Physiological and Behavioral Parameters in Beef Calves under Heat Stress. *Asian-Australas. J. Anim. Sci.* **2018**, *31*, 919–925. [CrossRef]
53. Olvera-Maneu, S.; Carbajal, A.; Serres-Corral, P.; López-Béjar, M. Cortisol Variations to Estimate the Physiological Stress Response in Horses at a Traditional Equestrian Event. *Animals* **2023**, *13*, 396. [CrossRef] [PubMed]
54. Michael, S.; Graham, K.S.; Davis, G.M. Cardiac Autonomic Responses during Exercise and Post-Exercise Recovery Using Heart Rate Variability and Systolic Time Intervals—A Review. *Front. Physiol.* **2017**, *8*, 301. [CrossRef] [PubMed]
55. von Borell, E.; Langbein, J.; Després, G.; Hansen, S.; Leterrier, C.; Marchant-Forde, J.; Marchant-Forde, R.; Minero, M.; Mohr, E.; Prunier, A.; et al. Heart Rate Variability as a Measure of Autonomic Regulation of Cardiac Activity for Assessing Stress and Welfare in Farm Animals—A Review. *Physiol. Behav.* **2007**, *92*, 293–316. [CrossRef] [PubMed]
56. Gormally, B.M.G.; Romero, L.M. What Are You Actually Measuring? A Review of Techniques That Integrate the Stress Response on Distinct Time-Scales. *Funct. Ecol.* **2020**, *34*, 2030–2044. [CrossRef]
57. Stucke, D.; Große Ruse, M.; Lebelt, D. Measuring Heart Rate Variability in Horses to Investigate the Autonomic Nervous System Activity—Pros and Cons of Different Methods. *Appl. Anim. Behav. Sci.* **2015**, *166*, 1–10. [CrossRef]
58. Stauss, H.M. Heart Rate Variability. *Am. J. Physiol. Integr. Comp. Physiol.* **2003**, *285*, R927–R931. [CrossRef] [PubMed]
59. Ille, N.; von Lewinski, M.; Erber, R.; Wulf, M.; Aurich, J.; Möstl, E.; Aurich, C. Effects of the Level of Experience of Horses and Their Riders on Cortisol Release, Heart Rate and Heart-Rate Variability during a Jumping Course. *Anim. Welf.* **2013**, *22*, 457–465. [CrossRef]
60. Turini, L.; Bonelli, F.; Lanatà, A.; Vitale, V.; Nocera, I.; Sgorbini, M.; Mele, M. Validation of a New Smart Textiles Biotechnology for Heart Rate Variability Monitoring in Sheep. *Front. Vet. Sci.* **2022**, *9*, 1018213. [CrossRef] [PubMed]
61. Ledowski, T. Objective Monitoring of Nociception: A Review of Current Commercial Solutions. *Br. J. Anaesth.* **2019**, *123*, e312–e321. [CrossRef] [PubMed]
62. Ledowski, T.; Bein, B.; Hanss, R.; Paris, A.; Fudickar, W.; Scholz, J.; Tonner, P.H. Neuroendocrine Stress Response and Heart Rate Variability: A Comparison of Total Intravenous Versus Balanced Anesthesia. *Anesth. Analg.* **2005**, *101*, 1700–1705. [CrossRef] [PubMed]
63. Jeanne, M.; Logier, R.; De Jonckheere, J.; Tavernier, B. Validation of a Graphic Measurement of Heart Rate Variability to Assess Analgesia/Nociception Balance during General Anesthesia. In Proceedings of the 2009 Annual International Conference of the IEEE Engineering in Medicine and Biology Society, Minneapolis, MN, USA, 2–6 September 2009; IEEE: Piscataway, NJ, USA, 2009; pp. 1840–1843.
64. Ruíz-López, P.; Domínguez, J.M.; Granados, M.d.M. Intraoperative Nociception-Antinociception Monitors: A Review from the Veterinary Perspective. *Vet. Anaesth. Analg.* **2020**, *47*, 152–159. [CrossRef] [PubMed]
65. Hernandez-Avalos, I.; Mota-Rojas, D.; Mora-Medina, P.; Martínez-Burnes, J.; Casas Alvarado, A.; Verduzco-Mendoza, A.; Lezama-García, K.; Olmos-Hernandez, A. Review of Different Methods Used for Clinical Recognition and Assessment of Pain in Dogs and Cats. *Int. J. Vet. Sci. Med.* **2019**, *7*, 43–54. [CrossRef] [PubMed]
66. Logier, R.; Jeanne, M.; De Jonckheere, J.; Dassonneville, A.; Delecroix, M.; Tavernier, B. PhysioDoloris: A Monitoring Device for Analgesia/Nociception Balance Evaluation Using Heart Rate Variability Analysis. In Proceedings of the 2010 Annual International Conference of the IEEE Engineering in Medicine and Biology, Buenos Aires, Argentina, 31 August–4 September 2010; IEEE: Piscataway, NJ, USA, 2010; pp. 1194–1197.
67. Mansour, C.; Merlin, T.; Bonnet-Garin, J.-M.; Chaaya, R.; Mocci, R.; Ruiz, C.C.; Allaouchiche, B.; Boselli, E.; Junot, S. Evaluation of the Parasympathetic Tone Activity (PTA) Index to Assess the Analgesia/Nociception Balance in Anaesthetised Dogs. *Res. Vet. Sci.* **2017**, *115*, 271–277. [CrossRef] [PubMed]
68. Mansour, C.; Mocci, R.; Santangelo, B.; Sredensek, J.; Chaaya, R.; Allaouchiche, B.; Bonnet-Garin, J.-M.; Boselli, E.; Junot, S. Performance of the Parasympathetic Tone Activity (PTA) Index to Predict Changes in Mean Arterial Pressure in Anaesthetized Horses with Different Health Conditions. *Res. Vet. Sci.* **2021**, *139*, 43–50. [CrossRef] [PubMed]
69. Yoshida, K.; Obara, S.; Inoue, S. Analgesia Nociception Index and High Frequency Variability Index: Promising Indicators of Relative Parasympathetic Tone. *J. Anesth.* **2023**, *37*, 130–137. [CrossRef] [PubMed]
70. Ruíz-López, P.; Morgaz, J.; Quirós-Carmona, S.; Navarrete-Calvo, R.; Domínguez, J.M.; Gómez-Villamandos, R.J.; Granados, M.M. Parasympathetic Tone Changes in Anesthetized Horses after Surgical Stimulation, and Morphine, Ketamine, and Dobutamine Administration. *Animals* **2022**, *12*, 1038. [CrossRef] [PubMed]
71. Leitão, C.J.; Lima-Rodríguez, J.R.; Ferreira, F.; Avelino, C.; Sánchez-Margallo, F.M.; Antunes, L. Parasympathetic Tone Activity Evaluation to Discriminate Ketorolac and Ketorolac/Tramadol Analgesia Level in Swine. *Anesth. Analg.* **2019**, *129*, 882–889. [CrossRef] [PubMed]
72. Hernández-Avalos, I.; Valverde, A.; Antonio Ibancovichi Camarillo, J.; Sánchez-Aparicio, P.; Recillas-Morales, S.; Rodríguez-Velázquez, D.; Osorio-Avalos, J.; Armando Magdaleno-Torres, L.; Chavez-Monteagudo, J.; Manuel Acevedo-Arcique, C. Clinical Use of the Parasympathetic Tone Activity Index as a Measurement of Postoperative Analgaesia in Dogs Undergoing Ovariohysterectomy. *J. Vet. Res.* **2021**, *65*, 117–123. [CrossRef] [PubMed]

73. Aguado, D.; Bustamante, R.; García-Sanz, V.; González-Blanco, P.; Gómez de Segura, I.A. Efficacy of the Parasympathetic Tone Activity Monitor to Assess Nociception in Healthy Dogs Anaesthetized with Propofol and Sevoflurane. *Vet. Anaesth. Analg* **2020**, *47*, 103–110. [CrossRef] [PubMed]
74. Mansour, C.; El Hachem, N.; Jamous, P.; Saade, G.; Boselli, E.; Allaouchiche, B.; Bonnet, J.-M.; Junot, S.; Chaaya, R. Performance of the Parasympathetic Tone Activity (PTA) Index to Assess the Intraoperative Nociception Using Different Premedication Drugs in Anaesthetised Dogs. *Int. J. Vet. Sci. Med.* **2020**, *8*, 49–55. [CrossRef] [PubMed]
75. Ruíz-López, P.; Domínguez, J.M.; Morgaz, J.; Quirós-Carmona, S.; Navarrete-Calvo, R.; Gómez-Villamandos, R.J.; Fernández-Sarmiento, J.A.; Granados, M.M. Evaluation of the Averaged Parasympathetic Tone Activity and Its Dynamic Variation to Assess Intraoperative Nociception in Relation to Hemodynamic Changes in Dogs. *Vet. Q.* **2023**, *43*, 1–9. [CrossRef] [PubMed]
76. Lima, L.; Dos-Santos, J.D.; Ribeiro, L.; Cabral, P.; Colaço, B.; Martins, J. The Performance of Using the Parasympathetic Tone Activity (PTA) Index to Assess Intraoperative Nociception in Cats. *Vet. Sci.* **2024**, *11*, 121. [CrossRef] [PubMed]
77. Domínguez-Oliva, A.; Hernández-Avalos, I.; Olmos-Hernández, A.; Verduzco-Mendoza, A.; Mota-Rojas, D. Thermal Response of Laboratory Rats (*Rattus norvegicus*) during the Application of Six Methods of Euthanasia Assessed by Infrared Thermography. *Animals* **2023**, *13*, 2820. [CrossRef] [PubMed]
78. Oka, T. Stress-Induced Hyperthermia and Hypothermia. In *Thermoregulation: From Basic Neuroscience to Clinical Neurology, Part II*; Romanovsky, A., Ed.; Elsevier: Amsterdam, The Netherlands, 2018; Volume 157, pp. 599–621. ISBN 9780444640741.
79. Mota-Rojas, D.; Olmos-Hernández, A.; Verduzco-Mendoza, A.; Lecona-Butrón, H.; Martínez-Burnes, J.; Mora-Medina, P.; Gómez-Prado, J.; Orihuela, A. Infrared Thermal Imaging Associated with Pain in Laboratory Animals. *Exp. Anim.* **2021**, *70*, 1–12. [CrossRef] [PubMed]
80. Hernández-Avalos, I.; Flores-Gasca, E.; Mota-Rojas, D.; Casas-Alvarado, A.; Miranda-Cortés, A.E.; Domínguez-Oliva, A. Neurobiology of Anesthetic-Surgical Stress and Induced Behavioral Changes in Dogs and Cats: A Review. *Vet. World* **2021**, *14*, 393–404. [CrossRef] [PubMed]
81. Travain, T.; Valsecchi, P. Infrared Thermography in the Study of Animals' Emotional Responses: A Critical Review. *Animals* **2021**, *11*, 2510. [CrossRef] [PubMed]
82. Blenkuš, U.; Gerós, A.F.; Carpinteiro, C.; de Castro Aguiar, P.; Olsson, I.A.S.; Franco, N.H. Non-Invasive Assessment of Mild Stress-Induced Hyperthermia by Infrared Thermography in Laboratory Mice. *Animals* **2022**, *12*, 177. [CrossRef] [PubMed]
83. Kohlhause, S.; Hoffmann, K.; Schlumbohm, C.; Fuchs, E.; Flügge, G. Nocturnal Hyperthermia Induced by Social Stress in Male Tree Shrews: Relation to Low Testosterone and Effects of Age. *Physiol. Behav.* **2011**, *104*, 786–795. [CrossRef] [PubMed]
84. Verduzco-Mendoza, A.; Bueno-Nava, A.; Wang, D.; Martínez-Burnes, J.; Olmos-Hernández, A.; Casas, A.; Domínguez, A.; Mota-Rojas, D. Experimental Applications and Factors Involved in Validating Thermal Windows Using Infrared Thermography to Assess the Health and Thermostability of Laboratory Animals. *Animals* **2021**, *11*, 3448. [CrossRef] [PubMed]
85. Bienboire-Frosini, C.; Wang, D.; Marcet-Rius, M.; Villanueva-García, D.; Gazzano, A.; Domínguez-Oliva, A.; Olmos-Hernández, A.; Hernández-Avalos, I.; Lezama-García, K.; Verduzco-Mendoza, A.; et al. The Role of Brown Adipose Tissue and Energy Metabolism in Mammalian Thermoregulation during the Perinatal Period. *Animals* **2023**, *13*, 2176. [CrossRef] [PubMed]
86. Andrade, D.V. Thermal Windows and Heat Exchange. *Temperature* **2015**, *2*, 451. [CrossRef] [PubMed]
87. Mota-Rojas, D.; Pereira, M.F.A.; Wang, D.; Martínez-Burnes, J.; Ghezzi, M.; Hernández-Ávalos, I.; Lendez, P.; Mora-Medina, P.; Casas, A.; Olmos-Hernández, A.; et al. Clinical Applications and Factors Involved in Validating Thermal Windows in Large Rumiants to Assess Health and Productivity. *Animals* **2021**, *11*, 2247. [CrossRef] [PubMed]
88. Gómez-Prado, J.; Pereira, A.M.F.; Wang, D.; Villanueva-García, D.; Domínguez-Oliva, A.; Mora-Medina, P.; Hernández-Avalos, I.; Martínez-Burnes, J.; Casas-Alvarado, A.; Olmos-Hernández, A.; et al. Thermoregulation Mechanisms and Perspectives for Validating Thermal Windows in Pigs with Hypothermia and Hyperthermia: An Overview. *Front. Vet. Sci.* **2022**, *9*, 1023294. [CrossRef]
89. Soroko, M.; Howell, K.; Zwyrzykowska, A.; Dudek, K.; Zielińska, P.; Kupczyński, R. Maximum Eye Temperature in the Assessment of Training in Racehorses: Correlations with Salivary Cortisol Concentration, Rectal Temperature, and Heart Rate. *J. Equine Vet. Sci.* **2016**, *45*, 39–45. [CrossRef]
90. Shu, H.; Li, Y.; Fang, T.; Xing, M.; Sun, F.; Chen, X.; Bindelle, J.; Wang, W.; Guo, L. Evaluation of the Best Region for Measuring Eye Temperature in Dairy Cows Exposed to Heat Stress. *Front. Vet. Sci.* **2022**, *9*, 857777. [CrossRef] [PubMed]
91. Ghezzi, M.D.; Napolitano, F.; Casas-Alvarado, A.; Hernández-Ávalos, I.; Domínguez-Oliva, A.; Olmos-Hernández, A.; Pereira, A.M.F. Utilization of Infrared Thermography in Assessing Thermal Responses of Farm Animals under Heat Stress. *Animals* **2024**, *14*, 616. [CrossRef] [PubMed]
92. Idris, M.; Uddin, J.; Sullivan, M.; McNeill, D.M.; Phillips, C.J.C. Non-Invasive Physiological Indicators of Heat Stress in Cattle. *Animals* **2021**, *11*, 71. [CrossRef] [PubMed]
93. Arfuso, F.; Acri, G.; Piccione, G.; Sansotta, C.; Fazio, F.; Giudice, E.; Giannetto, C. Eye Surface Infrared Thermography Usefulness as a Noninvasive Method of Measuring Stress Response in Sheep during Shearing: Correlations with Serum Cortisol and Rectal Temperature Values. *Physiol. Behav.* **2022**, *250*, 113781. [CrossRef] [PubMed]
94. Stewart, M.; Stafford, K.J.; Dowling, S.K.; Schaefer, A.; Webster, J. Eye Temperature and Heart Rate Variability of Calves Disbudded with or without Local Anaesthetic. *Physiol. Behav.* **2008**, *93*, 789–797. [CrossRef] [PubMed]

95. Chikkagoudara, K.P.; Pawan, S.D.; Deepandita, B.; Surender, S.L.; Ninad, B. Eye Temperature, an Indicator for Stress Levels in Young Buffalo Bulls—A Case Study of Micro-Environment Modification. *J. Agrometeorol.* **2021**, *22*, 266–273. [CrossRef]
96. Sutherland, M.A.; Worth, G.M.; Dowling, S.K.; Lowe, G.L.; Cave, V.M.; Stewart, M. Evaluation of Infrared Thermography as a Non-Invasive Method of Measuring the Autonomic Nervous Response in Sheep. *PLoS ONE* **2020**, *15*, e0233558. [CrossRef] [PubMed]
97. Casas-Alvarado, A.; Mota-Rojas, D.; Hernández-Ávalos, I.; Martínez-Burnes, J.; Rosas, M.E.; Miranda-Cortés, A.; Domínguez-Oliva, A.; Mora-Medina, P. Assessment of Thermal Response, Cardiorespiratory Parameters and Post-Operative Analgesia in Dogs Undergoing Ovariohysterectomy with Different Combinations of Epidural Analgesia and Isoflurane. *J. Anim. Behav. Biometeorol.* **2023**, *11*, e2023009. [CrossRef]
98. Bergamasco, L.; Edwards-Callaway, L.N.; Bello, N.M.; Mijares, S.H.; Cull, C.A.; Rugan, S.; Mosher, R.A.; Gehring, R.; Coetzee, J.F. Unmitigated Surgical Castration in Calves of Different Ages: Cortisol Concentrations, Heart Rate Variability, and Infrared Thermography Findings. *Animals* **2021**, *11*, 2719. [CrossRef] [PubMed]
99. Luck, J.C.; Kunselman, A.R.; Herr, M.D.; Blaha, C.A.; Sinoway, L.I.; Cui, J. Multiple Laser Doppler Flowmetry Probes Increase the Reproducibility of Skin Blood Flow Measurements. *Front. Physiol.* **2022**, *13*, 876633. [CrossRef]
100. Zherebtsov, E.A.; Zherebtsova, A.I.; Doronin, A.; Dunaev, A.V.; Podmasteryev, K.V.; Bykov, A.; Meglinski, I. Combined Use of Laser Doppler Flowmetry and Skin Thermometry for Functional Diagnostics of Intradermal Finger Vessels. *J. Biomed. Opt.* **2017**, *22*, 040502. [CrossRef] [PubMed]
101. Merla, A.; Di Donato, L.; Romani, G.L.; Proietti, M.; Salsano, F. Comparison of Thermal Infrared and Laser Doppler Imaging in the Assessment of Cutaneous Tissue Perfusion in Scleroderma Patients and Healthy Controls. *Int. J. Immunopathol. Pharmacol.* **2008**, *21*, 679–686. [CrossRef] [PubMed]
102. Galen, R.S.; Gambino, S.R. *Beyond Normality: The Predictive Value and Efficiency of Medical Diagnoses*; Wiley: New York, NY, USA, 1975.
103. Kesztyüs, D.; Brucher, S.; Wilson, C.; Kesztyüs, T. Use of Infrared Thermography in Medical Diagnosis, Screening, and Disease Monitoring: A Scoping Review. *Medicina* **2023**, *59*, 2139. [CrossRef] [PubMed]
104. Berckmans, D. *Advances in Precision Livestock Farming*, 1st ed.; Burleigh Dodds Science Publishing: London, UK, 2022.
105. Fuentes, S.; Gonzalez Viejo, C.; Chauhan, S.S.; Joy, A.; Tongson, E.; Dunshea, F.R. Non-Invasive Sheep Biometrics Obtained by Computer Vision Algorithms and Machine Learning Modeling Using Integrated Visible/Infrared Thermal Cameras. *Sensors* **2020**, *20*, 6334. [CrossRef] [PubMed]

animals

MDPI

Article

The Potential of Infrared Thermography for Early Pregnancy Diagnosis in Nili-Ravi Buffaloes

Umair Riaz [1,2,3], Musadiq Idris [4], Mehboob Ahmed [1,2], Farah Ali [3], Umer Farooq [4] and Liguo Yang [1,2,*]

1 Hubei Hongshan Laboratory, Wuhan 430070, China; umair.riaz@iub.edu.pk (U.R.); mehboob.alyani@gmail.com (M.A.)
2 National Center for International Research on Animal Genetics, Breeding and Reproduction (NCIRAGBR), Ministry of Science and Technology, Huazhong Agricultural University, Wuhan 430070, China
3 Department of Theriogenology, Faculty of Veterinary Science, The Islamia University of Bahawalpur, Bahawalpur 63100, Pakistan; drfarahiub@gmail.com
4 Department of Physiology, Faculty of Veterinary and Animal Sciences, The Islamia University of Bahawalpur, Bahawalpur 63100, Pakistan; musadiq.idris@iub.edu.pk (M.I.); umer.farooq@iub.edu.pk (U.F.)
* Correspondence: ylg@mail.hzau.edu.cn

Simple Summary: Early diagnosis of pregnancy in dairy buffaloes is of critical importance for harnessing their maximum production potential. Existing methods of early pregnancy diagnosis are highly technical, expensive, and laborious. Infrared thermography is emerging as a user-friendly, less laborious, and relatively simple technique that is capable of detecting variations in surface temperature associated with physiological changes in the animal's body. The present study evaluated the potential of infrared thermography to detect thermal changes associated with early pregnancy in different anatomical regions (vulva, eyes, muzzle, flanks) in buffaloes. The results of the present study revealed that body-surface temperature as detected by infrared was significantly higher for the left flank (maximum, average, and minimum) and left eye and vulva (maximum and average gradients). The muzzle temperature at the maximum gradient was significantly higher for the non-pregnant as compared to pregnant buffaloes. The results indicated that various regions of the body tend to show differences in the surface temperature after pregnancy has been established and that infrared thermography could serve as a tool to measure such pregnancy-associated thermal changes.

Abstract: This study was designed to explore the potential of infrared thermography (IRT) as an alternate approach for early pregnancy diagnosis in buffaloes. The surface temperature (ST) of different regions (eyes, muzzle, flanks, and vulva) was determined in 27 buffaloes using IRT from the day of artificial insemination (AI; Day 0), and measurement was repeated every fourth day until Day 24 post-AI. From all regions, the ST in each thermograph was recorded at three temperature values (maximum, average, minimum). Pregnancy status was confirmed through ultrasonography on Day 30, and animals were retrospectively grouped as pregnant or non-pregnant for analysis of thermographic data. In pregnant buffaloes, all three values of ST were significantly greater ($p \leq 0.05$) for the left flank, while, in the left eye and vulva, only the maximum and average values were significantly greater. By contrast, the maximum ST of the muzzle was significantly lower ($p \leq 0.05$) in pregnant buffaloes compared to non-pregnant buffaloes. However, the ST of the right eye and right flank did not show significant temperature variation at any value. These findings suggest that IRT has the potential to identify thermal changes associated with pregnancy in buffaloes at an early stage.

Keywords: buffalo; early pregnancy; infrared thermography; surface temperature

Citation: Riaz, U.; Idris, M.; Ahmed, M.; Ali, F.; Farooq, U.; Yang, L. The Potential of Infrared Thermography for Early Pregnancy Diagnosis in Nili-Ravi Buffaloes. *Animals* **2024**, *14*, 1966. https://doi.org/10.3390/ani14131966

Academic Editor: Daniel Mota-Rojas

Received: 10 June 2024
Revised: 26 June 2024
Accepted: 28 June 2024
Published: 2 July 2024

1. Introduction

Buffaloes hold significant importance as dairy animals, being the second-largest contributor to global milk production. To adequately harness the production potential of buffaloes, early identification of non-pregnant animals and their re-insemination are of paramount importance [1]. Delay in detecting the non-pregnant animals significantly escalates managerial and monetary inputs, jeopardizing the economic viability of rearing dairy buffaloes [2]. The significance of early pregnancy diagnosis is translated into improved reproductive efficiency, enhanced productivity, and competent farm economics, especially in the context of dairy-buffalo operations [1,3].

Currently, conventional methods employed for pregnancy diagnosis in buffaloes include rectal palpation [4], progesterone quantification in milk/blood [5,6], and ultrasonography [7]. These methods efficiently detect non-pregnant buffaloes around and after 25 days post-breeding, which is later than the next natural breeding opportunity, i.e., estrus (after an average of 21 days post-breeding). In addition to these traditional approaches, the detection of several biomarkers in maternal fluids such as pregnancy-associated glycoproteins [8], early pregnancy factors [9], urinary metabolites [10], and circulatory microRNAs [11,12] can be helpful in early pregnancy diagnosis in buffaloes. Unfortunately, these techniques demand a high level of technical expertise and are very expensive to use at a herd or farm level, limiting their adoption by farmers. Scientists are continuously working to find a practically convenient, relatively cheaper, non-invasive, and user-friendly method for early pregnancy estimation in buffalo.

Infrared thermography (IRT) is emerging as a potential tool for reproductive management, including estrus monitoring [13] and early pregnancy diagnosis [14] in farm animal species. IRT is handy for detecting and quantifying heat radiation emitting from the surface of an object. Various physiological (muscular activity, estrus, pregnancy, etc.) and pathological (stress, inflammations, pain, etc.) changes are associated with changes in the thermal status of the animal and warrant the use of this technique in veterinary practice [15–19]. These changes in body temperatures can be monitored through an infrared thermal camera. Major reproductive events like estrus and pregnancy are accomplished by the interplay of steroid hormones. These are responsible for changes in the blood flow to and from the genital organs. Progesterone has been reported to have a thermogenic property and, similarly, estrogen is known to induce hyperactivity and consequent temperature rise [13]. Considering these properties, IRT has been evaluated for estrus alerts [20], estimating the time of the luteinizing hormone (LH) surge and ovulation in animals [21,22]. Likewise, various studies have been conducted in different wild [23,24] and domesticated species to detect pregnancy and evaluate the progression of pregnancy [25–28]. Olğaç et al. [14] reported that the use of IRT is very promising for early pregnancy diagnosis in cattle. Buffaloes, due to their black skin color and less hairy coat, are assumed to possess better emissivity, as dark and smooth surfaces have a better ability to absorb and emit infrared radiation to maintain body temperature [13].

Therefore, it is hypothesized that IRT can be used to determine temperature variation arising possibly due to the difference in steroid-hormone profile and blood flow between pregnant and non-pregnant buffaloes at the time of insemination and during early pregnancy.

The present study is a pilot trial designed to evaluate the potential of IRT to detect surface-temperature variation at various anatomical regions (vulva, eyes, muzzle, flanks) between pregnant and non-pregnant buffaloes at the time of insemination and in early pregnancy.

2. Materials and Methods

2.1. Animals

The present study was conducted during the breeding season of the buffalo raised by small-scale livestock keepers (December and January 2021) in Bahawalpur, Pakistan. Twenty-seven 3.5-to-7-year-old lactating buffaloes (postpartum days, mean ± SE = 147.4 ± 17.5) of mixed parity that were clinically healthy with a body condition score of 3 ± 0.7 measured on a scale of 1 to 5 (1 = emaciated and 5 = obese) were selected as subjects in this study. Before enrollment in the trial, all buffaloes were examined by a trained veterinary professional and declared free of palpable abnormalities in their reproductive tract.

2.2. Experimental Design and Infrared Thermography

Following the occurrence of spontaneous estrus, all buffaloes were artificially inseminated by a trained technician with frozen semen sourced from a bull of known fertility. The day of artificial insemination (AI) was designated as 'Day 0' of the experiment. In all animals, before the AI, IRT was performed following the instructions and precautions explained earlier [13]. The infrared thermographic imaging continued at an interval of four days until 'Day 24' of the experiment in all animals. Thermal images were taken using an infrared thermal camera (Fluke PTi120 Pocket Thermal Imager with IR resolution 120 × 90, 10,800 pixels and temperature range −20 to 400 °C; Fluke Corporation, Everett, WA, USA) with emissivity set at 0.98. The anatomical regions of interest, specifically the vulva, eyes, muzzle, and flanks, were selected for taking thermal images, as they are prominent on the animal body surface and easily accessible for IRT imaging, having sufficient superficial blood circulation to dissipate metabolic heat to maintain homeostasis. It was ensured that the body parts of the animal subjected to thermographic imaging had clean surfaces without mud or dung present. Thermal images were taken for the above-mentioned anatomical regions in the morning at the same time every day, with the camera placed at an angle of 90° at a distance of approximately 1 m from the surface. All images were taken in the shade, and special attention was paid to avoiding physical activity by the animals before the thermal imaging. Images were transferred to the computer and were analyzed using software named 'Fluke Connect (version 1.1.551.0)'. The maximum, average, and minimum temperatures were analyzed.

2.3. Pregnancy Diagnosis

Pregnancy status was confirmed 30 days after artificial insemination with the help of transrectal B-mode ultrasonography. A linear probe having a frequency of 7.5 MHz fitted with an ultrasound unit 'Honda HS 200' (Honda Electronics, Toyohashi, Japan) was used to scan the uterine horns and body transrectally. Buffaloes found to have an embryonic heartbeat ($n = 11$) were declared pregnant, while those with absence of an embryo ($n = 16$) were recorded as non-pregnant. Retrospectively, the thermographic data of the animals were grouped and analyzed separately based on pregnancy status.

2.4. Statistical Analyses

The data obtained from 27 Nili-Ravi buffaloes to explore the potential of IRT as an alternate approach for diagnosing early pregnancy were analyzed using the statistical software Minitab 19 (Minitab® 19.1 Inc., Chicago, IL, USA). All data were analyzed for normality using the Anderson–Darling normality test. The data were analyzed using a generalized linear model (GLM) with the fixed factors Group, Day, and interaction between fixed factors (Group × Day), and post hoc comparisons were made with Tukey's 95% test. The animals' identification numbers were taken as a random factor. The statistical model can be described by the equation below:

$$Y_{\text{IRT(Body Surface)}} = \mu + \text{Group} + \text{Day} + \text{ID} + (\text{Group} \times \text{Day}) + e \quad (1)$$

Here, $Y_{\text{IRT(Body Surface)}}$ is the expected value of infrared body surface temperature; μ is the expected mean value for response variables, equal to zero, Group and Day are fixed factors, and e is the random error associated with experimental observations.

3. Results

The current study was planned to assess the potential of IRT as an alternate/quick approach for early pregnancy detection in buffaloes. The data representing variations in the body-surface temperature measured through IRT at different anatomical regions (left eye, right eye, muzzle, left flank, right flank, and vulva) of the pregnant and non-pregnant buffaloes are given in Table 1.

Table 1. Infrared surface temperature of different anatomical regions of pregnant and non-pregnant buffaloes.

Parameter	Surface Temperature (°C)		SED	F-Value	*p*-Value		
	Pregnant	Non-Pregnant			Group	Day	Group × Day
Left eye maximum	35.8 [a]	35.2 [b]	0.61	7.40	0.007	0.686	0.190
Left eye average	33.2 [a]	32.7 [b]	0.50	6.74	0.010	0.613	0.550
Left eye minimum	30.5	30.1	0.59	2.63	0.107	0.504	0.954
Right eye maximum	35.8	35.8	0.66	0.01	0.920	0.730	0.539
Right eye average	33.1	33.1	0.51	0.06	0.806	0.669	0.606
Right eye minimum	30.5	30.4	0.57	0.44	0.509	0.005	0.667
Muzzle maximum	34.6 [b]	35.2 [a]	0.64	4.94	0.027	0.060	0.002
Muzzle average	32.2	32.5	0.59	1.88	0.172	0.002	0.009
Muzzle minimum	29.7	29.8	0.68	0.09	0.764	0.000	0.049
Left flank maximum	35.6 [a]	34.9 [b]	0.65	6.24	0.013	0.000	0.009
Left flank average	32.7 [a]	32.2 [b]	0.56	6.74	0.010	0.000	0.011
Left flank minimum	29.9 [a]	29.4 [b]	0.67	3.72	0.055	0.003	0.088
Right flank maximum	35.6	35.4	0.65	0.56	0.457	0.035	0.767
Right flank average	32.8	32.4	0.69	1.81	0.180	0.027	0.948
Right flank minimum	30.1	29.56	0.87	2.52	0.115	0.072	0.954
Vulva maximum	36.5 [a]	35.3 [b]	0.64	23.82	0.000	0.001	0.994
Vulva average	33.9 [a]	33.1 [b]	0.54	11.64	0.001	0.000	0.960
Vulva minimum	31.3	31.0	0.59	1.43	0.233	0.000	0.388

SED: standard error of the difference between two means; Means with different superscripts differ significantly ($p \leq 0.05$) by Fisher's pairwise comparisons.

The maximum ($p = 0.007$) and average ($p = 0.010$) surface temperatures of the left eye were significantly higher in pregnant buffaloes compared to non-pregnant ones. However, the minimum surface temperature of the left eye did not vary significantly between groups ($p = 0.107$) (Table 1). No surface-temperature value of the right eye (maximum, average, or minimum) was statistically different between pregnant and non-pregnant buffaloes. Overall, the surface temperatures of the right and left eye did not differ significantly between the experimental groups on different experimental days (Group × Day; Table 1).

The maximum surface temperature of the muzzle was significantly higher ($p = 0.027$) in the non-pregnant buffaloes than in the pregnant animals; this result is opposite that observed for the left eye (Table 1). By contrast, the average and minimum surface temperatures of the muzzle exhibited no significant differences between the pregnant and non-pregnant buffaloes.

The interaction of the fixed variables (Group × Day) for the muzzle region revealed significantly higher surface-temperature maximum ($p = 0.002$), average ($p = 0.009$), and minimum ($p = 0.049$) values on the day of artificial insemination (Day 0) in pregnant animals compared to non-pregnant ones (Figure 1). However, during the remaining days of the experiment, i.e., Days 4, 8, 12, 16, 20, and 24, the difference in the infrared surface temperature of the muzzle between the two experimental groups remained non-significant.

The infrared surface temperature of the left flank was significantly higher for pregnant buffaloes as compared to non-pregnant buffaloes at maximum ($p = 0.013$) and average ($p = 0.010$) values (Table 1). However, the minimum infrared surface temperature of the left flank was marginally higher ($p = 0.055$). The maximum surface temperature of the left flank was significantly higher on Day 0. A similar trend was observed in the average value on Day 0 and Day 4 in pregnant compared to non-pregnant buffaloes (Figure 2). The infrared surface temperature of the right flank was not statistically significant between experimental groups and on different experimental days between pregnant and non-pregnant animals (Table 1).

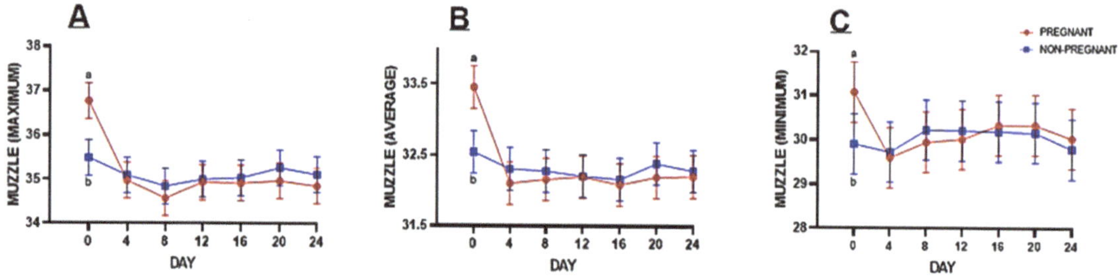

Figure 1. Fisher's pairwise comparison of infrared surface temperature for the muzzle region at all three values, i.e., maximum (**A**), average (**B**), and minimum (**C**), in pregnant and non-pregnant buffaloes on different experimental days.

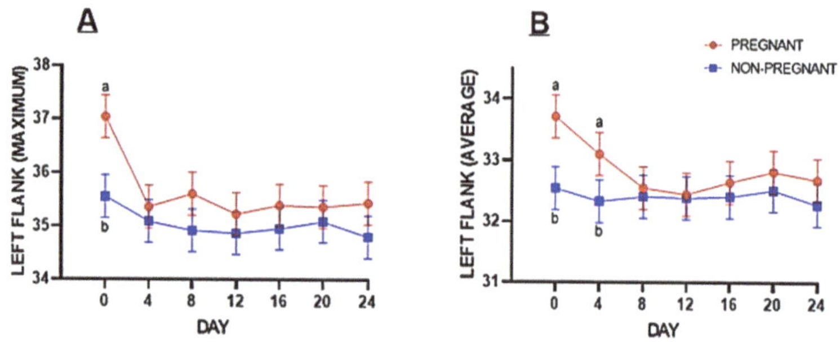

Figure 2. Fisher's pairwise comparison of maximum (**A**) and average (**B**) infrared surface temperature of the left flank region in pregnant and non-pregnant buffaloes on different experimental days.

The maximum ($p = 0.000$) and average ($p = 0.001$) infrared surface temperatures at the vulva measured through IRT were significantly higher in pregnant buffaloes compared to the non-pregnant group. The minimum vulvar surface temperature did not show a significant difference between the two groups (pregnant and non-pregnant) of buffalo. However, the interaction between pregnancy group and the day of experiment was non-significant at all temperature values for the vulvar region.

4. Discussion

In buffaloes, detection of non-pregnant individuals at an early stage is necessary for maximum harnessing of the production potential of animals. This is a basic need for a dairy-based enterprise to make it economically viable [29]. Existing conventional methods give results after the next natural breeding opportunity (estrus) has passed, resulting in the wastage of crucial time and valuable resources [10]. Other identification techniques using early biomarkers are expensive and involve a lot of sampling; they are thus laborious and cause significant discomfort for operators and animals at the farm level. There is a need for a user-friendly, non-invasive, and less expensive technique that is helpful in identifying pregnancy status before the next natural breeding time.

Infrared thermography (IRT) is a non-invasive and user-friendly technique that is emerging as a potential tool for reproductive management. It is helpful in generating estrus alerts [13] and estimation of early pregnancy in dairy animals [14]. The main reproductive functions, including estrus and pregnancy, are controlled by the interplay of steroid hormones. These hormones are responsible for changes in the blood flow to and from the genital organs. Likewise, various studies have been conducted in different species to detect pregnancy and evaluate the progression of pregnancy [25–27]. The use of IRT is very promising for early pregnancy diagnosis in cattle [14]. The present study is a pilot trial designed to evaluate the potential of IRT to detect temperature differences between pregnant and non-pregnant buffaloes at the time of insemination and in early pregnancy.

In the current study, IRT was hypothesized to have greater potential to identify early pregnancy through thermal variation associated with pregnancy in buffaloes compared to cattle owing to their dark color and less hairy coat, which result in greater emissivity [13]. For this purpose, various anatomical regions of interest (vulva, eyes, muzzle, flanks) were selected, as they are prominent and easily accessible for thermal imaging on the animal body, having sufficient superficial blood circulation to dissipate metabolic heat to maintain homeostasis.

In the current study, the infrared thermal temperature of eyes (right and left) was observed, and the results indicated that the maximum and average surface temperatures of the left eye were significantly higher in pregnant buffaloes than in non-pregnant ones. On the contrary, no variation in mean surface temperature was observed for any value in the right eye. Shu et al. [30] stated that IRT is more reliable and repeatable in the left eye. Moreover, the surface temperature of the left eye has been reported to have a strong correlation with the core body temperature in cattle. In various reports, a higher core body temperature has been reported in pregnant animals than in non-pregnant animals. Gill et al. [31] reported that at the time point corresponding to embryo entry into the uterus, during the earlier stages of pregnancy, the core body temperature increased in pregnant cattle compared to non-pregnant cattle. Another study by Scanavez et al. 2017 [32] 2017 also reported an increase in vaginal temperature with the advancement of pregnancy in cattle. In the current study, the higher surface temperature of the left eye in pregnant buffaloes could be attributed to high core body temperature, as the left eye has been reported to be a better indicator of variation in the core body temperature compared to the right eye [30]. Physiological stresses like pregnancy, as well as the hypothalamic-pituitary axis, which controls the reproductive hormones, are governed by the right hemisphere of the brain. The left eye is connected to and represents the information processed in the right hemisphere [30]. Therefore, compared to the right eye, the left eye is assumed to better represent the thermographic changes associated with pregnancy.

In the present study, the maximum surface temperature of the muzzle was significantly lower in the pregnant group of buffaloes than in the non-pregnant group. This could be due to the maintenance of high progesterone levels in the buffaloes who conceived. In addition to its main function of pregnancy maintenance, progesterone is responsible for enhancing minute ventilation, predominantly due to an increase in tidal volume in pregnant animals [33]. Progesterone also causes sensitization of the primary respiratory system of the dam to carbon dioxide during pregnancy, leading to ventilatory changes [34,35]. A temperature rise due to increased progesterone and enhanced metabolic demand may cause an increase in respiration rate to dissipate internal heat [36,37]. Moreover, progesterone plays a crucial role in thermoregulation by modulating TRPV receptors. Specifically, progesterone has been shown to downregulate the expression and activity of TRPV4 channels, as demonstrated in human-airway and mammary-gland epithelial cells, as well as in vascular smooth-muscle cells [38]. This downregulation reduces calcium influx and channel currents, thereby affecting the heat-dissipation mechanisms. Consequently, the elevated progesterone levels during pregnancy could lead to higher localized temperatures in body parts such as the vulva because TRPV4 activity is suppressed, reducing the body's ability to dissipate heat effectively. This mechanism supports our thermographic findings of increased body temperature in pregnant animals [38]. All these factors lead to moisturization of the muzzle surface due to evaporation and cause a drop in surface temperature at the muzzle. These explanatory factors were not considered practically in the present study and may be verified in the future in dedicated studies.

The results of the current study revealed that the surface temperature of the left flank was higher in pregnant compared to non-pregnant buffaloes, although the right flank region did not exhibit any difference in mean surface temperature between groups. This could be attributed to the rumen, which is a major organ under the left flank and is prone to temperature variations arising due to different physiological stages like estrus, the luteal phase, and pregnancy caused by the thermogenic property of progesterone [37]. In an earlier study, the temperature of the rumen in pregnant cows during the luteal phase (early pregnancy) was reported to be higher compared to that in non-pregnant cyclic cows as a result of enhanced metabolism, physiological changes to support fetal development, and increased feed intake [39,40].

The variation in the maximum and average surface temperatures of the vulva was greater in pregnant animals than in non-pregnant animals. This higher surface temperature could probably be due to the increase in blood supply to the reproductive tract that follows conception to deal with the increased needs of the dam to accommodate growing embryonic needs [41]. The blood flow to the genital organ increases during pregnancy owing to oxidative stress and enhanced metabolic demand [42]. The surface temperature of the tissue is representative of the underlying metabolic activity and changes in the blood flow that can be detected using IRT [43]. During estrus, an increase in vulvar temperature detected by IRT was reported to be associated with increased blood flow and was regarded as a very important factor in predicting estrus and/or ovulation [21,26]. There are reports that conclude, using Doppler ultrasonography, that blood flow to the genital tissues increases significantly during early pregnancy [44,45]. Therefore, the probable cause of increased vulvar temperature associated with the establishment of pregnancy in the current study might be associated with enhanced blood circulation to genital organs.

There was a significant increase in the surface temperature at the muzzle and left flanks on the day of insemination (Day 0) of animals subsequently identified as pregnant compared to those who failed to conceive (Figure 1). These findings are consistent with those reported earlier by Liles et al. [46], which stated that increased temperature at the time of insemination is positively correlated with pregnancy afterward in cattle. The chances of pregnancy increased 1.45 times with a 1 °C rise in body temperature. An elevated body temperature during estrus is important for resumption of meiotic activity in oocytes, which has functional significance during fertilization and pregnancy [46].

Previously various studies suggested the vulva and ocular region as good representatives of core body temperature during different physiological stages. However, the muzzle and flank were reported to be less well correlated with the core body temperature. Therefore, following the previous literature on preferred sites for IRT to monitor reproductive stages and the findings of the current pilot study, a systematic experiment is recommended.

The current study is a pilot study to evaluate the potential of infrared thermography as a tool to assess pregnancy-associated changes during early pregnancy in buffaloes. The study has limitations regarding the number of samples, and it is suggested that future studies should be conducted on a larger number of animals with more rigorous grouping. The current study made a comparison between pregnant and non-pregnant animals. However, all animals were inseminated, and it should be considered that possible cases of early/late embryonic death may have influenced the data gathered. Therefore, a more reliable control group should be created by including non-inseminated animals in the study.

5. Conclusions

This study revealed that the infrared body surface temperature was significantly higher for the left flank (maximum, average, and minimum) and left eye and vulva (maximum and average). The maximum muzzle temperature was significantly higher for the non-pregnant as compared to pregnant buffaloes. The results indicated that various regions of the body tend to show differences in the surface temperature after pregnancy has been established and that IRT could serve as a complementary tool to measure such pregnancy-associated thermal changes.

Author Contributions: Conceptualization, L.Y. and M.I.; methodology, U.R. and M.I.; Statistical analysis, M.I. and U.R.; Experimentation, U.R. and M.I.; writing—original draft preparation, U.R., M.A., F.A.; writing—review and editing, M.A. and M.I., U.F.; supervision, L.Y.; funding acquisition, L.Y. All authors have read and agreed to the published version of the manuscript.

Funding: This work was supported by the National Natural Science Foundation of China (32172731).

Institutional Review Board Statement: To ensure the welfare of all animals enrolled in the current trial, all possible measures were adopted in compliance with the standards of the Animal Ethics Board of Huazhong Agricultural University; Approval no. (HZAUBU-2017-001), Wuhan, China.

Informed Consent Statement: Not applicable.

Data Availability Statement: Data are contained within the article.

Acknowledgments: The authors wish to thank Hafiz Muhammad Tahir, research associate, and Abdul Salam, artificial insemination technician, for their help during the course of experimentation.

Conflicts of Interest: The authors declare no conflicts of interest.

References

1. Balhara, A.K.; Gupta, M.; Singh, S.; Mohanty, A.K.; Singh, I. Early pregnancy diagnosis in bovines: Current status and future directions. *Sci. World J.* **2013**, *2013*, 958540. [CrossRef] [PubMed]
2. Yadav, R.; Tripathi, H.; Kumar, P.; Ramesh, N. Assessing the economic losses to buffalo owners due to late diagnosis of pregnancy in their milch buffaloes. *AJAEES* **2019**, *33*, 1–6. [CrossRef]
3. Barile, V.L.; Menchetti, L.; Casano, A.B.; Brecchia, G.; Melo de Sousa, N.; Zelli, R.; Canali, C.; Beckers, J.F.; Barbato, O. Approaches to identify pregnancy failure in buffalo cows. *Animals* **2021**, *11*, 487. [CrossRef] [PubMed]
4. Karen, A.M.; Darwish, S.; Ramoun, A.; Tawfeek, K.; Van Hanh, N.; de Sousa, N.M.; Sulon, J.; Szenci, O.; Beckers, J.-F. Accuracy of transrectal palpation for early pregnancy diagnosis in Egyptian buffaloes. *Trop. Anim. Health Prod.* **2011**, *43*, 5–7. [CrossRef]
5. Perera, B.; Pathiraja, N.; Abeywardena, S.; Motha, M.; Abeygunawardena, H. Early pregnancy diagnosis in buffaloes from plasma progesterone concentration. *Vet. Rec.* **1980**, *106*, 104–106. [CrossRef] [PubMed]
6. Singh, A.; Puthiyandy, R. Estimation of progesterone in buffalo milk and its application to pregnancy diagnosis. *Reproduction* **1980**, *59*, 89–93. [CrossRef] [PubMed]
7. Karen, A.; Darwish, S.; Ramoun, A.; Tawfeek, K.; Van Hanh, N.; De Sousa, N.; Sulon, J.; Szenci, O.; Beckers, J.-F. Accuracy of ultrasonography and pregnancy-associated glycoprotein test for pregnancy diagnosis in buffaloes. *Theriogenology* **2007**, *68*, 1150–1155. [CrossRef]
8. Barbato, O.; Barile, V.L. Pregnancy-associated glycoproteins in buffalo: Origins, functions and clinical application for pregnancy follow-up. *Rev. Cient. Fac. Vet.* **2023**, *33*, 75. [CrossRef]
9. Mavedati, O.; Rastegarnia, A.; Habibian, R.; Bari, Y.N.; Bandarian, E. Early pregnancy diagnosis in water buffalo by early pregnancy factor measurement using rosette inhibition test. *Glob. Vet.* **2013**, *10*, 391–393. [CrossRef]
10. Sarangi, A.; Ghosh, M.; Sangwan, S.; Kumar, R.; Balhara, S.; Phulia, S.; Sharma, R.; Sahu, S.; Kumar, S.; Mohanty, A. Exploration of urinary metabolite dynamicity for early detection of pregnancy in water buffaloes. *Sci. Rep.* **2022**, *12*, 16295. [CrossRef]
11. Guelfi, G.; Stefanetti, V.; De Luca, S.; Giontella, A.; Barile, V.L.; Barbato, O. Serum microRNAs in buffalo cows: Potential biomarkers of pregnancy. *Res. Vet. Sci.* **2017**, *115*, 294–300. [CrossRef] [PubMed]
12. Sarwalia, P.; Raza, M.; Soni, A.; Dubey, P.; Chandel, R.; Kumar, R.; Kumaresan, A.; Onteru, S.K.; Pal, A.; Singh, K. Establishment of Repertoire of Placentome-Associated MicroRNAs and Their Appearance in Blood Plasma Could Identify Early Establishment of Pregnancy in Buffalo (*Bubalus bubalis*). *Front. Cell Dev. Biol.* **2021**, *9*, 673765. [CrossRef]
13. Riaz, U.; Idris, M.; Ahmed, M.; Ali, F.; Yang, L. Infrared Thermography as a Potential Non-Invasive Tool for Estrus Detection in Cattle and Buffaloes. *Animals* **2023**, *13*, 1425. [CrossRef] [PubMed]
14. Olğaç, K.T.; Yazlık, M.O.; Kaya, U.; Özkan, H.; Tırpan, M.B. The thermographic monitoring in early pregnancy detection in Holstein cows and heifers. *Anim. Reprod. Sci.* **2023**, *256*, 107317. [CrossRef] [PubMed]
15. Rekant, S.I.; Lyons, M.A.; Pacheco, J.M.; Arzt, J.; Rodriguez, L.L. Veterinary applications of infrared thermography. *Am. J. Vet. Res.* **2016**, *77*, 98–107. [CrossRef] [PubMed]
16. Zhang, C.; Xiao, D.; Yang, Q.; Wen, Z.; Lv, L. Application of Infrared Thermography in Livestock Monitoring. *Trans. ASABE* **2020**, *63*, 389–399. [CrossRef]
17. Ghezzi, M.D.; Ceriani, M.C.; Domínguez-Oliva, A.; Lendez, P.A.; Olmos-Hernández, A.; Casas-Alvarado, A.; Hernández-Avalos, I. Use of Infrared Thermography and Heart Rate Variability to Evaluate Autonomic Activity in Domestic Animals. *Animals* **2024**, *14*, 1366. [CrossRef]
18. Whittaker, A.L.; Muns, R.; Wang, D.; Martínez-Burnes, J.; Hernández-Ávalos, I.; Casas-Alvarado, A.; Domínguez-Oliva, A.; Mota-Rojas, D. Assessment of pain and inflammation in domestic animals using infrared thermography: A narrative review. *Animals* **2023**, *13*, 2065. [CrossRef] [PubMed]

19. Casas-Alvarado, A.; Mota-Rojas, D.; Hernández-Ávalos, I.; Mora-Medina, P.; Olmos-Hernández, A.; Verduzco-Mendoza, A.; Reyes-Sotelo, B.; Martínez-Burnes, J. Advances in infrared thermography: Surgical aspects, vascular changes, and pain monitoring in veterinary medicine. *J. Therm. Biol.* **2020**, *92*, 102664. [CrossRef]
20. Marquez, H.P.; Ambrose, D.; Schaefer, A.; Cook, N.; Bench, C. Evaluation of infrared thermography combined with behavioral biometrics for estrus detection in naturally cycling dairy cows. *Animal* **2021**, *15*, 100205. [CrossRef]
21. Talukder, S.; Kerrisk, K.; Ingenhoff, L.; Thomson, P.; Garcia, S.; Celi, P. Infrared technology for estrus detection and as a predictor of time of ovulation in dairy cows in a pasture-based system. *Theriogenology* **2014**, *81*, 925–935. [CrossRef] [PubMed]
22. Marquez, H.P.; Ambrose, D.; Schaefer, A.; Cook, N.; Bench, C. Infrared thermography and behavioral biometrics associated with estrus indicators and ovulation in estrus-synchronized dairy cows housed in tiestalls. *J. Dairy Sci.* **2019**, *102*, 4427–4440. [CrossRef] [PubMed]
23. Hilsberg-Merz, S. Infrared thermography in zoo and wild animals. *Zoo Wild Anim. Med. Curr. Ther.* **2008**, *6*, 20–33.
24. Krueger, F.; Knauf-Witzens, T.; Getto, S. New approach in thermal pregnancy diagnosis: Teat's heating in babirusa (*Babyrousa babyrussa*). *Theriogenology* **2019**, *133*, 144–148. [CrossRef]
25. Domino, M.; Borowska, M.; Kozłowska, N.; Zdrojkowski, Ł.; Jasiński, T.; Smyth, G.; Maśko, M. Advances in thermal image analysis for the detection of pregnancy in horses using infrared thermography. *Sensors* **2021**, *22*, 191. [CrossRef]
26. Radigonda, V.L.; Pereira, G.R.; da Cruz Favaro, P.; Barca Júnior, F.A.; Borges, M.H.F.; Galdioli, V.H.G.; Júnior, C.K. Infrared thermography relationship between the temperature of the vulvar skin, ovarian activity, and pregnancy rates in Braford cows. *Trop. Anim. Health Prod.* **2017**, *49*, 1787–1791. [CrossRef]
27. Bowers, S.; Gandy, S.; Anderson, B.; Ryan, P.; Willard, S. Assessment of pregnancy in the late-gestation mare using digital infrared thermography. *Theriogenology* **2009**, *72*, 372–377. [CrossRef]
28. Kul, H.; Erdoğan, G. Evaluation of infrared thermography findings in pseudopregnant rabbit. *Anim. Health Prod. Hyg.* **2022**, *11*, 12–20. [CrossRef]
29. Cardoso Consentini, C.E.; Wiltbank, M.C.; Sartori, R. Factors that optimize reproductive efficiency in dairy herds with an emphasis on timed artificial insemination programs. *Animals* **2021**, *11*, 301. [CrossRef] [PubMed]
30. Shu, H.; Li, Y.; Fang, T.; Xing, M.; Sun, F.; Chen, X.; Bindelle, J.; Wang, W.; Guo, L. Evaluation of the best region for measuring eye temperature in dairy cows exposed to heat stress. *Front. Vet. Sci.* **2022**, *9*, 857777. [CrossRef]
31. Gil, Z.; Kural, J.; Szarek, J.; Wierzchoś, E. Increase in milk and body temperature of cows as a sign of embryo entry into the uterus. *Theriogenology* **2001**, *56*, 685–697. [CrossRef]
32. Scanavez, A.; Fragomeni, B.; Rocha, L.; Voelz, B.; Hulbert, L.; Mendonça, L. Association between 4-day vaginal temperature assessment during the dry period and performance in the subsequent lactation of dairy cows during the warm season. *J. Anim. Sci.* **2017**, *95*, 5208–5217. [CrossRef] [PubMed]
33. Keith, I.; Bisward, G.; Manohar, M.; Klein, J.; Bullard, V. Respiratory effects of pregnancy and progesterone in Jersey cows. *Respir. Physiol.* **1982**, *50*, 351–358. [CrossRef] [PubMed]
34. LoMauro, A.; Aliverti, A. Respiratory physiology of pregnancy: Physiology masterclass. *Breathe* **2015**, *11*, 297–301. [CrossRef] [PubMed]
35. Bayliss, D.A.; Millhorn, D.E.; Gallman, E.A.; Cidlowski, J.A. Progesterone stimulates respiration through a central nervous system steroid receptor-mediated mechanism in cat. *Proc. Natl. Acad. Sci. USA* **1987**, *84*, 7788–7792. [CrossRef]
36. Noya, A.; Casasús, I.; Ferrer, J.; Sanz, A. Long-term effects of maternal subnutrition in early pregnancy on cow-calf performance, immunological and physiological profiles during the next lactation. *Animals* **2019**, *9*, 936. [CrossRef]
37. Suthar, V.; Burfeind, O.; Bonk, S.; Dhami, A.; Heuwieser, W. Endogenous and exogenous progesterone influence body temperature in dairy cows. *J. Dairy Sci.* **2012**, *95*, 2381–2389. [CrossRef] [PubMed]
38. Jung, C.; Fandos, C.; Lorenzo, I.M.; Plata, C.; Fernandes, J.; Gené, G.G.; Vázquez, E.; Valverde, M.A. The progesterone receptor regulates the expression of TRPV4 channel. *Pflügers Arch.-Eur. J. Physiol.* **2009**, *459*, 105–113. [CrossRef]
39. Kendall, P.; Webster, J. Season and physiological status affects the circadian body temperature rhythm of dairy cows. *Livest. Sci.* **2009**, *125*, 155–160. [CrossRef]
40. Kim, D.H.; Ha, J.J.; Yi, J.K.; Kim, B.K.; Kwon, W.-S.; Ye, B.-H.; Kim, S.H.; Lee, Y. Differences in ruminal temperature between pregnant and non-pregnant Korean cattle. *J. Anim. Reprod. Biotechnol.* **2021**, *36*, 45–50. [CrossRef]
41. Abouelela, Y.S.; Yasin, N.A.; Khattab, M.A.; El-Shahat, K.; Abdelnaby, E.A. Ovarian, uterine and luteal hemodynamic variations between pregnant and non-pregnant pluriparous Egyptian buffalos with special reference to their anatomical and histological features. *Theriogenology* **2021**, *173*, 173–182. [CrossRef]
42. Elmetwally, M.A.; Elshopakey, G.E.; Eldomany, W.; Eldesouky, A.; Samy, A.; Lenis, Y.Y.; Chen, D.b. Uterine, vaginal and placental blood flows increase with dynamic changes in serum metabolic parameters and oxidative stress across gestation in buffaloes. *Reprod. Domest. Anim.* **2021**, *56*, 142–152. [CrossRef] [PubMed]
43. Nääs, I.A.; Garcia, R.G.; Caldara, F.R. Infrared thermal image for assessing animal health and welfare. *J. Anim. Behav. Biometeorol.* **2020**, *2*, 66–72. [CrossRef]
44. Hassan, M.; Arshad, U.; Bilal, M.; Sattar, A.; Avais, M.; Bollwein, H.; Ahmad, N. Luteal blood flow measured by Doppler ultrasonography during the first three weeks after artificial insemination in pregnant and non-pregnant Bos indicus dairy cows. *J. Reprod. Dev.* **2019**, *65*, 29–36. [CrossRef]

45. Samir, H.; Kandiel, M.M. Accuracy of subjective evaluation of luteal blood flow by color Doppler ultrasonography for early diagnosis of pregnancy in Egyptian buffalo. *Anim. Reprod. Sci.* **2019**, *208*, 106129. [CrossRef] [PubMed]
46. Liles, H.L.; Schneider, L.G.; Pohler, K.G.; Oliveira Filho, R.V.; Neal Schrick, F.; Payton, R.R.; Rhinehart, J.D.; Thompson, K.W.; McLean, K.; Edwards, J.L. Positive relationship of rectal temperature at fixed timed artificial insemination on pregnancy outcomes in beef cattle. *J. Anim. Sci.* **2022**, *100*, skac100. [CrossRef] [PubMed]

animals

MDPI

Article

Effect of Density and Lineage on Dorsal Surface Temperature, Performance, and Carcass Condemnation of Broiler Grillers

Iara Cristina Marins [1,2], Bruno Giacomelli [2], Bruna Correia [3], Débora Cristina Olsson [4], Fabiana Moreira [1,3], Juahil Martins de Oliveira Júnior [3], Ivan Bianchi [1,3], Elizabeth Schwegler [1,3], Candice Bergmann Tanure [5], Monike Quirino [1], Tiago do Prado Paim [6], Natalia Nogueira Fonseca [6], Betina Raquel Cunha dos Santos [1] and Vanessa Peripolli [1,3,*]

[1] Programa de Pós-Graduação em Produção e Sanidade Animal, Instituto Federal Catarinense, Campus Araquari, Araquari 89245-000, SC, Brazil; iaramarins.zoo@gmail.com (I.C.M.); fabiana.moreira@ifc.edu.br (F.M.); ivan.bianchi@ifc.edu.br (I.B.); elizabeth.schwegler@ifc.edu.br (E.S.); quirinomonike@gmail.com (M.Q.); cunhabrs@yahoo.com.br (B.R.C.d.S.)
[2] Seara Alimentos Ltda, Itajaí 88305-030, SC, Brazil; giacomelli@vetanco.com.br
[3] Núcleo de Pesquisa, Ensino e Extensão em Produção Animal, Instituto Federal Catarinense, Campus Araquari, Araquari 89245-000, SC, Brazil; b-ru-correia@hotmail.com (B.C.); juahil.oliveira@ifc.edu.br (J.M.d.O.J.)
[4] Instituto Federal Catarinense, Campus Concórdia, Concórdia 89703-720, SC, Brazil; debora.olsson@ifc.edu.br
[5] Lohmann Breeders Canada, Brantford, ON N3S 7W4, Canada; c.bergmann@gmail.com
[6] Instituto Federal Goiano, Campus Rio Verde, Rio Verde 75901-970, GO, Brazil; pradopaim@hotmail.com (T.d.P.P.); natalia.nogueira@ifgoiano.edu.br (N.N.F.)
* Correspondence: vanessa.peripolli@hotmail.com or vanessa.peripolli@ifc.edu.br

Citation: Marins, I.C.; Giacomelli, B.; Correia, B.; Olsson, D.C.; Moreira, F.; de Oliveira Júnior, J.M.; Bianchi, I.; Schwegler, E.; Tanure, C.B.; Quirino, M.; et al. Effect of Density and Lineage on Dorsal Surface Temperature, Performance, and Carcass Condemnation of Broiler Grillers. *Animals* **2024**, *14*, 2195. https://doi.org/10.3390/ani14152195

Academic Editor: Daniel Mota-Rojas

Received: 10 July 2024
Revised: 24 July 2024
Accepted: 26 July 2024
Published: 28 July 2024

Simple Summary: Thermographic monitoring of the back surface of broiler grillers and the environment are accurate indicators of thermal comfort, allowing production to be enhanced and ensuring the quality of the final product. In this scenario, infrared thermography was used to evaluate chickens' thermophysiological state related to bird density, lineage, age, and time of day, isolated or integrated, on dorsal surface temperature, performance and carcass condemnations of broiler grillers reared in a Dark House system. The main results revealed that the dorsal surface temperature in the broiler grillers was affected by the combined effects of age and time of day, lineage and density, and lineage and age. The increase in the density of broiler grillers did not negatively influence mortality, average daily weight gain, and carcass condemnation. However, partial condemnations and arthritis condemnations were influenced by the lineage.

Abstract: The aims of this study were (i) to evaluate the effect of density, lineage, age, and time of day on dorsal surface temperature and (ii) to evaluate the effect of density and lineage on performance and carcass condemnations in broiler grillers. The evaluations were carried out in barns with the Dark House system, with two densities, 17 and 19 chickens/m^2 and two lineages, Cobb and Ross. The dorsal surface temperature of the chickens was measured by infrared thermography at 7, 14, 21, 23, 25 and 27 days of age, four times a day. The average daily weight gain, feed conversion, mortality, partial carcass condemnations, as well as those due to arthritis and dermatosis were also evaluated. The highest dorsal surface temperatures were observed in Cobbs housed at a density of 17 chickens/m^2, and in Ross housed at a density of 19 chickens/m^2. Cobbs housed at a 17 chickens/m^2 density showed the lowest feed conversion compared to Ross at the same density. Ross showed higher dorsal surface temperatures when compared to Cobbs at 14, 21, and 27 days. Cobbs showed higher percentages of partial carcass condemnation and arthritis compared to Ross. The higher density of broiler grillers in the Dark House system does not influence the dorsal surface temperature, performance, dermatosis, arthritis, and partial carcass condemnations.

Keywords: ambience; chicken; dark house; dermatosis; thermography

1. Introduction

The poultry sector aims to maximize results and reduce production costs at all stages of the production process, thus optimizing investments for better financial returns. For this reason, the density of chickens housed per square meter (m^2) becomes a relevant factor for production viability. However, the increase in this density contributes to the appearance of skin lesions, which influences carcass quality, impacting the financial return [1]. The quality of the carcass of chickens has become a challenge within the production process, especially for the production of low-body-weight chickens called broiler grillers commonly exported whole to international markets [2].

Skin diseases are frequent causes of chicken carcass condemnation [3,4]. To ensure a carcass without or at acceptable levels of skin lesion, the chickens must be comfortable, with low-stress level, and exposed to an appropriate environment. The Dark House system uses high technology, controlling all ambient factors, including luminosity control [5]. In this system, chickens are raised with controlled lighting from the beginning to the end of the flock, which allows greater control over the stress of the chickens, allowing an increase in bird density without increasing the incidence of dermatosis [6].

Therefore, thermographic monitoring of the body surface and the environment are accurate indicators of thermal comfort, enhancing production and ensuring the final product's quality [7]. In birds, variations in body surface temperature depend on peripheral blood flow, which indicates their active efforts to regulate core body temperature [8]. In this scenario, infrared thermography has been used to evaluate the thermophysiological state of chickens [7], with the head, wings, back, and feet being the areas commonly evaluated [9]. Measuring the temperature on the back (dorsal) surface of chickens is preferred because this area is less influenced by external factors such as environmental temperature fluctuations or direct sunlight exposure, providing a more accurate reflection of the bird's core body temperature [8,10]. The lineage used is also an important point to be considered due to its intrinsic characteristics of animal performance, carcass, productivity, bone development, and profitability [11], as well as the feather growth characteristics, structure, and moulting patterns that are important in commercial environments [12]. Therefore, we hypothesized that density, lineage, age and time of day, isolated or integrated, affect dorsal surface temperature, performance and carcass condemnations of broiler grillers. Thus, the aims of this study were (i) to evaluate the effect of density, lineage, age, and time of day on dorsal surface temperature and (ii) to evaluate the effect of density and lineage on the performance and carcass condemnations of broiler grillers.

2. Materials and Methods

2.1. Experimental Design

The experiment was carried out in the city of Seara (latitude 27°08′58″ south, longitude 52°18′38″ west and altitude of 550 m), located in Santa Catarina state, Brazil, where the climate is mesothermic, humid type (Cfa) according to the Köppen and Geiger [13] classification with an annual mean temperature of 19 °C.

A total of 423,360 female broilers, from both Cobb and Ross lineages were housed in four Dark House barns system. This system uses high technology, controlling all ambient factors, including luminosity control [5]. The chickens were housed by lineage and at two densities, 17 chickens/m^2 and 19 chickens/m^2, not exceeding 30 kg/m and observed from housing until slaughter at 28 days of age, being the average age necessary for the chickens to reach an average weight of 1.5 kg targeted by this griller-type broilers of rearing. Three consecutive flocks were evaluated (replications in time) with a mean interval of 15 days between flocks.

Dark House barns measuring 14× 140 m were used, totalling 1960 m^2 with the same conditions of ambiance, temperature curve, ventilation time, airspeed, static pressure, use of heaters, lighting, equipment adjustment, environmental conditions, and the number of litters. Chickens were raised with controlled lighting from the beginning to the end of the flock. A conventional lighting schedule of 23 h of light and 1 h of darkness was used in the first week, followed by 20 h of light and 4 h of darkness in the following weeks. The luminosity used was 40 Lux in the first week, 25 in the second week, 15 in the third, and 10 in the fourth week. In addition, a standard schedule was used to control temperature, humidity, ventilation, luminosity, cooling, and heating for the three flocks. This schedule considers the installation's surface in cubic meters, equipment capacity and the needs of the birds for each phase according to the lineages' guidelines.

Feed and water were provided ad libitum. The chickens received an isoenergetic and isoproteic diet formulated to meet the nutritional requirements of each period.

2.2. Performance Evaluation

The number of dead chickens was counted daily from housing to slaughter and used to calculate the mortality rate of the flock, considering the total number of chickens housed.

The feed conversion index was calculated from the volume of feed consumed by the chickens during the evaluation period, which was from housing to slaughter, divided by the weight gain of the flock in the same period.

2.3. Evaluation of the Dorsal Surface Temperature of Chickens

Thermal images were captured on the dorsal (back cape) surface of the chickens, in the upper portion of the wing due to the lower influence of external factors in this body area [8,10] at 7, 14, 21, 23, 25, and 27 days of age as shown in Figure 1. The temperature was randomly performed on three chickens in three areas inside each barn (at the entrance after passing the central door, in the central region where the feeder lines are located at the end of the barn, next to the exhaust fans), at four times: 4:00, 08:00, 14:00 and 20:00 h for each age evaluated.

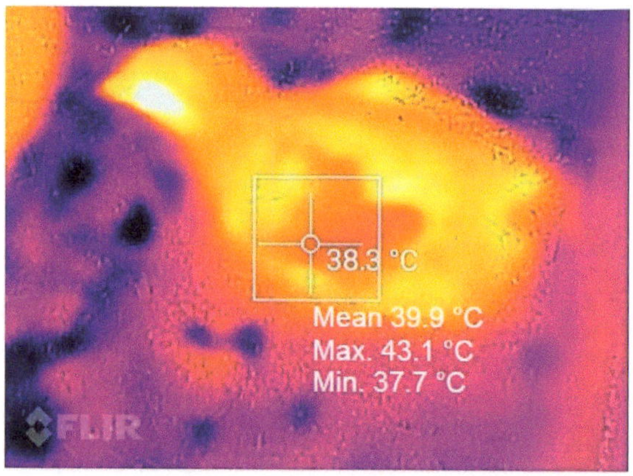

Figure 1. Thermographic image of the dorsal surface temperature of the chickens.

The sample size used to evaluate the dorsal surface temperature of the chickens was not determined according to the number of chickens in barns but by the expected variation considering the 95% confidence level and expected margin of error of 5% concerning the experimental data according to Cangar et al. [14].

The dorsal temperature of the chickens was measured using a Tg165X infrared thermographic camera (FLIR Systems Inc., Wilsonville, OR, USA). This camera had high precision (±0.1 °C) with an infrared resolution of 320 × 240 pixels and a thermal sensitivity of <0.05 °C at an ambient temperature of 30 °C and an accuracy ±2 °C. The emissivity coefficient (ε) used was 0.94 as suggested by Nääs et al. [10]. An image of each chicken was captured at a distance of 50 cm, using a ruler as a reference, to fill the image in its viewing angles.

View of the back cape region of the broiler where the surface temperature values were extracted.

The square represents the delimitation of the thermal window.

Since the waves of magnetic radiation propagate in a straight line, refraction can occur due to angle distortion generating a change in wavelength. To minimize the error which is almost insignificant for rough surfaces such as the body of the animals, the image was recorded using a 90-degree angle of the chickens' dorsal surface [9].

Simultaneously, the ambient temperature inside the barns was also measured, based on the readings taken by the controller installed internally.

2.4. Evaluation of Carcass Condemnations

The chickens of both lineages were slaughtered following the rules of Brazilian legislation and animal welfare, with a mean live weight of 1500 ± 100 g. In the slaughter line, the chickens were evaluated by the Federal Inspection Service (FIS), where carcass evaluations were carried out, and information on partial condemnations, and condemnations due to arthritis and dermatosis were collected. The evaluation of the carcasses was carried out individually in all chickens during slaughter through a macroscopic visual examination and, when necessary, palpation and cuts were performed.

Condemned chickens due to dermatosis presented skin lesions, and thus could not be packed as whole carcasses, but could be used for other purposes, such as cut-up parts, or processed meat for use in sausages. In condemnations due to arthritis, chickens with hemorrhagic lesions in the joints, which result in the condemnation of the thigh, were considered. The assessment criterion for both evaluations was based on the visual aspect, according to the FIS.

2.5. Statistical Analysis

The experimental design was completely randomized with four treatments in a 2 × 2 factorial arrangement, with two chicken densities (17 and 19 chickens/m^2) and two lineages (Cobb and Ross) with three replications in time (flocks).

Each Dark House barn was considered an experimental unit for evaluations of performance, dermatosis, and arthritis. Each chicken was considered an experimental unit for evaluation of dorsal surface temperature.

The performance, dermatosis, and arthritis data were tested for normality of distribution and homogeneity of residues using Shapiro Wilk and Levene tests, respectively, subjected to Person correlation analysis (PROC CORR) and analyzed using the MIXED procedure in a model that included density and lineage as a fixed effect and flock as a random effect, using the barn as the subject. Interactions between density and lineage were tested. Using Akaike's information criterion, the variance components (VC) structure was considered the best model for the residual covariance structure. The following statistical model was used:

$$Yijk = \mu + Di + Lj + DLij + Bk + \varepsilon ijk \quad (1)$$

where Yijk represents dependent variables; μ is the overall mean of the observations; Di is the fixed effect of the density (i = 17 and 19 chickens/m^2); Lj is the fixed effect of the lineage (j = Cobb and Ross); DLij is the interaction between density and lineage; Bk is the random effect of the flock (k = 1 to 3); and εijk is the random residual experimental error.

The dorsal surface temperature data were tested for normality of distribution and residue homogeneity using Shapiro Wilk and Levene tests, respectively, and analyzed using the MIXED procedure in a model that included density, lineage, age, and time as a fixed effect and flock as a random effect, using the barn as the subject. Interactions between density, lineage, age, and time were tested. Using Akaike's information criterion, the variance components (VC) structure was considered the best model for the residual covariance structure. The following statistical model was used:

$$Yijklm = \mu + Di + Lj + Ak + Tl + DLij + DAik + DTil + LAjk + TLjk + ATkl + Bm + \varepsilon ijklm \quad (2)$$

where Yijklm represents dependent variables; μ is the overall mean of the observations; Di is the fixed effect of the density (i = 17 and 19 chickens/m^2); Lj is the fixed effect of the lineage (j = Cobb and Ross); Ak is the fixed effect of the age (7, 14, 21, 23, 25, and 27 days); Tl = is the effect of the time (4:00, 08:00, 14:00 and 20:00 h); DLij is the interaction between density and lineage; DAik is the interaction between density and age; DTil is the interaction between density and time; LAjk is the interaction between lineage and age; LTjl is the interaction between lineage and time; ATkl is the interaction between age and time; Bm is the random effect of the flock (k = 1 to 3); and εijkm is the random residual experimental error.

Analyzes were performed using the Statistical Analysis System program (SAS Inst. Inc., Cary, NC, USA, version 9.3), and statistically significant differences were considered when $p < 0.05$.

3. Results

There was an interaction between age and time on broiler grillers' dorsal surface temperature ($p = 0.0052$) (Table 1). The dorsal surface temperatures decreased with the increasing age of the chickens. However, at 21 days, at 20:00, the temperature did not show the expected decline pattern according to the age of the chickens (Table 1).

Table 1. Effect of the interaction between age and time on dorsal surface temperature (°C) of broiler grillers.

Age, Days	Time, hh:mm				Mean	SEM
	04:00	08:00	14:00	20:00		
7	40.20 ^{Aa}	40.04 ^{Aa}	40.67 ^{Aa}	40.27 ^{Aa}	40.29	0.4523
14	38.26 ^{Ba}	38.02 ^{Ba}	38.82 ^{Ba}	38.39 ^{Ba}	38.37	0.4523
21	37.77 ^{Bb}	38.19 ^{Bb}	39.37 ^{Ba}	39.40 ^{Aa}	38.68	0.4520
23	36.48 ^{Cb}	35.97 ^{Cb}	38.37 ^{Ba}	38.22 ^{Ba}	37.26	0.4516
25	35.55 ^{Ca}	35.86 ^{Ca}	35.48 ^{Ca}	36.01 ^{Ca}	35.72	0.4520
27	35.57 ^{Ca}	35.98 ^{Ca}	35.47 ^{Ca}	36.10 ^{Ca}	35.78	0.4558
Mean	37.30	37.34	38.03	38.06		
SEM	0.4400	0.4398	0.4401	0.4399		
p-value						
Age × time	0.0052					

p-value: probability; SEM: standard error of the mean; ^{A,B,C} Different uppercase letters in columns and ^{a,b} different lowercase letters in rows differ by Tukey test at 5% ($p < 0.05$).

There was an interaction between density and lineage on dorsal surface temperature ($p < 0.0001$) and feed conversion ($p = 0.0235$) of broiler grillers (Table 2). The density of Cobb chickens at 17 chickens/m^2 and Ross chickens at 19 chickens/m^2 had the highest dorsal surface temperatures (Table 2). Cobb chickens at a housed density of 17 chickens/m^2 had a better feed conversion than Ross chickens at the same density (Table 2).

Table 2. Effect of the interaction between density and lineage on dorsal surface temperature (°C) and feed conversion of broiler grillers.

Lineage	Density		Mean	SEM
	17 Chickens/m^2	19 Chickens/m^2		
Back temperature				
Cobb	39.56 ^{Aa}	35.73 ^{Bb}	37.65	0.7254
Ross	37.84 ^{Ba}	38.94 ^{Aa}	37.94	0.7270
Mean	38.70	36.88		
SEM	0.7376	0.7297		
p-value				
Density × lineage	<0.0001			

Table 2. *Cont.*

Lineage	Density		Mean	SEM
	17 Chickens/m^2	19 Chickens/m^2		
Feed conversion				
Cobb	1.35 ^{Ba}	1.45 ^{Aa}	1.41	0.0195
Ross	1.47 ^{Aa}	1.38 ^{Aa}	1.42	0.0195
Mean	1.41	1.42		
SEM	0.0201	0.0201		
p-value				
Density × lineage	0.0235			

p-value: probability; SEM: standard error of the mean; ^{A,B} Different uppercase letters in columns and ^{a,b} different lowercase letters in rows differ by Tukey test at 5% ($p < 0.05$).

An interaction between lineage and age significantly influenced the back surface temperature of the broiler grillers ($p = 0.0003$) (Table 3). Ross exhibited higher dorsal surface temperatures than Cobb on days 14, 21, and 27 (Table 3).

Table 3. Effect of the interaction between lineage and age on dorsal surface temperature (°C) of broiler grillers.

Age, Days	Lineage		Mean	SEM
	Cobb	Ross		
7	40.20 ^{Aa}	40.62 ^{Aa}	40.41	0.7397
14	38.03 ^{BCb}	38.94 ^{Ba}	38.48	0.7397
21	38.39 ^{Bb}	39.20 ^{Ba}	38.79	0.7395
23	37.70 ^{Ca}	36.94 ^{Ca}	37.32	0.7392
25	36.13 ^{Da}	35.57 ^{Ca}	35.85	0.7395
27	35.43 ^{Db}	36.36 ^{Da}	37.90	0.7421
Mean	37.65	37.94		
SEM	0.7254	0.7270		
p-value				
Lineage × age	0.0003			

p-value: probability; SEM: standard error of the mean; ^{A,B,C,D} Different uppercase letters in columns and ^{a,b} different lowercase letters in rows differ by Tukey test at 5% ($p < 0.05$).

There was an effect of the lineage on arthritis condemnation ($p = 0.0084$) and partial carcass condemnation ($p = 0.0175$), where Cobbs had higher percentages for both condemnations compared to Ross (Table 4).

Table 4. Effect of density and lineage on performance variables, dermatosis, arthritis, and partial condemnation of broiler grillers.

Trait	Density		Lineage		Mean	SEM	*p*-Value		
	17 Chickens/m^2	19 Chickens/m^2	Cobb	Ross			Density	Lineage	Density × Lineage
Mortality, n	4.97	3.92	5.29	3.60	4.44	0.5498	0.3874	0.1920	0.9491
ADG, g/day	52.15	51.84	51.60	52.40	52.00	0.5846	0.8279	0.5801	0.9328
Dermatosis, %	2.08	2.64	3.10	1.62	1.98	0.2815	0.5698	0.1716	0.5115
Arthritis, %	1.75	1.15	2.20 ^{a}	0.69 ^{b}	1.43	0.2894	0.2154	0.0084	0.2948
Partial condemnations, %	1.16	4.32	5.59 ^{a}	2.89 ^{b}	4.24	0.5030	0.8319	0.0175	0.9365

p-value: probability; SEM: standard error of the mean; ADG: average daily weight gain; ^{a,b} Different lowercase letters in rows differ by Tukey test at 5% ($p < 0.05$).

Considering both densities, there was a strong positive correlation between the percentage of dermatosis and partial carcass condemnation. In addition, the percentage of arthritis showed a moderate positive correlation with the mortality rate in chickens (Table 5). Considering the density of 19 chickens/m^2, there was a strong positive correlation between feed conversion and mortality, between arthritis and feed conversion, and between dermatosis and partial condemnations (Table 5).

Table 5. Correlation between performance variables, dermatosis, arthritis, and partial condemnation of broiler grillers.

	ADG	FC	Dermatosis	Arthritis	PC
		General			
Mortality	0.03521	−0.31205	−0.01326	**0.66733**	0.28105
ADG	-	−0.22276	0.09458	−0.06804	0.05852
FC	−0.22276	-	0.01022	−0.22737	−0.10804
Dermatosis	0.09458	0.01022	-	−0.04805	0.84695
Arthritis	0.66733	−0.06804	−0.22727	-	0.46889
PC	0.05852	−0.10804	**0.84695**	0.46889	-
		17 chickens/m^2			
Mortality	0.41428	−0.80861	0.00198	0.55180	0.29707
ADG	-	−0.00056	0.04911	0.05201	0.00718
FC	−0.00056	-	−0.21792	−0.77565	−0.62271
Dermatosis	0.04911	−0.21792	-	0.12324	0.76456
Arthritis	0.05201	−0.77565	0.12324	-	0.72912
PC	0.00718	−0.62271	0.76456	0.72912	-
		19 chickens/m^2			
Mortality	−0.68482	**0.95953**	0.09813	0.86053	0.37988
ADG	-	−0.52962	0.15082	−0.30487	0.10784
FC	−0.52962	-	0.19722	**0.84843**	0.51176
Dermatosis	0.15082	0.19722	-	−0.07926	**0.91839**
Arthritis	−0.30487	**0.84843**	−0.07926	-	0.28611
PC	0.10784	0.51176	**0.91839**	0.28611	-

ADG: average daily weight gain; FC: feed conversion; and PC: partial condemnation. Bold numbers are statistically significant at 5% ($p < 0.05$).

4. Discussion

The Dark House system has greater potential to produce broilers, presenting superior performance when compared to conventional systems [5]. Barns that use the Dark House system typically have a higher density of broiler chickens, resulting in more contact between chickens [15]. At high densities, chickens need greater heat dissipation, adopting strategies such as opening their wings and ruffling feathers, generating greater movement to maintain higher contact with the air, and ensuring faster cooling, predisposing to a significant increase in carcass condemnations. The Dark House system has reduced condemnation rates for dermatosis caused by scratches [15]. This is due to the high control of the ambient conditions that ensure thermal comfort to the chickens and the greater ease of handling in the barns, especially at the time of catching by controlling the luminosity [15].

Carcass condemnations are significant in the poultry sector, and factors that may contribute to the increase in these condemnations should be minimized. In the present study, the increase in the density of broilers did not cause an increase in mortality and partial carcass condemnations, as well as arthritis and dermatosis, corroborating with previous studies [16–20]. However, there was an interaction between density and lineage on dorsal surface temperature and feed conversion, which can be explained by the performance curve of each lineage, where broilers from commercial lineages presented different genetic potentials due to selection pressure applied to the characteristics of economic interest, such as weight gain, feed conversion, and resistance, resulting in differences in the lineage's profile, making each lineage present a specific performance and behave differently when subjected to different densities [21].

Among commercial lineages, Cobb presents better productive performance, among commercial lineages, due to better weight gain, as it has excellent feed conversion, high muscle deposition capacity, greater rusticity, and resistance to handling, regardless of density [22]. Marcato et al. [23] also observed that Cobb had a higher growth rate and nutrient deposition, reaching a higher slaughter weight when compared to Ross.

In the present study, the lineages did not affect the average daily weight gain, consistent with Lara et al.'s [22] findings of Ross showing similar productivity to Cobb. However, the former has lower initial growth, with a compensatory gain after 21 days of age, obtaining high final weight gain, generally superior to that of Cobb. For the same authors, Ross showed better viability, with a lower percentage of mortality and better use at slaughter due to lower carcass condemnations, which was also observed in the present study.

The dorsal surface temperature was higher in chickens reared at lower density, which demonstrates a greater ability to control the environment even at high densities in the Dark House system. This system allows the use a higher density of chickens per square meter in the barns due to the use of technology to control the environment, allowing lower percentages of dermatosis, better feed conversion, and greater average daily weight gain [15]. Thus, the ambient control and the technology of the Dark House system used in the present study allowed maintenance of the mortality rate, the average daily weight gain, and the carcass quality of the chickens even when raised in higher density.

In the present study, Cobb presented higher percentages of partial condemnations and arthritis. According to Lara et al. [22], Cobb has a higher capacity for muscle deposition and better feed conversion, however, this rapidly developing musculature can make the skin more exposed and susceptible to scratches, resulting in higher percentages of dermatitis, pododermatitis, and inflammation.

In commercial lineages, one of the determining factors for carcass quality is the rate of feathering. The earlier the onset of feathering, the lower the percentage of dermatosis-related carcass condemnation, since the feathers act as a protective barrier and help avoid the appearance of skin lesions in commercial broilers raised at high densities. The feathering process involves complex physiological mechanisms, influenced by nutritional, hormonal, genetic, and environmental factors, as well as by the interaction between them [24]. Some broiler lineages with slower feathering rates tend to have a higher incidence of skin lesions [12]. Since Cobb has a faster feathering rate when compared to Ross [12], the former was expected to have lower percentages of partial carcass condemnations. However, the early feathering trait alone was not enough to minimize partial carcass condemnations, and the fast-growing trait should be considered when choosing the ideal lineage to produce broilers in a Dark House system.

Very high densities compromise chicken welfare, and performance, increasing surface temperatures [25]. Therefore, it is crucial to know the factors that influence the surface temperature, and the thermoregulation mechanisms of chickens to minimize the impacts of increased density [26]. Infrared thermography is a noninvasive remote sensing method used to measure changes in heat transfer and blood flow in humans and animals by detecting small variations in body temperature, providing valuable information [10,27]. Abudabos et al. [25] reported that birds housed in medium and high densities exhibited higher body, head, neck, wing, and shin temperatures compared to those in lower densities. The authors concluded that the increased density negatively impacted the welfare and performance of the birds. Therefore, knowing the surface temperature allows adjustment of the environmental temperature according to the needs of the chickens. In the present study, the dorsal surface temperature of the chickens was influenced by the integrative effect between age and time, between lineage and density, and between lineage and age.

It is important to consider that the surface temperature varies in response to environmental and physiological conditions [10]. The feathered parts of the body have a lower response to environmental temperature, as opposed to the parts with no feathers, thus highlighting the impact of the environmental temperature [8]. Thus, assessments performed on the dorsal area of the chicken represent a slower response to the environment [8,10].

The decrease in body temperature with increasing age should be gradual and points to the comfort condition of the chickens [14]. In the present study, there was a gradual decrease in the dorsal surface temperatures as the chickens got older, stabilizing at 23 days of age at all times of the day. However, at 21 days, at 20:00, the temperature did not show the same temperature decline pattern considered normal according to the age of the chickens.

At the beginning of life, chickens have higher difficulty in thermal regulation, being greatly influenced by the environment temperature. The high temperatures in the first weeks of life, as observed in the present study, may be related to environmental factors, which directly reflect on the surface temperature of young chickens. In addition, environmental humidity can influence the surface temperature of one-week-old birds, redistributing heat, and generating changes between body and surface temperature [28].

The gradual decrease in temperature up to 21 days coincided with the period of development of the thermoregulatory system and with the feathering of the chickens [29]. Thus, it is recommended to use heaters for at least 21 days, even in summer, because only after this period will the chickens have a fully developed thermoregulatory system [30].

Considering the evaluation of both densities, a moderately positive correlation was observed between arthritis and mortality. As for the 19 chickens/m^2 density, a strong positive correlation was observed between feed conversion and arthritis. This is because arthritis causes pain and discomfort to chickens, decreasing their performance, which worsens feed conversion [20–31]. In addition, arthritis can be fatal for chickens, which explains the positive correlation with mortality.

Thermography can be utilized to assess temperature changes, providing valuable insights into housing conditions and management practices in poultry farming. Furthermore, studying thermoregulatory patterns in chickens through thermography could contribute to genetic selection programs to breed more heat-tolerant and resilient poultry lineages.

5. Conclusions

The dorsal surface temperature of broiler grillers in a Dark House system is influenced by the integrative effect of age and time of day, lineage and density, and lineage and age.

The increase in the density of broiler grillers does not negatively influence mortality, average daily weight gain, and carcass condemnation. However, partial condemnations and arthritis condemnations are influenced by lineage.

Author Contributions: I.C.M.—conceptualization, methodology, investigation, data curation, writing—original draft, project administration. B.G.—conceptualization, methodology, investigation. B.C.—conceptualization, methodology, writing—original draft. D.C.O.—conceptualization, methodology, writing—review and editing. F.M.—conceptualization, methodology, writing—review and editing. J.M.d.O.J.—conceptualization, methodology, formal analysis. I.B.—conceptualization, methodology, writing—review and Editing. E.S.—conceptualization, methodology. C.B.T.—conceptualization, methodology, writing—review and editing. M.Q.—methodology, writing—review and editing. T.d.P.P.—methodology, writing—review and editing. N.N.F.—methodology, writing—review and editing. B.R.C.d.S.—methodology, writing—review and editing. V.P.—conceptualization, methodology, investigation, formal analysis, supervision, writing—review and editing, project administration. All authors have read and agreed to the published version of the manuscript.

Funding: Instituto Federal Goiano supported the article processing charge (APC).

Institutional Review Board Statement: It was approved by the Ethics Commission for Animal Use (CEUA), protocol no. 2019-24, of Instituto Federal Catarinense, Brazil.

Informed Consent Statement: Not applicable.

Data Availability Statement: All data generated and analyzed during the current study are available from the corresponding author upon reasonable request.

Acknowledgments: The authors are grateful to the National Council for Scientific and Technological Development (CNPq) for a scholarship, Instituto Federal Goiano for financial support, and SEARA Alimentos for providing the chicks, and facilities.

Conflicts of Interest: Authors Iara Cristina Marins and Bruno Giacomelli were employed by the company Seara Alimentos Ltda. Author Candice Bergmann Tanure was employed by the company Lohmann Breeders Canada. The remaining authors declare that the research was conducted in the absence of any commercial or financial relationships that could be construed as a potential conflict of interest.

References

1. Santana, A.P.; Murata, L.S.; Freitas, M.G.; Delphino, M.K.; Pimentel, C.M. Causes of condemnation of carcasses from poultry in slaughterhouses located in State of Goiás Brazil. *Cien. Rural.* **2008**, *38*, 2587–2592. [CrossRef]
2. Vieira, S.L. *Qualidade Visual das Carcaças de Frango de Corte*, 2nd ed.; Editora Rede: São Paulo, Brazil, 2012; 104p.
3. Ebling, P.D.; Basurco, V. Análise das perdas econômicas oriundas da condenação de carcaças nos principais estados brasileiros produtores de frangos de corte. *Cien. Agrovet. Aliment.* **2016**, *1*, 1–11.
4. Muchon, J.L.; Garcia, R.G.; Gandra, E.R.S.; Assunção, A.S.A.; Komiyama, C.M.; Caldara, F.R.; Nääs, I.A.; Santos, R.A. Origin of broiler carcass condemnations. *R. Bras. Zootec.* **2019**, *48*, e20180249. [CrossRef]
5. Carvalho, R.H.; Soares, A.L.; Grespan, M.; Spurio, R.S.; Coró, F.A.G.; Oba, A.; Shimokomaki, M. The effects of the dark house system on growth, performance and meat quality of broiler chicken: Dark house and broiler meat. *Anim. Sci. J.* **2015**, *86*, 189–193. [CrossRef] [PubMed]
6. Andreazzi, M.A.; Pinto, J.S.; Santos, J.M.G.; Cavalieri, F.L.B.; Matos, N.C.S.; Barbieri, I.O. Desempenho de frangos de corte criados em aviário convencional e Dark-House. *R. U. Vale Rio Verde* **2018**, *16*, 1–6. [CrossRef]
7. Caniatto, A.R.M.; Carão, A.C.P.; Tonetti, P.A. Da Guerra para a Granja: Tecnologia de luz Infravermelha no Controle da Temperatura. Revista Avisite. 2013. Available online: https://www.avisite.com.br/da-guerra-para-a-granja-tecnologia-de-luz-infravermelha-no-controle-da-temperatura/#gsc.tab=0 (accessed on 20 January 2024).
8. Damane, M.M.; Barazandeh, A.; Mokhtari, M.S.; Esmaeilipour, O.; Badakhshan, Y. Evaluation of body surface temperature in broiler chickens during the rearing period based on age, air temperature and feather condition. *Iran. J. Appl. Anim. Sci.* **2018**, *8*, 499–504.
9. Dahlke, F.; Gonzales, E.; Gadelha, A.C.; Maiorka, A.; Borges, S.A.; Rosa, P.S.; Furlan, R.L. Empenamento, níveis hormonais de triiodotironina e tiroxina e temperatura corporal de frangos de corte de diferentes genótipos criados em diferentes condições de temperatura. *Cien. Rural.* **2005**, *35*, 664–670. [CrossRef]
10. Nääs, I.A.; Romanini, C.E.B.; Neves, D.P.; Nascimento, G.R.; Vercellino, R.A. Broiler surface temperature distribution of 42 day old chickens. *Sci. Agric.* **2010**, *67*, 497–502. [CrossRef]
11. Da Costa, M.J.; Zaragoza-Santacruz, S.; Frost, T.J.; Halley, J.; Pesti, G.M. Straight-run vs. sex separate rearing for 2 broiler genetic lines Part 1: Live production parameters, carcass yield, and feeding behavior. *Poult. Sci.* **2017**, *96*, 2641–2661. [CrossRef]
12. Api, I.; Takahashi, S.E.; Mendes, A.S.; Paixão, S.J.; Refati, R.; Restelatto, R. Efeito da sexagem e linhagens sobre o desempenho e rendimento de carcaça de frangos de corte. *Cien. Anim. Bras.* **2017**, *18*, e-32691. [CrossRef]
13. Köppen, W.; Geiger, R. Klimate der Erde. Wall-Map 150 cm × 200 cm; Verlag Justus Perthes: Gotha, Germany, 1928.
14. Cangar, O.; Aerts, J.-M.; Buyse, J.; Berckmans, D. Quantification of the spatial distribution of surface temperatures of broilers. *Poult. Sci.* **2008**, *87*, 2493–2499. [CrossRef]
15. Gallo, B.B. Dark House: Manejo × Desempenho Frente ao Sistema Tradicional. 2009. Available online: https://pt.engormix.com/avicultura/artigos/dark-house-manejo-t36773.htm (accessed on 1 July 2023).
16. Feddes, J.; Emmanuel, E.; Zuidhoft, M. Broiler performance, body weight variance, feed and water intake, and carcass quality at different stocking densities. *Poult. Sci.* **2002**, *81*, 774–779. [CrossRef] [PubMed]
17. Moreira, J.; Mendes, A.A.; Roça, R.O.; Garcia, E.A.; Naas, I.A.; Garcia, R.G.; Paz, I.C.L.A. Effect of stocking density on performance, carcass yield and meat quality in broilers of different commercial strains. *R. Bras. Zootech.* **2009**, *33*, 1506–1519. [CrossRef]
18. Sirri, F.; Minelli, G.; Folegatti, E.; Lolli, S.; Meluzzi, A. Foot dermatitis and productive traits in broiler chickens kept with different stocking densities, litter types and light regimen. *Ital. J. Anim. Sci.* **2007**, *6*, 734–736. [CrossRef]
19. Meluzzi, A.; Fabbri, C.; Folegatti, E.; Sirri, F. Survey of chicken rearing conditions in Italy: Effects of litter quality and stocking density on productivity, foot dermatitis and carcase injuries. *Br. Poult. Sci.* **2008**, *49*, 257–264. [CrossRef] [PubMed]
20. Lima, R.C.; Freitas, E.R.; Gomes, H.M.; Cruz, C.E.B.; Fernandes, D.R. Performance of broiler chickens reared at two stocking densities and coir litter with different height. *R. Cien. Agronon.* **2018**, *49*, 519–528. [CrossRef]
21. Tavárez, A.A.; Santos, F.S. Impact of genetics and breeding on broiler production performance: A look into the past, present, and future of the industry. *Anim. Front.* **2016**, *6*, 37–41. [CrossRef]
22. Lara, L.J.C.; Baião, N.C.; Rocha, J.S.R.; Lana, A.M.Q.; Cançado, S.V.; Fontes, D.O.; Leite, R.S. Influência da forma física da ração e da linhagem sobre o desempenho e rendimento de cortes de frangos de corte. *Arq. Bras. Med. Vet. Zootech.* **2008**, *60*, 970–978. [CrossRef]
23. Marcato, S.M.; Sakomura, N.K.; Fernandes, J.B.K.; Siqueira, J.C.; Dourado, L.R.B.; Freitas, E.R. Crescimento e deposição de nutrientes nos órgãos de frangos de corte de duas linhagens comerciais. *R. Bras. Zootech.* **2010**, *39*, 1082–1091. [CrossRef]
24. Garcia, R.G.; Almeida Paz, I.C.L.; Caldara, F.R.; Nääs, I.A.; Bueno, L.G.F.; Freitas, L.W.; Graciano, J.D.; Sim, S. Litter materials and the incidence of carcass lesions in broilers chickens. *Braz. J. Poult. Sci.* **2012**, *14*, 27–32. [CrossRef]

25. Abudabos, A.M.; Samara, E.M.; Hussein, E.O.S.; Al-Ghadi, M.Q.; Al-Atiyat, R.M. Impacts of stocking density on the performance and welfare of broiler chicken. *Ital. J. Anim. Sci.* **2013**, *12*, 66–71. [CrossRef]
26. Nascimento, S.T.; Silva, I.J.O.; Maia, A.S.C.; Castro, A.C.; Vieira, F.M.C. Mean surface temperature prediction models for broiler chickens—A study of sensible heat flow. *Int. J. Biometeorol.* **2013**, *58*, 1–7. [CrossRef] [PubMed]
27. McManus, C.; Tanure, C.B.; Peripolli, V.; Seixas, L.; Fischer, V.; Gabbi, A.M.; Menegassi, S.R.O.; Stumpf, M.T.; Kolling, G.J.; Dias, E.; et al. Infrared thermography in animal production: An overview. *Comput. Electron. Agric.* **2016**, *123*, 10–16. [CrossRef]
28. Lin, H.; Zhang, H.F.; Du, R.; Gu, H.X.; Zhang, Z.Y.; Buyse, J.; Decuypere, E. Thermoregulation responses of broiler chickens to humidity at different ambient temperatures. I. One week of age. *Poult. Sci.* **2005**, *84*, 1166–1172. [CrossRef] [PubMed]
29. Abreu, V.M.N.; Abreu, P.G. Os desafios da ambiência sobre os sistemas de aves no Brasil. *R. Bras. Zootech.* **2011**, *40*, 1–14.
30. Macari, M.; Mendes, A.A. *Manejo de Matrizes de Corte*; Facta: Campinas, Brazil, 2005; 421p.
31. Sellers, H.S. Current limitations in control of viral arthritis and tenosynovitis caused by avian reoviruses in commercial poultry. *Vet. Microbiol.* **2017**, *206*, 152–156. [CrossRef]

animals

MDPI

Article

The Relationship between the Infrared Eye Temperature of Beef Cattle and Associated Biological Responses at High Environmental Temperatures

Musadiq Idris [1,*], Megan Sullivan [2], John B. Gaughan [2] and Clive J. C. Phillips [3,4]

1 Department of Physiology, Faculty of Veterinary and Animal Sciences, The Islamia University of Bahawalpur, Bahawalpur 63100, Pakistan
2 School of Agriculture and Food Sciences, Gatton Campus, The University of Queensland, Gatton, QLD 4343, Australia; megan.sullivan12@outlook.com (M.S.); j.gaughan@uq.edu.au (J.B.G.)
3 Curtin University Sustainability Policy (CUSP) Institute, Faculty of Humanities, Curtin University, Perth, WA 6845, Australia; clive.phillips@curtin.edu.au
4 Institute of Veterinary Medicine and Animal Sciences, Estonian University of Life Sciences, Kreutzwalki 1, 51014 Tartu, Estonia
* Correspondence: musadiq.idris@iub.edu.pk

Citation: Idris, M.; Sullivan, M.; Gaughan, J.B.; Phillips, C.J.C. The Relationship between the Infrared Eye Temperature of Beef Cattle and Associated Biological Responses at High Environmental Temperatures. *Animals* **2024**, *14*, 2898. https://doi.org/10.3390/ani14192898

Academic Editor: Daniel Mota-Rojas

Received: 28 July 2024
Revised: 23 September 2024
Accepted: 3 October 2024
Published: 8 October 2024

Simple Summary: Cattle are increasingly exposed to high temperatures with accelerating climate change, presenting the risks of heat stress, poor welfare, and even death. In Australia, the sudden onset of extreme conditions means that deaths are not uncommon. Understanding the impact of high environmental temperatures on cattle behaviour and physiology is critical. We aimed to investigate changes in infrared eye temperature and their association with behavioural responses, such as changes in body postures, alterations in feeding behaviour, and body self-care. Twenty-four Angus steers were individually housed to assess heat load responses when they were fed either a standard finisher diet based on cereal grains or an alternative diet with 8% of the grains replaced by an isoenergetic amount of lucerne hay. No differences were observed between different dietary cohorts. The infrared eye temperature (IRT-Eye), of the steers exposed to high environmental temperatures was increased, and this had a strong association with behaviour and physiology. Cattle with increased IRT-Eye temperature showed stress-related behavioural responses, including a downward-facing head, ears directed backwards, and other indicators of heat stress such as increased panting, standing, and increased rumen temperature. The strong association of IRT-Eye with behaviour and physiology suggests that it can be used to detect heat stress in cattle.

Abstract: Cattle in regions with high ambient temperatures are at risk of heat stress. Early detection is important to allow action to be taken to minimise the risks to cattle exposed to thermal stress. This study aimed to investigate the impact of heat stress on IRT-Eye temperature and its association with the behavioural and physiological responses of heat-stressed Angus steers (n = 24) on finisher and or substituted diets. Overall, 2 cohorts of 12 Angus steers were individually housed in a climate-controlled facility to examine responses to heat stress when fed on a standard finisher diet, based on a high percentage of cereal grains, and on a substituted diet in which 8% of the grains were replaced by an isoenergetic amount of lucerne hay. Exposing feedlot cattle to hot environmental conditions increased IRT-Eye temperature, which had a strong association with behaviour and physiology. There was no evidence of differences between the different dietary cohorts. The cattle with increased IRT-Eye temperature showed stress-related responses, including a downward-facing head, ears directed backwards, and other indicators of heat stress such as increased panting, standing, and increased rumen temperature. The strong association of IRT-Eye temperature with stress-related behaviours, as well as with rumen temperature and panting behaviour, highlights the potential for IRT-Eye to be utilised as a non-invasive tool to assess cattle responses in hot conditions.

Keywords: beef cattle; eye temperature; heat stress; heat load condition; infrared thermography

1. Introduction

Hot environmental conditions are a significant welfare and production problem for livestock, especially in beef farming industries. This is especially true in tropical and subtropical regions of the world, where animals may be constantly exposed to hot conditions for prolonged periods [1–3]. An increased core body temperature is a major indicator of heat load [3–5], with cattle attempting to maintain homeostasis through physiological and behavioural responses [1]. Their core body temperatures range from 38.0 to 38.5 °C under thermoneutral conditions [2]. Body temperature is usually recorded from the rectum, rumen, vagina, or tympanic membrane of the ear [6–8], all differing in their accuracy and degree of invasiveness [6]. Some methods may require the handling of animals in confined places, which may result in stress to the animals and increased body temperatures [9,10]. In feedlots in Australia and the US, the death of cattle is associated with heat waves that occur during the summer period (e.g., [11]).

Infrared thermography (IRT) is widely used as a non-invasive, passive, and remote method to assess animal responses during stressful conditions. Practically, IRT has proved to be a useful non-invasive tool in veterinary and animal science for measuring fluctuations in surface temperature that may correlate with core body temperatures [1,10]. Infrared thermography has also been used as a tool for the detection of febrile conditions related to inflammation [12] and for early disease detection [13,14]. At high environmental temperatures, the animal sees increased heat loss from the skin surface through vasodilatation and increased peripheral circulation [1,15]. Infrared thermography can detect these changes [16]. The measurement of eye temperature using infrared thermography is a useful indicator of pain and stress in animals [1] during dehorning [17], bovine viral diarrhoea [13], castration [18], foot and mouth disease [19], and periods of anxiety [20]. This may be due to the position of the eye, being closer to the emotional processing centres of the brain hemispheres. Secondly, eyes have rich capillary beds, supplying blood, that are highly innervated by the sympathetic nervous system. These directly influence body surface temperature [18]. Infrared thermography can be a promising and reliable technique for assessing cattle responses during heat load conditions, with the added advantages of minimum stress to cattle and quick measurement [21,22].

Finisher or grain-based diets are usually fed to beef cattle as they are high in energy, promoting rapid weight to achieve high production levels in the beef industry. However, in heat-stressed cattle, grain-based diets may reduce rumen pH, increasing stress. In extreme cases, cattle become lame, ataxic, and uncoordinated [23,24]. Fiber-based diets are thought to improve rumen health, assisting in efficient digestion. The presence of adequate amount of fibre in diets is necessary to maintain optimal rumen pH by stimulating salivation during bolus chewing [25,26] in cattle, reducing heat load.

To define the relationship between infrared eye temperature and the animal's responses to heat load conditions, the impact of heat load on the biological responses of feedlot cattle was studied to investigate the hypothesis that hyperthermic conditions will influence eye temperature, and that this will be associated with the behaviour and physiology of feedlot cattle. Following adverse reactions to heat stress in the first cohort, the diet was changed for the second cohort, allowing us to explore whether a substituted diet (grain substituted with forage) influenced cattle responses to heat stress, as evidenced by changes in infrared eye temperature.

2. Materials and Methods

2.1. Animal Housing

In the initial feedlot phase, animals in each cohort were kept in a feedlot pen of 162 m^2 (27 × 6 m), with an east–west alignment. Each contained a concrete feed bunk, water troughs, and a shaded area of 1.3 m^2/animal at midday. Animals were then randomly assigned to individual pens (each 10 × 3.4 m) in an outdoor facility containing a concrete apron, a feed bunk, water troughs, and a shaded area. In the next phase, animals were housed in two climate-controlled rooms (each 10 × 10 m), separated by a concrete wall. Each climate-controlled room was further divided into two chambers by a clear Perspex wall, with 3 pens per chamber. The ambient dry-bulb temperature, humidity cycles, and lighting schedule in each room were programmed automatically. The ventilation rate inside each CCR was maintained with a centrally controlled air-conditioning system (Table 1). Oxygen, ammonia and carbon dioxide were monitored at 0600, 1200 and 1800 h (and at midnight on HOT days) each day, using a hand-held gas meter (Gas Micro Alert, BW technologies, Honeywell, Charlotte, NC, USA). Lighting was set at 10% of maximum from 1900 to 0500 h, and to maximum at all other times. The individual pens (2.5 × 2.5 m) each had a rubber mat floor over a steel grill, which facilitated the drainage and cleaning of the pens. Pens were cleaned daily at 0630–0730 h, prior to feeding at 0900 h, by hosing all excrement from the mats and pen flooring. All animals were provided with individual access to a water trough and a feed trough (each 500 × 500 × 500 mm) inside the climate-controlled facility. The climate-controlled facility was equipped with two cameras (CW214H; K-guard Info. Co., Ltd., New Taipei City, Taiwan) above each pen that were connected to a digital video recorder (LG, XQ-L900H; T.O.M Technology Co., Ltd., Yeouido-dong, Seoul, Republic of Korea) for the surveillance of the animals.

Table 1. The ambient temperature, relative humidity, and temperature humidity index for cattle (n = 12) when they were in the climate control facility.

Day	Treatment Phase	Min T_A (°C)	Max T_A (°C)	Mean T_A (°C)	Min RH (%)	Max RH (%)	Mean RH (%)	Min THI	Max THI	Mean THI
0	ACC	19.7	21.0	20.1	60.9	90.6	66.2	65.5	69.1	66.3
1	ACC	19.7	21.0	20.1	60.9	90.6	66.2	65.5	69.1	66.3
2	ACC	19.5	21.6	20.0	60.0	89.3	67.1	65.3	68.9	66.2
3	TN	19.5	20.8	19.9	61.3	90.1	67.9	65.3	68.8	66.1
4	TN	19.6	24.0	20.2	60.0	89.1	68.2	65.5	72.0	66.4
5	TP1	19.9	40.5	33.2	42.9	88.4	66.1	66.3	92.6	84.9
		Finisher dietary cohort—transition to 30 °C from 00.00 h on day 5								
		Substituted dietary cohort—transition to 30 °C from 21.00 h on day 5								
6	HOT	28.4	40.2	33.0	43.3	82.8	65.8	80.5	91.8	84.8
7	HOT	28.4	38.1	32.1	42.3	84.2	63.7	78.3	89.1	83.0
8	HOT	24.9	34.3	28.7	44.3	82.0	65.9	73.6	85.0	78.4
9	HOT	22.6	34.4	28.0	45.81	79.5	66.2	69.9	85.7	77.5
10	HOT	20.6	30.3	24.3	54.4	80.5	66.7	67.2	80.0	72.3
11	HOT	20.4	30.4	24.2	45.3	80.6	65.8	67.1	79.2	72.0
12	HOT	19.7	21.3	20.3	50.0	90.5	64.6	65.8	68.8	66.4
13	TP2	19.7	20.7	20.1	56.4	91.3	65.5	65.6	68.6	66.2
14	Recovery	19.7	21.4	20.1	58.1	89.0	66.7	65.6	69.6	66.2
15	Recovery	19.6	20.5	19.9	58.4	90.3	66.4	65.6	68.2	66.0
16	Recovery	19.4	25.0	20.5	57.8	93.5	66.4	65.2	73.2	66.8
17	Recovery	19.3	23.7	21.1	58.1	69.0	61.9	64.9	71.1	67.5

T_A: ambient temperature (°C); RH: relative humidity; THI: temperature humidity index; ACC: acclimatisation to climate-controlled facility; TN: thermoneutral conditions before high-temperature treatment; TP1 and 2: transition phases to and from hot conditions; HOT: high-temperature treatment; recovery: thermoneutral conditions after high-temperature treatment as a recovery period.

2.2. Animal Management

Steers were immunised ten days before entry to the experimental feedlot facility for tick fever (trivalent chilled, 3 germ; *Anaplasma centrale*, *Babesia bovis* and *Babesia bigemina*; Department of Agriculture, Fisheries and Forestry (Biosecurity Queensland, Longreach, Australia)), clostridial diseases (5-in-1 enterotoxaemia, [pulpy kidney disease], malignant oedema, Black's disease, tetanus, and blackleg; Pfizer Animal Health, Sydney, Australia), and for external and internal parasites (150 g/L hydrocarbon liquid; Cydectin, 5 g/L moxidectin solvent, Fort Dodge Australia P/L, Baulkham Hills, NSW, Australia). We administered bovine respiratory disease vaccination (Bovillis MH, inactivated *Mannheimia haemolytica*; Coopers Animal Health, Macquarie Park, Australia) and bovine ephemeral fever vaccination (BEFV) on d 20 of the feedlot phase and administered a second dose 24 days later before animals left the feedlot pen. Animals were inserted with a hormonal growth promoter (HGP) implant (Synovex® containing trenbolene acetate, Zoetis, Parsippany, NJ, USA) on entry into the feedlot. Steers in cohort 1 were offered a starter diet of concentrates for the first 8 d in the climate-controlled facility and then an intermediate diet for 6 d. They then transitioned to a finisher diet over the next 3 d, which was used until the end of trial (Table 2). Due to the adverse heat stress response of some cattle in the first cohort, the second cohort was fed a substituted (high-fibre) diet from the second day of the HOT period and were transitioned back to the finisher diet over four days during the recovery thermoneutral period. Individually housed cattle were fed their diet at 2.5% of body weight on a DM basis, with refusals removed and weighed each morning prior to the provision of 50% of the ration at 0900 h and the remainder at 1300 h. The feed dry matter content was determined by oven drying. The animals were provided with ad libitum water during the study, and water consumption in the CCR was recorded at the time of each observation using endurance multi-jet turbine water meters (RMC Zenner, Eagle Farm, Australia).

Table 2. Diet ingredients and nutrient composition for starter, intermediate, finisher and substituted diets fed to cattle.

Item	Starter	Intermediate	Finisher	Substituted
Ingredients, % of diet				
Grain-based mix *	62.1	74.5	86.8	78.7
Whole cottonseed	9.0	16.5	9.0	9.0
Lucerne hay	28.9	9.0	4.2	12.3
Nutrient composition				
DM, g/kg fresh weight	880	893	887	886
ADF, g/kg DM	263	257	119	177
NDF, g/kg DM	404	375	229	253
NE_g, MJ/kg DM	29	29	30	30
ME, MJ/kg DM	116	119	132	131
DE, MJ/kg DM	143	147	163	162
Crude fibre, g/kg DM	218	197	87	124
Nitrogen-free extract, g/kg DM	503	548	678	685
Fat, g/kg DM	46	43	46	43
Feed digestibility, g/kg DM	768	791	861	868
Digestible DM, g/kg DM	676	707	763	769
Digestible protein g/kg DM	133	125	130	131
Starch, g/kg DM	229	218	432	432

* grain-based mix: steam-rolled barley 89.2%, feedlot pellet 9.2%, and vegetable oil 1.6%; feedlot pellet: wheat brain, aleurone, germ, and pollard (Millrun Wheat, Riverina, Brisbane, Australia) 55.9%, ammonium sulphate 2.6%, dry-rolled wheat 12.5%, calcium carbonate (limestone) 15.6%, Rumensin 100 0.3%, magnesium oxide 0.7%, zinc supplement (Availa zinc 100) 0.34%, vegetable oil 3.1%, salt, (NaCl) 2.8%, urea 5.7%, vitamin A 500 0.009%, vitamin E 0.057%, and mineral supplement (XFE-Select L) 0.385%.

2.3. *Animals and Treatments*

Twenty-four (24) healthy yearling Black Angus steers were procured for the study from a commercial property in Armidale, New South Wales, and transported to The University of Queensland (UQ), Gatton Campus' Queensland Animal Science Precinct (QASP) (27.56° S, 152.34° E; 99 m elevation above sea) during the Southern Hemisphere's summer. The steers had an initial live weight (non-fasted) of 493 ± 6.8 kg. During summer, the climate at the location is hot, humid, and sub-tropical. The maximum and minimum ambient temperatures recorded during the time of the study at the location were 30.98 ± 1.48 to 16.4 ± 2.42, respectively. A randomised design was used for the study, with the steers randomly divided into 2 cohorts of 12 animals to run separately through the climate control room, which only had capacity for this number of animals. The first cohort was fed a standard feedlot ration based on cereal grains (the finisher diet), and the second received the same diet but with 8% of the grains by mass replaced by an isoenergetic amount of lucerne hay (the substituted diet). The detailed composition of the diet is presented in Table 2.

Each cohort of 12 steers was kept for 50 d in a feedlot pen and then for 10 d in individual outdoor pens to allow acclimatisation to the handling and feeding management practices that they would receive in the climate rooms. The animals were moved to the experimental facility where cattle were kept in two climate-controlled rooms (CCR) for 18 days. The climatic conditions were changed over five different periods:

1. Thermoneutral conditions before hot conditions (TN; d 3–4)
2. Transition period to HOT (TP1; d 5)
3. Hot conditions (HOT; d 6–12)
4. Transition period to recovery (TP2; d 13)
5. Thermoneutral conditions after hot treatment (recovery; d 14–17)

In the initial thermoneutral (TN) and recovery, the ambient dry-bulb temperature (T_A) and relative humidity (RH) in CCR were maintained at 20 °C and 65%, respectively. The thermoneutral period transitioned to the hot climatic phase (HOT) over 1 d in TP1. In the HOT, T_A and RH increased each day from 0700 h to reach a maximum at 1100 h, which was maintained until 1600 h and then decreased hourly from 1600 h to the daily minimum T_A and RH at 2000 h. The ambient temperature declined over the HOT period before cattle entered the recovery period (Table 1). Animals were continuously monitored for real-time body temperature, panting score, respiration rate, and animal behaviour to assess the impact of a high heat load. An action plan was devised with a specific parameter set to identify animals that were not adequately coping with the conditions, as outlined below.

Ethical approval (AEC Approval No. SAFS/460/16) was obtained from The University of Queensland Animal Ethics Committee for this study, which was opportunistically added to another study of metabolic responses to heat stress. An intervention protocol was in place to mitigate severe heat stress based on animal responses to the heat stress. Monitoring was conducted every hour throughout the hot period, including the entire night. If animals exhibited signs of severe distress (constantly excessive panting, open-mouth breathing, elevated core body temperature, etc.), mitigating measures were immediately applied, including providing access to ventilation and water spray. Despite these mitigating measures, there was one sudden death on the first day of the hot period in the first cohort of animals. A second animal from this cohort was closely monitored for the whole of the first night of the hot period. Then it was removed from the study and euthanised to prevent further suffering. Following the change in ambient temperature for the second cohort, two animals were removed from the experiment due to their adverse responses to hot conditions, but their conditions were not serious and they did not have to be euthanised.

2.4. Behavioural and Other Key Observations

Respiratory behaviour was recorded 7 times a day, every 2 h from 0600 to 1800 h during TN and recovery and every hour over 24 h during the HOT period. A team of five trained observers recorded respiration rate and panting score (PS), with only one observer recording measurements at any one time (Table 3). The respiration rate (RR) was determined by timing the time taken for the animal to take ten breaths, as observed from flank movements. This was then converted into breaths per minute. The panting scores (PSs) of animals were visually scored using a modified scale from 0 to 4.5, where 0 indicates no heat stress and 4.5 represents a severely heat-stressed animal [27–31].

From the video recordings, standing, lying, the stepping of each limb, eating, ruminating, grooming, and scratching were continuously recorded using event-logging software (BORIS v. 6.0.4 [32]) for 24 h (5 min/h) on d 3 (TN), d 6–12 (HOT), and d 14–16 (recovery) (Table 3). Head, ear, and tail positions were also recorded at 5-min intervals every h for 24 h. Not all behaviours could be observed at all times: the accurate recording of ear position was not possible when the animal was ruminating, eating, drinking, scratching, or grooming; the recording of head position was not possible during eating, drinking, grooming, or scratching; and tail position could not be recorded during defecation, urination, eating, drinking, grooming, or scratching.

Table 3. Ethogram for recorded behaviours.

Item	Description
Standing	Animal standing with limb positioned upright
Lying	Animal resting on the floor with their limb laterally or sternally recumbent
Eating	Animal consuming feed at the trough
Rumination	Animal chewing the cud or regurgitating bolus
Respiration rate	Number of breaths in a 1 min period
Panting score	Animal visually scored for the extent of panting based on a 0 to 4.5 score scale
Grooming	Animal licking any part of the body or striking one part with another part of the body
Scratching	Animal rubbing or striking any part of the body against the fixture of the pen
Ear positions	
Ear raised	Both ears being held upright above the neck with the ear pinnae facing forwards or to the side
Ear forward	Both ears' pinnae directed forwards in front of the focal animal and held horizontally
Ear backward	Both ears being held backwards on the focal animal's head
Ear downward	Both ears being loosely hung downwards, falling perpendicular to the head
Ear specific	Both ear pinnae (right and left) being oriented in opposite directions, or perpendicular to head rump axis, thus failing to satisfy raised, forward, backward, or downward ear positions
Head positions	
Head raised	The head held upright above withers or body top-line
Head neutral	The head held horizontally at the level of withers or body top-line
Head downwards	The head held downwards below withers or body top-line
Stepping	
Front right (FR) limb	Animal raising a front right limb and replacing it forthwith on the surface of the pen
Front left (FL) limb	Animal raising a front left limb and replacing it forthwith on the surface of the pen
Back right (BR) limb	Animal raising a back right limb and replacing it forthwith on the surface of the pen
Back left (BL) limb	Animal raising a back left limb and replacing it forthwith on the surface of the pen
Tail positions	
Tail raised	Tail held in a fixed position, held at than 45 degrees from the vertical
Tail vertical	Tail hanging downward from the vertical line of body and relaxing with no movements
Tail swishing	Swift movement of the tail in any direction around the hind quarters from its base in a side-to-side flicking manner
Tail tucked	Tail held tightly pressed in a fixed position against the rump, with the tip of the tail tucked in behind the hind limb

Adapted from Idris et al. [30,31,33].

2.5. Body Surface Temperature

Body surface temperature was measured by using an infrared thermometer (TN410LCE, Zytemp, HsinChu, Taiwan) on the head (midway between the eyes), shoulder (cranio-lateral surface of the thoracic cavity at the site of attachment of the forelimb), lower leg (halfway between the fetlock and hock joint), and rump (the dorso-caudal aspect of the body corresponding to the gluteal muscles) of each animal at a 1 m distance.

2.6. Rumen Temperature

Rumen temperature (T_{RUM})-monitoring boluses (Smartstock, Pawnee, OK, USA) were inserted orally using a customised applicator on d 0 upon entry to the feedlot pens. Before administration, the rumen boluses were calibrated for 24 h by placing in a water bath at 39 °C. The boluses contained an active radio frequency identification (RFID) transmitter operating at 915–928 MHz, detectable up to 90 m. The real-time radio transmissions were communicated by a Yagi antenna to a base station and were then transformed into a database via proprietary software (TechTrol Inc., Pawnee, OK, USA). Individual internal body temperature T_{RUM} recordings were transmitted to an on-site computer at 10 min intervals throughout the study. At each transmission, the previous 11 data points (i.e., the previous 110 min) were also collected in order to minimise data loss due to missed transmission. Individual measurements for the internal body temperature T_{RUM} recordings of each animal were recorded at the time of the infrared thermal imaging of the cattle's eyes.

2.7. Infrared Thermography (IRT)

Images of both eyes (IRT-Eye) of each animal were taken using an infrared thermal imager (FLK-Ti25 9 Hz, Fluke Corporation, Everett, WA, USA) at 0600, 1200, and 1800 h daily on d 3 (TN), d 6–12 (HOT), and d 14–16 (recovery). The images of the animals were taken whilst located in the pen inside climate control rooms at approximately 1 m distance from the eye. In order to maximise the usefulness of thermal images, the IRT-Eye images were captured at a perpendicular angle to the infrared thermal camera. Only usable infrared thermal images (which comprised 94% of all images) were selected for analysis. We used an emissivity value of 0.98; non-usable images (6%) were excluded (Figure 1). Minimum, maximum, and mean eye temperatures from both right and left eyes were determined using a zone analysis marker (Figure 2); however, only the maximum eye temperatures within the zone analysis were used for further analysis, as this represented the best image with which to determine stress in cattle [16,20]. Mean eye temperature was calculated from the infrared thermal eye temperatures of both eyes ((right eye + left eye)/2). The infrared thermography images obtained were analysed using proprietary software (Fluke Smart View Software, version 4.3, Fluke Corporation, Everett, WA, USA).

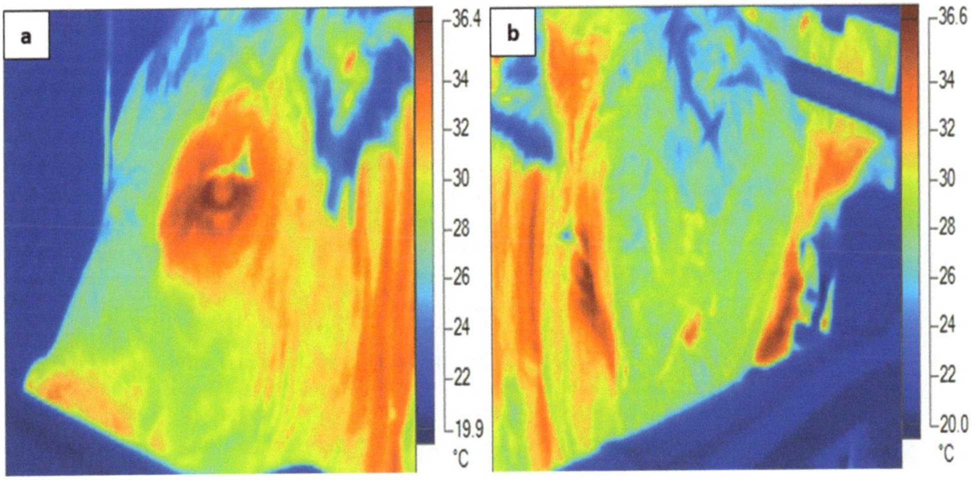

Figure 1. Examples of (**a**) useable and (**b**) unusable infrared thermal images of the eye.

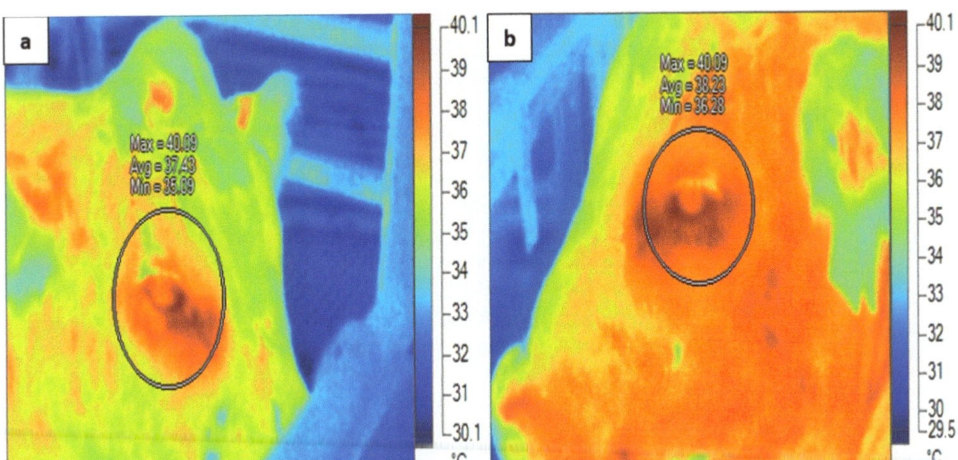

Figure 2. Zone analysis of the eye: (**a**) right eye; (**b**) left eye.

2.8. Climatic Data

Ambient temperature (T_A) and relative humidity *(RH)* values were maintained inside the climate-controlled room using a cyclic air-conditioning system. These were recorded at 10 min intervals using two temperature and humidity data loggers per room (HOBO UX100-011, Onset, MA, USA), which were installed on the pen at 1.5 m from the ground. A temperature humidity index *(THI)* was calculated using the following equation, adapted from Thom [34]:

$$THI = (0.8 \times T_A) + \{[(RH/100) \times (T_A - 14.4)\} + 46.4 \quad (1)$$

where RH = relative humidity in % and T_A = ambient temperature in °C

2.9. Statistical Analyses

In total, 4 animals (2 from each cohort) were removed from the experiment due to their adverse response to the heat load at the start of the HOT period. Overall, there were 2 in the first cohort, with one of that died and a second that required euthanasia, and two in the second cohort that were just removed from the experiment. Therefore, in each dietary cohort, the data obtained from the 10 steers were analysed using the statistical software Minitab 17 (Minitab® 17.3.1 Inc. Chicago, IL, USA) and Minitab 18 (Minitab® 18.1 Inc. Chicago, IL, USA) for Windows. The transition days provided time for metabolic changes at the start of the hot conditions after the acclimatisation process and similarly at the start of the thermoneutral conditions after the hot conditions (TP1 and 2, respectively), and this was reflected in different levels of metabolic heat production due to the altered metabolic rate. Therefore, the data obtained during these transition phases to and from hot conditions were not included in our statistical analyses.

Two separate models were used to describe the data. The first made a comparison between the TN and the HOT period, accounting for diet (*D*; finisher and substituted diet), treatment periods [*P*; HOT and TN], and interactions. We used diet × treatment period as a fixed factor, and the animal identification (*ID*) was a random factor.

Secondly, linear relationships between climatic conditions and the infrared eye temperature (°C) of the feedlot cattle were determined using a mixed-effects model, with the animal identification (*ID*) acting as a random factor, and we used the following fixed factors: dietary cohort (*D*; finisher and substituted diets), treatment period [P; HOT and recovery], and the time (*t*) while the day (*d*) of the experiment was nested within the treatment. We also examined the interactions of diet × treatment period and diet × day. Data from the TN were used as covariate *(Cov)* values. The equation for the analysis was as follows:

$$Y = \mu + D + P + t + d\,(P) + ID + (D \times P) + (D \times d\,(P)) + Cov + e \quad (2)$$

where Y is the expected value for biological response variables; μ is the expected mean value for response variables when input variables = zero and the factors are as described above; and e is the random error associated with experimental observations. Pairwise comparisons between treatment means were performed using Fisher's test ($p \leq 0.05$). When necessary, the Kolmogorov–Smirnov test was used to test for normal distributions of residuals. The second comparison, between the hot and recovery periods, was made to assess how quickly cattle return to thermoneutrality after experiencing heat stress and to evaluate the persistence of the effects of heat stress on eye temperature. This comparison provides insight into the animals' recovery capacity and whether their physiological responses, including eye temperature, fully normalise after the heat stress period. The thermoneutral period was included as a covariate in the statistical models to account for baseline physiological conditions, allowing for the more accurate detection of changes specifically occurring due to heat stress and recovery. By comparing the hot and recovery periods, we could better understand the animals' resilience and how quickly they recover, which is crucial for managing heat stress in livestock.

Spearman's rank correlation coefficients (r_S) with a two-tailed level of significance ($p < 0.05$) were determined. The data were examined using a multivariate general linear model with forward-selection stepwise regression (α to enter variable = 0.25) in order to investigate the association of infrared eye temperature with the differential means (HOT-TN) of the respiration rate, the panting score (PS), the rumen temperature, the surface temperature (ST) of the body (head, shoulder, leg, and rump), the stepping, the posture (standing, lying, head, ear and tail positions), the self-care (grooming and scratching), and the activity (eating, drinking and rumination) of animals.

3. Results

3.1. Initial Thermoneutral Period vs. Hot Period

The infrared eye temperatures of the feedlot cattle ($n = 20$), exposed first to TN and then to the high-temperature environment, are compared in Table 4. The infrared thermographic temperature (°C) of right, left, and mean infrared eye temperatures of both eyes in the feedlot cattle was greater in the HOT period than in the TN. No difference in infrared eye temperature (°C) was observed in the different dietary cohorts during the HOT treatment period. No difference was observed in the infrared R/L eye ratio (°C) of feedlot cattle during different treatment periods and diets, or interaction between the treatment period and diets.

Table 4. Infrared eye temperature (IRT-Eye) of cattle on a finisher or substituted diet exposed to high temperatures (HOT) or an initial thermoneutral period (TN).

Parameters	Finisher Diet		Substituted Diet		SED	F-Value (1, 18; d.f. †)	*p*-Value		
	TN	HOT	TN	HOT			Period (P)	Diet (D)	D × P
Mean eye temperature (°C)	36.75 ^{b}	38.73 ^{a}	36.91 ^{b}	38.52 ^{a}	0.123	429	≤0.001	0.800	0.050
Right eye temperature (°C)	36.73 ^{b}	38.72 ^{a}	36.91 ^{b}	38.50 ^{a}	0.114	492	≤0.001	0.900	0.020
Left eye temperature (°C)	36.76	38.74	36.93	38.51	0.149	287	≤0.001	0.800	0.070
Right/left eye ratio (°C)	0.999	0.999	0.999	1.00	0.0016	0.11	0.800	0.800	0.970

IRT: infrared thermography; SED: standard error of the difference between two means; HOT: high-temperature treatment period; day 6–12; TN: initial thermoneutral period before high-temperature treatment; day 3; † treatment, error degrees of freedom; D: diet; P: period. Superscripts ^{a,b}; Means within each variable with different superscripts are different at $p \leq 0.05$ by Fisher's test.

3.2. Hot Period vs. Recovery Period

The infrared eye temperature in °C (right eye, left eye and mean infrared eye) of the cattle was greater in the HOT period than in the recovery period (Table 5). A significantly lower infrared eye temperature (right eye, left eye and mean infrared eye) was observed at 0600 h, increasing to a maximum value at 1200 h and then declining at 1800 h, reflecting changes in the ambient temperature (Figure 3). No significant difference was observed in the infrared eye temperature of the right eye and left eye or in the mean infrared eye temperature (°C) of beef cattle on different diets or subjected to variations in period and day. The R/L eye ratio was not affected by period, time, diet or the relevant interactions.

Table 5. Infrared eye temperature (IRT-Eye) of feedlot cattle on finisher or substituted diets exposed to high temperatures (HOT treatment) or to a recovery period.

Parameters	Finisher Diet		Substituted Diet		SED	F-Value (d.f. †)	*p*-Value				
	HOT	Recovery	HOT	Recovery			Period (P)	Diet (D)	Time (t)	D × P	D × d
Mean Eye Temperature (°C)	38.73	36.58	38.52	36.49	0.241	1758 (1, 540)	≤0.001	0.300	≤0.001	0.200	0.900
Right Eye Temperature (°C)	38.73	36.59	38.50	36.52	0.247	1545 (1, 511)	≤0.001	0.300	≤0.001	0.100	0.700
Left Eye Temperature (°C)	38.72	36.57	38.52	36.43	0.242	1486 (1, 484)	≤0.001	0.300	≤0.001	0.600	0.900
Right/Left Eye ratio (°C)	1.00	1.00	1.00	1.00	0.0023	2.09 (1, 459)	0.149	0.782	0.366	0.993	0.580

IRT: infrared thermography; SED: standard error of the difference between two means; HOT: high-temperature treatment period; day 6–12; recovery: thermoneutral period after high-temperature treatment; d 14–16); † treatment, error degrees of freedom; D: diet; d: day; P: period.

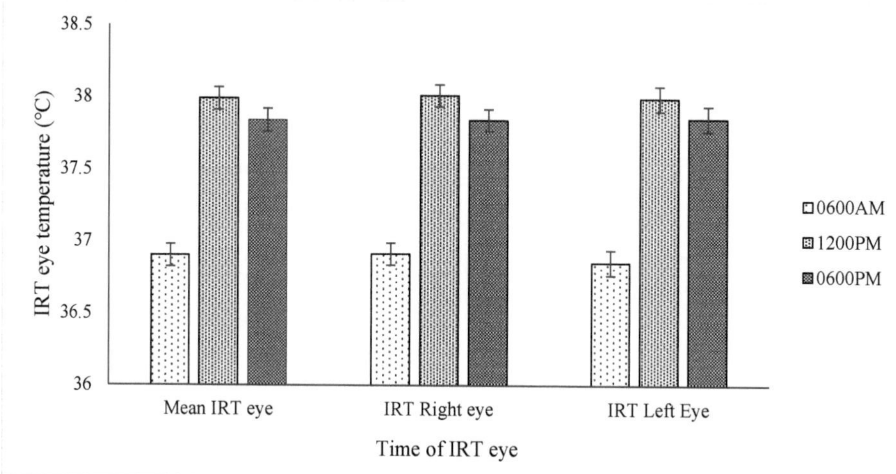

Figure 3. Infrared thermographic eye temperature (°C) of both eyes, the right eye, and the left eye at different times of the day ($p \leq 0.001$).

Using stepwise regression analysis, it was determined that the differentials in infrared eye temperature (HOT—TN) were positively correlated with HOT—TN differentials in panting score, standing, eating, vertical tail, backward and downward ear positions, and downwards and neutral head positions. They were negatively correlated with HOT—TN differentials in head and shoulder surface temperatures, scratching, stepping by the back left and front right limb, forward ear position, and raised head position.

$$\begin{aligned}&\textit{Infrared eye temperature } (^\circ\text{C}) \textit{ differential} = 0.830\ (\pm 0.0025; p = 0.002) + 0.360\ PS\ (\pm 0.00081; p = 0.001) + 0.88\\ &HPD\ (\pm 0.0082; p = 0.006) + 2.22\ HPN\ (\pm 0.0070; p = 0.002) + 8.00\ \textit{Eating time}\ (\pm 0.017; p = 0.001) + 2.00\ \textit{Standing}\\ &(\pm 0.0063; p = 0.002) + 0.370\ EPB\ (\pm 0.0045; p = 0.008) + 16.00\ EPD\ (\pm 0.056; p = 0.002) + 7.60\ TV\ (\pm 0.0088;\\ &p = 0.001) - 6.00\ EPF\ (\pm 0.0028; p \leq 0.001) - 0.19\ BL\ (\pm 0.00044; p = 0.002) - 0.11\ FR\ (\pm 0.00043; p = 0.002) -\\ &14.70\ \textit{Scratching}\ (\pm 0.027; p = 0.001) - 0.081\ \textit{Shoulder temperature}\ (\pm 0.00017; p = 0.001) - 0.010\ \textit{Head temperature}\\ &(\pm 0.00018; p = 0.01) - 854\ HPR\ (\pm 0.75; p = 0.001)\end{aligned} \tag{3}$$

where *PS* = panting score; *HPD* = head downwards; *HPN* = head neutral; *EPB* = backward ear; *EPD* = ear position downward; *TV* = tail vertical; *EPF* = forward ear, *BL* = back left stepping; *FR* = front right stepping; and *HPR* = raised head.

The quadratic relationship of rumen temperature with IRT-Eye temperature (°C) is expressed by the equation below and presented in Figure 4:

$$\textit{Rumen temperature } (^\circ\text{C}) = 88.3 - 3.2\ \textit{IRT-Eye temperature } (^\circ\text{C}) + 0.05\ \textit{IRT-Eye temperature}^2\ (^\circ\text{C})\ (r^2_{adj} = 70.7\%;\ F_{value} = 672; p \leq 0.001) \tag{4}$$

To determine if there were any other non-linear relationships, several non-linear terms were tested, but they were not significant and/or the r^2 values were not increased.

Spearman's rank correlations of the differences in the HOT—TN means of infrared eye temperature (°C) with differences in the HOT—TN means of behaviours were used to determine the behavioural impact when IRT cattle responded the most. Change in IRT-Eye temperature was positively associated with change in head downward position (correlation coefficient (CC) 0.481; $p = 0.03$), ear backward (CC 0.54; $p = 0.01$), tail vertical (CC 0.68; $p \leq 0.001$), standing (CC 0.51; $p = 0.02$), front right limb stepping (CC 0.503; $p = 0.05$), panting (CC 0.51; $p = 0.02$); and rumen temperature (CC 0.56; $p = 0.01$). It was negatively correlated with scratching (CC—0.60; $p = 0.01$) and grooming (CC—0.44; $p = 0.05$).

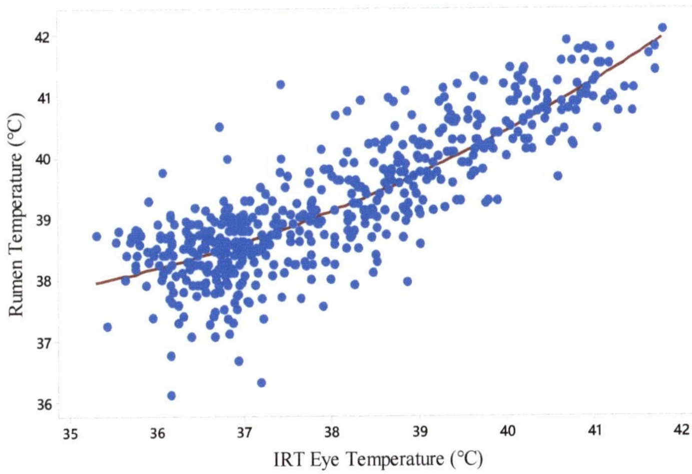

Figure 4. Quadratic relationship between rumen temperature and infrared thermographic eye temperature (°C). {Rumen temperature (°C) = 88.3 − 3.2 IRT-Eye temperature (°C) + 0.05 IRT-Eye temperature^2 (°C) (r^2_{adj} = 70.7%; F_{value} = 672; $p \leq 0.001$)}.

4. Discussion

The results for the behavioural responses of beef cattle to hot conditions are detailed in the work of Idris et al. [31]. In this manuscript, the impact of heat load on biological responses of feedlot cattle was studied to investigate the hypothesis that hyperthermic conditions influence eye temperature, which is thought to be associated with the behaviour and physiology of feedlot cattle. We further explored whether a substituted diet (grain substituted with forage) would influence cattle responses to heat stress through any changes in the infrared eye temperature. Behavioural responses that were related to the animals' ability to cope with the hot conditions were associated with increased IRT-Eye temperature (°C), reflecting its importance for possible applications in feedlot industry.

Hot climatic conditions can serve as a potential stressor to beef cattle, sometimes leading to low production and high mortality rates in feedlots. Cattle are susceptible to hyperthermia during hot environmental conditions and when their thermoregulatory responses of animals are insufficient to maintain thermoneutrality, resulting in a rise in core body temperature [35], a major indicator of hyperthermic conditions. Earlier efforts to use IRT to assess heat stress in feedlot cattle through the association of the infrared body surface temperature with the rumen temperature failed to predict feedlot cattle responses to hot conditions [10]. However, the association of infrared eye temperature (°C) with rumen temperature in the current study demonstrates the value of IRT-Eye measurements as an alternative method for measuring core body temperature (Figure 4). Consequently, infrared thermography may be a promising and reliable technique for assessing cattle responses and well-being [21,22] during hot days, with the additional advantage of minimum stress and quick measurement.

The rise in core body temperature during heat stress provokes animals to increase their peripheral circulation [15] to dissipate accumulated heat from the body. This change in the blood circulation leads to increased heat dissipation from the body's surfaces, including the eyes. This is especially the case around the *caruncula lacrimalis*, which is innervated by rich capillary beds [16]. Not all surface temperatures are correlated with body temperature (e.g., limb surface temperature); however, infrared eye temperature has been shown to have a correlation with core body temperature [35,36]. Therefore, infrared thermography can determine changes in peripheral circulation through changes in heat loss mechanisms as a non-invasive, passive, remote, and quick tool for assessing animal responses.

In our cattle, increases in the infrared thermographic temperature (°C) of the right eye and left eye and in the mean eye temperature of both eyes were observed in the HOT period compared with the TN period and recovery periods (Tables 4 and 5). Similarly, IRT-Eye was at its maximum at 1200 h when the dry-bulb ambient temperature was at its daily maximum limit during hot days and dropped following a decline in ambient temperature (Table 4; Figure 3). The rise in body temperatures (rectal and rumen temperature) in cattle exposed to hot environmental conditions is well established (e.g., Gaughan [15]; Brown-Brandl et al. [37]; Lees [38]). Stewart et al. [39] assessed IRT-Eye temperature in dairy cattle as a measure of pain and stress during handling. Further, Church et al. [40] studied the impact of environmental factors on the IRT-Eye temperature of cattle. In other studies, the infrared thermography of eyes has widely been reported as a tool for the detection of thermal conditions related to inflammation [13], dehorning [19], castration [20] and disease detection [13,14]. Recently, Shu et al. [41] reported the eye to be the best region for measuring thermal variation in order to assess the heat stress response in dairy cattle. Similarly, in our study, the rise in the IRT-Eye temperature in the cattle during the HOT period confirms the potential of the IRT-Eye to be used as a useful tool (in addition to rectal and rumen temperature) for assessing thermal variations in cattle experiencing hot environmental temperatures.

No difference in infrared eye temperature (in °C) was observed in different dietary cohorts during the HOT treatment period. The absence of significant differences in the IRT response between the two dietary cohorts could be for several reasons. One possibility is that, although the diets differed in composition and were nutritionally important, the variation might not have been large enough to impact the animals' ability to regulate their body heat. Both diets provided energy and or fibre levels that provided for the animals' requirements. Secondly, infrared thermography could only measure the body surface temperature [1] and it was unable to capture the internal adaptations inside the body, especially minor perturbations. In summary, we suspect that the dietary differences between cohorts were not enough to result in physiological changes that would result in differences in body surface temperature that could be measured through IRT.

In the current study, a negative association of IRT-Eye temperature differentials existed with temperature differentials in head and shoulder surfaces, demonstrating less difference in body surface temperatures with increasing IRT-Eye temperature. Thus, cattle that had the biggest IRT-Eye temperature response saw the smallest change in peripheral circulation in hot conditions. This probably reflects natural variation in the ability of cattle to respond to heat through evaporative cooling from the skin's surface, enabling them to dissipate accumulated heat during hot conditions [42,43]. Those with a limited ability to do this had the greatest increase in core body temperature, as evidenced by increased IRT-Eye values. No evidence of infrared eye temperature or lateralised IRT-Eye responses was obtained, despite the fact that cattle show lateralised visual preferences during various stressful conditions [44,45]. However, using an eye more does not mean that its temperature is expected to increase. In the current study, cattle exhibited compensatory behaviours (e.g., increased standing time, reduced activity such as grooming, scratching) that might have minimised the impact of diet on thermoregulation, resulting in similar eye temperatures across both groups.

During hot environmental conditions, cattle attempted to cope with hot conditions by reducing the body heat load through various behavioural and physiological adjustments in order to maintain a constant body temperature [46]. Adjustments made to maintain body temperature apparently lag three to four hours behind changes in daily dry-bulb ambient temperature [15,47]. However, the exact delay can vary; in some cases, respiratory signs might appear sooner if the animal is already in a compromised state, or later if the animal has better coping mechanisms. This time lag in apparent core body temperature may delay signs of hyperthermia until it is too late to provide support to animals [47]. Monitoring coping behaviours associated with immediate increases in body temperature can be helpful in providing necessary support to animals suffering from hot conditions. Cattle with greater increases in the IRT-Eye (°C) had smaller changes in scratching and stepping behaviour (front right and back left stepping). A decline in scratching behaviour [48] and eating [49] has been reported in dairy cows in response to increased body temperatures. Animals with higher differential of IRT-Eye measures had a greater decline in eating time, as reported earlier in our detailed behaviour studies, published as Idris [30] and Idris et al. [31]. A reduced eating time reflects a decline in feed intake to reduce the heat of digestion. This could be due to a decline in body energy that might have affected the animals' chewing ability [49–51]. Increased back left and front right limb stepping have been reported to be indications of stress-related responses in sheep during simulated sea motion [52]. However, the increase in back left and front right limb stepping with increasing IRT-Eye could be an effort to attempt to escape from the stressful condition [53,54]. Ultimately, the decline in these energy-dependent activities such as scratching can be attributed to a decline in body energy resources [55] and/or to a reduction in heat production [56] in response to hot environmental conditions.

The regression analysis shows that, during the HOT period, the cattle with the biggest increase in IRT-Eye temperature were more likely to spend more time with a vertical tail, hanging stationary. A vertical tail is the most observed tail position during routine daily activities, such as walking, standing and lying, and is sometimes taken as an indication of calm and relaxed behaviour in cattle [54,57]. However, in the hot conditions of this study, the involvement of cattle in protective behaviours such as tail swishing (the opposite of vertical stationary tail), scratching, and stepping potentially helped them to cope with the conditions by creating air movement (convection) over exposed surfaces. Convection depends on the air flow and contributes about 15% to heat loss compared to conduction, which accounts for only 5% of heat loss [58]. In this study, tail swishing allows for more significant heat loss by convection associated with the vulvar region, which tends to be under the effect of extensive peripheral vasodilation. Thus, cattle not engaging in these protective behaviours had the highest IRT-Eye increases.

Among various signs of heat stress, increased body temperature (rectal and rumen), panting, and standing time have all been reported as early signs of heat stress, as cattle attempt to maintain constant body temperature by dissipating body heat [59,60]. Lees et al. [9] reported that during hot environmental conditions, cattle accumulate heat from the surrounding environment, increasing the rumen temperature which must be dissipated from the body. The animal increases heat loss from the skin surface through vasodilatation and increased peripheral circulation [1,15]. This includes the eyes, especially around the *caruncula lacrimalis,* which is innervated by rich capillary beds [16]. Eyes are not usually influenced by ambient temperature due to the rich blood supply and high level of innervation by the sympathetic nervous system, which directly influence body surface temperature. There is a strong correlation between infrared eye surface temperature and core body temperature [16,18]. In the current study, the quadratic relationship between rumen temperature and eye temperature reinforces earlier findings of a strong correlation of IRT-Eye with core body temperature (i.e., rumen temperature in our study). Furthermore, rumen temperature increases rapidly at high core body temperatures, probably due to the high environmental temperatures. Regarding the relationship between eye temperature and rumen temperature, it is important to acknowledge the limitations of rumen temperature as a proxy for core temperature. Rumen temperature can be influenced by factors like feed intake and water consumption, which may cause discrepancies when compared to rectal or vaginal temperature measurements [16,18]. IRT-Eye measurements have an advantage, being non-invasive and minimising stress; however, they may not always outperform easily observable indicators like respiration rate (RR) or panting score (PS) in field settings where equipment may not be available. The importance of infrared thermography cannot be neglected in the industry for the continuous monitoring of large herds via the installation of automated thermal cameras in designated areas.

Standing helps by increasing the surface area of the body for maximum evaporation from the skin surface [33], particularly the naked skin underneath the torso, and similarly panting increases heat dissipation through respiratory channels and the evaporation of saliva from the tongue surface, which is highly innervated with blood vessels [37]. In the current study, the association of the infrared eye temperature of cattle with rumen temperature (Figure 4), standing, and panting during hot conditions further supports the inclusion of IRT-Eye temperature in the list of the tools that can be used as major indicators of heat stress in beef cattle.

The cattle with the most increased infrared eye temperatures in the current study showed a greater increase in the time spent with their heads downward, more ears backward, showed a greater decline in neutral head position, and had ears downwards during hot conditions. There was less change in the time spent with a raised head and forward ear position. Increased head downward [54,59] and backward ears [61,62] have been previously reported in cattle and sheep during various stressful activities, as compared with normal behaviours, which result in a neutrally positioned head and forward, downward or axial ear positions. The behavioural changes in the orientation of the ears, head, and tail have already been reported in Idris et al. [31]. This downward head position (below the withers) may indicate discomfort [52,59], while holding the head above or at the level of the withers indicates healthy, active, and vigilant animals [63]. Cattle spend an increased amount of time with their head in a downwards position following castration [64] and when handling stress [59]. Cattle respond to heat with more time spent with their head downward [30,31]. The greater downward ear position response in the current study could be attributed to low energy reserves in animals due to heat stress. There are inconsistent reports regarding ear positions, as backward and forward ear position have also been reported during feeding and grooming (brushing) activities [55]. Similarly, forward ears have also been reported as the predominant ear position, suggesting reactive/anxious behaviour [65]. These inconsistent results for ear positions can probably be attributed to different arousal levels in animals and their association with both positive and negative valence [61].

The association of IRT-Eye temperature with previously reported stress-related responses, such as downward head [52,59], backward ears [60,61], and other indicators of heat stress—increased panting [15], standing, and rumen temperature [9]—rather than normal relaxed cattle behaviours (such as scratching), shows the importance of IRT-Eye measurements as non-invasive tools to assess cattle responses to hyperthermia. The current study was performed in controlled environmental conditions where the major confounding factors reported for the use of infrared thermography, including camera settings (object localisation, distance, humidity, and emissivity), solar radiations, and air velocity, were closely controlled. However, in the field conditions, these can be overcome by the installation of automated thermal cameras in areas with no direct exposure of solar radiations and minimum air velocity [16]. The automated thermal imagers with an ability to localise the eye [66] would best for the measurement of the infrared eye temperature of cattle in field conditions. Future studies should explore the effects of dietary interventions on a larger and more diverse sample of cattle across varying climatic conditions. Moreover, investigating other physiological and behavioural responses to heat stress would help validate the efficacy of different dietary strategies to mitigate heat stress. Additionally, while eye surface temperature was used as an indicator of thermal status, other regions of interests for infrared thermography, such as the animal's flank or using body surface temperature at multiple sites, could provide a more comprehensive assessment. Future studies should explore the effects of dietary interventions on a larger and more diverse sample of cattle across varying climatic conditions, but with careful control of the onset of hot conditions. Moreover, investigating other physiological and behavioural responses to heat stress would help validate the efficacy of different dietary strategies to mitigate heat stress.

5. Conclusions

Cattle suffering from heat stress had higher mean infrared eye temperatures of both eyes, with no difference in response between those fed finisher and substituted diets. Infrared eye temperature (°C) was strongly associated with already validated stress-related behaviours and reported indicators of heat stress in cattle. There was no evidence of lateralised eye temperature responses in cattle. Based on changes in infrared eye temperature (°C) during hot conditions and their association with the rumen temperature and panting behaviour, as well as with other coping behaviours, infrared eye temperature could be combined with these related behaviours and utilised as a non-invasive tool in the feedlot industry during hot environmental conditions. The strong association of IRT-Eye with behavioural and physiological responses suggests that IRT-Eye can be used as an indicator to recognise heat stress in beef cattle.

Author Contributions: Conceptualization, M.I. and C.J.C.P.; methodology, M.I., M.S., J.B.G. and C.J.C.P.; image analysis, M.I.; statistical analysis, M.I. and C.J.C.P.; investigation, M.I.; resources, M.S., J.B.G. and C.J.C.P.; data curation, M.I.; writing—original draft preparation, M.I.; writing—review and editing, M.I., M.S., J.B.G. and C.J.C.P. All authors have read and agreed to the published version of the manuscript.

Funding: The authors are grateful to University of Queensland, Australia, for funding the research experiments conducted to fulfil the requirements of a doctoral degree of corresponding author Musadiq Idris.

Institutional Review Board Statement: Ethical approval for the study was obtained from The University of Queensland Production and Companion Animal Ethics Committee (SAFS/460/16).

Informed Consent Statement: Not applicable.

Data Availability Statement: The data presented in this study are available on request from the corresponding author.

Acknowledgments: The authors are grateful to the staff at the Queensland Animal Science Precinct for their assistance in this study. The authors also acknowledge that the manuscript is a section of a PhD work that was already completed and submitted by the corresponding author to the University of Queensland t bvhesis repository to fulfil the requirements of a doctoral degree [30].

Conflicts of Interest: Clive Phillips is a Director of Humane Society International, a member of the Voiceless Grant Applications committee, was a hearing expert to the European Food Safety Authority in 2021/2, and is chairperson of the Queensland government's animal welfare advisory committee. None of these organisations had any involvement in this manuscript.

References

1. Idris, M.; Uddin, J.; Sullivan, M.; McNeill, D.M.; Phillips, C.J.C. Non-Invasive physiological indicators of heat stress in cattle. *Animals* **2021**, *11*, 71. [CrossRef] [PubMed]
2. Sjaastad, O.V.; Hove, K.; Sand, O. *Physiology of Domestic Animals*; Scandinavian Veterinary Press: Oslo, Norway, 2010.
3. Hempel, S.; Menz, C.; Pinto, S.; Galán, E.; Janke, D.; Estellés, F.; Müschner-Siemens, T.; Wang, X.; Heinicke, J.; Zhang, G.; et al. Heat stress risk in European dairy cattle husbandry under different climate change scenarios–uncertainties and potential impacts. *Earth Syst. Dyn.* **2019**, *10*, 859–884. [CrossRef]
4. Yue, S.; Qian, J.; Du, J.; Liu, X.; Xu, H.; Liu, H.; Zhang, J.; Chen, X. Heat Stress Negatively Influence Mammary Blood Flow, Mammary Uptake of Amino Acids and Milk Amino Acids Profile of Lactating Holstein Dairy Cows. *Pak. Vet. J.* **2023**, *43*, 73–78.
5. Li, J.B.; Wang, X.; Sun, A.; Li, H.B.; Luo, Y.; He, F.; Huan, C.; Zhou, X.; Li, C.J.; Zhang, B.Z.; et al. Comparative Transcriptomic Analysis of Spermatozoa from Xiangxi and Simmental Bulls under Heat Stress: Implications for Fertility Prediction. *Pak. Vet. J.* **2023**, *43*, 184–188.
6. Hillman, P.; Gebremedhin, K.; Willard, S.; Lee, C.; Kennedy, A. Continuous measurements of vaginal temperature of female cattle using a data logger encased in a plastic anchor. *Appl. Eng. Agric.* **2009**, *25*, 291–296. [CrossRef]
7. Reuter, R.; Carroll, J.; Hulbert, L.; Dailey, J.; Galyean, M. Development of a self-contained, indwelling rectal temperature probe for cattle research. *J. Anim. Sci.* **2010**, *88*, 3291–3295. [CrossRef]
8. Lee, C.; Gebremedhin, K.; Parkhurst, A.; Hillman, P. Placement of temperature probe in bovine vagina for continuous measurement of core-body temperature. *Int. J. Biometeorol.* **2015**, *59*, 1201–1205. [CrossRef]

9. Lees, A.; Lees, J.; Lisle, A.; Sullivan, M.; Gaughan, J. Effect of heat stress on rumen temperature of three breeds of cattle. *Int. J. Biometeorol.* **2018**, *62*, 207–215. [CrossRef]
10. Lees, A.M.; Lees, J.; Sejian, V.; Wallage, A.; Gaughan, J. using infrared thermography as an in situ measure of core body temperature in lot-fed Angus steers. *Int. J. Biometeorol.* **2018**, *62*, 3–8. [CrossRef]
11. Vogel, E.; Meyer, C.; Eckardt, R. Severe Heatwaves Show the Need to Adapt Livestock Management for Climate. The Conversation, 27 February 2017. Available online: www.theconversation.com/severe-heatwaves-show-the-need-to-adapt-livestock-management-for-climate-73447#:~:text=During%20the%20recent%20heatwave%20in%20New%20South%20Wales,,the%20likelihood%20of%20this%20kind%20of%20record-breaking%20heatwave (accessed on 1 February 2020).
12. Mccafferty, D.J. The value of infrared thermography for research on mammals: Previous applications and future directions *Mammal Rev.* **2007**, *37*, 207–223. [CrossRef]
13. Schaefer, A.; Cook, N.; Tessaro, S.; Deregt, D.; Desroches, G.; Dubeski, P.; Tong, A.; Godson, D. Early detection and prediction of infection using infrared thermography. *Can. J. Anim. Sci.* **2004**, *84*, 73–80. [CrossRef]
14. Schaefer, A.; Cook, N.; Bench, C.; Chabot, J.; Colyn, J.; Liu, T.; Okine, E.; Stewart, M.; Webster, J. The non-invasive and automated detection of bovine respiratory disease onset in receiver calves using infrared thermography. *Res. Vet. Sci.* **2012**, *93*, 928–935 [CrossRef] [PubMed]
15. Gaughan, J.B. Respiration Rate and Rectal Temperature Responses of Feedlot Cattle in Dynamic, Thermally Challenging Environments. Ph.D. Thesis, The University of Queensland, St Lucia, QLD, Australia, 2002.
16. Stewart, M.; Webster, J.; Verkerk, G.; Schaefer, A.; Colyn, J.; Stafford, K. Non-invasive measurement of stress in dairy cows using infrared thermography. *Physiol. Behav.* **2007**, *92*, 520–525. [CrossRef] [PubMed]
17. Stewart, M.; Stookey, J.; Stafford, K.; Tucker, C.; Rogers, A.; Dowling, S.; Verkerk, G.; Schaefer, A.; Webster, J. Effects of local anesthetic and a nonsteroidal antiinflammatory drug on pain responses of dairy calves to hot-iron dehorning. *J. Dairy Sci.* **2009** *92*, 1512–1519. [CrossRef]
18. Stewart, M.; Verkerk, G.; Stafford, K.; Schaefer, A.; Webster, J. Non-invasive assessment of autonomic activity for evaluation of pain in calves, using surgical castration as a model. *J. Dairy Sci.* **2010**, *93*, 3602–3609. [CrossRef]
19. Gloster, J.; Ebert, K.; Gubbins, S.; Bashiruddin, J.; Paton, D.J. Normal variation in thermal radiated temperature in cattle Implications for foot-and-mouth disease detection. *BMC Vet. Res.* **2011**, *7*, 73. [CrossRef]
20. Uddin, J.; Phillips, C.J.C.; Goma, A.A.; McNeill, D.M. Relationships between infrared temperature and laterality. *Appl. Anim Behav. Sci.* **2019**, *220*, 104855. [CrossRef]
21. Stewart, M.; Webster, J.; Schaefer, A.; Cook, N.; Scott, S. Infrared thermography as a non-invasive tool to study animal welfare *Anim. Welfare* **2005**, *14*, 319–325. [CrossRef]
22. Stewart, M.; Webster, J.; Verkerk, G.; Colyn, J.; Schaefer, A. Infrared thermography as a non-invasive measure of stress in dairy cows. *J. Anim. Sci.* **2005**, *83*, 374.
23. Mishra, M.; Martz, F.; Stanley, R.; Johnson, H.; Campbell, J.; Hilderbrand, E. Effect of diet and ambient temperature-humidity on ruminal pH, oxidation reduction potential, ammonia and lactic acid in lactating cows. *J. Anim. Sci.* **1970**, *30*, 1023–1028. [CrossRef]
24. Shearer, J. Rumen acidosis, heat stress and laminitis. In Proceedings of the 4th Annual Arizona Dairy Production Conference, Tempe, AZ, USA, 11 October 2005.
25. Conte, G.; Ciampolini, R.; Cassandro, M.; Lasagna, E.; Calamari, L.; Bernabucci, U.; Abeni, F. Feeding and nutrition management of heat-stressed dairy ruminants. *Ital. J. Anim. Sci.* **2018**, *17*, 604–620. [CrossRef]
26. Erdman, R.A. Dietary buffering requirements of the lactating dairy cow: A review. *J. Dairy Sci.* **1988**, *71*, 3246–3266. [CrossRef]
27. Brown-Brandl, T.M.; Nienaber, J.A.; Eigenberg, R.A.; Mader, T.L.; Morrow, J.; Dailey, J. Comparison of heat tolerance of feedlot heifers of different breeds. *Livest. Sci.* **2006**, *105*, 19–26. [CrossRef]
28. Mader, T.L.; Davis, M.; Brown-Brandl, T. Environmental factors influencing heat stress in feedlot cattle. *J. Anim. Sci.* **2006**, *84*, 712–719. [CrossRef]
29. Gaughan, J.B.; Mader, T.L.; Holt, S.M.; Lisle, A. A new heat load index for feedlot cattle. *J. Anim. Sci.* **2008**, *86*, 226–234. [CrossRef]
30. Idris, M. Behavioural and Physiological Responses of Beef Cattle to Hot Environmental Conditions. Ph.D. Thesis, The University of Queensland, Gatton Campus, QLD, Australia, 2020.
31. Idris, M.M.; Gaughan, J.B.; Phillips, C.J.C. Behavioural responses of beef cattle to hot conditions. *Animals* **2024**, *14*, 2444. [CrossRef]
32. Friard, O.; Gamba, M. BORIS: A free, versatile open-source event-logging software for video/audio coding and live observations. *Methods Ecol. Evol.* **2016**, *7*, 1325–1330. [CrossRef]
33. Idris, M.; Gay, C.C.; Woods, I.G.; Sullivan, M.; Gaughan, J.B.; Phillips, C.J.C. Automated quantification of the behaviour of beef cattle exposed to heat load conditions. *Animals* **2023**, *13*, 1125. [CrossRef]
34. Thom, E.C. The discomfort index. *Weatherwise* **1959**, *12*, 57–61. [CrossRef]
35. Mota-Rojas, D.; Pereira, A.M.; Wang, D.; Martínez-Burnes, J.; Ghezzi, M.; Hernández-Avalos, I.; Lendez, P.; Mora-Medina, P.; Casas, A.; Olmos-Hernández, A.; et al. Clinical applications and factors involved in validating thermal windows used in infrared thermography in cattle and river buffalo to assess health and productivity. *Animals* **2021**, *11*, 2247. [CrossRef]
36. Macmillan, K.; Colazo, M.G.; Cook, N.J. Evaluation of infrared thermography compared to rectal temperature to identify illness in early postpartum dairy cows. *Res. Vet. Sci.* **2019**, *125*, 315–322. [CrossRef] [PubMed]
37. Brown–Brandl, T.M.; Yanagi, T.; Xin, H.; Gates, R.S.; Bucklin, R.A.; Ross, G.S. A new telemetry system for measuring core body temperature in livestock and poultry. *Appl. Eng. Agric.* **2003**, *19*, 583.

38. Lees, A.M. Biological Responses of Feedlot Cattle to Heat Load. Ph.D. Thesis, The University of Queensland, St Lucia, QLD, Australia, 2016.
39. Stewart, M.; Webster, J.; Schaefer, A.; Stafford, K. Infrared thermography and heart rate variability for non-invasive assessment of animal welfare. *ANZCCART Hum. Sci. News* **2008**, *21*, 1–4.
40. Church, J.S.; Hegadoren, P.; Paetkau, M.; Miller, C.; Regev-Shoshani, G.; Schaefer, A.; Schwartzkopf-Genswein, K. Influence of environmental factors on infrared eye temperature measurements in cattle. *Res. Vet. Sci.* **2014**, *96*, 220–226. [CrossRef]
41. Shu, H.; Li, Y.; Fang, T.; Xing, M.; Sun, F.; Chen, X.; Bindelle, J.; Wang, W.; Guo, L. Evaluation of the best region for measuring eye temperature in dairy cows exposed to heat stress. *Front. Vet. Sci.* **2022**, *9*, 857777. [CrossRef] [PubMed]
42. Silanikove, N. Effects of heat stress on the welfare of extensively managed domestic ruminants. *Livest. Prod. Sci.* **2000**, *67*, 1–18. [CrossRef]
43. Schumacker, P.; Rowland, J.; Saltz, S. Glossary of terms for thermal physiology Third Edition revised by The Commission for Thermal Physiology of the International Union of Physiological Sciences (IUPS Thermal Commission). *JPN J. Physiol.* **2001**, *51*, 245–280.
44. Robins, A.; Phillips, C. Lateralised visual processing in domestic cattle herds responding to novel and familiar stimuli. *Laterality* **2010**, *15*, 514–534. [CrossRef]
45. Kappel, S.; Mendl, M.T.; Barrett, D.C.; Murrell, J.C.; Whay, H.R. Lateralized behaviour as indicator of affective state in dairy cows. *PLoS ONE* **2017**, *12*, e0184933. [CrossRef]
46. Vermunt, J.J.; Tranter, B.P. Heat stress in dairy cattle–A review, and some of the potential risks associated with the nutritional management of this condition. In *Review of AVA QLD Division Conference 25-27/3/10*; Australian Veterinary Association: St Leonards, Australia, 2011; pp. 212–221.
47. Brown-Brandl, T.; Eigenberg, R.; Nienaber, J.; Hahn, G.L. Dynamic response indicators of heat stress in shaded and non-shaded feedlot cattle, Part 1: Analyses of indicators. *Biosyst. Eng.* **2005**, *90*, 451–462. [CrossRef]
48. Fogsgaard, K.K.; Røntved, C.M.; Sørensen, P.; Herskin, M.S. Sickness behavior in dairy cows during Escherichia coli mastitis. *J. Dairy Sci.* **2012**, *95*, 630–638. [CrossRef] [PubMed]
49. Karimi, M.; Ghorbani, G.; Kargar, S.; Drackley, J.K. Late-gestation heat stress abatement on performance and behavior of Holstein dairy cows. *J. Dairy Sci.* **2015**, *98*, 6865–6875. [CrossRef]
50. Alam, M.; Hashem, M.; Rahman, M.; Hossain, M.; Haque, M.; Sobhan, Z.; Islam, M. Effect of heat stress on behavior, physiological and blood parameters of goat. *Progress. Agric.* **2011**, *22*, 37–45. [CrossRef]
51. Soriani, N.; Panella, G.; Calamari, L. Rumination time during the summer season and its relationships with metabolic conditions and milk production. *J. Dairy Sci.* **2013**, *96*, 5082–5094. [CrossRef] [PubMed]
52. Robins, A.; Berthoux, G.; Santurtun, E.; Navarro, G.; Phillips, C.J. Sheep Quickstep while the Floor Rock and Rolls: Visuomotor Lateralization during Simulated Sea Travel. *Animals* **2019**, *9*, 700. [CrossRef] [PubMed]
53. Rousing, T.; Bonde, M.; Badsberg, J.H.; Sørensen, J.T. Stepping and kicking behaviour during milking in relation to response in human–animal interaction test and clinical health in loose housed dairy cows. *Livest. Prod. Sci.* **2004**, *88*, 1–8. [CrossRef]
54. Goma, A.A.; Pearce, G.P.; Uddin, J.; Rimon, E.; Davies, H.; Phillips, C.J.C. A forced lateralisation test for dairy cows and its relation to their behaviour. *Appl. Anim. Behav. Sci.* **2018**, *207*, 8–19. [CrossRef]
55. Mandel, R.; Whay, H.R.; Nicol, C.J.; Klement, E. The effect of food location, heat load, and intrusive medical procedures on brushing activity in dairy cows. *J. Dairy Sci.* **2013**, *96*, 6506–6513. [CrossRef]
56. Mader, T.L.; Griffin, D. Management of cattle exposed to adverse environmental conditions. *Vet. Clin. N. Am.-Food Anim. Prac.* **2015**, *31*, 247–258. [CrossRef]
57. de Oliveira, D.; Keeling, L.J. Routine activities and emotion in the life of dairy cows: Integrating body language into an affective state framework. *PLoS ONE* **2018**, *13*, e0195674. [CrossRef]
58. Duke, J. Temperature Disturbances. In *Anesthesia Secrets*, 4th ed.; Mosby Elsevier: Philadelphia, PA, USA, 2011; pp. 217–221. ISBN 9780323065245. [CrossRef]
59. Nienaber, J.; Hahn, G. Livestock production system management responses to thermal challenges. *Int. J. Biometeorol.* **2007**, *52*, 149–157. [CrossRef] [PubMed]
60. Hemsworth, P.H.; Rice, M.; Karlen, M.G.; Calleja, L.; Barnett, J.L.; Nash, J.; Coleman, G.J. Human–animal interactions at abattoirs: Relationships between handling and animal stress in sheep and cattle. *Appl. Anim. Behav. Sci.* **2011**, *135*, 24–33. [CrossRef]
61. Boissy, A.; Aubert, A.; Désiré, L.; Greiveldinger, L.; Delval, E.; Veissier, I. Cognitive sciences to relate ear postures to emotions in sheep. *Anim. Welfare* **2011**, *20*, 47. [CrossRef]
62. Gleerup, K.B.; Andersen, P.H.; Munksgaard, L.; Forkman, B. Pain evaluation in dairy cattle. *Appl. Anim. Behav. Sci.* **2015**, *171*, 25–32. [CrossRef]
63. Kluever, B.M.; Breck, S.W.; Howery, L.D.; Krausman, P.R.; Bergman, D.L. Vigilance in cattle: The influence of predation, social interactions, and environmental factors. *Rangeland Ecol. Manag.* **2008**, *61*, 321–328. [CrossRef]
64. de Oliveira, F.A.; Luna, S.P.L.; do Amaral, J.B.; Rodrigues, K.A.; SantAnna, A.C.; Daolio, M.; Brondani, J.T. Validation of the UNESP-Botucatu unidimensional composite pain scale for assessing postoperative pain in cattle. *BMC Vet. Res.* **2014**, *10*, 200. [CrossRef]

65. Reefmann, N.; Bütikofer Kaszàs, F.; Wechsler, B.; Gygax, L. Ear and tail postures as indicators of emotional valence in sheep. *Appl. Anim. Behav. Sci.* **2009**, *118*, 199–207. [CrossRef]
66. Jaddoa, M.A.; Al-Jumaily, A.; Gonzalez, L.; Cuthbertson, H. Automatic eyes localization in thermal images for temperature measurement in cattle. In Proceedings of the 12th International Conference on Intelligent Systems and Knowledge Engineering (ISKE), Nanjing, China, 24–26 November 2017; pp. 1–6.

MDPI AG
Grosspeteranlage 5
4052 Basel
Switzerland
Tel.: +41 61 683 77 34

Animals Editorial Office
E-mail: animals@mdpi.com
www.mdpi.com/journal/animals

www.ingramcontent.com/pod-product-compliance
Lightning Source LLC
Chambersburg PA
CBHW041419210926
51850CB00013B/3313
* 9 7 8 3 7 2 5 8 3 3 5 7 3 *